AF334463

Growth and Form
Nonlinear Aspects

NATO ASI Series

Advanced Science Institutes Series

A series presenting the results of activities sponsored by the NATO Science Committee, which aims at the dissemination of advanced scientific and technological knowledge, with a view to strengthening links between scientific communities.

The series is published by an international board of publishers in conjunction with the NATO Scientific Affairs Division

A	**Life Sciences**	Plenum Publishing Corporation
B	**Physics**	New York and London
C	**Mathematical and Physical Sciences**	Kluwer Academic Publishers
D	**Behavioral and Social Sciences**	Dordrecht, Boston, and London
E	**Applied Sciences**	
F	**Computer and Systems Sciences**	Springer-Verlag
G	**Ecological Sciences**	Berlin, Heidelberg, New York, London,
H	**Cell Biology**	Paris, Tokyo, Hong Kong, and Barcelona
I	**Global Environmental Change**	

Recent Volumes in this Series

Volume 270—Complexity, Chaos, and Biological Evolution
edited by Erik Mosekilde and Lis Mosekilde

Volume 271—Interaction of Charged Particles with Solids and Surfaces
edited by Alberto Gras-Martí, Herbert M. Urbassek,
Néstor R. Arista, and Fernando Flores

Volume 272—Predictability, Stability, and Chaos in N-Body Dynamical Systems
edited by Archie E. Roy

Volume 273—Light Scattering in Semiconductor Structures and Superlattices
edited by David J. Lockwood and Jeff F. Young

Volume 274—Direct Methods of Solving Crystal Structures
edited by Henk Schenk

Volume 275—Techniques and Concepts of High-Energy Physics VI
edited by Thomas Ferbel

Volume 276—Growth and Form: Nonlinear Aspects
edited by M. Ben Amar, P. Pelcé, and P. Tabeling

Series B: Physics

Growth and Form
Nonlinear Aspects

Edited by

M. Ben Amar

Ecole Normale Supérieure
Paris, France

P. Pelcé

Université de Provence–St. Jerome
Marseille, France

and

P. Tabeling

Ecole Normale Supérieure
Paris, France

Plenum Press
New York and London
Published in cooperation with NATO Scientific Affairs Division

Proceedings of a NATO Advanced Study Institute on
Nonlinear Phenomena Related to Growth and Form,
held July 17–29, 1990,
in Cargèse, France

Library of Congress Cataloging in Publication Data

NATO Advanced Study Institute on Nonlinear Phenomena Related to Growth and
Form (1990: Cargèse, France)
    Growth and form: nonlinear aspects / edited by M. Ben Amar, P. Pelcé, and P.
Tabeling.
         p.   cm.—(NATO ASI series. Series B, Physics; v. 276)
    "Proceedings of a NATO Advanced Study Institute on Nonlinear Phenomena
Related to Growth and Form, held July 17–29, 1990, in Cargèse, France"—T. p.
verso.
    "Published in cooperation with NATO Scientific Affairs Division."
    Includes bibliographical references and index.
    ISBN 0-306-44046-6
    1. Solid state physics—Congresses. 2. Crystals—Growth—Congresses. 3.
Nonlinear theories—Congresses. 4. Dentritic crystals—Congresses. 5. Solidifi-
cation—Congresses. I. Ben Amar, M. II. Pelcé, Pierre, date. III. Tabeling, P. IV.
North Atlantic Treaty Organization. Scientific Affairs Division. V. Title. VI. Series.
QC176.A1N324   1990   91-28344
530.4′1—dc20   CIP

ISBN 0-306-44046-6

© 1991 Plenum Press, New York
A Division of Plenum Publishing Corporation
233 Spring Street, New York, N.Y. 10013

SPECIAL PROGRAM ON CHAOS, ORDER, AND PATTERNS

This book contains the proceedings of a NATO Advanced Research Workshop
held within the program of activities of the NATO Special Program on
Chaos, Order, and Patterns.

Growth and Form is the title of a famous book written by D'Arcy Thomson at the beginning of the century. It relates a large number of problems of shapes of bodies either in the physical world or the biological realm. Keywords in this field are shapes, spirals, growth law, gravity field, surface tension, scaling laws, diffusion and mechanical efficiency. This field is the source of a considerable amount of work, even today, and this conference was a place where some of this work was discussed. Except for a few contributions with biophysical inspiration, the main part of the conference was devoted to physical problems related to growth and form and especially to the problem of the motion of interfaces under various nonequilibrium conditions. Even with this restriction, this field is huge, from the more applied area (combustion, metallurgy) to the more fundamental (singularities in the complex plane, solvability conditions). One day, at dinner time, in a restaurant with a good view of the corsica sea, W. Kurz from Lausanne told us about teleferique cables and the kind of material which was necessary to build them. Considering the important abyss between this kind of concept and for instance, the huge formalism involving Green functions used to find operating points for dendritic growth, we immediatelty had the giggles for five minutes. This large domain was the occasion to confront many scientists from different areas (physicists, applied mathematicians, specialists of combustion, metallurgists and geologists).

The simplest problem for interface propagation is viscous fingering. The modelling of this configuration in terms of mechanics of continuous media is completely understood now (Tanveer, Brener et al.). What remains to be understood are the mathematical properties of the time-dependent solutions (Kadanoff et al.) and the connections that exist between this continuous model and the connected statistical problem posed by diffusion limited aggregation that we will present below.

The modelling of free dendritic growth in a pure undercooled melt is almost understood too. The solutions of these models are not yet completely. Considering dendritic growth in a pure melt, there are two different theories, not necessarily exclusive. One theory considers anisotropic surface tension of the liquid-solid interface and obtains needle crystals growing with stationary shape and constant velocity. Another theory neglects this effect and considers time-dependent solutions for dendritic growth with an operating poiunt associated with the marginal stability of these solutions. Theses two theories give two different values for the well known $\rho^2 v$ and experimenters have difficulties in clarifying the two theories (Gollub, Bilgram and Hurliman). More complex situations remain to be understood: rapid dendritic growth (Herlach and Eckler), dendritic growth in a forced flow (Bouissou et al.), and dendritic growth in a more exotic systems like Langmuir monolayers (Muller and Gallet).

One of the most interesting parts of the session was certainly the one on directional solidification. The session started with J. Hunt on numerical integration of the equations. When he suggested that dendritic growth could occur without crystalline anisotropy he was immediately attacked verbally by many people of the audience. Then came the controversy on the oscillatory instability of deep cells. Karma and Pelce proposed that an oscillatory instability of deep cells appears at a sufficiently large velocity. Kessler and Levine did not find it in their numerical integration of the equation. Then came the problem of the two branches of solution (Van Saarloos and Weeks) which possibly can solve the previous discrepancy. If some years ago, the small Péclet number studies in directional solidification were thought to be a breakthrough, things now appear more difficult when comparisons with experiments are performed (Billia et al., Kurowski et al.). Thus many questions remain open: selected state (marginal stability or minimum undercooling), oscillatory instability of deep cells, is growth at a small Péclet number possible, how many branches of solutions are there for the cells. Here too, more complex situations remain to be understood, like the directional solidification of a facetted material (Adda Bedia and BenAmar) or the directional solidification of smectic material (Oswald).

Consider another kind of interface, the flames. This system is interesting since it is relatively complex (many chemical reactions), but presents well reproducible phenomena. The way of thinking here can help the understanding of the complex situations that can occur in biophysics, for instance. We present here mainly a collaborative work from the Laboratoire de Recherche en Combustion on the problem of interaction of flames with acoustics. This problem is rich: generation of sound (Clavin, Pelce, and Rochwerger), and parametric instabilities (Searby).

Still reaction-diffusion but in another context, the excitable medium is a subject that is of importance nowdays because of the discovery that cardiac fibrillation and chemical waves have a similar origin. The more important structure in this kind of media is the rotor, i.e. a spiral wave rotating at uniform angular velocity. Important progress was made recently in determining this rotational frequency as a function of the control parameters characterizing the excitable medium (Pelcé and Sun). A problem which remains to be understood is the possible instability of the spiral wave, namely, the meandering process, i.e. an almost erratic motion of the spiral tip and the interaction between spirals. Numerical simulations of simple reaction-diffusion systems show that such behavior is associated with a Hopf bifurcation (Karma). Another interesting problem concerns the interaction between spirals (Meron).

Even if the main part of the conference was devoted to nonlinear aspects of growth and form, an important part was concerned with the statistical aspects of the problem. The more important problem in this field is the one called diffusion-limited-aggregation (DLA). It concerns the growth of clusters formed by aggregation of small spheres coming from outside with a brownian motion. Despite very clever attempts to understand the structure of these clusters (Hakim), the problem remains unsolved. Having the idea to superimpose many of these clusters growing in a channel, Arneodo et al. found that the envelope of the clusters is a Saffman-Taylor shape, making a beautiful connection between a statistical problem and a nonlinear one. Other well studied statistical problems of growth concern the growth of thin films (Sander and Yan) or the inhomogeneopus growth of rough surfaces (Villain, Wolf and Tang, Tang et al.).

Finally, some miscellaneous subjects, in general related to the main subject of the conference. One can mention for instance the problem of chiral structures in condensed matter (Pomeau), growth and form of disliocation patterns (Walgraef and Ghoniem), and problems coming from the geological (Brandeis) or from the biological context (Sawada et al.).

From many sources this conference was greatly appreciated, and therefore it is a pleasure to thank all the people who helped us to organize this conference; M.Gillino,M.F.Hanseler and all the staff of Cargèse. We also thank the organisations which financed the project: NATO,CEE,DRET,CNES,CNRS, and the Université de Provence.

M.Ben Amar, P.Pelcé, P.Tabeling

CONTENTS

VISCOUS FINGERING

DENDRITIC GROWTH

VISCOUS FINGERING

SINGULARITIES IN COMPLEX INTERFACE DYNAMICS

Wei-shen Dai, Leo P. Kadanoff, Su-min Zhou

The James Franck Institute
The University of Chicago
5640 South Ellis Avenue
Chicago Illinois 60637, USA

ABSTRACT

The motion of the interface between two fluids in a quasi two-dimensional geometry is studied via simulations. We consider the case in which a zero-viscosity fluid displaces one with finite viscosity, and compare the interfaces which arise with zero surface tension with those which occur when the surface tension is small, but finite.

The interface dynamics can be analyzed in terms of a complex analytic function which maps the unit circle into the interface between the fluids. The physical region of the domain is the exterior of the circle, which then maps into the region occupied by more viscous fluid. In this physical region, the mapping is analytic and its derivative is never zero.

At zero surface tension we have an integrable problem. The derivative of the mapping function, $g(\omega,t)$, then necessarily has all its zeros and poles within the unit circle. If $g(\omega,t)$ is a rational function, then the integrable dynamics simply describes the motion of these singularities. The analysis fails at a critical time at which one of the singularities hits the unit circle.

This paper focuses upon the determination of the nature of g and of the interface when the surface tension is small. Two cases are considered: In case A in which the t=0 interface is described by a g with only zeros in the unit circle; in case B in which the singularities closest to the unit circle are instead poles. In case B, the motion is qualitatively similar with and without surface tension: the singularities move outward and asymptotically approach the

Growth and Form, Edited by M. Ben Amar *et al.*
Plenum Press, New York, 1991

circle. In case A, for zero surface tension, the zeros move outward and hit the interface after a finite time, whereupon the solution breaks down. But, for finite surface tension, a zero disappears and is replaced by a pair of pole-like excitations which again seem to approach the unit circle asymptotically.

I. INTRODUCTION

Bubble growth in a Hele-Shaw cell has drawn a lot of attention recently. Here, two closely-spaced glass plates contain two fluids. For this idealized case, one fluid is viscous and incompressible, while the other has zero viscosity. The latter fluid is a bubble in an infinite sea composed of the more viscous fluid. The area of zero viscosity fluid grows at a steady rate. The interface separating the two fluids is described by a surface tension. This and similar systems have been studied experimentally and various growth features have been observed[1].

For many initial conditions, the zero surface tension case can be solved analytically[2,3,4,5]. These solutions show that for a large range of initial conditions the interface will develop cusps after a finite time interval. After this critical time, the analysis is not meaningful.

Starting with the work of Saffman and Taylor[6], there has been considerable discussion of the effect of the surface tension upon the interface motion in a Hele-Shaw cell. Work on this problem has shown that the surface tension is a singular perturbation so that the solutions with and without surface tension may be qualitatively different[7]. Since the bubble growth problem is the simplest one in this general class, the question whether the presence of a small surface tension will qualitatively change the solution is of great interest.

In this paper, we shall first give a mathematical formulation of the bubble growth problem using the formulation of Shraiman and Bensimon[8] and that of Tanveer[9]. In this formulation, the interface is described by a function $f(\omega,t)$, where ω is complex and t (the time) is real. Here $f(\omega,t)$ maps the exterior of the unit circle onto the region of the viscous fluid. The derivative of f:

$$g(\omega,t) = \frac{\partial}{\partial \omega} f(\omega,t) \qquad (1)$$

is analytic and has no zeros outside the unit circle of ω. We shall use the word "singularities" to describe zeros, poles or other points of non-analyticities of g. Our results will be described in terms of the motion of these singularities.

We will study this motion through simulations. We shall show both the shape of the actual bubble in comparison with those of the zero surface tension case and the comparative motion of the singularities. For some initial conditions, a small surface tension does change the shape of the interface significantly and the existence of this small surface tension enables us to carry our calculation longer than the critical time mentioned above. But in some other cases, the small surface tension does not seem to affect the solution very much.

4

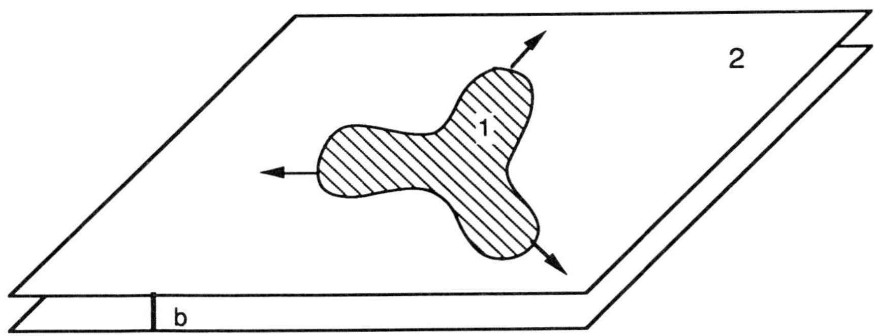

Fig. 1. In a Hele-Shaw cell, a bubble containing fluid 1 with very small viscosity grows into a fluid 2 with larger viscosity.

II. MATHEMATICAL FORMULATION

The system has two kinds of fluids. They are confined between two parallel glass plates which are kept very close to each other. (See Fig.1) The interface between the two fluids is bubble-shaped. The fluid inside the bubble (fluid 1) has a very small viscosity and is kept at a constant pressure. The fluid outside the bubble (fluid 2) has a larger viscosity and is incompressible.

For the fluid outside the bubble, we can use Darcy's law:

$$\mathbf{v} = -\frac{b^2}{12\,\mu}\,\nabla p \qquad (2)$$

where $\mathbf{v}, p, \mu$ are the velocity, pressure and viscosity of fluid 2, and b is the spacing between the two plates. From the condition of incompressibility, we have $\nabla \cdot \mathbf{v} = 0$. Therefore the pressure field satisfies the Laplace equation:

$$\nabla^2 p = 0 \qquad (3)$$

The pressure is constant inside the bubble and has a jump at the interface which is equal to surface tension, τ, times the local curvature, κ. Since the constant added to the pressure has no dynamical effects, we can write, as our boundary condition:

$$p\big|_{\text{interface}} = \tau\,\kappa \qquad (4)$$

At infinity,

$$p \to \frac{1}{2\pi}\frac{dS}{dt}\ln r \qquad (5)$$

where r is the distance from the injection point.

The boundary condition (4) is not always a fully correct description of the situation in real, three dimensional fluid cells[10]. In this paper we nonetheless use Eq. (4) in part because it provides an interesting mathematical problem.

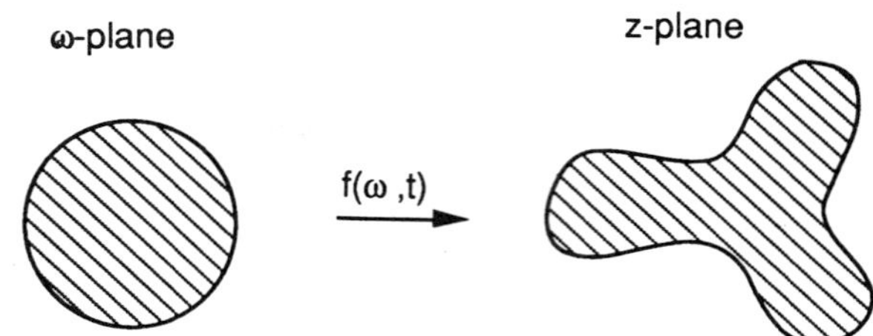

Fig. 2. A conformal map $f(\omega, t)$ maps a unit circle in ω-plane onto the real physical interface and the outside of the circle into the out side of the bubble.

A hodographic method is used to solve the equations (see for example references 2, 5 and 8). A conformal mapping $f(\omega,t)$ is used to map the unit circle in the ω-plane onto the interface in the z-plane and the exterior of the unit circle into the exterior of the bubble. (See Fig.2) So in the ω-plane, the outside of the unit circle $|\omega|\geq 1$ corresponds to the physical domain and the inside of the unit circle $|\omega|<1$ corresponds to the unphysical domain. All the zeroes and poles of g have to remain inside the unit circle to make the solution meaningful. Using this formulation and rescaling the variables in the problem, we can reduce the above equations into one equation describing the time evolution of the mapping $f(\omega,t)$

$$\frac{\partial f}{\partial t} = \omega \frac{\partial f}{\partial \omega} \hat{A} \left(\frac{1- d_0 \, \mathrm{Re}\left(\omega \frac{\partial \hat{A}\{\kappa\}}{\partial \omega}\right)}{\left|\frac{\partial f}{\partial \omega}\right|^2} \right) \tag{6}$$

This holds for the situation in which the area of the bubble grows linearly with time, in this case, the time derivative of the area is 2π. In this equation, $\hat{A}$ is an operator which acts on functions which are analytic in a strip about the unit circle. For a function $F(\omega)$ that can be expressed as:

$$F(\omega) = \sum_{n=-\infty}^{\infty} a_n \, \omega^n \tag{7}$$

we define:

$$\widehat{A}\{F(\omega)\} = a_0 + 2 \sum_{n \le -1} a_n \, \omega^n \tag{8}$$

The curvature, κ, can be expressed in term of $f(\omega, t)$ as:

$$\kappa = -\text{Im}\left(i \, \frac{\dfrac{\partial f}{\partial \omega} + \omega \dfrac{\partial^2 f}{\partial \omega^2}}{\dfrac{\partial f}{\partial \omega} \left|\dfrac{\partial f}{\partial \omega}\right|} \right) \tag{9}$$

d_0 is dimensionless a parameter proportional to the surface tension in the system. In terms of the parameters of the system, d_0 can be expressed as:

$$d_0 = \frac{\pi^2 \, \tau \, b^2}{3 \, \mu \, \dfrac{d\,S}{dt} \, A(0)} = \frac{\pi \, \tau \, b^2}{6 \, \mu \, A(0)} \tag{10}$$

where dS/dt is the time derivative of the bubble's area, $A(0)$ is the zeroth Fourier coefficient of the initial condition of the function $g(\omega,0)$[11].

Following the work in Ref.8, 2, 5, we use the singularities in the unphysical domain to describe the analytic structure of the mapping f and the physical quantities of the bubble. We know that this method works when the surface tension is zero, and hope that the study of the singularities can give an effective treatment of the problem when the surface tension is not zero.

III. BRIEF REVIEW OF THE ZERO SURFACE TENSION CASE

Before moving on to the simulation results for the non-zero surface tension case, we give a brief review of some previous work on the bubble growth at zero surface tension[2,3,5,8]. We choose our initial condition for $g(\omega,t)$ to have the zero or pole structure:

$$g(\omega,t) = A(t) \, \frac{\displaystyle\prod_{i=1}^{m} \left(1 - \frac{z_i(t)}{\omega^q}\right)}{\displaystyle\prod_{j=1}^{n} \left(1 - \frac{p_j(t)}{\omega^q}\right)} \tag{11}$$

For $m \ge n$, this analytic structure will be preserved and the original partial differential equation is reduced to some ordinary differential equations governing the evolution of the zeroes $z_i(t)$, the poles $p_j(t)$, and the number $A(t)$.

For zero surface tension, the work of Sarkar[5] has given a good description of what happens when all the singularities of g are, in fact, zeros. Then it is almost always true that one or more of the zeros will hit the unit circle. At that critical time, the interface will usually develop cusps and the solution will break down.

A particularly simple situation arises when we have q zeros symmetrically placed within the unit circle. We call this the 'one-zero' case. The solution has the structure:

$$g(\omega,t) = A(t) \left(1 - \frac{z(t)}{\omega^q}\right) \tag{12}$$

Since we have a q-fold symmetry here, we shall not distinguish between the different qth roots of $z(t)$ or $p(t)$. So when we talk about a singularity, it should be understood that there are actually q symmetrically placed singularities.

The equations of motion for $A(t)$ and $z(t)$ are:

$$\frac{d\,A(t)}{d\,t} = \frac{1}{A(t)\,(1-z(t)^2)} \tag{13}$$

$$\frac{d\,z(t)}{d\,t} = \frac{(q-2)\,z(t)}{A^2(t)\,(1-z^2(t))} \tag{14}$$

For q=3 they have the solution:

$$A(t) = \frac{A(0)}{z(0)} \left\{ 1 - \left[1 - 2\,z^2(0) + z^4(0) - 4\,t\,\frac{z^2(0)}{A^2(0)} \right]^{\frac{1}{2}} \right\}^{\frac{1}{2}} \tag{15}$$

$$z(t) = \frac{z(0)}{A(0)}\,A(t) \tag{16}$$

Where $A(0)$ and $z(0)$ define the initial conditions. Notice that when t reaches the critical time t_c,

$$t_c = \left(\frac{A(0)\,(1-z(0)^2)}{2z(0)}\right)^2 \tag{17}$$

the solution becomes singular. It then fails to make sense for $t>t_c$. This t_c is the critical time we mentioned at the beginning.

In contrast, we can consider cases in which the solution remains valid for all time. We call the simplest of these the 'one-pole' case. Here there are q symmetrically-placed poles and zeros. The solution has the structure

$$g(\omega,t) = A(t) \frac{1 - \dfrac{z(t)}{\omega^q}}{1 - \dfrac{p(t)}{\omega^q}} \qquad (18)$$

Here $z(t)$ and $p(t)$ respectively describe the positions of the zeros and poles. We shall consider only the case when z and p are real and obey $0 < z(0) < p(0) < 1$. Then the solution will be one in which the poles and zeros move asymptotically toward the unit circle but never get there. Furthermore for all the later time we have $z(t) < p(t)$. So the solution exists for all times. For very large times, the asymptotic solution is

$$A(t) = (2\ t)^{1/2} \qquad (19)$$

$$p(t) = 1 - c_2 \exp\left(-q\ \frac{t^{1/2}}{c_1} \right) \qquad (20)$$

$$z(t) = 1 - \frac{c_1}{t^{1/2}} \qquad (21)$$

where c_1 and c_2 are two constants of integration.

IV. SIMULATION RESULTS FOR THE NON-ZERO SURFACE TENSION CASE

Following Bensimon[8], we use a pseudo-spectral method in our calculation. Instead of solving Eq.(5), we solve the equation for $g(\omega,t)$, which is,

$$\frac{\partial g}{\partial t} = \frac{\partial}{\partial \omega}\left(\omega\ g\ \widehat{A}\left(\frac{1 - d_0\omega\dfrac{\partial \widehat{A}\{\kappa\}}{\partial \omega}}{|g|^2} \right) \right) \qquad (22)$$

We can make a Fourier expansion of the function $g(\omega,t)$. Considering the q-fold symmetry on the unit circle in the ω-plane, this expansion looks like:

$$g(\omega,t) = \sum_{n=0}^{\infty} g_n(t)\ \omega^{-n\ q} \qquad (23)$$

Using a second-order Runge-Kutta method we can calculate the coefficients $g_n(t)$. The expansion is truncated at a certain number of terms N, beyond which all g_n fall into round-off error.

In our simulations, we considered in detail two cases which have initial conditions of the 'one-zero' and 'one-pole' cases described above.

A. "One-Zero" Initial Condition

For the analog of the one-zero case, we used $d_0 = 10^{-3}$, and the initial condition

$$g(w,t)|_{t=0} = 1 - \frac{0.5}{\omega^3} \tag{24}$$

Thus, in the previous notation, $A(0)=1.0$, $z(0)=0.5$, $q=3$.

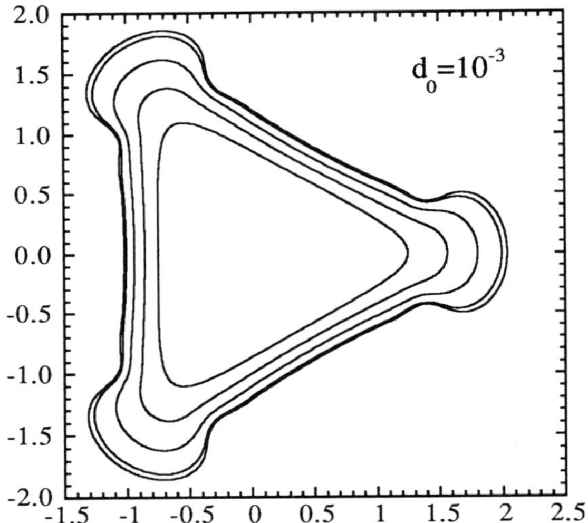

Fig. 3(a) Evolution of a bubble for $d_0 = 10^{-3}$ for a one-zero initial condition with $z=0.5$. The innermost bubble is our initial condition, the time for the other bubbles are 0.18, 0.36, 0.54, 0.59 from inside out. Notice here $t_c = 0.56...$ and we have calculated beyond that time.

Fig.3 shows the interface for $d_0 = 10^{-3}$ (Fig.3(a)) together with that for $d_0 = 0$ (Fig.3(b)). Here $t_c = 0.5625...$. We can see that for $d_0 = 0$ case the tips of the interface develop into cusps when $t=t_c$, while for $d_0 = 10^{-3}$ case the tips look much rounder and we can carry the calculation beyond t_c. In Fig.3(c), the two cases are shown together.

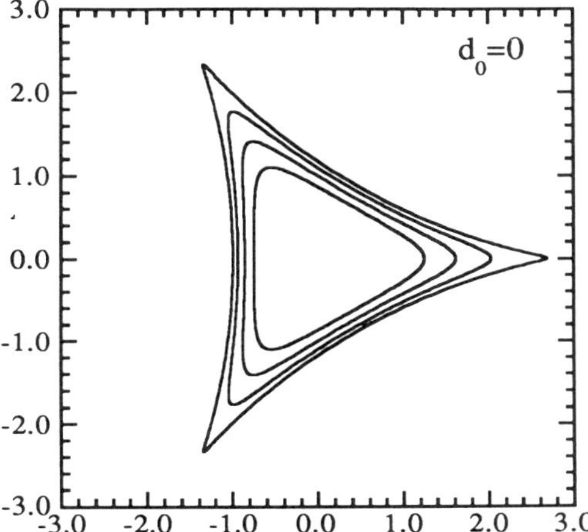

Fig. 3(b) Evolution of a bubble for d0=0 for the one one-zero case with initial condition z=0.5. The corresponding time from inside out is: 0.0, 0.18, 0.36, 0.54. The tips are developing into cusps.

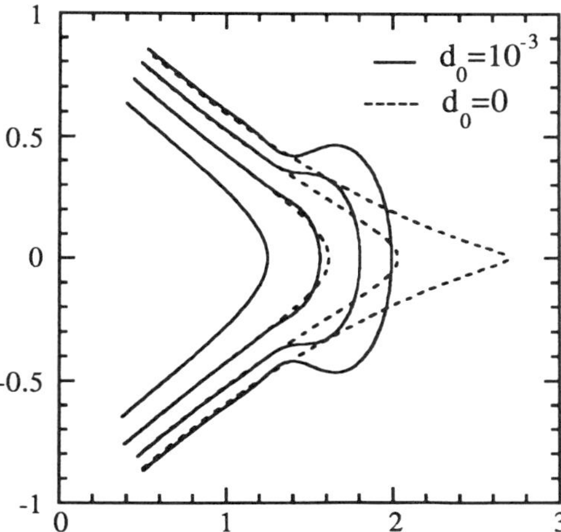

Fig. 3(c) Fig. 3(a) and Fig. 3(b) are displayed together on the same scale for times 0, 0.18, 0.36, 0.45, 0.54. Only one third of the whole interface is shown here. The small surface tension changes the shape of the interface significantly.

From our simulation results, $g(\omega,t)$ can, for $t \gtrsim 0.005$, be reasonably well fit by the structure:

$$g(\omega,t) = H\left(\frac{1}{\omega^3}\right) + A(t)\left(\frac{1}{1 - \dfrac{p(t)}{\omega^3}} + \frac{1}{1 - \dfrac{p(t)^*}{\omega^3}}\right) \tag{25}$$

where $H(x)$ only contains low powers of x. Instead of the initial zero, we now see two complex conjugate poles. For high frequencies, we expect g_n is of the form $\exp(-n|p|)\cos(n\theta + \theta_0)$.

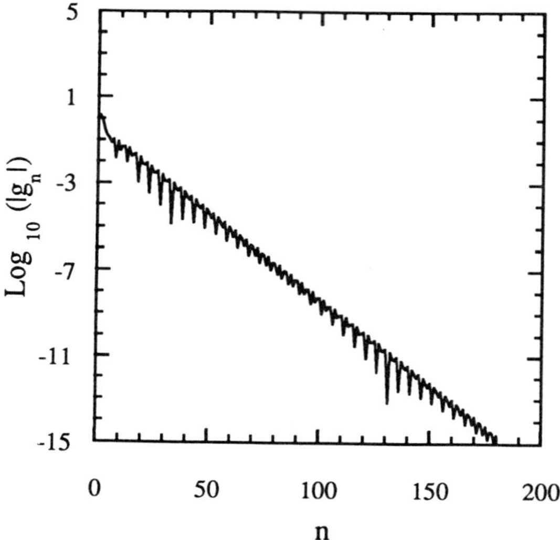

Fig. 4(a) $\mathrm{Log}_{10}(g_n)$ vs n at time 0.36 for the initial one-zero case. The periodic bumps on the top of a straight line suggests a two-pole structure.

In Fig.4(a), $\mathrm{Log}_{10}(g_n)$ vs n at time=0.36 is plotted. By ignoring the bumps, we fit the data by a straight line to get an estimate of $|p|$. Then we apply Fourier transform to the series $g_n \exp(n|p|)$ to get an estimate of θ. Then we change $|p|$, θ and θ_0 around their estimated values to get the best fit. If the structure in Eq.(24) is true, $g_n/[(|p'|)g_{n+1} + \exp(-|p'|)g_{n-1}]$ vs n should be a horizontal line. Furthermore, this ratio is $2/\cos(\theta)$ and $|p'|=|p|$. In Fig.4(b) $g_n/[\exp(|p'|)g_{n+1} + \exp(-|p'|)g_{n-1}]$ is plotted against n and for the high frequencies, they do look like horizontal lines. The values of p' are generally different from those of p by roughly 1 part in 10^3 and θ we get using these two different methods are different by roughly 1 part in 100. Given fits of this quality, it seems reasonable to call $p(t)$ and $p^*(t)$ the positions of the singularities in ω^q.

Fig.5 shows how these singularities move inside the unit circle. Each of the original three zeros split into two poles, which then move towards the unit circle. Fig.6(a) shows how the quantity (1-|p(t)|), which is a measure of the distance between the singularities and the unit circle, changes as a function of time. The distance is plotted on a logarithmic scale. It suggests that the singularities might asymptote the unit circle exponentially in time. Fig.6(b) shows the phase of p(t) as a function of time.

These singularities may not be poles. The basic equations do not seem to admit pole solutions[16], but nonetheless the singularities generated by the numerical work look very much like poles.

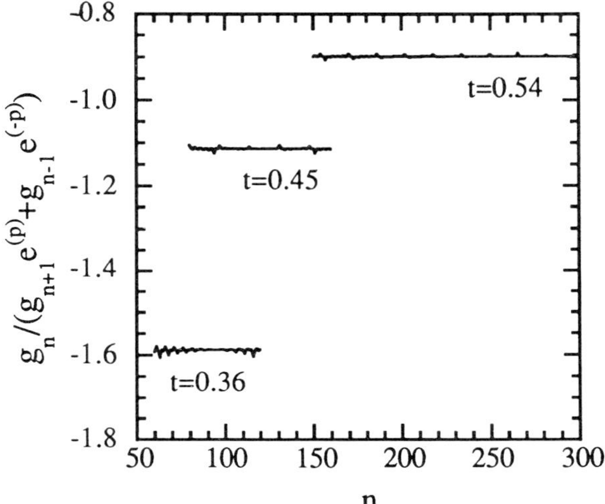

Fig. 4(b) $g_n/[\exp(|p'|)g_{n+1}+\exp(-|p'|)g_{n-1}]$ is plotted against n for high frequencies for times 0.36, 0.45, 0.54. The apparent horizontal lines confirm the previous observations that the leading singularity structure looks like two poles.

B. "One-Pole" Initial Condition

A simpler solution is presented in the case in which the initial condition is taken to be an analog of the one-pole case for zero surface tension. We use again $d_0 = 10^{-3}$, and take as the initial condition

$$g(w,t)|_{t=0} = \frac{1}{1 - \dfrac{0.5}{\omega^3}} \tag{26}$$

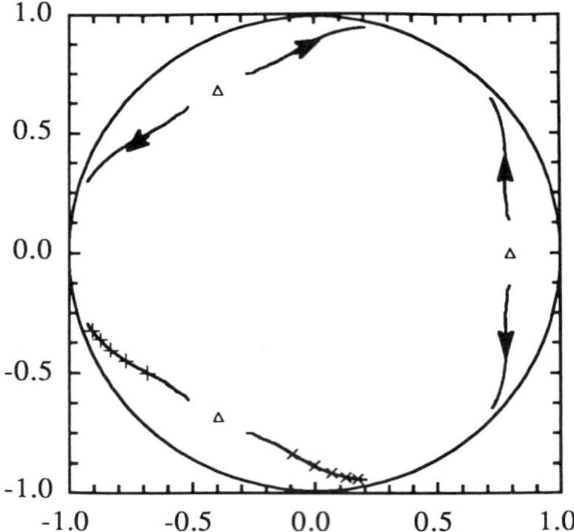

Fig. 5 The motion of the singularities inside the unit circle for $d_0 = 10^{-3}$ and the one-zero initial condition. The initial zeroes which are cubic roots of 0.5 are denoted by 'Δ'. Each of them splits into two poles which move towards the unit circle. On the bottom lines, towards the unit circle, the symbols "x" mark the position of the poles on those two lines for times 0.1, 0.2, 0.3, 0.4, 0.5.

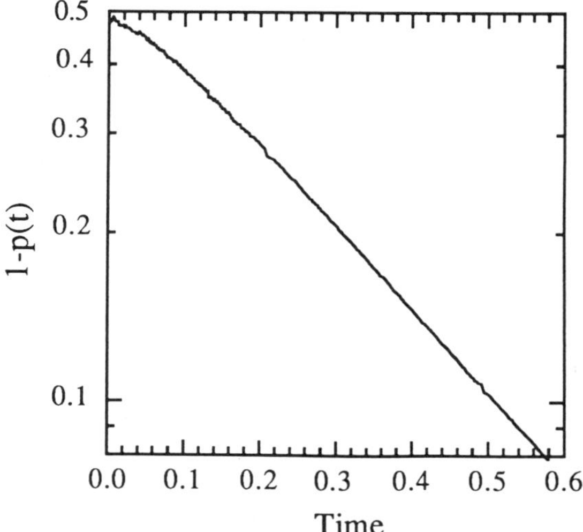

Fig. 6(a) The distance between the singularities and the unit circle from our fitting results is plotted in logarithmic scale against time. It looks like a quite good straight line for large time which suggests that the poles might asymptote the unit circle exponentially in time.

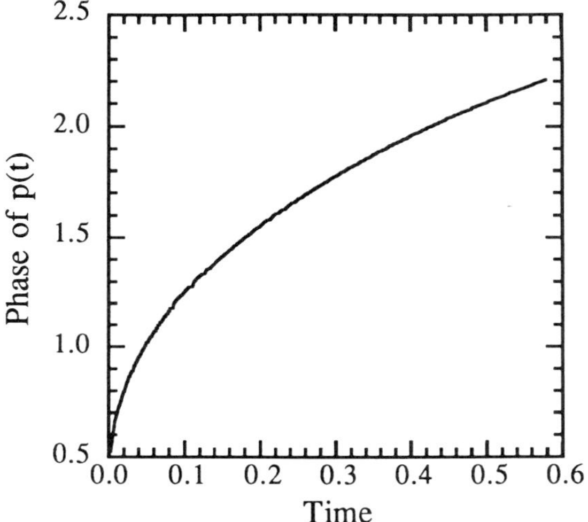

Fig. 6(b) The phase of p(t) vs time.

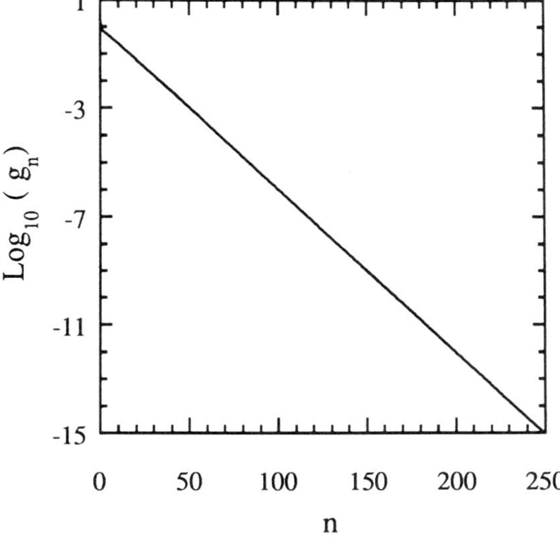

Fig. 7(a) Fourier coefficients in logarithmic scale for $d_0 = 10^{-3}$ and the one-pole initial condition. The exponential behavior shown by the coefficients suggests a pole structure for g.

In our previous notation, A(0)=1.0, z(0)=0.0, p(0)=0.5, q=3.

Fig.7(a) shows the Fourier coefficients generated by the simulation at time=0.6 when this initial condition is used. The good fit to a straight line on this log linear plot indicates that one can get a good fit to the data by taking the leading singularity to be a pole. Hence we try a fit in the form:

$$g(w,t)= H(1/w^3) + \frac{B(t)}{1 - \dfrac{p(t)}{\omega^3}} \qquad (27)$$

Here again, the power series expansion of H(x) converges quite rapidly. Our simulation results thus suggest that a pole structure is quite a good representation of the data. By measuring slop of the curve on this log linear plot, we can find the quantity p(t) which is plotted in Fig.9. In Fig.7(b) g_{n+1}/g_n is plotted against 1/n for time=0.6. This ratio is almost a constant, but it does show a small and significant variation. This variation indicates that the singularity structure is not simply a single pole. Nonetheless, when we try to extrapolate the data to $n \to \infty$, we can get a number which is different from p by 1 part in 10^3. Therefore we'll still use the pole to describe the structure.

Fig.8(a) is the interface for $d_0 = 10^{-3}$. In Fig.8(b) the zero surface tension case is carried to t=20. For time smaller than t=0.6, the interfaces for the two cases are almost identical to each other. We haven't been able to carry the calculation for the non-zero surface tension case long enough to see if there arises the interesting shape exhibited by the zero surface tension case. In Fig.9 where p(t)'s for both cases are plotted, we can see some small changes caused by the surface tension. Notice that the poles move more slowly towards the unit circle for the non-zero surface tension than when surface tension is zero. This also suggests that even for non-zero surface tension the singularities asymptote the unit circle as time goes to infinity.

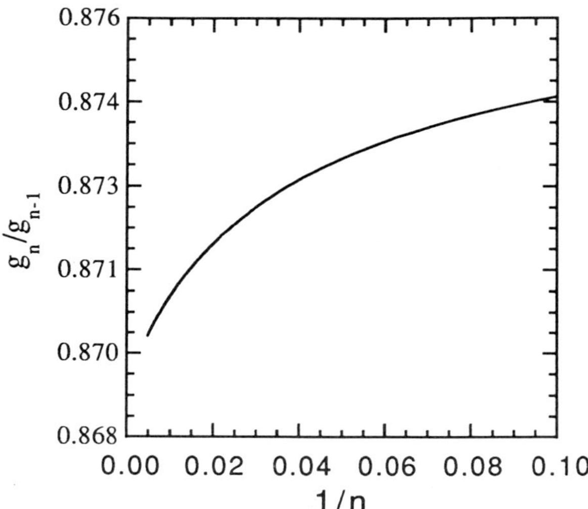

Fig. 7(b) Here, g_{n+1}/g_n doesn't show a linear behavior in 1/n for large n. This indicates that a single branch cut cannot describe all the singularity structure of g.

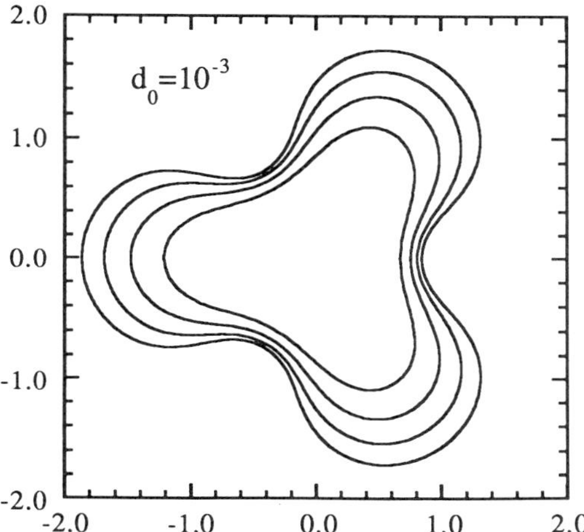

Fig. 8(a) Fourier coefficients in logarithmic scale for $d_0 = 10^{-3}$ and the one-pole initial condition. The exponential behavior shown by the coefficients suggests a pole structure for g.

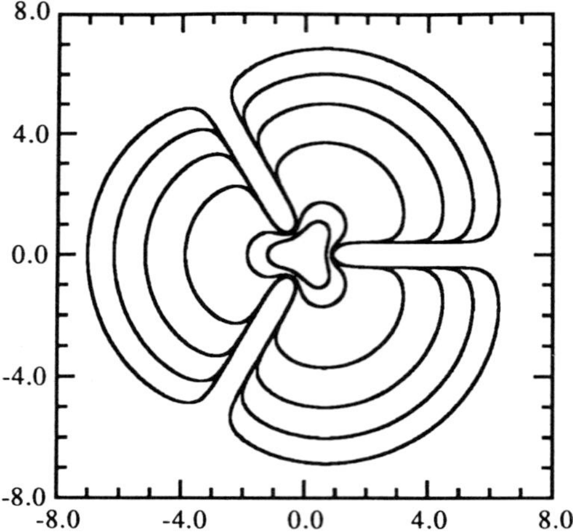

Fig. 8(b) For the same initial condition as in Fig.8(a), $d_0=0$, and the times: 0.0, 0.6, 5.0, 10.0, 15.0, 20.0. We know that at least up to t=0.6, the effect of the surface tension on the shape of the interface is rather small. We are not sure whether this property will be kept for some much later time.

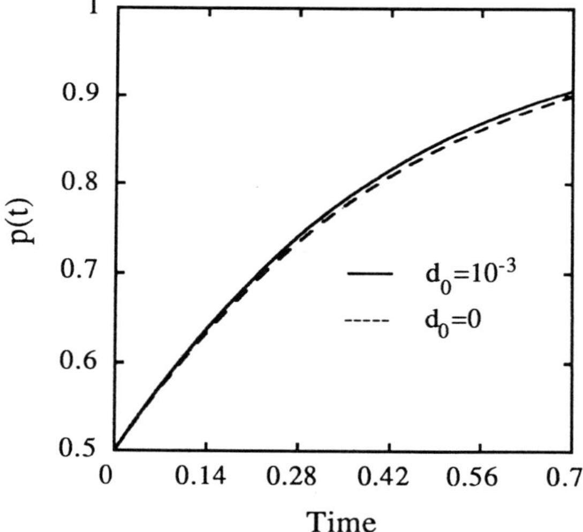

Fig. 9 p(t), the position of the singularity, is plotted against time. The solid line is the $d_0=10^{-3}$ case and the dotted line is the $d_0=0$ case. We can see some small changes caused by the surface tension.

V. CONCLUSIONS

We calculated the bubble growth in the Hele-Shaw cell for the non-zero surface tension case and compared them with the corresponding zero surface tension case. We notice that a pole-like singularity seems to maintain its form even with non-zero surface tension. Conversely, an initial condition with a zero in g is quite unstable against the 'perturbation' caused by surface tension. We speculate that isolated pole-like structures might be a good solution to the equations even for non-zero surface tension, and that these singularities might maintain their form even at very large time, when the singularity position asymptotically approaches the unit circle.

VI. ACKNOWLEDGEMENT

The authors would like to thank the following people for helpful discussions: P. Constantin, B. Shaw, M. Shelley, S. Tanveer.

This work is supported by ONR and the University of Chicago Materials Research Laboratory. We would also like to acknowledge the support of CNSF for computational facilities.

REFERENCES

[1] L. Paterson, J. Fluid Mech. 113, 513 (1981); L. Paterson, Phys. Fluids 28, 26 (1985); S.N. Rauseo, P.D. Barnes, Jr., and J.V. Maher, Phys. Rev. A35, 2840 (1987).

[2] B. Shraiman and D. Bensimon, Phys. Rev. A30, 2840 (1984).

[3] S.D. Howison, J. Fluid Mech. Vol.167, 439 (1986).

[4] D. Bensimon, P. Pelce, Phys. Rev. A33, 4477 (1986).

[5] S. Sarkar and M. Jensen, Phys. Rev. A35, 1877 (1987).

[6] P.G. Saffman and G.I. Taylor, Prog. Roy. Soc. A245: 312 (1958).

[7] In : "Dynamics of Curved Fronts " P. Pelce ed., Academic Press, Inc., San Diego (1988) and references therein.

[8] D. Bensimon, L.P. Kadanoff, S. Liang, B.I. Shraiman and C. Tang, Rev. Mod. Phys. 58, 977 (1986).

[9] S. Tanveer, Phys. Fluids 30, 1589 (1987).

[11] C.-W. Park & G. M. Homsy, J. Fluid Mech. 139, 291 (1984) and J. W. MacLean & P. G. Saffman, J. Fluid Mech. 102, 455 (1981).

[12] In the situation of a channel flow, a dimensionless capillary number similar to $1/d_0$, Ca, can be defined. Following Meiburg and Homsy[13],

$$ Ca = \frac{3\mu u L^2}{\tau b^2} \, , $$

where u is the mean velocity with which fluid 1 is injected and L is the width of the channel. So μL in the channel is the area growth rate dS/dt in the bubble growth case. If we take A(0) to be 1 and L to be 2π as in Ref. 13, the number of d_0 is the number of π^2/Ca and correspondingly, the number of $4\pi^2$ times the dimensionless surface tension in the calculation of Tryggvason and Aref[14] and Degregoria and Schwartz[15].

[13] E. Meiburg and G. M. Homsy, <u>Phys. Fluids</u> **31**, 429 (1988).

[14] G. Tryggvason and H. Aref, <u>J. Fluid Mech.</u> **136**, 1 (1983) and <u>J. Fluid Mech.</u> **154**, 287 (1985).

[15] A. J. Degregoria and L.W. Schwartz, <u>Phys. Rev. Lett.</u> **58**, 1742 (1987).

[16] L. Kadanoff, work in progress.

SAFFMAN TAYLOR FINGER PROBLEM WITH THIN FILM EFFECTS

S. Tanveer

Department of Mathematics
Ohio State University
Columbus, OH 43210

I. INTRODUCTION

A Hele-Shaw cell is a pair of parallel very long plates each of width $2a$ that are separated by a small gap b such that $b << a$. The motion of a less viscous fluid displacing a more viscous fluid in this geometry is the simplest of a heirarchy of increasingly complex but related problems in pattern formation that include dendritic crystal growth and directional solidification (see Pelce[1] and Kessler, Koplik & Levine[2] for recent reviews).

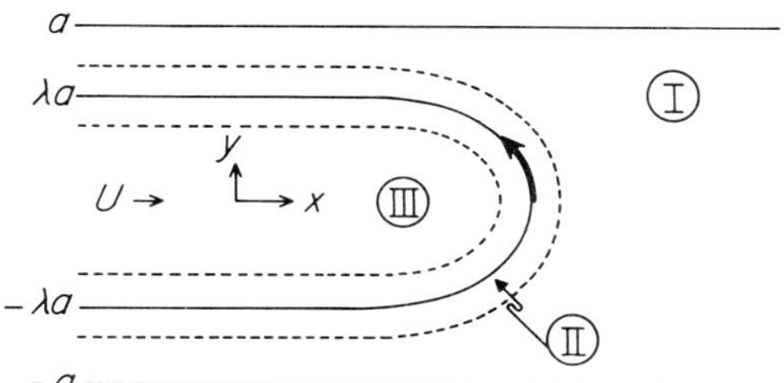

Fig.1   Lateral frame finger view. Flow regimes II and III relevant only
with thin viscous film in the narrow gap as in Fig. 3. The dotted
lines bounds region II of width $O(b)$. The arrow indicates direction of increasing ν (used in eqn.(10)) on the interface

It is well known that when a viscous fluid of viscosity μ filling the cell is displaced by a less viscous fluid under the action of a pressure gradient, a planar interface is unstable[3] and that through a process of competition, a single steady finger emerges that translates down the channel with constant velocity U . Using the idealization of an infinitely long steady finger (lateral $x - y$ plane view in Fig. 1) completely expelling the viscous fluid in the thin gap region with no interfacial tension, a two parameter family of exact solutions due to Saffman & Taylor[3,4] exist for which both the asymptotic finger width λ relative to the channel and the relative position of the finger with respect to channel centerline remain arbitrary (Fig. 1 shows only symmetric finger). Experiment[3], on the otherhand, showed that fingers were symmetric about the channel centerline and for capillary number $Ca = \frac{\mu U}{T}$ (T is the

Growth and Form, Edited by M. Ben Amar *et al.*
Plenum Press, New York, 1991

interfacial tension) not too small, the agreement with the $\lambda = \frac{1}{2}$ symmetric theoretical finger was close.

In recent years, considerable theoretical work, both numerical[5-8] and analytical[9-14] has been focussed on the singular effect of surface tension that breaks the continuum of Saffman-Taylor finger solutions under the assumption that the viscosity of the displacing fluid is negligible and that gravity or any other body forces are absent. As with the case of Saffman-Taylor[3,4] theoretical work, it is assumed that the more viscous fluid is completely expelled by the advancing finger and that the shape of the interface is planar (Fig. 2) in the narrow gap region.

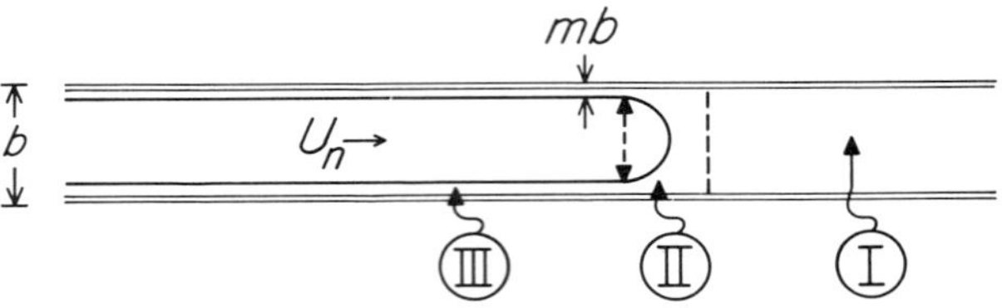

Fig. 2   The interface shape in the narrow gap direction according to MS theory. Arrow indicates direction of advance of the less viscous fluid.

The above simplifying assumptions leads to a set of relatively simple boundary conditions on the flow variables at the interface and will be referred to as the Mclean-Saffman[5] (MS) boundary conditions henceforth. In this case, the pressure field is strictly two dimensional right upto the interfacial boundary and the gap averaged flow in the laboratory frame is determined by the pressure gradient. From the incompressibility of the fluid flow, the pressure satisfies 2-D Laplace's equation. Further there exists a single non-dimensional control parameter $\mathcal{G} = \frac{\epsilon^2 \pi^2}{12 Ca}$, where $\epsilon = \frac{b}{2a}$ is the gap width to channel width ratio. It may be noted that through a mere transformation of parameters[3,4], the MS boundary conditions can easily be extended to include a finite viscosity ratio between the displacing and displaced fluids, a finite film thickness m that does not change at various points of the lateral interface or gravity acting along the channel (x -axis). The analytical[9-11] and numerical evidence[5-8] to date suggests that with the MS boundary conditions, for $\mathcal{G}$ over some range, a discrete set of steady solutions is possible, each characterized by a different λ . For small $\mathcal{G}$, the finger width on each branch $\lambda = \frac{1}{2} + k \, \mathcal{G}^{2/3}$ where k depends on the branch of solution and is the smallest for the finger solution of Mclean & Saffman[5], which is the only one that is linearly stable[15-17]. However, since all the solutions coalesce as $\mathcal{G} \rightarrow 0$, even the Mclean-Saffman branch is nonlinearly unstable[16] with a threshold amplitude shrinking to zero as $\mathcal{G} \rightarrow 0$.

The MS theory prediction of $\lambda = \frac{1}{2}$ in the limit of $\mathcal{G} \rightarrow 0$ is in rough conformity with experiment[3,18] though Tabeling et al[18] have found fingers with width slightly less than a half. However, the actual dependence of λ on $\mathcal{G}$ for larger $\mathcal{G}$ is not consistent. Detailed experiment[18] with the finger suggest that there isn't a single control parameter $\mathcal{G}$ as would be the case if the MS theory were accurate. To a significant extent, the gap to width ratio ϵ and the capillary number Ca are two independent control parameters. The experiment[18] also shows that the interfacial shape in the narrow gap direction is as given in Fig. 3, with the relative film thickness m depending on ν , a parameter describing the on the interfacial location in Fig. 1, and is greater at locations where the normal velocity U_n is larger. Experiment[18] also shows that the finger is unstable for small $\mathcal{G}$. However, for a given level of level of noise in the experiment, a sharp critical $\mathcal{G}$ for instability did not appear to exist; the critical value appearing to depend on Ca as well. Thus the mechanism for nonlinear stability, as originally suggested by Bensimon[16], needs significant modifications if it is to be valid at all for the real finger problem.

More realistic boundary conditions that include the thin film effects have been formulated, we call them the Saffman-Park-Homsy-Reinelt (SPHR) conditions because of the contributions of Saffman[19], Park & Homsy[20] and Reinelt[21] in the development of this condition. Numerical calculations[22] based on these and their simplifications[23,24] suggest that the discrepancy between experiment and the theory is significantly reduced. However, the numerical calculations[22] become unreliable as $\mathcal{G} \rightarrow 0$ because of the degeneracy of λ at

$\mathcal{G} = 0$. Further, since there are two control parameters, it is difficult to guess any scaling laws from the numerics[22−24].

That the MS boundary conditions is deficient in some respects comes as no surprise. What is surprising is that it works as well as it does in predicting a finger width of one half for small $\mathcal{G}$. Indeed, an order of magnitude estimate based on the SPHR conditions would seem to suggest that the MS theory (with an adjustment of parameter $\mathcal{G}$ by a factor of $\frac{\pi}{4}$) should be valid when $Ca << \mathcal{G}^3 << 1$, a range outside most of the experimental data. Indeed, it would seem that for experimental range of control parameters Ca and ϵ (or equivalently $\mathcal{G}$), the lateral curvature term, which is included in the MS boundary conditions, is much smaller than terms due to thin film effects in the SPHR boundary conditions. Our initial interest in this problem arose in trying to understand if $\lambda = \frac{1}{2}$ had any specific significance in the limit of $\mathcal{G} \rightarrow 0$ for the realistic SPHR boundary conditions. We also wanted to know if fingers on different branches of solutions have the same asymptotic width as $\mathcal{G} \rightarrow 0$, a crucial requirement for the Bensimon[16] mechanism for nonlinear stability to be applicable.

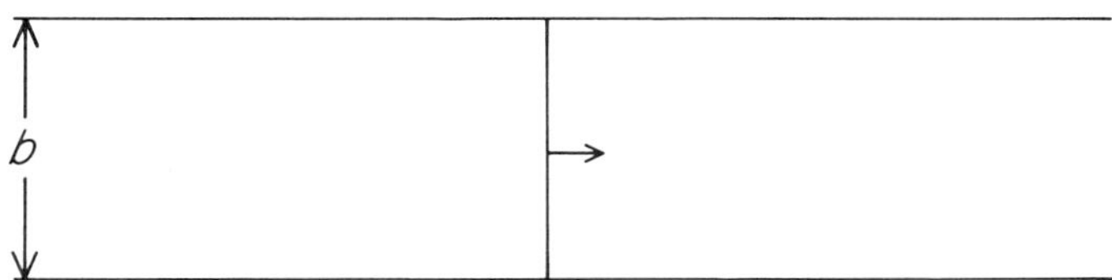

Fig. 3   Realistic interface shape in the narrow gap direction. U_n denotes the normal velocity of advance of the interface. A transverse view of regions I, II and III of Fig. 1 is also given.

This paper is arranged as follows: In section II, we briefly discuss the SPHR boundary conditions that has been developed in its current form by Reinelt[21,22]. In section III, we discuss the method of solution, the similarities and differences with the theory for the MS boundary conditions are highlighted. In Section IV, we discuss the results and comment on agreement with experimental results. The details of the analytical method have already appeared elsewhere[25]

II. SPHR BOUNDARY CONDITIONS

The fluid flow in a Hele-Shaw cell is modelled as a two-dimensional problem by averaging the Stokes equation for low Reynolds number flow across the narrow gap and the thin film in regions I and III respectively. The averaged fluid velocity in the laboratory frame is proportional to the gradient of a two dimensional harmonic pressure field. In region II with width is of order b , the three dimensionality of the flow field is important. However, in the Hele-Shaw limit $\frac{b}{a} \rightarrow 0$, one can solve[21] the region II equations to determine the relative asymptotic film thickness m and the difference of pressure Δp between regions I and III as a function of the normal interfacial velocity U_n and curvature $\frac{1}{R}$ in the lateral plane (Fig. 1). These provide the boundary conditions in the $x - y$ plane linking flow regions I and III. These boundary conditions are then applied at the leading edge of the interface since region II is of negligible thickness compared to the lateral length scale a . It is shown that m and Δp nondimensionalized by $4\mu aU/b^2$ are given by

$$m = m^0(\mu U_n/T) + m^1(\mu U_n/T)\,\epsilon/R \tag{1}$$

$$\Delta p = \frac{\epsilon}{3Ca} \left[\kappa^0(\mu U_n/T) + \kappa^1(\mu U_n/T)\,\epsilon/R \right] \tag{2}$$

The four functions m^0 , m^1 , κ^0 and κ^1 are in general determined numerically solving the region II. However, in the limit of small Ca , following Bretherton's[26] analysis for the motion of a bubble in a tube, Park & Homsy[20] and Reinelt[21] find that

$$\kappa^0(Ca) = -1 - 3.878\,Ca^{2/3} + ... \tag{3}$$

$$\kappa^1(Ca) \;=\; -\pi/4 \;+\; 4.153\, Ca^{2/3} \;+\; ... \tag{4}$$

$$m^0(Ca) \;=\; 1.3375\, Ca^{2/3} \;+\; ... \tag{5}$$

$$m^1(Ca) \;=\; -1.3375\,\frac{\pi}{4}\, Ca^{2/3} \tag{6}$$

In region III, to the leading order, the flow velocity in the laboratory frame is 0 and so the pressure in that region equals the pressure inside the finger which is a constant that can be chosen as 0 without any loss of generality. Thus the pressure at the boundary of region III applied at the finger interface is

$$p \;=\; \frac{\epsilon}{3Ca}\left[\kappa^0(\mu U_n/T) \;+\; \kappa^1(\mu U_n/T)\,\epsilon/R\right] \tag{7}$$

Once again using the continuity of fluid flux between regions I and III one finds that at the finger boundary

$$\frac{\partial p}{\partial n} \;=\; -n_x\,[1 \,-\, m] \tag{8}$$

where ∂n denotes the outward normal derivative and n_x is the component of the unit normal in the x direction (equals one at the finger tip).

Introducing a velocity potential ϕ, which is harmonic, normalizing all velocities by finger velocity U and lengths by channel half width a, and using $\mathcal{G}$ rather than ϵ, the boundary conditions in the frame of the steady finger are[22]:

$$\phi \,+\, x \;=\; -\frac{4\mathcal{G}}{\pi^2}\left[\sqrt{\frac{\pi^2}{12}}\, Ca^{-1/2}\, \mathcal{G}^{-1/2}\,\kappa^0(Ca\ n_x) \;+\; \kappa^1(Ca\ n_x)\frac{1}{R}\right] \tag{9}$$

$$\psi_\nu \;=\; -\left[m^0(Ca\ n_x) \;+\; m^1(Ca\ n_x)\sqrt{\frac{12}{\pi^2}}\,\mathcal{G}^{1/2}\, Ca^{1/2}\,\frac{1}{R}\right] y_\nu \tag{10}$$

where ψ is the harmonic conjugate of ϕ and ν is a parametrizes finger boundary along the arrow direction in Fig. 1.

The MS boundary conditions correspond to replacing m^0, m^1, κ^0 by zero and κ^1 by -1. It is obvious that this simplifies the mathematical problem considerably. For $Ca \ll \mathcal{G}^3 \ll 1$, it is clear from a comparison of magnitude of various terms that to the leading order, SPHR conditions (9) and (10) reduce to the MS boundary conditions once the parameter $\mathcal{G}$ is adjusted by a factor of $\frac{\pi}{4}$. Note the constant -1 in (3) plays no role in (9) as it can be absorbed as part of x by redefining the origin. However, this range of parameter space is outside the range of all experiments to date. Thus, for a quantitatively correct theory explaining the experimental observations, we are forced to deal with the more complicated SPHR conditions.

III DESCRIPTION OF THE METHOD OF SOLUTION

Our interest in the problem was in the limit of $\mathcal{G} \to 0$. since this range has not been accessed by the numerics[22-24]. Detailed analysis has appeared elsewhere[25] and here we restrict ourselves to a qualitative description of the steps involved.

For arbitary Ca, when $\mathcal{G} = 0$, numerical computation of solutions[25] based on equations (9) and (10) show that λ can be arbitrary and so there is a selection problem as for the MS theory without surface tension (i.e. Saffman-Taylor solutions[3,4]). However, note that the commonly used statement 'Surface tension causes selection' would not make sense in the context of this problem since there are terms involving Ca in (10) that are nonzero for $\mathcal{G} = 0$. What really causes selection of finger width λ is the presence of the lateral curvature term $\frac{1}{R}$ in (9). Note that for $Ca = O(1)$, the Saffman-Taylor exact solutions is not the solution to the problem when $\mathcal{G} = 0$, since they are obtained by dropping every term on the right hand side of (9) and (10). However, when $Ca \ll 1$, it is clear from (3), (4), (5) and (6) that one can further ignore every term on the righthand of (9) and (10) since the constant -1 in (3) can be eliminated by redefining the origin of x or ϕ. This is the reason why the Saffman-Taylor experimental data for for small $\mathcal{G}$ and Ca shows good shape agreement with their exact solution provided the proper λ is chosen. Also, note that for fixed ϵ as in the Saffman-Taylor experiment[3], when Ca is reduced sufficiently, $\mathcal{G}$ does not stay small any more and so the Saffman Taylor shape is in disagreement with the experimental shape for very small Ca.

A regular perturbation expansion in powers of $\mathcal{G}$ appears to be possible for arbitary Ca using numerical techniques without any constraints on λ. This series can be obtained explicitly for small Ca without any condition on λ, the relative ordering of Ca and $\mathcal{G}$ playing an important part in determining the complete asymptotic series. For simplicity, we restricted ourselves to fingers that were symmetric about the channel centerline; though we expect that for the nonsymmetric finger both the finger width and the location of the finger relative to the channel centerline will remain arbitary in a regular pertubation analysis as for the MS boundary conditions[5,12]. Thus, one is led to terms beyond all orders in $\mathcal{G}$ to look for possible constraints in λ arising from the condition of a smooth finger tip as for the MS problem[8-14]. The leading order term beyond all orders has been calculated and the constraint on λ obtained explicitly only for small Ca, though we have shown what the form of the constraint would be for arbitrary Ca. To complete the explicit calculation of λ for arbitrary Ca, one needs some subsidiary numerical calculations of the functions m^0 and κ^0 for complex arguments and this is not yet available.

Here, we will only describe the calculations for small Ca. First, we summarize the ideas behind the methods used by other investigators for the relatively simple MS boundary conditions. This provides us with a framework to compare and contrast the SPHR analytic calculations[25] with that based on the MS boundary conditions[9-14]. This comparison also shows whey why some of the methods of other investigators are inapplicable for the SPHR conditions.

First, comes the mathematical formulation. Owing to the invariance of Laplace's equation with respect to conformal mapping, a hodograph representation is helpful and has been used in all analytical studies[3,4,9-14,25] of the Saffman-Taylor problem. Instead of solving for $\phi(x,y)$ in an unknown domain, most of these studies[9-14] look $z = x + iy$ as a function of $W = \phi + i\psi$ or alternately as a function of a known map of W. For the steady finger problem, with the MS boundary conditions, the finger boundary is is characterized by $\psi = Im\,W = 0$. Thus the hodograph mapping allows conversion of an unknown domain problem into a fixed domain problem. Mclean & Saffman[5] have used such a hodograph representation in their numerical problem as well. However, for the SPHR problem, a variation of the above method is needed as (10) implies that the finger boundary is not a streamline. Following the earlier introduction of a semi-circle[12] as our work plane, (call it the ζ plane) for the MS problem, we notice that the arc of the unit circle can always represent the finger boundary provided we relax the requirement that ζ is a known transformation of W. We have to treat both W and z as functions of independent variable ζ. This proved to be convenient way to analytically determine the linear stability of the MS finger[17] as well. Despite the convenience of the hodograph transformation, I do not consider it crucial and I believe that the same results can be obtained using a physical plane representation, though not as conveniently.

Once the equations have been obtained in the hodograph plane, there is once again a difference in the solution procedures that different investigators have employed. For the MS problem one procedure, originally used by Shraiman[10], is to linearize about the $\mathcal{G} = 0$ solution. One obtains a inhomogeneous linear equation. By requiring the Fredholm alternative condition that the right hand side is orthogonal to the solutions of the adjoint linear homogeneous equations one obtains integral constrains that determine λ. In evaluating the integral constraint for small $\mathcal{G}$ via steepest descent, one finds that most of the contribution comes from the neighborhood of some singular point in the complex plane where the linearized equations are no longer valid. However, for the MS problem all the scaling information is contained in the form of the integrand near the singular point where most of the contribution is picked up. Thus the results are qualitatively correct for these problems and, surprisingly, the scaling constants are not very different than what one would get with a fully nonlinear analysis[9,12,13,14]. This provides a conceptually simple way of obtaining quick results and has been used in context of dendritic crystal growth[27] as well. The same results can be obtained by the use of variation of parameters[11,12] on a linearized equation analytically continued to the unphysical domain. The requirement of a unique analytically continued function, as would be the case for a smooth finger boundary, determines the finger width.

A more intuititive and rigorous approach is to do an inner-outer matching where the outer solution is the regular perturbation solution in powers of $\mathcal{G}$ and the inner region consists of neighborhoods of one or more critical points where the outer perturbation expansion is invalid. Kruskal & Segur[28] introduced this procedure in the context of a model nonlinear problem. It was first used in the MS problem by Combescot et al[9]. In the physical region, the outer expansion is uniformly valid. (Strictly speaking this is not true near the tails, where a secondary expansion can be made to match[5]. We ignore this complication in our description). The nonuniformity of the perturbation expansion occur on analytic continuation to the unphysical plane in the neighborhood of one or more critical points. Introducing appropriate scalings in an 'inner region', one obtains the inner equation, which is nonlinear even to the leading order. Solution to this equation is sought that match with the outer solution in an overlap domain. In terms of the inner independent variable this overlap domain does not extend to every direction in the complex plane. The Stokes lines play a crucial role in deciding which directions the inner solution should be matched to the outer solutions. These lines are found by seeking a WKB type of solution to the linearized equation for small $\mathcal{G}$ and determining were the WKB solutions change from being exponentially small to exponentially large. The Stokes lines emanating from the critical points divide up the complex plane into Stokes sectors and in carrying out the inner-outer matching, one has to ensure that the matching occurs in Stokes sector that continue all the way to the physical domain. The matching condition provides a constraint on λ .

In studying the inner equations for the MS problem[9,12] or for the problem of crystal growth in the presence of crystalline anisotropy[29] , one finds just a single parameter that needs to be determined. If one simplifies the nonlinear equation and replaces it by a linear equation, the asymptotic form of the solution in the overlap region between outer and inner expansion remains unchanged to the leading order. Thus the only thing that is affected in solving the linear problem is the numerical value of the single parameter which contains all the scaling information. Detailed analytical investigation by Combescot et al[14] using Borel summation technique has revealed the structure of the solution of the inner problem. In particular, they have also shown why the scaling constants obtained by solving the the nonlinear inner equation is the same as that for the linearized inner equation for asymptotically large values of n , the integer indexing the different branches of solution in the order of increasing width.

For our problem (SPHR conditions), we follow the Kruskal-Segur approach. The Shraiman approach[10] or any other linear approach would fail to even give the scaling law as will be discussed shortly. Compared to the application of the Kruskal-Segur method to the MS problem[9,12,13,14], this problem introduces new difficulties. First, the the structure of the inner region can be very much more complicated because of the nested inner regions and rather complicated structure of the Stokes lines in these inner regions. For instance, when $Ca\, \mathcal{G}^{-1/2}\, (1 - 2\,\lambda)^{-1}\, \lambda^2\ >>\ 1$, a case where the analytical evidence suggests the possibility of successful inner-outer matching for appropriate values of λ , one finds nested inner regions in the neighborhood of the critical point $\zeta\ =\ \frac{i}{q}$, where $q\ =\ (1 - 2\,\lambda)^{1/2}$ (See Fig. 4). This critical point is located outside the unit semi-circle (unphysical region) corresponding to the analytic continuation of equations across the arc of the unit semi-circle, i.e. the finger boundary). The outer most inner region centered at $\zeta\ =\ \frac{i}{q}$ is of relative size $Ca^{2/3}$ (Fig. 4). The inner independent variable in this neighborhood is denoted by ξ_1 . In the transformation from ζ to ξ_1 the imaginary ζ axis is transformed to the real ξ_1 axis.

The leading order inner equation in this region is nonlinear and involves only the kinematic leakage term m^0 in (9). If we construct an asymptotic expansion in this region, we find that it is not uniformly valid in the neighborhood of some point $\xi_1\ =\ \xi_{1_0}$ and its complex conjugate though the figure only shows ξ_{1_0} . It is worthwhile noting that this nonuniformity would be absent if the leading order equation for $\xi_1\ =\ O(1)$ is replaced by a linearized equation as one effectively does using Shraiman's method. The nonuniformity at ξ_{1_0} introduces the necessity of an inner neighborhood around ξ_{1_0} of size $O(\mathcal{G}^{1/3}\, (1 - 2\lambda)^{2/3}\, \lambda^{-4/3})$ where the independent variable $\xi_2\ =\ O(1)$. The leading order equation in this neighborhood includes the dynamic κ^0 term in (9) besides the effect of m^0 . However the lateral curvature term $\frac{\kappa^1}{R}$ does not appear at this order. Once again the asymptotic series expansion for $\xi_2\ =\ O(1)$ becomes invalid in a small region shown around a set of points which are zeros of the Airy function $Ai(\xi_2)$. We only show one of these points (shown by a triangle) and the innermost of the inner regions surrounding this point. In this inner-inner region, the lateral curvature term comes into play, however the deviation of κ^1 from $-\frac{\pi}{4}$ and m^1 are unimportant to the leading order.

Now comes the question of matching. In the outer most of the inner regions, we clearly want an asymptotic solution that matches with the Saffman-Taylor solutions (since we are talking only about $Ca\ <<\ 1$) in the Stokes sectors that adjoin the arc of the unit ζ semicircle (i.e. the physical domain). The structure of this outer Stokes line for $Ca\ <<\ 1$ is exactly the same as for the MS problem[12] Such a solution as $\xi_1\ \rightarrow\ \xi_{1_0}$ has to be matched with the middle-solution as $\xi_2\ \rightarrow\ \infty$ again for Stokes sectors that are a continuation of each of the outer Stokes sectors. A similar matching needs to be performed between the inner-inner regions and the middle inner region. Throughout the matching procedure, it is important to take into account the Stokes line structure in order to ascertain the range of argument of the complex independent variable for which matching has to be accomplished to the next outer region.

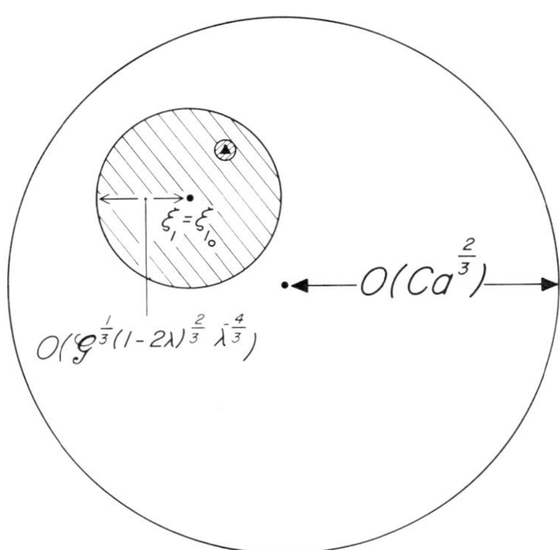

Fig. 4   Nested inner regions near $\zeta = \frac{i}{q}$ (center of largest circle), where $q = (1 - 2\lambda)^{1/2}$. The inner most region is shown as a small circle around a triangle

In the end, the matching condition provides the desired relation between λ , Ca and $\mathcal{G}$.

IV RESULTS AND COMPARISON WITH EXPERIMENTS

We obtained new scalings for the dependence of λ on Ca and $\mathcal{G}$. We find that for $\mathcal{G} << Ca << 1$ fingers with asymptotic widths in the interval $(0, \frac{1}{2})$ are possible

$$\frac{\lambda^2 (1 - \lambda)}{(1 - 2\lambda)} = \bar{k} \frac{\mathcal{G}^{1/2}}{Ca} \tag{11}$$

where $\bar{k}$ depends on the branch of solution and equals to $3.061, 4.878$ for the 1st two branches of solution. Asymptotically for the n -th branch, for large n ,

$$\bar{k} = 2.06(n + 0.279)$$

This formula is pretty accurate even for $n = 2$. In deriving the scaling law (11), we noticed a very crucial role played by the thin film leakage term m^0 and the dynamic term κ^0 but none so ever by m^1 or the deviation of κ^1 from $-\frac{\pi}{4}$ (atleast not to the leading order). That the thinfilm leakage is crucial in obtaining skinny fingers was noticed by Reinelt[22] in his numerical calculations and his comparison with Degregoria & Schwartz's[23] calculation which does not include the leakage term and gives worse agreement with experiment for than the MS theory in the small $\mathcal{G}$ range.

Experimentally, it may be more convenient to think of gap to width ratio ϵ and Ca as the two control parameters, in which case (10) is equivalent to the prediction that for $\epsilon << Ca << 1$,

$$\frac{\lambda^2 (1 - \lambda)}{(1 - 2\lambda)} = k \frac{\epsilon}{Ca^{3/2}} \tag{12}$$

where $k = 2.776$ for the 1st branch. Unfortunately, due to difficulties in numerics, the Reinelt[22] calculations do not extend to sufficiently small $\mathcal{G}$ to quantitatively verify (12). Table 1 shows a comparison between the Tabeling[18] experimental data and formula (12) which is in general qualitative agreement. Note that while conditions $\epsilon << 1$ and $\epsilon << Ca$ are satisfied by the Tabeling data[18] $Ca << 1$ condition is not strictly satisfied as would be necessary for equation (12) to be valid. This may explain the limited agreement as we see in the table.

Table 1   Comparison with Tabeling et al[18] experiment in the small ϵ small Ca range. Also, the values of experimental λ quoted here are eyeball readings from Fig. 8 of theirpaper[18].

ϵ	Ca	$1/B$	$\mathcal{G}$	$3.061 \frac{\mathcal{G}^{1/2}}{Ca}$	Expt. λ	Theor. λ
1/112.5	0.0395	6000	0.165 x 10^{-2}	3.14	0.48	0.481
1/65	0.118	6000	0.165 x 10^{-2}	1.05	0.47	0.447
1/112.5	0.0263	4000	0.247 x 10^{-2}	5.77	0.49	0.489
1/65	0.0789	4000	0.247 x 10^{-2}	1.93	0.48	0.470

Since it is known that the Bretherton approximation holds only for really small Ca , the accuracy of (12) is restricted to a small range of Ca . Notice that for the Tabeling et al data in Table 1 and indeed for all available data we could find when Ca was small, ϵ was not small enough to be $O(Ca^{3/2})$ or smaller. Experiments in this range of parameter space would provide a quantitative test to (12), though my understanding is that experiments are difficult to perform in this range because enormous pressure differences tends to bend the parallel plates of the Hele-Shaw cell. Table 2 shows comparison with the Kopf-Sill-Homsy[30] experiments, where skinny fingers were obtained. We see very little agreement. This is not unexpected since in their experiment, even the SPHR boundary conditions will be invalid in some small neighborhood near the tip, because the product of the gap width and the tip curvature is estimated to be of order unity. It is clear that in the derivation[21] of the SPHR conditions, one assumes that the product of the lateral curvature and the gap width is small everywhere. However, this product at points away from the tip is small for the Kopf-Sill-Homsy experiment and so the SPHR conditions hold except right near the tip. We suggest that this is in some sense equivalent to a tip perturbation on the regular Saffman-Taylor finger and that might explain the similarity of the observed features with that of Couder et al[31] or of Zocchi et al[32] where small tip perturbations in the form of a small bubble or a piercing needle dramatically affects the width selection.

Table 2   Comparison of Kopf-Sill Homsy[30] data with theory

ϵ	Ca	$\mathcal{G}$	$3.061 \frac{\mathcal{G}^{1/2}}{Ca}$	Expt. λ	Theor. λ
0.38 x 10^{-2}	0.25	0.49 x 10^{-4}	0.089	0.15	0.24
0.38 x 10^{-2}	0.054	0.22 x 10^{-3}	0.85	0.20	0.44
0.38 x 10^{-2}	0.0128	0.95 x 10^{-3}	7.35	0.25	0.49
0.38 x 10^{-2}	0.0036	0.33 x 10^{-2}	48.44	0.31	0.50

Notice several interesting things about (12). When $\epsilon \ll Ca^{3/2}$ we have fingers with very narrow widths $\lambda \ll 1$. This is true for every branch for which the right hand side of (12) is small. On the otherhand when $\epsilon = O(Ca^{3/2})$, λ on different branches have different values, each being in the interval $(0, \frac{1}{2})$. This means that if one branch is stable and others unstable (as our preliminary calculations on stability suggest), there is no mechanism for nonlinear instability as suggested by Bensimon when $\epsilon = O(Ca^{3/2})$. Similar to fingers one obtains through external perturbation of the flow field near the tip, one would expect these solutions which are naturally skinny fingers will be very stable to tip breaking disturbances. Only, in the case when $\epsilon \gg Ca^{3/2}$ (Note $\epsilon \ll Ca \ll 1$ as well), do we find an asymptotic finger width close to one half. Thus the approximate agreement with the experimental data[3,18] with the Mclean-Saffman calculations of $\lambda = \frac{1}{2}$ as $\mathcal{G} \to 0$ is fortuitous; an artifact of the choice of the range of experimental parameters than anything fundamental.

An exhaustive study for all possible orderings of the two small parameters Ca and $\mathcal{G}$ has not been done. However, for $1 \gg \mathcal{G}^{7/3} \gg Ca$, analytical evidence suggests that solutions with

$$\lambda = \frac{1}{2} + k \, \mathcal{G}^{2/3} \tag{13}$$

calculated originally using the MS boundary conditions persist. This range appears to be outside all experimental data that we know of. However, we like to comment on its theoretical validity since it may have some relevance to other problems. As we mentioned in section I, a simple order of magnitude argument based on size of different terms in the SPHR boundary conditions conditions would seem to suggest that SPHR conditions go over to the MS conditions (with a multiplicative adjustment of $\mathcal{G}$) when $1 >> \mathcal{G}^3 >> Ca$ which is a smaller range than what we found for validity of (13). Thus for $\mathcal{G}^3 << Ca << \mathcal{G}^{7/3}$, we have an unusual situation where to the leading order, the solutions on the MS theory gives the correct finger shape and the width scaling with $\mathcal{G}$ even when the 3-D terms in the SPHR conditions far exceeds the lateral curvature term on the finger boundary. The reason for this unexpected validity of the MS theory results is, first, that both Ca and $\mathcal{G}$ are small so that the deviation from the Saffman-Taylor finger solutions is actually small. Indeed, the role of terms such as lateral curvature and transverse curvature is not so much to change the Saffman-Taylor shapes as to determine the finger width which is arbitrary to the zeroth order. Second, the finger width is being determined by transcendentally small terms in $\mathcal{G}$ in the physical domain which can only be dtermined by analytic continuation of the equations to the neighborhood of some point in the unphysical plane that is the source of the transcendentally small correction. The relative size of lateral, transverse curvature and the thin film leakage terms in an 'inner' region near this point is rather different from what they are in the physical domain; yet it is this relative size which determines the finger width. It turns out that the lateral curvature is far bigger in the 'inner' region than the thin film leakage or the transverse curvature when $Ca << \mathcal{G}^{7/3} << 1$ and that explains the unexpected range of validity of (13).

V. CONCLUSION

We have described the calculations of finger width analytically when realistic thin film effects are included in the Saffman-Taylor fingering. We have obtained new scaling laws which have important consequences on the question of nonlinear stability. We have also shown that the agreement of the special value $\lambda = \frac{1}{2}$ of the theory neglecting thin film effects with Saffman-Taylor experiment has more to do with the choice of control parameters than anything fundamental.

This work was supported by the National Science Foundation (DMS-8713246).

VI. REFERENCES

1. P. Pelce, 1988, "Dynamics of Curved Front", Academic Press.

2. D. Kessler, J. Koplik, & H. Levine, Patterned Selection in fingered growth phenomena, Advances in Physics, 37:255 (1988).

3. P.G. Saffman, & G.I. Taylor, The penetration of a fluid into a porous medium of Hele-Shaw cell containing a more viscous fluid, Proc. R. Soc. London Ser. A 245:312 (1958).

4. G.I. Taylor & P.G. Saffman, A note on the motion of bubbles in a Hele-Shaw cell and porous medium, Q. Jour. Mech. Appl. Math 12:265 (1959).

5. J. W. McLean, & P.G. Saffman, The effect of surface tension on the shape of fingers in a Hele-Shaw cell, J. Fluid Mech. 102:455 (1981).

6. L. A. Romero, Ph.d thesis, Department of Applied Math, California Institute of Technology, 1982.

7. J.M. Vanden-Broeck, Fingers in a Hele-Shaw cell with surface tension, Phys. Fluids, 26:2033 (1983).

8. D. Kessler & H. Levine, The theory of Saffman-Taylor finger, Phys. Rev. A 32:1930 (1985).

9. R. Combescot, T. Dombre, V. Hakim, Y. Pomeau, & A. Pumir, Shape selection for Saffman-Taylor fingers, Phys. Rev Lett., 56:2036 (1986).

10. B.I. Shraiman, On velocity selection and the Saffman-Taylor problem, Phys. Rev.'Lett., 56:2028 (1986).

11. D.C. Hong, & J.S. Langer, Analytic theory for the selection of Saffman-Taylor finger, Phys. Rev. Lett., 56:2032 (1986).

12. S. Tanveer, Analytic theory for the selection of symmetric Saffman-Taylor finger, Phys. Fluids 30:1589 (1987).

13. A.T. Dorsey & O. Martin, Saffman Taylor fingers with anisotropic surface tension, Phys. Rev. A 35: 3989 (1987).

14. R. Combescot, T. Dombre, V. Hakim, Y. Pomeau, & A. Pumir, Analytic theory of the Saffman-Taylor fingers, Phys. Rev. A 37: 1270 (1987).

15. D. Kessler & H. Levine, Stability of finger patterns in Hele-Shaw cells, Phys. Rev. A 33: 2632 (1986).

16. D. Bensimon, Stability of viscous fingering, Phys. Rev. A 33: 1302 (1986).

17. S. Tanveer, Analytic theory for the linear stability of Saffman-Taylor finger, Phys. Fluids 30: 2318 (1987).

18. P. Tabeling, G. Zocchi, & A. Libchaber, An experimental study of the Saffman-Taylor instability, J. Fluid Mech., 177: 67 (1987).

19. P.G. Saffman, 1982, "Fingering in Porous Medium", Lecture Notes in Physics, pp 208, Ed. Burridge et al, ed. , Springer Verlag.

20. C. W. Park, & G. M. Homsy, Two-phase displacement in Hele-Shaw cells: theory., J. Fluid Mech., 139: 291, (1985).

21. D. A. Reinelt, Interface conditions for two-phase displacement in Hele-Shaw cells, J. Fluid Mech, 183: 219 (1987).

22. D. A. Reinelt, The effect of thin film variations and transverse curvature on the shape of fingers in a Hele-Shaw cell, Phys. Fluids 30: 2617 (1987).

23. L.W. Schwartz, & A. J. Degregoria, Simulation of Hele-Shaw cell fingering with finite capillary number effects included, Phys. Rev. A, 35:276, (1987).

24. S. Sarkar, & D. Jasnow, 1987, Quantitative test of solvability theory for the Saffman-Taylor problem, Phys. Rev. A, 35:4900 (1987).

25. S. Tanveer, Analytic theory for the selection of Saffman-Taylor finger in the presence of thin-film effects, Proc. Roy. Soc. A 428: 511 (1990).

26. F.P. Bretherton, The motion of long bubbles in tubes, J. Fluid Mech, 10:166 (1961).

27. A. Barbeiri, D.C. Hong, and J. Langer, Velocity selection in the symmetric model of dendritic crystal growth, Phys. Rev. A, 35: 1802 (1986).

28. M. Kruskal, & H. Segur, Asymptotics beyond all orders, Aeronautical Res. Associates of Princeton, Technical Memo 85-25 (1986).

29. M. Ben Amar & Y. Pomeau, Theory of dendritic growth in a weakly undercooled melt, Europhys. Lett., 2: 307 (1986).

30. A. Kopf-Sill, & G. M. Homsy, Narrow fingers in a Hele-Shaw cell, Phys. Fluids 30: 2607 (1987).

31. Y. Couder, N. Gerard & M. Rabaud, Narrow fingers in the Saffman-Taylor instability, Phys. Rev. A, 34: 5175 (1986).

32. G. Zocchi, B. Shaw, A. Libchaber & L. Kadanoff, Finger narrowing under local perturbations in the Saffman-Taylor problem, Phys Rev. A, 36:1894 (1987).

Growth of Non-Reflection Symmetric Patterns

Efim Brener[1] , Herbert Levine and Yuhai Tu

Department of Physics and
Institute for Nonlinear Science
University of California, San Diego
La Jolla, CA 92093

1 INTRODUCTION

In recent years, patterns formed by instabilities in propagating interfaces between different phases have received considerable attention[1]. Two of the best known examples of this type of system are dendritic growth[2] and the Saffman-Taylor finger[3]. It has become clear that the degeneracy of the macroscopic problem (family of Ivantsov parabolas[4] and family of Saffman-Taylor Solutions) are lifted by surface tension acting as a singular perturbation. Most surprisingly, this selection mechanism is beyond all orders of perturbation theory; therefore, it requires a rather sophisticated analysis to reveal its workings.

For the dendritic growth problem, one outcome of "microscopic solvability" theory is the critical importance of crystalline anisotropy. Most of the work to date has focused on the simple case of surface tension anisotropy in a cubic, reflection-symmetric crystal. The theory then predicts that dendrites will grow along the minimum surface tension (maximum surface energy) direction, with a velocity controlled by the magnitude of the anisotropy. This picture has been verified experimentally, both quantitatively[5] and qualitatively[6].

Another possible source of anisotropy is the kinetic coefficient which relates interfacial undercooling to local growth velocity. Again, let us first discuss cubic reflection-symmetric crystals. Then, we must distinguish between two possibilities; direction of minimum kinetic coefficient might coincide with that of minimal surface tension, or alternatively, might be at a relative angle of $\pi/4$. For the first case, kinetics merely enhance the effect of surface tension anisotropy and no qualitative change ensues.

The case of nonzero relative angle, i.e. that of competing anisotropies, has been studied in several papers[7,8], all within the limitation of imposing reflection- symmetry. It has been shown that there is a first order transition from surface tension dendrites to kinetic dendrites, as the relative strength of the kinetic term is increased. These dendrites are distinguished by their growth directions, being determined by surface tension and kinetics respectively.

---

[1]Permanent address: Institute for Solid-State Physics, Academy of Sciences of the U.S.S.R., 142434 Chernogolovka, Moscow Dist. USSR

Growth and Form, Edited by M. Ben Amar *et al.*
Plenum Press, New York, 1991

In this paper, we extend the study of dendritic crystals to non-reflection-symmetric anisotropy. The motivation for this is both theoretical and experimental. Theoretically, we will see that this problem requires the extension of the selection method, from a problem of fixing one eigenvalue (velocity), to one of fixing two (velocity and angle). This extension comes about via the breaking of reflection-symmetry for the local equation found – a la Kruskal-Segur[9] – by restricting attention to the neighborhood of the singular point.

Experimentally, there has been recent interest in patterns seen during the diffusion-limited growth of condensed phase phospholipid monolayers[10]. Many of these systems are composed of helical molecules and, hence, do not possess reflection symmetry. Furthermore, there appears to be evidence of the importance of microscopic handedness for macroscopic structures, especially for the spiral dendrite[11]. A preliminary explanation[12] of these spirals explicitly requires non-symmetric growth.

In the case of Saffman-Taylor finger, besides the original family of solutions found by Saffman and Taylor, the family was generalized in a later paper to a double continuum family , parametrized by the finger width λ and the position of the finger x_0 (see figure 1). Based on the same selection mechanism as in the dendrite problem, following picture has emerged for the selection problem of Saffman-Taylor finger. At any fixed value of the surface energy, only a discrete set of finger solutions exist. These solutions are labelled by their asymptotic widths λ and only the smallest width finger is linearly stable. As the surface energy is decreased, the finger width asymptotically approaches one half the channel width. Without any non-symmetric forcing, all selected fingers are reflection symmetric, i.e., $x_0 = 0$[13]. This picture is in good agreement with experimental findings[14], once complications due to thin films on the two plates are accounted for[15].

In this paper, we generalize this family of solutions, as well as the selection mechanism, via inclusion of non-symmetric forcing. In particular, we assume that there exists a gravitational force which acts on the viscous fluid of density ρ. If the cell is rotated away from horizontal by angle ω, we find a two parameter family of exact solutions with zero surface tension for arbitrary ω, which reduces to the known results at $\omega = 0$, because of the non-symmetric forcing, when the surface tension is taken into account, the selected finger, therefore, will be non-symmetric. And the problem, just like the asymmetric dendrites, will be one of fixing two eigenvalue, i.e., λ and x_0.

2 DENDRITES

The basic equations governing dendritic growth are well known[16]. We will describe and do our calculations for the thermal diffusion-limited growth of a crystal from a pure melt.

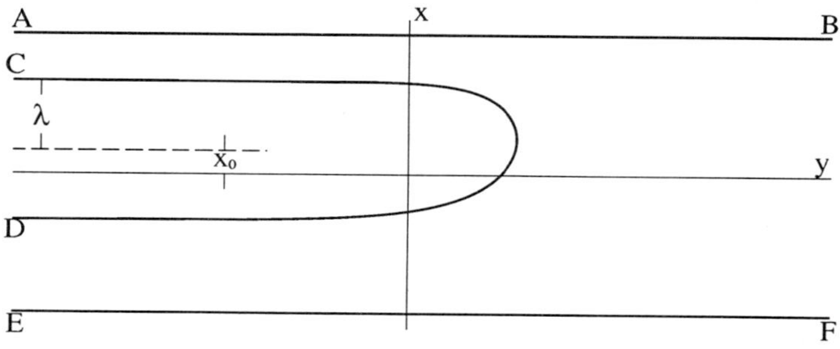

1. Schematic picture of the asymmetrical Saffman-Taylor finger.

The changes necessary to study the impurity diffusion-limited process, which presumably governs the phospholipid experiment, are understood not to alter any of the essential features of the problem.

Let us normalize temperature by L/c_p for latent heat L, specific heat c_p. Then, we must solve

$$D\nabla^2 T = \frac{\partial T}{\partial t} \tag{1}$$

with boundary conditions

$$T(\vec{x} \to \infty) = -\Delta \tag{2a}$$

$$\hat{n} \cdot \nabla T|_{liq} - \hat{n} \cdot \nabla T|_{sol} = \frac{-v_n}{D} \tag{2b}$$

$$T_{sol} = T_{liq} = -d_0(\theta)\kappa - \beta(\theta)v_n \tag{2c}$$

Here D is the thermal diffusivity (assumed equal in liquid and solid), Δ is the dimensionless undercooling $c_p(T_m - T_\infty)/L$, (T_m is melting temperature) and v_n is the local growth velocity. Eq. (2b) is just the Stefan condition relating interface motion to sources for the heat equation. Finally, the anistropy enters in the two functions d_0 (surface tension, which multiplies curvature κ) and β, (kinetic coefficient), which depend on the angle θ between the interface normal and the underling lattice.

The integro-differential shape equation arises via assuming steady-state motion in some direction $\hat{g}$. Since we will restrict our attention here to the two dimensional problems (for simplicity and because the experiments on monolayers are indeed two dimensional), we will label this direction by the angle $\theta^* \equiv \cos^{-1}(\hat{g} \cdot \hat{y})$ where $\hat{y}$ will be the direction of minimal surface energy (see later). Solving for the temperature field and imposing the condition (2c), we find

$$\Delta - \tilde{d}_0(\theta)\kappa - \tilde{\beta}(\theta)\cos(\theta - \theta^*) = \int ds'\, G(\vec{x}(s) - \vec{x}(s'))\cos(\theta - \theta^*) \tag{3}$$

where G is the Green's function

$$\frac{1}{\pi}K_0\left(\sqrt{(\vec{x} - \vec{x}\,')^2}\right)e^{-\hat{g}\cdot(\vec{x} - \vec{x}\,')}$$

where length is measured in units of $v/2D$ and $\tilde{d}_0, \tilde{\beta}$ are the dimensionless analogues of the previously defined objects.

Let us recall the results of investigating the above equation in the case of reflection-symmetric anisotropies; for example, imagine that

$$\tilde{d}_0(\theta) = \bar{d}_0(1 - \epsilon\cos 4\theta)$$

$$\tilde{\beta}(\theta) = \bar{\beta}(1 - \alpha\cos 4\theta)$$

Then, we have the following results:[1,17]

a) At $\bar{d}_0 = \bar{\beta} = 0$, there is a doubly infinite family of Ivantsov parabola solutions, corresponding to rotations of $y(x) = \frac{-x^2}{2p}$, $\Delta = \frac{e^p}{\sqrt{\pi p}}\mathrm{erfc}(\sqrt{p})$, through any angle θ^*.

b) At any non-zero $\bar{d}_0$, if $\epsilon = \alpha = 0$, there are no solutions at all.

c) Solutions exist in the presence of non-zero ϵ and/or α and fall into two categories. Note that our definition of $\hat{y}$ requires positive ϵ. The first possibility is that $\alpha \geq 0$; then, solutions must have $\theta^* = 0$.

d) If $\alpha < 0$, we can have surface tension dendrites ($\theta^* = 0$) or kinetic dendrites ($\theta^* = \pi/4$). The latter branch exists only for large velocity and emerges discontinuously. The actual system has been conjectured to undergo a first order transition to this new branch as velocity is increased. There is some evidence for this scenario in the growth of ammonium bromide[18] and also in fluid mechanical analogues[6].

Following the discovery of the selection mechanism in numerical work, analytic methods were devised to solve this problem. The most logically appealing and rigorously valid methodology is based on the Kruskal- Segur method[9], originally devised for the geometrical model; the method works here because the integro-differential shape equation can be replaced by a differential equation near the relevant singularity in the complex x plane. Here, we review this method as it applies to general forms of surface tension and kinetic anisotropies.

For small $\bar{d}_0, \bar{\beta}$, the solution is close to the Ivantsov one. Hence, we can drop all terms which are products of the above small parameters and the interface shift z. This allows us to linearize and, thereafter, evaluate the integral term in the shape equation. The terms on the left hand side, which involve derivatives, cannot be linearized. Using the rotated coordinate system (and dropping the tildes and the $*$ superscript on the coordinates), the left hand side becomes

$$\frac{-d_0(\theta + \theta^*)(z'' + 1/p)}{\left(1 + (\frac{x}{p} + z')^2\right)^{3/2}} - \frac{\beta(\theta + \theta^*)}{\left(1 + (\frac{x}{p} + z')^2\right)^{1/2}}$$

where $\theta = \tan^{-1}\frac{x}{p} + z'$ and $'$ denotes $\frac{d}{dx}$.

The most important feature of these terms is the singularity near $x = ip$. As shown first by Ben-Amar and Pomeau[19], the leading term in the (linearized) integral, becomes local near this point. The details of how to do the re-scaling have been discussed by Ben-Amar and Pomeau and, for arbitrary undercooling and for the case of kinetic terms, by Brener and Meln'kov[17]. Let us pick the forms

$$d_0(\theta) = \bar{d}_0(1 - \epsilon \cos 4\theta)$$

$$\beta(\theta) = \bar{\beta}(1 - \alpha \cos 4(\theta - \theta_\beta)) \tag{7}$$

Then, the final equation takes the form

$$\frac{d^2 F}{du^2} - \frac{\sqrt{2}\lambda\tau^{3/2} F}{1 - 2e^{4i\theta^*}/\tau^2} =$$

$$-1 - 2\mu\tau\frac{\tau^2 - 2\nu e^{4i(\theta^* - \theta_\beta)}}{\tau^2 - 2e^{4i\theta^*}} \tag{8}$$

In this equation, $F(u)$ is the unknown function of u which measures the rescaled shift form $x = ip$ and $\tau = u + \frac{dF}{du}$. The velocity eigenvalue appears in the combination $\lambda = \epsilon^{7/4}p^2/\bar{d}_0$ and, of course, the unknown growth direction is given by θ^*. Finally, the parameters ν and μ are α/ϵ and $\left(p\bar{\beta}/\bar{d}_0\right)\sqrt{\epsilon}$ respectively. The latter parameter governs the relevant importance of kinetic effects in the system – not surprisingly, it is directly proportional to the Peclet number p.

Before analyzing the general case, let us first recall the picture when $\theta_\beta = 0$ or $\pi/4$. We must choose the solution which suppresses the possible growth

$$F \sim e^{|u|^{7/4}}$$

along the rays arg $u = \pm 4\pi/7$. This fixes both constants of integration in the general solution and leaves us with some fixed function of the solution along the ray arg $u = 0$, which in general will grow exponentially. Therefore, we must demand that λ and θ^* be chosen to suppress this unwanted behavior. Setting $\theta^* = 0$ results in one (real) condition for one unknown , which can be readily determined either numerically or via WKB. At non-zero growth angle, there are two conditions (one complex coefficient equals zero) and two variables; however, it has been shown that there are no nontrivial solutions at

34

$\theta^* \neq 0, \pi/4$. Thus, for reflection-symmetric cubic crystals, growth is either along $0°$ or $45°$.

Let us now study the situation where θ_β is an arbitrary angle[20]. For simplicity, we will fix ν to equal one and consider the behavior at variable μ.

Let us first consider what happens for small kinetic coefficient, $\mu = .1$. Here, we find that there is a solution branch which starts, of course, at zero growth angle. As θ_β is increased, θ^* increases. However, as θ_β approaches $\pi/4$, we recover reflection-symmetry and θ^* decreases back to zero. The velocity and growth of this surface tension dendrite are shown in Fig 2. This type of solution is expected to persist for all values of μ. There are also secondary branches corresponding to slower velocities which are all unstable and, hence, do not appear as physical patterns.

A second set of runs were done at $\mu = .8$, i.e. at large kinetic coefficient. The surface tension dominated branch is still present, but there is a second possibility. At $\theta_\beta \simeq .634$, a new kinetic dendrite branch appears; the velocity for this solution is given in Fig. 3a. The upper part of the solution branch is unstable, since it corresponds to velocity decreasing as the driving is increased. The selected angle for the stable lower part of the branch is depicted in Fig. 3b. Note that as $\theta_\beta \to \pi/4$, the selected angle also approaches $\pi/4$, becoming the kinetic dendrite solution seen in previous work.

How are the above two results connected? As μ is lowered from .8, the onset of the new branch occurs at larger θ_β. Finally, we reach a critical μ for which $\theta^* = \theta_\beta$ is the only possible kinetic dendrite. This value of the kinetic coefficient corresponds to the critical value previously found for the reflection- symmetric problem[8]. For smaller kinetic term, the branch disappears.

We have shown how to find the direction of dendritic growth for systems without reflection-symmetry. The results indicate that this direction cannot be determined a priori, but instead emerges as an eigenvalue of the shape equation.

As mentioned in the introduction, one possible experimental realization of this hypothesized system might be growth of phospholipid monolayers. Specifically, Weis and McConnell[11] found that right-handed dipalmitoyl phosphatidylcholine (R-DPPC) molecules form chiral domains which are absent when the mixture is racemic. Pomeau[12] has attributed this shape to a bending of the crystal axis, caused by nonsymmetric incorporation of impurities, which introduce elastic strain. Our paper shows that non-symmetric dendrites are indeed the select pattern for crystals with no reflection-symmetry and, hence, make the elastic bending scenario possible.

Perhaps, the simplest prediction amendable to testing is the fact that the selection angle will, in general, depend on velocity through the competition between surface tension and anisotropy. One must grow a dendrite in, say, R-DPPC, and measure the relative angle between some static crystal feature (found by diffraction, e.g.) and the growth direction. This offset angle should be a continuous function of velocity. This is in contrast to more common cubic materials where the growth choices are limited to (100), (110), and (111).

3 Non-Symmetric Saffman-Taylor finger

Let us quickly review and extend the equations at motion for fluid displacements in a Hele-Shaw cell. Under the assumption of an exceedingly narrow gap[21] and neglecting all film effects[15], we have Darcy's law

$$\vec{v} = -\frac{k}{\mu}\,\vec{\nabla}p \qquad (9)$$

with $k = b^2/12$ for gap thickness b and fluid viscosity μ. Assuming incompressibility, the

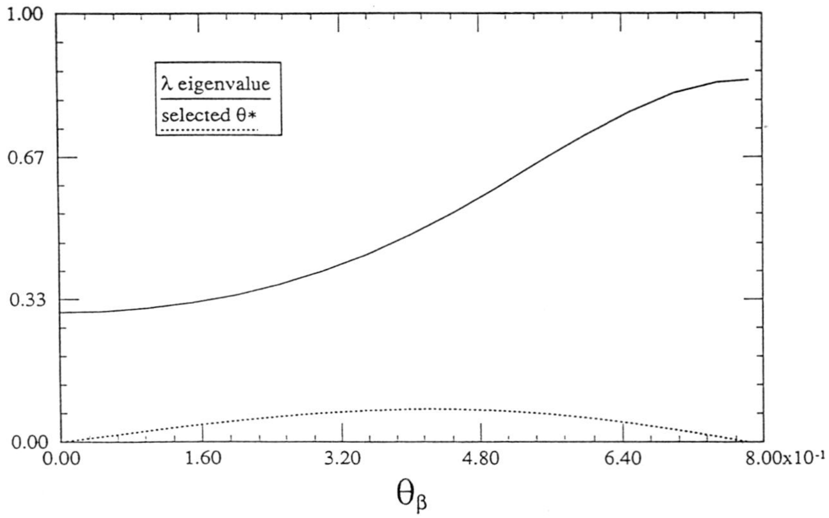

2. Velocity eigenvalue (λ) and selected angle (θ^*) vs. competition angle, small kinetic coefficient case

velocity potential $\phi = -\frac{k}{\mu}\,p$ satisfies Laplace's equation $\nabla^2\phi = 0$. There also exists a stream function ψ such that $w = \phi + i\psi$ is an analytic function of $z = ix + y$.

The boundary conditions are as follows. We fix the overall flow velocity at infinity, giving $\phi \sim Vy$ far ahead of the interface. At the walls $x = \pm 1$, $\psi = \pm V$. On the interface itself, we assume that we are dealing with a steady-state profile moving with velocity U. Therefore

$$U\cos\theta = \frac{\partial\phi}{\partial n} \equiv \frac{\partial\psi}{\partial s} \tag{10}$$

Using $\cos\theta = \frac{dx}{ds}$, we have

$$\psi = U(x - x_0) \tag{11}$$

where x_0 is arbitrary as long as $-1 < x_0 < 1$.

The final condition relates to the pressure on the surface. Neglecting the viscosity of the "pushing" fluid and setting the surface tension to zero, we have

$$\phi = \frac{k}{\mu}\,g\rho x \sin\omega \tag{12}$$

where ω is the tilt angle, g the gravitational acceleration and ρ the fluid density. If we rescale such that the asymptotic velocity V is set to one, we introduce the dimensionless parameter

$$\sigma = \frac{(g\rho\sin\omega)b^2}{12\mu V} \tag{13}$$

which governs the relative strength of the gravitational and pressure gradients.

Following the standard methodology[3,21], we work in the w plane and try to find the analytic function $z(w)$. And after using some conformal mapping in the w-plane and some algebra, the solutions are found to be[22]:

$$x = x_0 - \lambda + \frac{4\lambda\cos\epsilon\pi/2}{\pi}\int_0^\alpha (\tan\alpha')^\epsilon\,d\alpha' \tag{14}$$

$$y = \frac{2}{\pi}(1 - \lambda - x_0)\ln\cos\alpha + \frac{2}{\pi}(1 - \lambda + x_0)\ln\sin\alpha \tag{15}$$

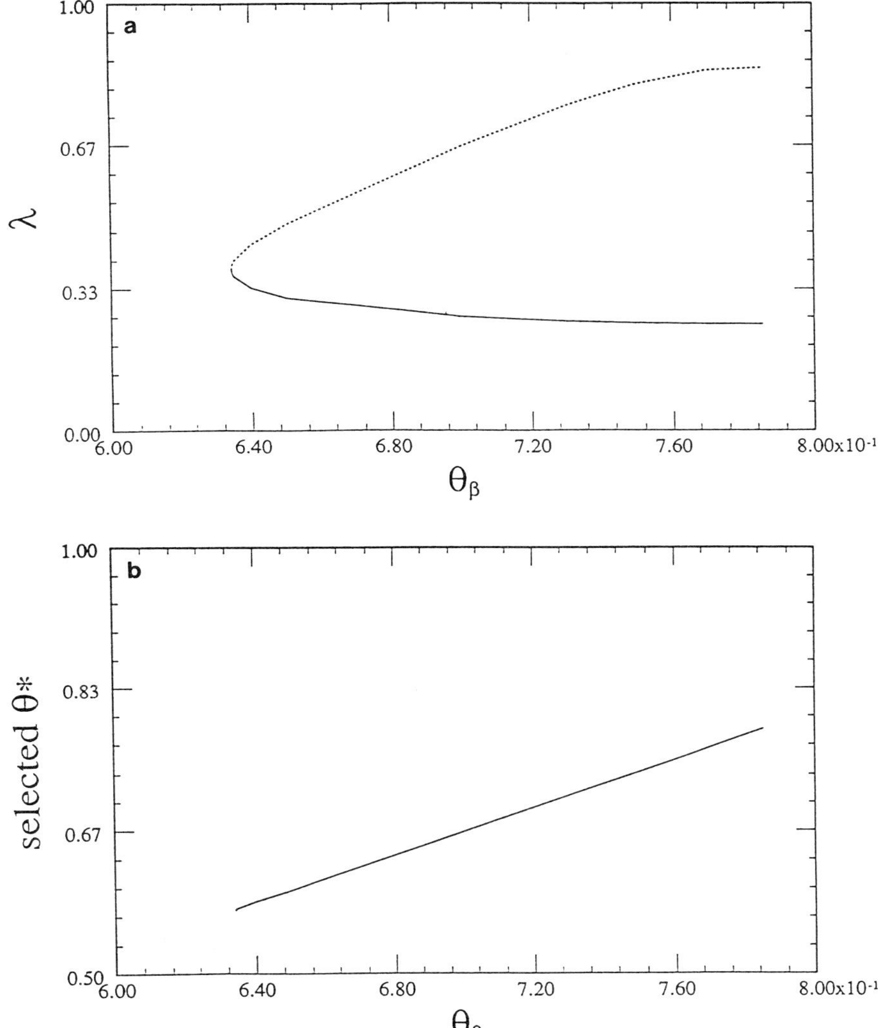

3. a) Eigenvalue versus competition angle, kinetic dendrite branch, $\mu = .8$ b) Selected angle for stable part of the kinetic branch

where the parameter α runs from 0 to $\pi/2$. For the special case $\epsilon = 0$, we have

$$x = x_0 + \lambda \left(\frac{4\alpha}{\pi} - 1 \right) \tag{16}$$

and we recover the previous solution set of Saffman and Taylor[21,23].

If we assume that the tilting angle ω is small, the the singular point in the surface tension will be very close to the point where $\partial y_0 / \partial x = i$. Following exactly the same steps as in the dendrite case, we can get the local equation around the singular point, let us define t via:

$$\frac{\partial y_0}{\partial x} = i\,(1-t) \quad t << 1 \tag{17}$$

The local equation have the form:

$$y_1''(t) + y_1'(t)\frac{\partial}{\partial x}\left(\ln \frac{\partial^2 y_0}{\partial x^2}\right) - \frac{y_1\, 2^{1/2}\, \tau^{3/2}}{\lambda\gamma \left(\frac{\partial^2 y_0}{\partial x^2}\right)^2} = \frac{1}{\frac{\partial^2 y_0}{\partial x^2}} \tag{18}$$

All primes denote derivatives with respect to t and all derivatives of y_0 with respect to x can be explicitly computed given the analytic expressions (14) and (15). Here $\tau = t - y_1' \frac{\partial^2 y_0}{\partial x^2}$, causing the equation to be non-linear in y_1. The same equation as derived in ref. 17 for the problem of dendritic growth in a channel geometry.

From the explicit expression for y_0, we have

$$\frac{\partial y_0}{\partial x} = \frac{1 - \lambda + x_0}{2\lambda \cos \frac{\epsilon\pi}{2}} (\tan \alpha)^{1-\epsilon} - \frac{(1 - \lambda - x_0)}{2\lambda \cos \frac{\epsilon\pi}{2}} (\tan \alpha)^{-1-\epsilon} \tag{19}$$

$$\frac{\partial^2 y_0}{\partial x^2} = -\left[\frac{(1 - \lambda + x_0)\pi}{8\lambda^2 \cos^2 \frac{\epsilon\pi}{2}} (\tan \alpha)^{-2\epsilon}(1 - \epsilon) + \frac{1 - \lambda - x_0}{8\lambda^2 \cos^2 \frac{\epsilon\pi}{2}} (\tan \alpha)^{-2-2\epsilon}(1 + \epsilon) \right] (1 + \tan^2 \alpha) \tag{20}$$

Our analytic approach is valid wherever γ and ϵ are both small. This was required for the derivation of the above equation (18) and also guarantees that the zeros of $\frac{\partial^2 y_0}{\partial x^2}$ are close to $t = 0$. In this case, we can express $\frac{\partial^2 y_0}{\partial x^2}$ as a function of t by inverting to find $\alpha(t)$; after a great deal of algebra, we have

$$\frac{\partial^2 y_0}{\partial x^2} \equiv g(t) = \frac{\pi}{2} \left(i\sqrt{2t - 2t_0} - \epsilon \right) \left(\frac{x_0}{\lambda^2} - \frac{1 - \lambda}{\lambda^2} i\sqrt{2t - t_0} \right) \left(\frac{1 - \lambda - x_0}{1 - \lambda + x_0} \right)^{\epsilon} \tag{21}$$

where $t_0 = \delta + i\pi\epsilon/2$ and

$$1 - \delta = \frac{((1 - \lambda)^2 - x_0^2)^{1/2}}{\lambda} \left(\frac{1 - \lambda - x_0}{1 - \lambda + x_0} \right)^{\epsilon/2} \tag{22}$$

Note that for the simplest Saffman-Taylor problem, $\epsilon = x_0 = 0$ and the two square roots combine to give a linear zero at real t, i.e., along the imaginary $\frac{\partial y_0}{\partial x}$ axis. For the general case, there are two square root zeros which are at complex values of t.

Knowing the above expression (21), we can derive scaling laws in two different limits. First, we consider the case when ϵ as the smallest parameter in the problem; the exact criterion for this will be derived self-consistently in what follows. We first start by setting $\epsilon = 0$ and employ the usual Saffman-Taylor rescaling

$$\begin{aligned} t &\rightarrow \delta\tilde{t} \\ y_1 &\rightarrow \delta\tilde{y}_1 \\ \gamma &\rightarrow \tilde{\gamma}\delta^{3/2} \end{aligned} \tag{23}$$

Substituting these into (18) with the solution at $x_0 = 0$ gives the usual eigenvalue problem which determines $\tilde{\gamma}$ as a function of $\delta_0 = \frac{1-\lambda}{\lambda}$.

We now include finite ϵ and the breaking of reflection symmetry via first order perturbation theory. To see the correct scaling of x_0, we explicitly expand $\partial^2 y_0 / \partial x^2$ in eq (21):

$$\frac{\partial^2 y_0}{\partial x^2} = \frac{\pi}{2} \frac{1-\lambda}{\lambda^2}(t - \delta_0) - \frac{\pi}{2}\left(\frac{1-\lambda}{\lambda^2}\right)\left(\frac{i\pi\epsilon}{2}\right) + \frac{\pi i}{2}\sqrt{2t - 2\delta_0}\,\frac{x_0}{\lambda^2} \tag{24}$$

The two imaginary terms (one due to x_0, one to ϵ) must be the same order of magnitude, and hence x_0 is order $\epsilon/\sqrt{\delta_0}$. For this expansion to be valid, we clearly must have $\epsilon << \delta_0$. The equation to first order then reduces to determining two eigenvalues, $\tilde{x}_0 = \frac{x_0\sqrt{\delta}}{\epsilon}$ and the first order in ϵ correction to $\tilde{\gamma}$.

The second case we consider is the opposite limit; whereas before $\epsilon << \delta$ and therefore $\epsilon << \gamma^{2/3}$, we now imagine $\epsilon >> \gamma^{2/3}$. Clearly, the lowest order scaling will now be changed from the previous case, since the finite angle effect cannot be treated perturbatively. Let us rescale δ and t by ϵ, defining

$$\delta \rightarrow p\epsilon \tag{25}$$
$$t \rightarrow \hat{t}\epsilon \tag{26}$$

The first implies that the position of t_0 is no longer purely real to zeroth order. We assume that $x_0 >> (1 - \lambda)\sqrt{\epsilon}$ which will be shown self-consistently below. From (18) using these rescalings, the term which depends on γ becomes

$$\frac{y_1\,\epsilon^{3/2}\,\hat{t}^{3/2}\,2^{1/2}}{\lambda\gamma\left(\hat{t} - (p + \frac{i\pi}{2})\right)\frac{\epsilon x_0^2}{\lambda^4}\left(\frac{1-\lambda-x_0}{1-\lambda+x_0}\right)^{2\epsilon}} \tag{27}$$

Comparing this to the first term which is order y_1/ϵ^2, we find the requirement

$$\frac{x_0^2}{\lambda^4}\left(\frac{1-\lambda-x_0}{1-\lambda+x_0}\right)^{2\epsilon} \sim \frac{\epsilon^{5/2}}{\lambda\gamma} \tag{28}$$

Along with the equation (22), this equation determines the scalings of λ and x_0 with ϵ and γ.

To find the final scalings, we must consider two subcases separately

(a) $\epsilon^{5/2} << \gamma << \epsilon^{3/2}$

We can now neglect the power terms in both (28) and (22); this is valid because they are of order $x_0\epsilon$ which always will be of lower order. With this assumption,

$$x_0 \sim \frac{\epsilon^{5/4}}{\gamma^{1/2}} \tag{29}$$

and from (23),

$$\frac{1}{2} - \lambda \sim x_0^2 \sim \frac{\epsilon^{5/2}}{\gamma} \tag{30}$$

Self consistency requires $x_0 >> (1 - \lambda)\sqrt{\epsilon}$ which is just the condition $\gamma << \epsilon^{3/2}$. The other condition $\gamma >> \epsilon^{5/2}$ implies that x_0 is small.

(b) $\gamma << \epsilon^{5/2}$

Equation (28) now implies that x_0^2/λ^4 is large. Since x_0 is limited in range, this implies that λ must go to zero. Requiring finite δ in (22) demands that $\lambda \sim 1 - x_0$, showing that the finger moves to the side-wall. Finally, (28) gives rise to scaling

$$\lambda \sim \frac{\gamma^{1/3}}{\epsilon^{5/6}} \tag{31}$$

This is how the finger width goes to zero as the finger approaches the side-wall. To properly interpret eq. (31), recall that the definition of ϵ gives $\epsilon \sim \frac{2}{\pi}\sigma\lambda$. This should be substituted into (31) to recover the scaling of the pattern variable λ with the experimentally tunable parameter σ.

Our discussion required neglect of the term

$$1 - \left(\frac{1 - \lambda - x_0}{1 - \lambda + x_0}\right)^{\epsilon/2} \tag{32}$$

As $x_0 \to 1$ this approximation breaks down if $\epsilon \geq \frac{1}{|\ln \sigma|}$. This is the limit of validity of our treatment; note that this is not a very stringent condition for physically meaningful γ. This completes our discussion of the analytic form of the scalings.

Based on our analytic results, we predict the following behavior of a Saffman-Taylor finger as a function of tilt angle. For small enough angle such that $\sigma << \gamma^{2/3}$, the finger will slightly deform, with the width staying at about 1/2 but slightly decreasing. At some intermediate angle, the rate of deformation and width shrinkage will accelerate. Finally, the width will become small and the finger will hug the bottom sidewall.

It is worthwhile to take some typical experimental numbers to get a rough estimate of the range of angle, necessary to see these effects. We use the experiment of Thome, et al[24] for which $T = 20.9 \times 10^{-3}$ n/m, $\mu = 96.5 \times 10^{-3}$ $\frac{kg-sec}{m}$, $2a = 12cm$ and $b = .1cm$. If we choose $\frac{T}{12\mu V}\left(\frac{b}{b}\right)^2 \simeq 10^{-3}$, and we use a density similar to that of water, we find $\sigma = 1.7(\sin\omega)$. This gives $\epsilon = .5\sin\sigma$. According to the previous theory, we should start seeing large effects when $\epsilon \sim \gamma^{2/5} \sim .06$. So, we predict significant narrowing of the finger and a significant deviation from the center at angles $\omega \sim 2^0$. This is easily measurable!

4   REFERENCES

1. For recent reviews, see D. Kessler, J. Koplik and H. Levine, *Adv. in Physics* **37**, 255 (1988).

2. See J. S. Langer in **Chance and Matter**, J. Souletie ed., North-Holland (1987).

3. For more details about the Saffman-Taylor system, see D. Bensimon, L. Kadanoff, S. Liang, B. I. Shraiman and L. Tang, *Reviews of Modern Physics* **58**, 977 (1986); G. M. Homsy *Ann. Review of Fluid Mech.* **19**, 271 (1987).

4. G. P. Ivantsov, *Dokl. Akad Nauk. SSSR* **58**, 567 (1947).

5. D. Kessler and H. Levine, *Acta Metall* **36**, 2693 (1988); A. Dougherty and J. Gollub, *Phys. Rev A***38**, 3043 (1988); H. Chou and H. Cummins, *Phys. Rev. Lett. 61*, 173 (1988).

6. E. Ben-Jacob, R. Godbey, N.D. Goldenfeld, J. Koplik, H. Levine, T. Mueller, and L.M. Sander, *Phys. Rev. Lett* **1315** (1985).

7. E. Ben-Jacob, P. Garik and D. Grier, *Superlattices and Microstructures* **3**, 599 (1987).

8. E. A. Brener, *Sov. Phys. JETP* **69**, 133 (1989).

9. M. Kruskal and H. Segur, *Physica* **28D**, 228 (1987); also see discussion of this method in refs. 1.

10. See for example, A. Miller and H Mohwald, *J. Chem Phys.* **86**, 4258 (1987).

11. R. Weis and H. M. McConnel, *Nature* **310**, 47 (1984).

12. Y. Pomeau, *Europhys. Lett.* **3**, 1201 (1987).

13. B. I. Shraiman, *Phys. Rev. Lett* **56**, 2028 (1986); D. Hong and J. S. Langer, *Phys. Rev. Lett* **56**, 2032 (1986); R. Combescot, T. Dombre, V. Hakim, Y. Pomeau and A. Pumir, *Phys. Rev. Lett* **56**, 2036 (1986) and *Phys Rev* **A37** 1270 (1988); S. Tanveer, *Phys. Fluids* **30**, 1589 (1987) and also R. Combescot and T. Dombre, *Phys. Rev* **A38**, 2573 (1988).

14. P. Tabeling, G. Zocchi and A. Libchaber, *J. of Fluid Mech* **177**, 67 (1987).

15. D. A. Reinelt, *Phys. Fluids* **30**, 2617 (1987) and *J. of Fluid Mech.* **183**, 219 (1987); S. Tanveer, *Proc. Roy. Soc. London* **A428**, 511 (1990).

16. D. P. Woodruff, **The Solid-Liquid Interface**, Cambridge. Univ. Press, 1973.

17. E. A. Brener and V. I. Mel'nikov, submitted to Adv. in Physics (1989).

18. S. K. Chan, H.-H. Reimer, M. Kahlwelt, *J. Cryst. Growth* **32**, 303 (1976).

19. M. Ben-Amar and Y. Pomeau, *Europhys. Lett* **2**, 302 (1986).

20. E. Berner and H. Levine, *Phys. Rev. A* **43**, 883(1991).

21. P. G. Saffman and G. I. Taylor, *Proc., Roy. Soc. London* **A245**, 312 (1958).

22. E. Berner, H.Levine and Y. Tu, *Phys. of Fluids A* **3**, 529(1991).

23. G. I. Taylor and P. G. Saffman, *Q. J. of Applied Mech. and Applied Math* **12**, 265 (1959).

24. H. Thome, M. Rabaud, V. Hakim and Y. Couder, *Phys. Fluids* **A1**, 224 (1989).

DENDRITIC GROWTH

AN EXPERIMENTAL ASSESSMENT OF CONTINUUM MODELS OF DENDRITIC GROWTH

J.P. Gollub

Physics Dept., Haverford College, Haverford, PA 19041 USA and
Physics Dept., University of Pennsylvania, Philadelphia, PA 19104 USA

Abstract

Experimental evidence pertinent to theories of needle crystals (dendrites) based on continuum models is reviewed and assessed critically. Some predictions, such as the dependence of the growth state on crystalline anisotropy, have not been convincingly demonstrated, and the models may not be appropriate in all cases, for example when kinetic effects are important. On the other hand, the continuum models provide an internally consistent explanation for many of the observations, including some related to sidebranching.

A. Introduction

The problem of explaining the needle crystals or dendritic growth patterns that occur during solidification has attracted much interest and effort for several distinct reasons. From a mathematical point of view, the dendrite is significant as a steady and stable solution of a nonlinear pattern-forming system. Furthermore, the solution exhibits scaling behavior, in the sense that dendrites for different specified boundary conditions (undercooling) are the same except for a change of scale. Another remarkable property is the fact that the macroscopic form of a dendrite is determined by microscopic parameters such as the length characterizing capillarity, which is of the order of Angstroms for typical materials. From a physical point of view, dendrites are significant because they are ubiquitous in nature, and because they represent an ordered non-equilibrium state. Dendrites are of particular interest to metallurgists in part because cast metals are often composed of them.

Recent theories of pattern selection in dendritic growth that have been summarized in reviews by Kessler et al. [1], Langer [2], and Pelcé [3] provide a mathematically consistent approach to predicting the shape and velocity of dendrites using continuum model equations. In brief, the theory shows that if the growth is diffusion-controlled, and if the crystal is symmetric about the growth axis, then the macroscopic transport equations allow only a single stable solution when anisotropic surface tension is taken into account. The term "microscopic solvability" is often used to describe the theory because the capillary length

Growth and Form, Edited by M. Ben Amar *et al.*
Plenum Press, New York, 1991

scale characteristic of surface tension is typically only a few Angstroms. The purpose of this article is to assess the experimental evidence for (or against) the applicability of this approach to real dendritic crystals.

B. Systems Displaying Dendritic Growth

Dendritic crystal growth occurs in a great variety of materials during solidification.[4] It has been studied in two quite different situations: growth into the melt (or solution) at constant undercooling (supersaturation); and directional solidification with an imposed temperature gradient. Examples of materials displaying dendritic growth include many alloys, transparent organic materials such as CBr_4 and succinonitrile (either pure or with added solute), inorganic salts such as NH_4Br and NH_4Cl, and rare gas solids such as He, Xe, and Kr. In fact, the dendritic growth form is a ubiquitous non-equilibrium stationary state whenever the interfaces are microscopically rough (not faceted). Rough interfaces can occur either as a result of thermal roughening, or as a result of the growth process itself.

An experimental example of free dendritic growth from supersaturated solution, as studied by Dougherty et al.[5] is shown in Fig. 1. The main features worth noting are: constant growth speed; a fixed approximately parabolic shape near the tip; and sidebranches that grow progressively with distance back from the tip. It is often useful to describe the growth using a frame of reference moving with the dendrite; in this frame the branches appear to be waves propagating backward at constant speed. All of these features were documented much earlier in a classic study by Huang and Glicksman.[6]

Dendritic growth also occurs in quite different circumstances that are apparently unconnected with solidification. For example, dendrites are found in the electrolytic deposition of metals in solution, as shown in Fig. 2, which is taken from Ref. [7]. On the other hand, modification of the electrolyte concentration leads instead to disordered growth that has been compared quantitatively to diffusion-limited aggregation.[8] Another unconventional example of dendritic growth occurs in the transition between certain liquid crystal phases.[9]

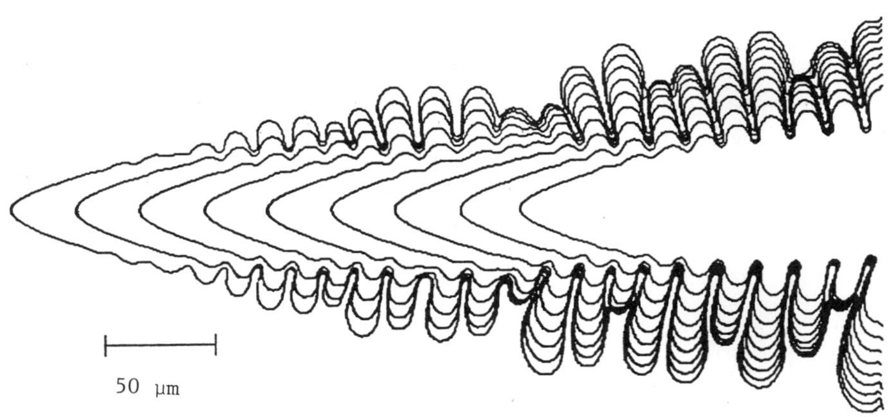

Fig. 1. Contours of an ammonium bromide dendrite growing from supersaturated solution, from Ref. [5]. The interval between contours is 20 s. Sidebranches also grow perpendicular to the plane of the diagram.

C. Mullins-Sekerka Instability

It is worthwhile to note briefly the basic instability that leads to the dendritic state. Interfaces whose rate of growth is controlled by diffusion were shown to be unstable with respect to sinusoidal perturbations by Mullins and Sekerka.[10] The growth rate of the instability rises with wavenumber k, reaches a maximum, and falls at high k due to surface tension. A lovely quantitative study of this process was performed by Chou and Cummins,[11] and an example from their work is displayed in Fig. 3. These authors showed that the early growth process is irregular, and can be adequately described by a stochastic equation of the following form:

$$dA_k/dt = \sigma(k)A_k + \text{Gaussian noise} \quad . \tag{1}$$

In this equation, A_k is the amplitude of the Fourier mode with wavenumber k, and $\sigma(k)$ is its growth rate. Whether the noise is due to thermal fluctuations or some other source is unclear.

The sensitivity of the early growth process to noise has been nicely illustrated in a numerical simulation of the unstable growth of a spherical nucleus by Pines et al.[12] The growth was constrained to be axisymmetric and random noise was provided as part of the initial condition. The radius is shown as a function of the polar angle in Fig. 4 with an exaggerated vertical scale. The dominant features can be traced back to very early times. Once started, these features persist and compete with each other for material.

D. Predictions for Steady State Dendritic Growth

The central theoretical prediction for axisymmetric dendritic growth is that surface tension enters the problem as a singular perturbation, and solutions for stable needle crystals depend on a small anisotropy ε of the capillary length d_0 (proportional to the surface energy). This effect is usually modeled (for cubic crystals) by a functional dependence of the form

$$d_0(q) = d_0(1 + \varepsilon\cos 4\theta) \quad , \tag{2}$$

where θ is the angle between the local normal to the surface and the (100) direction. The selected state of the dendrite is given for small undercooling by a parameter

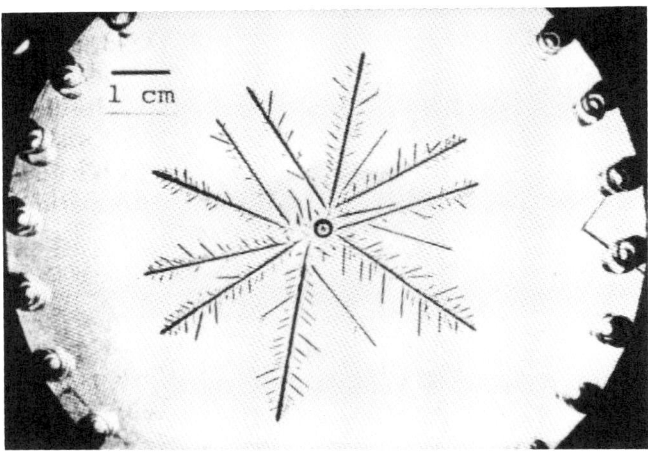

Fig. 2. Dendritic growth in the electrodeposition of zinc from solution, from Ref. [7].

$$\sigma^* = 2D/v\rho^2 \, , \tag{3}$$

where D is the diffusion constant (thermal or molecular) that controls the growth, ρ is the radius of curvature of the tip of the dendrite, and v is the speed of propagation of the interface. In addition, the product ρv is a function only of the undercooling (or supercooling) Δ for diffusion-controlled growth, so a unique state is predicted for each value of Δ. (The radius decreases while the speed increases as Δ is increased.) For axisymmetric growth, σ^* has been predicted to vary as $\varepsilon^{7/4}$ as $\varepsilon \to 0$. However, for the physically more relevant range $\varepsilon \sim 0.01$, the prediction [13][14] is closer to $\sigma \sim \varepsilon^1$. This prediction has been checked by two-dimensional numerical simulations. [15]

However, the theoretical situation is still apparently unresolved for non-axisymmetric three-dimensional growth [16]. In fact, the tip shape deviates significantly from a paraboloid somewhat before the sidebranches become prominent, so it is important to look at the experimental situation.

There is a long history of theoretical models of axisymmetric dendritic growth. Some references to the older approaches, such as marginal stability, are given in [17]. While these models give predictions that are in some cases not far from experimental results for steady state growth, their theoretical justification and mathematical consistency has been undermined by the microscopic solvability approach.

E. Measurements of Steady State Dendritic Growth from the Melt

Studies of dendritic growth radii and velocities have been undertaken by a number of investigators. However, a test of the theory also requires a quantitative measurement of the small parameter ε, which is quite difficult to perform accurately for small ε.

For the growth of highly purified material from the melt, Glicksman et al. [17] reported that succinonitrile and pure pivalic acid have nearly identical values of σ^* (0.0195 for SCN and 0.022 for PVA), while their anisotropies appear to differ substantially (0.005 for SCN and 0.05 for pure PVA). The anisotropies were determined from interfacial angular discontinuities at grain boundaries.

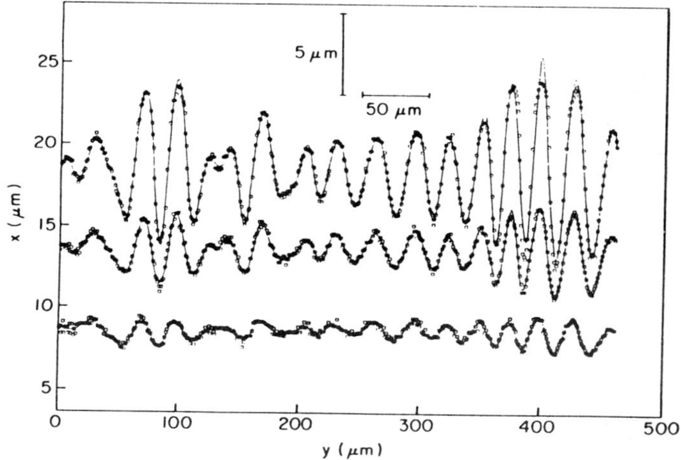

Fig. 3. Experimental measurement of the early stages of the Mullins-Sekerka instability, from Ref. [11].

Data on anisotropies and growth constants for these two materials is summarized in Table 1, along with corresponding data for solution growth to be discussed in the following section.

Studies of the dendritic solidification of krypton and zenon by Bilgram et al. [18] are particularly interesting because of the simplicity of rare gas solids. These authors found σ^* to be a function of supersaturation, something that would not be expected for small undercooling unless kinetic or impurity effects, or convection were important. The results are sufficiently surprising that further tests may be warranted.

F. Measurements of Steady State Dendritic Growth from Solution

The situation for growth from solution is somewhat different. It is perhaps worth indicating some of the experimental compromises that are involved in such measurements. It is desirable to use a growth cell larger than the diffusion length 2D/v to avoid finite size effects. On the other hand, the dendrites are small for solution growth, and adequate magnification can only be achieved with short focal length lenses. This unfortunately limits the cell thickness, but has the advantage of suppressing convection. Other difficulties are the achievement of spatially uniform solute concentration and adequate temperature control. Finally, one has to worry about possible optical distortions resulting from diffraction and the concentration gradients around the dendrite. These various constraints can be met in practice to a reasonable degree. Analogous problems affect studies of growth from the melt, but in different proportions.

Careful measurements for solution growth were undertaken by Dougherty and Gollub [19] for NH$_4$Br and later by Dougherty [20] for pivalic acid, using digital microscopy. The parameter ε was determined from the equilibrium radius R(θ) of a small spherical crystal (with mean radius roughly 40 μm) as a function of angle θ. This quantity has a fourfold component that gives ε directly.   To make these measurements, it was necessary to hold the small crystal

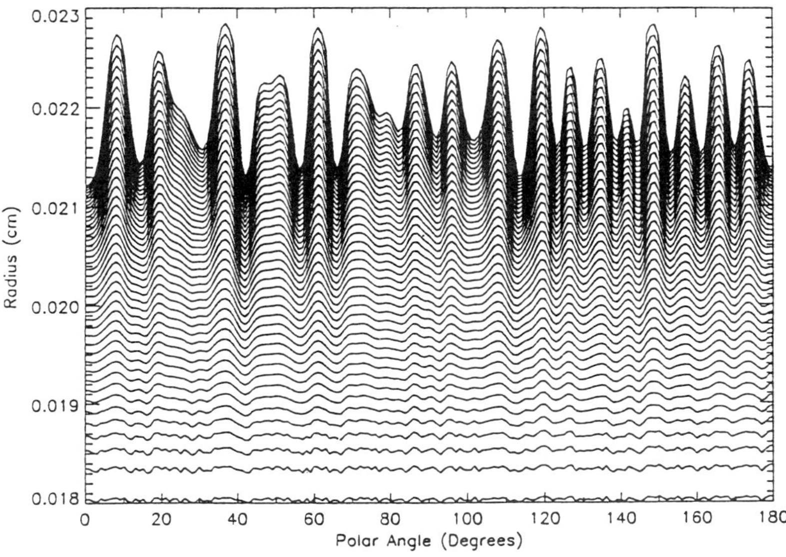

Fig. 4. Numerical computation of the axisymmetric growth of a spherical nucleus, illustrating the sensitivity of the structure to noise at early times, from Ref. [12].

Table 1. Comparison of predictions for the dendritic growth constant σ^* based on microscopic solvability (assuming axisymmetric growth) with various experiments. The data was taken from Ref. [17] for growth from the melt and Refs. [9], [19], and [20] for growth from solution. PVA is pivalic acid; HET is hexaoctyloxytriphenylene, a discotic liquid crystal; SCN is succinonitrile. These materials have fourfold anisotropy, except for HET, which is hexagonal. See the text for discussion.

	ε	(σ^*) from expt.	(σ^*) from theory
SCN (melt)	0.005	0.0195	0.01
PVA (melt)	0.05	0.022	~0.1
PVA (soln.)	0.006±0.002	0.05±0.02	0.022±0.007
NH$_4$Br (soln.)	0.016±0.004	0.081±0.02	0.083±0.025
HET (soln.)	0.003±0.001	0.038	0.040

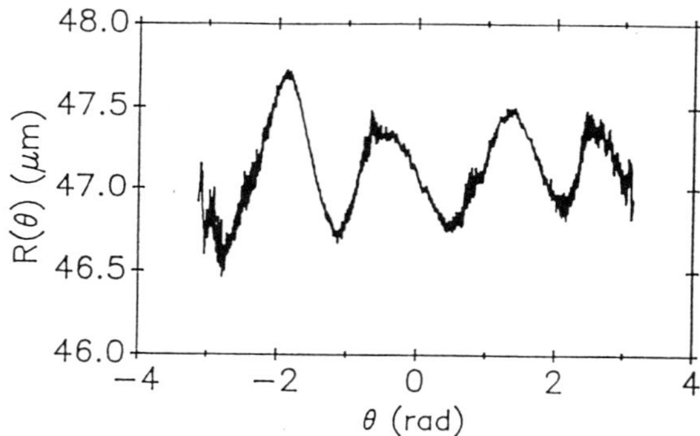

Fig. 5. Angular dependence of the equilibrium radius $R(\theta)$ of a small crystal of pivalic acid, from Ref. [20]. The fourfold component is proportional to the anisotropy ε.

in equilibrium for long periods of time to allow equilibration. However, the crystal is unstable, tending either to grow or shrink. To keep the mean crystal size fixed, Dougherty developed a novel temperature control system with optical feedback. An example of $R(\theta)$ from Ref. [20] for pivalic acid in ethanol solution is given in Fig. 5. The measured anisotropy (averaged over several crystals) is $\varepsilon=0.006\pm0.002$ for this material when grown from 1% ethanol solution.

A comparison of theory and experiment for the growth from solution of ammonium bromide and pivalic acid is contained in Table 1. The theoretical predictions for σ^* pertain to axisymmetric growth for the measured ε, and the errors quoted on the theoretical values stem from the uncertainty in ε. The experimental results on these two materials are consistent with the predictions, but the statistical errors are too large (or the anisotropies too similar) to allow a clear dependence on ε to be detected.

Maurer et al. [21] have reported a somewhat higher value ($\sigma^*=0.12\pm0.01$) for the growth of NH_4Br in a gel, where possible convective effects should be strongly suppressed. This latter measurement is somewhat more precise than those reported in the table, but one must hope that the gel does not modify the growth process.

Finally, we note that Oswald [9] studied interfacial growth (in a very thin layer) at the boundary between two liquid crystal phases: a columnar hexagonal liquid crystal mesophase, and the isotropic phase of the same material. In this case, the anisotropy is 6-fold and was measured to be about 0.003, while $\sigma^*=0.038$. The corresponding theoretical value [22] for two-dimensional 6-fold growth (adjusted for the different diffusivities for impurities in the two phases) is 0.040. There are many uncertainties in the comparison, but qualitatively at least, the experiment seems not inconsistent with the predictions.

Now examining all the data in Table 1, we note that the case of pivalic acid grown from the melt is the only one that appears to contradict the theory. On the other hand, the values of σ^* for PVA grown from solution and from the melt differ by just the factor of two expected for the change from one-sided diffusion of heat to two-sided diffusion of molecules, without any additional change that might be attributed to a modification of the surface tension anisotropy upon addition of the ethanol "solvent".[20] The internal consistency of the data thus suggests that the measurements of the growth constants are correct. It would therefore seem worthwhile to remeasure ε for pure PVA using digital imaging methods.

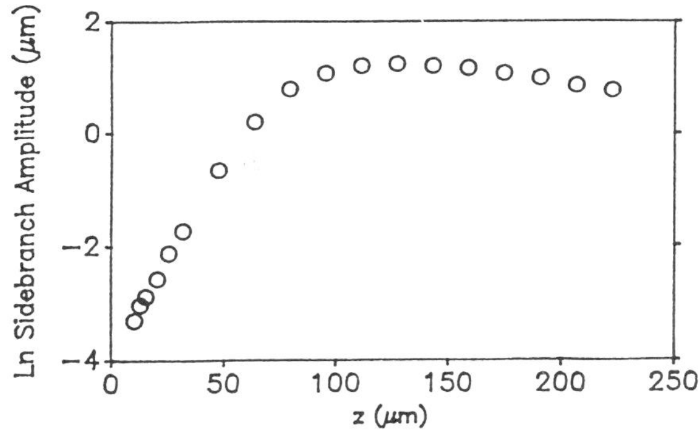

Fig. 6. Root-mean-square sidebranch amplitude as a function of distance from the tip, showing exponential growth. From Ref. 5.

G. Early Sidebranching

According to our present theoretical understanding, dendritic tips are stable with respect to variations of the radius of curvature at the tip. However, they are believed to be "convectively unstable" with respect to sidebranching waves. This means that a small perturbation can grow while propagating away from the tip (in the tip frame of reference), while leaving the tip unchanged. If this understanding is correct, one expects [32] that noise in the tip region would produce small wavepackets of mean wavelength

$$\lambda \sim 2\pi\rho(\sigma^*)^{1/2} \tag{4}$$

and amplitude A(s) as a function of arc length s given by

$$A(s) \sim \exp\{(\sigma^*)^{-1/2}(s/\rho)^{1/4}\} \quad . \tag{5}$$

Any source of noise might have this effect, including thermal noise or microscopic irregularities arising from atomic discreteness or defects in the crystal. The dendrite serves as a frequency-selective (or wavelength-selective) amplifier.

Something like this in fact seems to occur. For example, measurements of the logarithm of the root-mean-square sidebranch amplitude as a function of distance from the tip (from Ref. [5]) are shown in Fig. 6. An approximately exponential growth (which would be indistinguishable from the form given in (5) is seen to occur.

A more direct test has been made by Qian and Cummins,[23] who studied the response to a localized pulse of radiant energy sent through an optical fiber aimed at the tip. They indeed observed the expected wavepacket and have compared its form with the theoretical predictions.

The concept that the dendritic tip is a frequency-selective amplifier was verified by Bouissou et al. [24]. These authors subjected the growing dendrite to a small oscillatory flow. They found that the sidebranch structure appears to resonate when forced weakly at a particular frequency, at which very large and regular branches are found, as shown in Fig. 7.

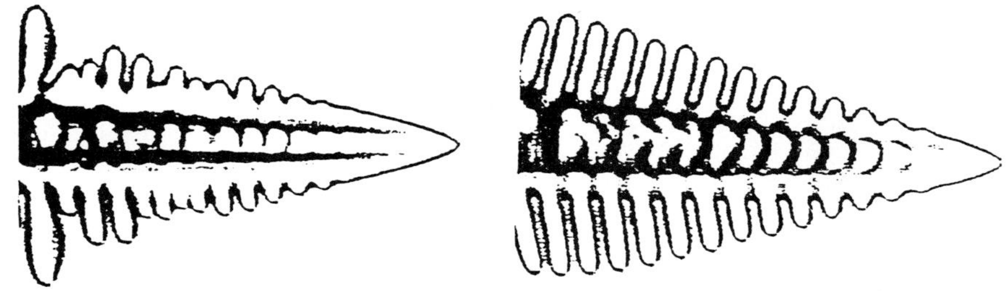

Fig. 7. Resonant response of a dendrite to a periodic perturbation, from Ref. [24]. Left, unforced; right, forced.

The mean sidebranch spacing has been predicted [1][14] based on the concept of noise-induced convective instability in the tip region. Though the predictions are somewhat model-dependent, the spacing should scale as $(\sigma^*)^{1/2}$. For ammonium bromide grown from solution, it is expected that λ/ρ should be roughly 3.7, while the experimental result [19] is somewhat larger, 5.2±0.8. For pure succinonitrile, both the prediction and the experimental observation [6] are lower than these values by a factor of 1.8, so the quantitative discrepancy is again about 30% of the experimental value. For pivalic acid grown from solution, the situation is worse, with the prediction and experimental value being about 3 and 6, respectively.[20] It is possible that the lack of axisymmetry of the real crystals plays a role in this quantitative disagreement.

The irregularity of the sidebranching pattern, as seen visually in Fig. 1, is affected eventually by local competition of adjacent branches for material. However, the irregularity is *also* seen quite close to the tip,[5] where the branches are barely visible in Fig. 1. This feature would also be a natural consequence of noise-induced convective instability in the tip region. The effect is qualitatively similar to the amplification of noise by the Mullins-Sekerka instability, as documented in Figs. 3 and 4. In a sense, the Mullins-Sekerka instability continues to operate, even in the steady state.

H. Kinetic Effects

One of the critical assumptions often made in describing dendritic growth is that departures from local equilibrium (or kinetic effects) can be neglected. However, experiments and theories have increasingly made efforts to deal with kinetic effects. For example, in the continuum theory of dendritic growth, one can allow the non-dimensional concentration at the interface to contain a term $-\beta(\theta)v$ that is linear in the velocity but depends on the angle of the local normal to a particular crystal axis. Such a model can explain the transition from (100) dendrites to (111) dendrites that was observed by Chan et al.[25] at high velocity long ago. For a recent example of the incorporation of kinetic effects into continuum growth models, see for example the discussion by Brener and Levine in Ref. [26].

Dendrites are sometimes partially faceted [27], and in this case kinetic effects must surely be taken into account. Theoretical predictions for such faceted dendrites have been proposed by Ben Amar and Pomeau.[28] Raz and Lipson [29] have noted that the dendritic growth of NH_4Cl shows a nonlinear dependence of the growth velocity on supersaturation, a likely hallmark of kinetic effects. In a later study, Raz et al. [30] discovered an interesting superlattice structure for dendritic growth in a very thin layer, which they attributed in part to kinetic effects.

In some cases, a crystal that is faceted in equilibrium will be roughened when grown even slowly. (For example, see Ref. [19].) This so-called "kinetic roughening" process is fairly complex. Phenomena of this type have been studied numerically by Xiao et al.[31] These authors used a modified Monte Carlo diffusion-limited aggregation model that includes fairly realistic boundary conditions. Surface diffusion prior to attachment, lattice anisotropy, and thermal fluctuations were included. At low temperatures, where the bond strength is much larger than the thermal energy kT, the faceted shape is preserved. However, at higher temperatures a clear transition from faceted to dendritic shapes is found in the numerical computations. Such simulations may be the only adequate way to treat these complex situations where kinetic effects are important.

J. Conclusion

A considerable body of experimental evidence now exists concerning the problem of pattern selection in dendritic growth. To what extent does it support the continuum models and their solution based on microscopic solvability? This is not an easy question to answer, and it is especially difficult to provide a summary that researchers in all of the relevant research disciplines (metallurgy and materials science, physics, and mathematics) will find equally acceptable, since they have divergent views as to the relevance of different types of evidence. The following is an effort, surely imperfect, to provide a balanced assessment.

First, there seems to be little doubt that the continuum models have been solved correctly and in a mathematically consistent way. The results have been checked by several independent investigators, and have also been compared to numerical computations.

Furthermore, the same methods have been found to produce correct predictions when applied to mathematically similar hydrodynamic situations, such as the viscous fingering problem.[1] As a solution to a mathematical problem of pattern formation within the framework of continuum models, the theory is impressive.

The real issue is that the models used to date are only approximate descriptions of growing crystals. First, the lack of axisymmetry of most three-dimensional dendritic crystals is obviously important, but still a subject of theoretical controversy [16] Second, kinetic effects are complex, and the simple addition of a linear term $(-\beta v)$ to the boundary condition may not provide a sufficient remedy. Third, microscopic effects, including both molecular discreteness and crystal defects, may in some cases be significant, especially for sidebranching phenomena.

Still, it is important to ask to what extent the continuum models of axisymmetric dendritic growth adequately describe the basic phenomena. The following elements are important in such a comparison:

-- The theory provides a mathematically consistent way to understand the existence of a unique radius and velocity at given undercooling, as is found experimentally.

-- The theory provides a natural way to understand sidebranches that grow continuously from the tip, are sensitive to noise in that region, and resonate with an external perturbation at a particular frequency.

-- The evidence concerning the dependence of the growth constant σ^* on anisotropy is inconclusive at present. For growth from solution, the trend seems correct but a sufficient range of ε is unavailable. Further work on the rare gas solids also seems desirable to settle the question of the constancy of σ^* for these materials. For pure systems, several cases are at least roughly consistent with the theory (including the liquid crystal example). However, the one material (pure pivalic acid) for which a large value of ε has been reported does not show the predicted large growth constant. It would be worthwhile to remeasure ε for this material using different methods, given its importance to testing the theory.

-- The fact that dendrites grow along crystal axes and do not generally grow in amorphous materials certainly indicates that anisotropy is of some importance. Also, in the mathematically analogous case of hydrodynamic viscous fingering, dendrites (needle-like fingers with sidebranches) can be produced by introducing anisotropy into the boundary conditions, and do not occur in isotropic situations. This connection to hydrodynamics shows that the mathematics does produce qualitatively reasonable predictions in a situation where anisotropy can be controlled more easily than it can in solidification.

-- Even if the theory is eventually found to be reasonably consistent with experiment, there remain other important aspects of interfacial growth, such as the statistical properties of the coarsening process in the region of strong competition far from the tip, that are still incompletely understood.

Acknowledgements

This work was supported primarily by the National Science Foundation Low Temperature Physics Program (DMR-8901869). Some facilities were provided by the W.M. Keck Foundation and by the University Research Initiative Program (DARPA/ONR-N00014-85-K0759). I am indebted to A. Dougherty (of Lafayette College), some of whose work is reviewed in this article, and also to H. Levine and M. Glicksman for helpful comments. The hospitality of the Institute for Nonlinear Science at The University of Califiornia at San Diego, where this article was written, is much appreciated.

References

[1] D.A. Kessler, J. Koplik, and H. Levine, Adv. Phys. **37**, 255 (1988).

[2] J.S. Langer, in *Chance and Matter*, ed. by J. Souletie, J. Vannimenus, and R. Stora (Elsevier, New York, 1987), p. 629; and Science **243**, 1150 (1989).

[3] P. Pelcé, *Dynamics of Curved Fronts* (Academic Press, New York, 1988).

[4] For a general review from a materials science perspective see W. Kurz and R. Trivedi, Acta Metall. Mater. **38**, 1 (1990).

[5] A. Dougherty, P.D. Kaplan, and J.P. Gollub, Phys. Rev. Lett. **58**, 1652 (1987).

[6] S.-C. Huang and M.E. Glicksman, Acta Metall. **29**, 701 and 717 (1981).

[7] Y. Sawada, A. Dougherty, and J.P. Gollub, Phys. Rev. Lett. **56**, 1260 (1986).

[8] F. Argoul, A. Arneodo, and G. Grasseau, Phys. Rev. Lett. **61**, 2558 (1988).

[9] P. Oswald, J. Phys. France **49**, 1083 (1988).

[10] W.W. Mullins and R.F. Sekerka, J. Appl. Phys. **34**, 323 (1963) and **35**, 444 (1964).

[11] H. Chou and H.Z. Cummins, Phys. Rev. Lett. **61**, 173 (1988).

[12] V. Pines, M. Zlatkowski, and A. Chait, Phys. Rev. A **42**, 6129 and 6137 (1990).

[13] D.A. Kessler and H. Levine, Phys. Rev. A **33**, 3352 (1986).

[14] A. Barbieri and J.S. Langer, Phys. Rev. A **39**, 5314 (1989); J.S. Langer, Phys. Rev. A **36**, 3350 (1987).

[15] Y. Saito, G. Goldbeck-Wood, and H. Müller-Krumbhaar, Phys. Rev. Lett. **58**, 1541 (1987) and Phys. Rev. A **38**, 2148 (1988).

[16] D. Kessler, in *Proceedings of the Conference on Asymptotics Beyond All Orders*, ed. by S. Tanveer (Plenum, 1991), to appear. A different approach to the effect of non-axisymmetry on state selection has been given by Y. Miata, S.H. Tirmizi, and M.E. Glicksman, "Dendritic growth with interfacial energy anisotropy," J. Cryst. Growth (1991), to appear.

[17] M.E. Glicksman and N.B. Singh, J. Cryst. Growth **98**, 277 (1989).

[18] J.H. Bilgram, M. Firmann, and E. Hürlimann, J. Cryst. Growth **96**, 175 (1989).

[19] A. Dougherty and J.P. Gollub, Phys. Rev. A **38**, 3043 (1988).

[20] A. Dougherty, Surface tension anisotropy and the dendritic growth of pivalic acid, J. Cryst. Growth (1991), in press.

[21] J. Maurer, B. Perrin, and P.Tabeling, Three dimensional structure of NH4Br dendrites growing out of a gel (1990), preprint.

[22] M. Ben Amar, private communication.

[23] X.W. Qian and H.Z. Cummins, Phys. Rev. Lett. **64**, 3038 (1990).

[24] P. Bouissou, A. Chiffaudel, B. Perin, and P. Tabeling, Europhys. Lett. **13**, 89 (1990).

[25] S.-K. Chan, H.-H. Reimer, and M. Kahlweit, J. Cryst. Growth **32**, 303 (1976).

[26] E. Brener and H. Levine, Phys. Rev. A **43**, 883 (1991).

[27] J. Maurer, P. Bouissou, B. Perrin, and P. Tabeling, Europhys. Lett. **8**, 67 (1988).

[28] M. Ben Amar and Y. Pomeau, Europhys. Lett. **6**, 609 (1988).

[29] E. Raz, S.G. Lipson, and E. Polturak, Phys. Rev. A **40**, 1088 (1989).

[30] E. Raz, S.G. Lipson, and E. Ben-Jacob, Meta-ordering observed during dendritic growth of ammonium chloride crystals in thin layers (1990), preprint.

[31] R.-F. Xiao, J.I.D. Alexander, and F. Rosenberger, Phys. Rev. A **38**, 2447 (1988); J. Cryst. Growth **100**, 313 (1990).

[32] R. Pieters and J.S. Langer, Phys. Rev. Lett. **56**, 1948 (1986); M. Barber, A. Barbieri, and J.S. Langer, Phys. Rev. A **36**, 3340 (1987); D.A. Kessler and H. Levine, Europhys. Lett. **4**, 215 (1987); B. Caroli, C. Caroli, and B. Roulet, J. Phys. Paris **48**, 1423 (1987).

DENDRITIC SOLIDIFICATION OF RARE GASES

J.H. Bilgram and E. Hürlimann

Laboratorium für Festkörperphysik
ETH
CH 8093 Zürich, Switzerland

ABSTRACT

Rare gases Kr and Xe have been used to study dendritic solidification
of a pure melt. The advantages of using single component systems in comparison to solution systems are: short relaxation times and good repro-
ducibility. The parameters characterizing the shapes of the dendrite tips
have been determined. The volume solidification rate of dendrites increases
with increasing supercooling of the melt, thus the stability-"constant" decreases with increasing supercooling. The volume of dendrites V has been determined as a function of the length L of the dendrite. $V \propto L^3$. The average
density of a dendrite V/L^3 is independent of L, dendrites are no fractals.
V/L^3 decreases with increasing supercooling. A model is presented, which describes the volume of the dendrites.

INTRODUCTION

Experiments on dendritic solidification have been performed as a part
of a project where dynamical processes during the freezing transition are
studied [1,2]. Data of dendritic solidification of rare gases are presented
in this paper. Together with other experimental results [3,4] these data
can be used to verify theories on dendritic solidification [5-7].

Several decisions have been made for the selection of the system:
1.) Metals or dielectric materials
2.) Growth from the solution or from the pure melt
3.) Organic liquids or simple liquids

1.) We want to study the tip parameters, the volume and the shape of the
side arms *in situ* during the growth of the dendrite into a metastable
liquid. Therefore we use transparent materials.

2.) The essential parameter in solidification experiments is the supercooling or the supersaturation of the liquid phase. The equilibrium conditions
of single component systems are well known. The triple point parameters of
rare gases can be used as fixed points in the temperature scale. The equilibrium conditions of solute systems are not known as well as the triple

Growth and Form, Edited by M. Ben Amar *et al.*
Plenum Press, New York, 1991

points of rare gases. Solute concentrations are difficult to control. Therefore we have selected single component systems. Another reason to use pure
systems is the relaxation time necessary for approaching a steady state.
Theories have been developed for steady states. The typical time scale of
a transient state in a diffusion process is $\propto l/D$. In a single component
system D corresponds to the thermal diffusivity $\alpha \simeq 10^{-3}$ cm^2/s. Solute
diffusion constants are at least by two orders of magnitude smaller and
thus relaxation times in solute systems are a factor 100 longer as relaxation times in single component systems.

3.) A short range order exists in organic liquids. This short range order
has to be destroyed before a molecule can be built into the crystal [1,8].
This step is not necessary in a simple liquid. A one component system has
to be a pure system and no impurities should be segregated at the solid-
liquid interface (s.-l.-interface). Organic liquids are always formed by
a chemical reaction. Therefore there are always small amounts of impurities
in an organic liquid. No chemical decomposition does exist for rare gases.
Purification of rare gases can be achieved to an impurity concentration below ppb levels by rare gas purifiers, which have been developed for semiconductor components production. Therefore we have selected rare gases for
our experiments.

Two aspects of dendritic solidification will be discussed in this
paper: The parameters of the dendrite tip in a 3 dimensional system and
the volume of a whole dendrite. The volume measurements give informations
about the whole dendrite and a global information on the later stages in
the development of a dendrite.

SOLIDIFICATION AND HEAT TRANSPORT

We want to compare various substances. Therefore it is necessary to
use reduced quantities such as the growth velocity of the tip V, the tip
radius ρ, the supercooling Δ, and the Péclet number p. These reduced quantities are related to the quantities measured in the experiment by

$$V = \frac{d_c}{2\alpha} \tag{1}$$

$$\rho = \frac{R}{d_o} \tag{2}$$

$$\Delta = \Delta T \ (c_p / L) \tag{3}$$

$$p = \frac{v_{tip} \ R}{2 \ \alpha} \tag{4}$$

where c_p, L, ΔT, d_o, v_{tip}, α, and R are the specific heat of the melt, the
latent heat of fusion, the supercooling of the melt, the capillarity, the
tip growth velocity, the thermal diffusivity and the radius of curvature
of the dendrite tip respectively. The supercooling is given by

$$\Delta T = T_\infty - T_m \tag{5}$$

where T_∞ is the temperature in the melt far away from the crystal and T_m is the equilibrium melting temperature of the crystal.

$$d_o = \frac{T_m \, \gamma \, c_p}{L^2} \tag{6}$$

here γ is the solid-liquid interface free energy.

An important goal of the theoretical treatment of dendritic solidification is the prediction of R and v_{tip} for a given ΔT. A first step to such a prediction is the Ivantsov solution [9]:

$$\Delta = p \, e^p \, E_1(p) \tag{7}$$

with

$$E_1(p) = \int_p^\infty \frac{e^{-y'}}{y'}$$

Horvay and Cahn [10] have shown that for small Péclet numbers

$$p \propto \Delta \tag{8}$$

Eq. 7 represents a solution of the Laplace equation, which describes the heat flow around a growing dendrite tip. More recent treatments of this problem are discussed in [6,7]. All the treatments of the diffusion of the latent heat away from the dendrite tip are based on a serie of assumptions. Some of these assumptions are listed here:
1.) a stationary state is considered
2.) the shape of the dendrite is a rotational paraboloide,
3.) the dendrite has infinite length,
4.) only thermal transport is rate limiting because,
5.) interface kinetics is fast.

The main shortcomming of the solutions of the heat flow problem is: it yields only a relation between the product v_{tip} R and ΔT, whereas the experiment yields simultaneously v_{tip} and R for a given supercooling of the melt. The experiment shows that there is a unique relation beween ΔT and v_{tip}. The inclusion of the Gibbs-Thomson effect leads to a maximum growth velocity for a given supercooling. For some time it has been assumed that the dendrites grow with the highest possible growth velocity. Experiments have shown that the maximum growth velocity hypothesis does not predict the growth velocities of the dendrites. In the "stability hypothesis" it is assumed that the quantity $v\sqrt{\rho^2}$ is a universal constant. Now experimental data are available to test this hypothesis. It is found that $\sqrt{\rho^2}$ varies from substance to substance [8] and depends on supercooling for some substances [8,11].

PROPERTIES OF THE DENDRITE TIP

Rare gas dendrites are growing into a volume of about 100 cm^3. This size of the growth vessel is sufficient to exclude any confinement effects. [8,11]. The crystal seed is introduced into the center of this volume by the capillary injection technique. The melt is thermostated by a liquid bath and the temperature can be stabilized within $\pm 10^{-4}$ K for several hours. Details about the experimental setup can be found in [11-13]. The shape of the dendrites is described in [14]. The dendrites do not have the

shape of a rotational paraboloide 4 fins run along the dendrites and side
arms grow at the backs of the fins. There are always 5 dendrites developing at the end at the injection capillary. The total length of the dendrites is typically a few mm. Thus for distances far away from the dendrites
the dendrites act like a point source of heat and the thermal field does
not have the shape of a rotational parabolide. At distances only a few tip
radii away from the dendrite tip the thermal field is stationary and can
be described in terms of parabolical coordinates [11].

In fig. 1 the growth rates of dendrites are plotted vs. the supercooling. We have used reduced quantities (1) and (3). The data of Xe and
Kr are very close together, therefore we have plotted the data points of
Xe dendrites and the Kr data are represented by the drawn line. All the
data on Xe and Kr can be found in ref. [11]. The line which is fitted to
the krypton data continues in the interrupted line. The data points of
succinonitrile (SCN) and pivalic acid (PVA) have been taken from ref.[4].
We find that the scaling according to eq. (1) and (3) leads to an excellent
agreement of the data. This scaling is based on the assumption that the
heat is transported away from the dendrites by thermal diffusion. Convection does not contribute significantly to the thermal transport [8,11].
The radius of curvature of the parabola fitted to the dendrite tips is
plotted vs. the supercooling in fig.2. again we use dimensionless units
eq.(2) and (3). The scaling of the data is not as perfect as in fig.1 but
within the errors of measurements it is acceptable. For a discussion on
the precision of the data see [11]. Here the question may be raised, whether
the good fit in fig. 1 is accidentially and whether d_o is the proper quantity to scale the tip radius. In fig.3 the Péclet number is plotted vs.Δ. For
Xe and Kr it is found:

$$2p = \Delta \tag{9}$$

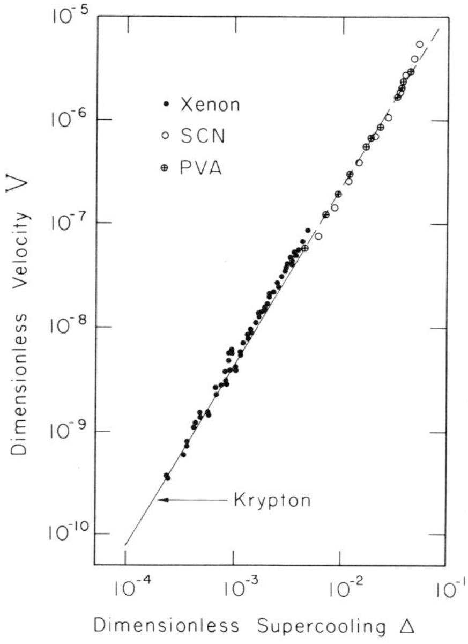

Fig.1 Dimensionless
velocity vs. dimension-
less supercooling. Rare
gas data [11] are com-
pared with data of SCN
and PVA [4].

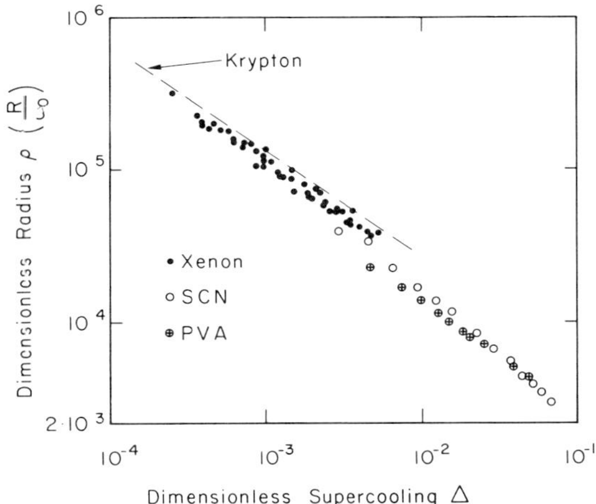

Fig.2 Tip radius vs. supercooling. Rare gas data [11] are compared with data of SCN and PVA [4].

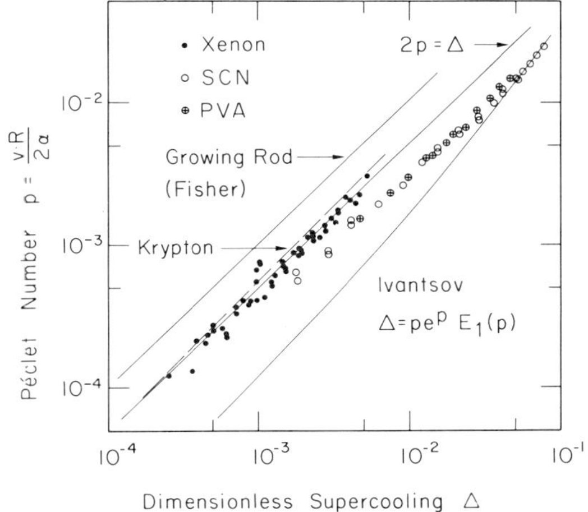

Fig.3 Péclet-number vs. supercooling.

in agreement with (8). The data of SCN and PVA do not show this proportionality. The stability constant is plotted as a function of supercooling in fig.4. In agreement with [8] we find that σ varies with substance. We find that σ decreases with temperature for Kr and Xe. The data for V and ρ can be represented by:

$$V(Kr) = (6.33 \pm 0.12) \times 10^{-4} \, \Delta^{1.729 \pm 0.007}$$

$$V(Xe) = (7.87 \pm 0.92) \times 10^{-4} \, \Delta^{1.745 \pm 0.017} \tag{10}$$

$$\rho(Kr) = (1.16 \pm 0.04) \times 10^{3} \, \Delta^{-0.683 \pm 0.010}$$

$$\rho(Xe) = (0.93 \pm 0.09) \times 10^{3} \, \Delta^{-0.685 \pm 0.014} \tag{11}$$

THE VOLUME OF DENDRITES

We have measured the volume of rare gas dendrites V as a function of the length L of the dendrite. For experimental details see [12,13]. We have developed a simple model to describe our experimental data.

We consider a stationary state. A dendrite grows with the velocity v_{tip}. This velocity does not depend on time. The length of the dendrite is L. $v_{tip} = dL/dt$. We consider a sphere with a diameter 2 L. All dendrite tips and the tips of the side branches grow inside this sphere. All latent heat which is produced by the dendrites has to pass through this sphere, if the dendrite grows in a stationary state. From continuity we deduce the change in volume of solidified material in time:

$$\frac{dV}{dt} = \frac{\lambda}{L} \int \nabla_L \, d\sigma \tag{12}$$

The integral has to be taken over the surface of a sphere with radius L. ∇_L is the thermal gradient at the surface of this sphere. ∇_L does not depend on time (stationary state) but it depends on the orientation in space, e.g. at the point where a dendrite touches the large sphere $\nabla_L = \nabla_R$ the thermal gradient at the dendrite tip, which is the maximum value of ∇_L. We define a normalized thermal gradient ∇_L^*:

$$\int \nabla_L d\sigma = 4\pi \, L^2 \, \nabla_L^* \, \Delta T \tag{13}$$

In (12) and (13) it is assumed that the latent heat diffuses through the sphere. The experiment shows that this is a good approximation. With (3), (9) and (4) we obtain

$$\frac{dV}{dt} = 4\pi \, \nabla_L^* \, L^2 \, v_{tip} \, R \tag{14}$$

or

$$V = \frac{4}{3}\pi \, \nabla_L^* \, L^3 \, R \tag{15}$$

For the filling factor of the sphere or the averaged density we obtain:

$$\frac{V}{\frac{4}{3}\pi L^3} = \frac{\nabla_L^*}{V/R} \tag{16}$$

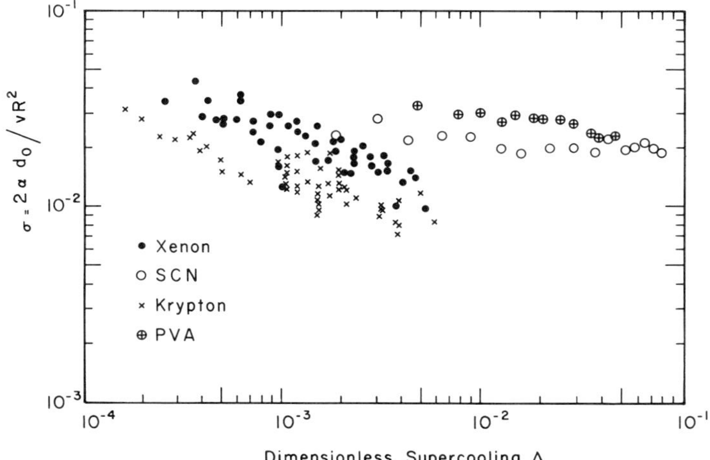

Fig.4 Stability constant vs. supercooling. Rare gas data [11] are compared with data of SCN [10] and PVA [4]. The results of the experiments with rare gases show that the volume solidification rate increases with increasing supercooling.

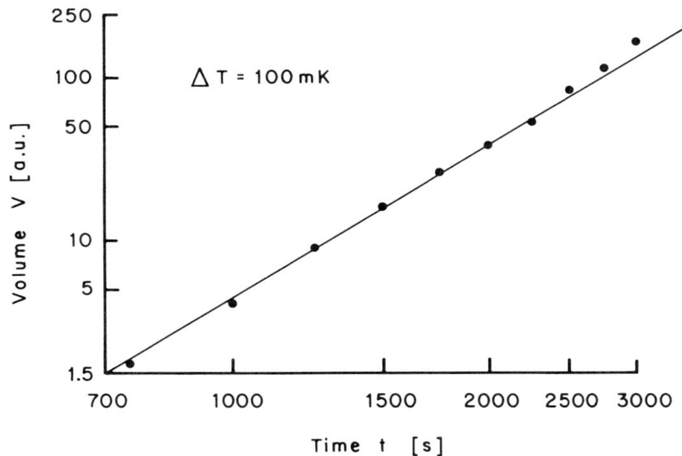

Fig.5 The volume of a dendrite vs. the time elapsed sice the dendrite has left the growth capillary. The drawn line corresponds to slope 3 according to eq.15.

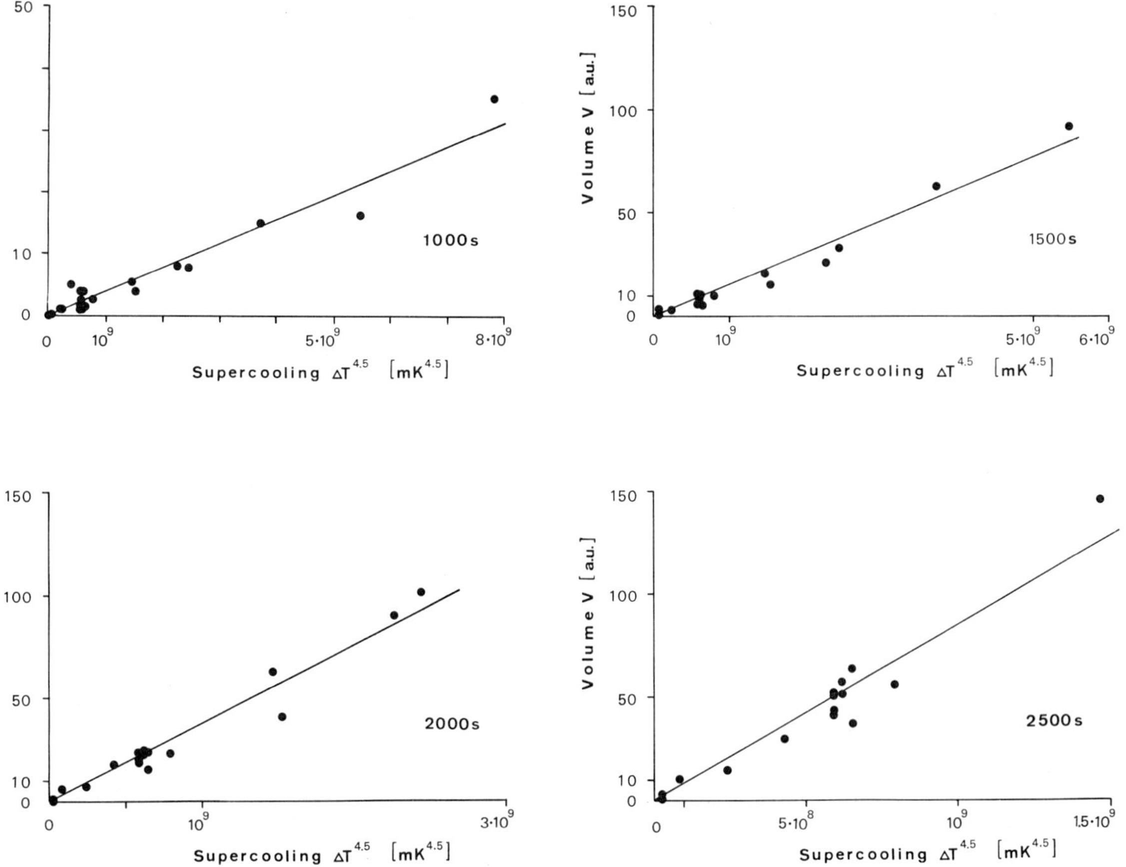

Fig.6 The temperature dependence of the volume solidification rate. The volume is measured at given times for various supercoolings. The volume increases with $\Delta T^{4.5}$ in agreement with the model.

The fillig factor (or averaged density) of the dendrite does not depend on
L. Dendrites are no fractals. The only quantity which determines the filling factor is the tip radius. We have used in this simple model the assumption of continuity of heat. We use the steady state condition v_{tip} = const.
We use two experimental results: Most of the latent heat liberated during
the growth of the dendrite has to diffuse through the sphere: We give an
example with numbers: ΔT = 83mK, L = 6 mm, V = 138 mm^3. The latent heat is
7.16 J, the heat which is necessary to heat the volume inside of the sphere
to T_m is 0.076 J. Thus 99% of the latent heat have to diffuse through the
sphere. We have used the experimental result (9).

Eq. (15) can be verified in several ways:
1.) The filling factor is independent of L, $V \propto L^3$. This is shown in fig.5.
2.) If the volumes of various dendrites growing at different supercoolings
    are measured after a given time, then this volume depends on supercool-
    ing according (15) i.e. the temperature dependences of V (eq.10) and R
    (eq, 11). This behavior is shown in fig.6, $V \propto \Delta T^{4,5}$.

DISCUSSION AND CONCLUSIONS

Systems with only one component have several advantages in comparison
to solute systems: Equilibrium conditions are better defined as in solute
systems. Steady state is reached much quicker in single component systems
as in solute systems.

The shape of a dendrite cannot be described by a rotational parabolide
but this shape is very usefull for the discussion of heat flow problems.
The assumptions that the latent heat is transported away from the dendrite
tip by diffusion leads to scaling laws (1-4), which have been verified for
the substances, where experimental data are available. The excellent scaling of V vs. Δ in fig. 1 for 4 substances over 4 orders of magnitude in Δ
and 5 orders of magnitude in V is gratifying. The experiments show that the
volume solidification rate of a dendrite increases with increasing supercooling. Thus the stability "constant" $\sigma = 2\alpha d_o/v_{tip} \cdot R^2$ decreases with increasing supercooling. This result is in good agreement with thermodynami-
cal principles, which state that volume solidification rate vanishes at
zero supercooling. This disagreement of the experiment with some theories
might be due to assumptions on which theories are based and which are not
given in the experiments:
1.) Although growth rate is constant, it might be that the dendrite tip is
not in a steady state when side branches are growing.
2.) The dendrite in the experiment does not have the shape of a rotational
parabolide.
3.) Dendrites with an infinite length cannot be studied in experiments. A
steady state temperature distribution can be found close to the dendrite
tip but not at distances which are large in comparison to the length of
the dendrite.
4.) Experiments with solute systems [8] as well as with metals [15] show
that kinetic effects cannot be neglected in dendritic solidification.

It has been claimed that solute systems show that $V\rho^2$ is indpendent
of supersaturation [16]. It has been mentioned in the introduction that
solute systems are difficult to characterize. In the experiment [16] growth
velocity has been varied by less than a factor of 2. If this variation is
compared with the range covered in fig.1, it is obvious that solute systems
cannot give information on the variation of σ upon supersaturation.

The volume of the dendrites can be described by a very simple model. This model is based on two assumptions: steady state and continuity. The model predicts: a): the fillig factor of the dendrites is independent of the size of the dendrites and b) the only relevant quantity is the tip radius. The fillig factor decreases with increasing supercooling. This predictions are in agreement with the experiment.

ACKNOWLEDGEMENTS

We thank Prof. H.-R. Ott for his support. This work was supported by the Swiss National Science Foundation.

REFERENCES

1. J.H. Bilgram, Phys. Reports 153,1(1987)
2. J.H. Bilgram and R. Steininger, J. Cryst. Growth 99,30(1990)
    R. Steininger and J.H. Bilgram, J. Cryst. Growth
3. S.-C. Huang and M.E. Glicksman, Acta Met. 29,701(1981)
4. M.E. Glicksman and N.B. Singh, J. Cryst. Growth 98,277(1989)
5. J.S. Langer, Rev. Mod. Phys. 52,1(1980)
6. J.S. Langer, Lectures in the theory of pattern formation, in: Chance and Matter, Proc. Les Houches Summer School, Session XLVI, 1986, Eds. J. Souletie, J. Vannimenus and R. Stora ,North-Holland, Amsterdam, 1987
7. Theory of dendritic solidification is covered by other papers of this conference
8. R. Trivedi and J.T. Mason, Met. Trans. in press
9. G.P. Ivantsov, in: Growth of Crystals, Vol.1 Eds.: A.V. Shubnikov and N.N. Sheftal (Consultants Bureau, New York, 1958) p.76
10. G. Horvay and J.W. Cahn, Acta Met. 9, 695(1961)
11. J.H. Bilgram, M. Firmann and E. Hürlimann, J. Cryst. Growth 96,175(1989)
12. J.H. Bilgram and E. Hürlimann, Proc. VIIth European Symp. on materials and fluid sciences in microgravity, ESA SP-295,173(1990)
13. E. Hürlimann, R, Trittibach and J.H. Bilgram, Helv. Phys. Acta 63,473 (1990)
14. E. Hürlimann and J.H. Bilgram, this conference
15. R. Willnecker, D.M. Herlach and B. Feuerbacher, Phys. Rev. Lett. 62, 2707(1989)
16. P. Bouissou, B. Perrin, and P. Tabeling, Phys. Rev. A 40,509(1989)

LIGHT SCATTERING EXPERIMENTS DURING THE FREEZING AND MELTING TRANSITION

J.H. Bilgram and R. Steininger

Laboratorium für Festkörperpysik
ETH
CH 8093 Zürich, Switzerland

ABSTRACT

Anomalous light scattering is observed at the solid-liquid interfaces of growing crystals. Two models for the description of the phenomena
are under discussion: 1.) Light is scattered by continuous fluctuations in
a mesophase layer; 2.) Light is scatterd by gas bubbles in front of the
growing crystal. In this paper it is shown that the bubbles model is not
compatible with the experiment. The model with continuous fluctuations is
compatible with the experimental results.

INTRODUCTION

The freezing transition is a first order phase transition. It is not
a bulk process but takes place at the interface only. First order phase
transitions cannot be studied at equilibrium conditions in contrary to
second order phase transitions. The latent heat has to be removed continuously during freezing. Therefore only little is known about the freezing
transition and the processes at the interface.

Light scattering is a unique technique for *in situ* studies of processes in a small volume [1]. Light scattering techniques were introduced
to study the solid-lquid (s.-l.) phase transition about 12 years ago [2].
Intense Rayleigh scattering at the s.-l. interface of growing ice crystals
was observed. The dynamics of the processes at the s.-l. interface can be
characterized by a single diffusion constant D_i, with $D_i \simeq 10^{-8}$ cm^2/s. This
enhanced Rayleigh scattering was shown to originate in an interface layer
[3]. This layer is located at the liquid side of the s.-l. interface, and
has a typical thickness of about 5 μm [3,4]. Enhanced Rayleigh scattering
experiments have been performed in several laboratories and at various
s.-l. interfaces: H_2O [5.6], D_2O [7], salol [8,9,10], cyclohexanol [11],
naphtalene [12], biphenyl [12], succinonitrile [13], and cyclohexane [14].
Enhanced Rayleigh scattering seems to be a general phenomenon of crystal
growth.

Growth and Form, Edited by M. Ben Amar *et al.*
Plenum Press, New York, 1991

Several models were developed to describe the phenomenon of diffusive
light scattering, for a review see [15]. A corrugated interface was discussed [2]. In this model it is assumed that light is scattered by surface
corrugations with fluctuating amplitudes. This model is a two-dimensional
model. The experiments of Böni et al.[3] have shown that the light is
scattered in a three dimensional layer. Thus a model using a 2 dimensional
corrugated layer is not compatible with the experiment. Experiments at the
s.-l. interface of ice [3] and salol [8,9] showed that light is scattered
at the liquid side of the s.-l. interface. The light scattering phenomena
take place in a medium with a mean index of refraction close to the one of
the melt. Therefore a model assuming light scattering to take place in the
solid phase is not compatible with the experiment.

The propagation of growth steps has been studied in light scattering
experiments at the s.-l. interface of salol crystals [8,9]. The velocity
of propagating steps was determined by means of Doppler velocimetry. Dürig
et al. [8] performed light scattering experiments with red and blue light.
They found that light with different wave-lengths is scattered at the
same scattering vectors but at different scattering angles. Thus they could
prove that light is scattered at a moving optical grating by Bragg-scatter-
ing. The light is not reflected by lateral moving macroscopic objects.
Crystal growth mediated by lateral moving steps is described by the theory
of Burton, Cabrera and Frank (BCF). Predictions of this theory allowed to
conclude that the aggregation of molecules at the growth steps is not determined by the constant of self diffusion in the bulk liquid, but by the
diffusion constant D_i which is measured in the light scattering experiments
at the s.-l. interface [8,15]. The interpretation of light scattering experiments, where the propagation of growth steps is studied, is generally
accepted. The interpretation of the so called diffusive light scattering
is a matter of controversy. In the present state, two different models to
describe diffusive light scattering at the s.-l.interface are under discussion, both assume that the light is scattered in a 3-dimensional layer
in front of the s.-l. interface.

i) It was proposed that a layer of a mesophase appears at the surface of
growing crystals [3]. Density fluctuations in this mesophase scatter the
light. This assumption is corroborated by experimental and theoretical
work. In a recent synchrotron radiation experiment a labile ordered layer
at the crystal/solution interface of ADP (ammonium dihydrogen phosphate)
was observed [16]. Molecular dynamics simulations showed the presence of a
quasiliquid layer at the s.-l. interface. Recent molecular dynamics crystal
growth simulations indicated that the [111] crystal face of a Lenard-Jones
crystal grows by addition of clusters of atoms rather than by single atom
addition [17]. Discrepancies remain: up to now molecular dynamics simulations give a thickness of a few atomic spacings, whereas we measure a
thickness of several microns. But so far it is not possible to model rea-
listically a s.-l. transition.

ii) Recently another model was proposed by Cummins et al. for the enhanced
scattering [18]. These authors assume that the scattering is caused by
small microbubbles in front of the s.-l. interface. This hypothesis is
supported by experiments, where the liquid was contaminatec with gas [6].

In this paper we will focus at the experimental evidences which favour
the one or the other of the two models for diffusive light scattering at
the s.-l. interface.

THE MESOPHASE MODEL

The correlation length

We imagine that the spatial density fluctuations inside the s.-l. interface layer can be characterized by an autocorrelation function $\gamma(r)$. In the first Born approximation, (the validity of which cannot be doubted in our case) the autocorrelation function $\gamma(r)$ and the scattered intensity $I(q)$ are Fourier transforms of each other [19]. If $\gamma(r)$ decays over a characteristic length ξ one speaks of the correlation length. Roughly, ξ is a measure of the mean distance between neighbouring high density or low density regions. Since the range of scattering vector q of the intensity-measurements at the s.-l. interface is not sufficient to perform a Fourier transform, we use autocorrelation functions, which have been successfully used in the description of related systems, and which are characterized by a single parameter ξ.

Debye and Bueche [20] found that the frozen-in density fluctuations responsible for the x-ray and light scattering in acrylic glass can be described by

$$\gamma(r) = e^{-r/\xi} \tag{1}$$

which leads to

$$I(q) \propto 8\pi\xi^3 \left(1 + q^2\xi^2 \right)^{-2} \tag{2}$$

As the enhanced scattering closely resembles critical opalescence, we also try Ornstein-Zernike [21] pair correlation function $\gamma(r)$:

$$\gamma(r) = \frac{1}{r} e^{-r/\xi} \tag{3}$$

which leads to

$$I(q) \propto 4\pi\xi^2 \left(1+q^2\xi^2 \right)^{-1} \tag{4}$$

The scattering diagrams calculated for both correlation functions are plotted in fig.1 together with data measured at the s.-l. interface of cyclohexane [14]. The calculated curves and the experimental points were arbitrarily shifted parallel to the intensity axis in such a way that they intersect for $\Theta = 35^{\circ}$. Both correlation functions are found to be compatible with the experimental data. For the correlation length ξ we deduced $\xi \approx 0.2$ µm using the Ornstein-Zernike pair correlation function and $\xi \approx 0.05$ µm using the Debye-Bueche pair correlation function.

Although we are not able to derive the exact form of the pair correlation function from the intensity data, it is interesting to note that $\xi \simeq 0.2$ µm deduced assuming an Ornstein-Zernike form of $\gamma(r)$ agrees with the length calculated from the measured diffusion constant $D_i = \Gamma/q^2$ using the Stokes-Einstein-Kawasaki relation [22]

$$D = \frac{k_B T}{6\pi\eta\xi} \tag{5}$$

and inserting the shear viscosity of the bulk liquid η. This relation was deduced using (3). It was shown in [23] that (5) is not very sensitive to a change of $\gamma(r)$. Thus the enhanced Rayleigh scattering at the s.-l. interface seems to resemble the critical slowing down occuring in second order phase transformations. However, this conclusion leads to contradictions: A long correlation length gives rise to an increase of the scattered intensity and favours forward scattering. Equ.(5) is deduced in the hydrodynamic limit and therefore is valid for $q\xi \lesssim 1$ [22]. For $q\xi \gtrsim 1$ deviations from $\Gamma \propto q^2$ have to be expected, where Γ stands for the linewidth of the light scattered in the interface layer by the fluctuations. In our experiments at $\lambda=488$ nm, q ranges from 10 μm^{-1} to 34 μm^{-1}, so that $2 \lesssim q\xi \lesssim 6.8$. We did never find any tendency for a deviation from $\Gamma = D_i q^2$ in all our experiments [2,3,8,14]. Therefore thermodynamic fluctuation theory for critical phenomena does not describe the s.-l. interface layer. If thermodynamic fluctuation theory and eq.(5) would be applied to our problem, the shear viscosity η in the interface layer would have to be significantly higher (at least 1 order of magnitude) than in the bulk liquid. In addition to linewidth measurements we performed intensity measurements, which provide a second independent means for the determination of the correlation length ξ. We find again values of ξ which are not compatible with the hydrodynamic limit. This result does not depend upon whether Debye-Bueche or the Ornstein-Zernike correlation function is chosen.

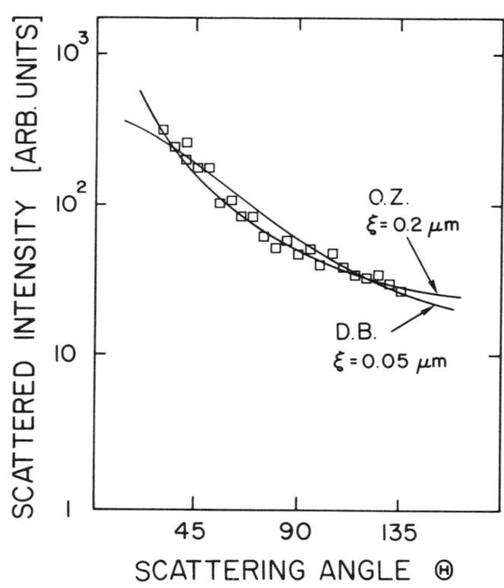

Fig.1 Comparison of the measured scattering diagram for $\lambda=488$ nm with the one calculated from the Ornstein-Zernike pair correlation function and the Debye-Bueche pair correlation function. The incident light is vertically polarized. The experimental data are not compatible with the scattering theorie of Mie.

Absolute Intensities of the scattered light

The scattering power is expressed by the Rayleigh ratio R

$$R = \frac{I}{I_o} R^2 \frac{1}{V} \tag{6}$$

I_o is the irradiance of the incoming light and I the irradiance of the scattered light at the distance R of the test substance. For pure liquids R is proportional to the isothermal compressibility. Two types of techniques are used to determine R. Absolute measurements require precise knowledge of the geometrical factors and lead directly to the scattering powers. A simpler technique is the relative one, which measures the scattering power in comparison to some reference substance with tabulated scattering power. An additional problem in our experiments is that the volume of the scattering layer is much smaller than the scattering volume imaged on the sensitive spot of the photomultiplier. For the experiments with cyclohexane we found after careful evaluation of geometrical factors [14] that the scattering power of the interface layer is enhanced by a factor between 40 and 900 relative to the scattering power of the liquid. An enhancement in the same range has been found for H_2O [3]. For the following discussion of the s.-l. interface of cyclohexane we use

$$R_{layer} \simeq 100 \, R_{cyclohexane} \tag{7}$$

The local dielectric constant ε of a medium with fluctuations $\Delta\varepsilon$ around the mean dielectric constant ε_0 can be written as

$$\varepsilon = \varepsilon_0 + \Delta\varepsilon \tag{8}$$

The intensity scattered by these fluctuations is [20]

$$I = I_0 \, \frac{\pi^2}{\lambda^4 R^2} \, \frac{\overline{\Delta\varepsilon}^2}{\varepsilon_0^2} \, V \int \gamma(r) \, e^{iqr} \, dr \tag{9}$$

where the dielectric correlation function $\gamma(r)$ has been introduced.

$$\gamma(r) = \frac{1}{\overline{\Delta\varepsilon}^2} \int \Delta\varepsilon(r) \, \Delta\varepsilon(r-r') \, dr' \tag{10}$$

where $\overline{\Delta\varepsilon}^2$ is the mean square of the dielectric fluctuation $\Delta\varepsilon$. The integral term in eq.(9) is responsible for the angular dependence of the intensity of the scattered light if the correlation length of the fluctuation is large. This was used in the preceeding paragraph to calculate the correlation length.

It is well known that the amplitude of thermodynamic fluctuations can be related to the compressibility or the correlation length by the use of theories of the type of the Ornstein-Zernike theory. We showed that these fluctuation theories cannot be applied to the enhanced Rayleigh scattering. Therefore we try another approach. We insert Debye-Bueche correlation function into eq.9. Together with eq.6 one obtains:

$$R = \frac{\pi^2}{\lambda^4} \, \frac{\overline{\Delta\varepsilon}^2}{\varepsilon_0^2} \, 8\pi\xi^3 \, (\, 1+q^2\xi^2)^{-2} \tag{11}$$

This new expression represents the Rayleigh factor for the enhanced scattered intensities in terms of ξ, ε_0 and $\overline{\Delta\varepsilon^2}$. Inserting the known values R_{layer}, λ, ε_0, ξ and q one obtains from eq. 11

$$\overline{\Delta\varepsilon}^2 = 1.1 \cdot 10^{-6} \tag{12}$$

and with $n^2 = \varepsilon$, $\Delta n \simeq 2.5 \cdot 10^{-4}$
This is the root mean square value of the fluctuations of the index of refraction. The same procedure using the Ornstein Zernike function leads to $\Delta n \simeq 3 \cdot 10^{-8}$.

THE GAS BUBBLES MODEL

Two basic experimental facts are:
1.) The autocorrelation function of the photoncounts can be fitted over
   more than two orders of magnitude by a single exponential (fig.2).
2.) The linewidth of the scattered light increases with the square of the
   scattering vector.
These observations imply: There are many inhomogeneities which scatter the
light. The dynamics of these inhomogeneities can be described by one well
described diffusion constant. This means in terms of the gas bubbles model:
There are many gas bubbles, and all of them have the same size. Cummins and
Mesquita and coworkers [18, 12, 24] discussed in some detail the production
of monodisperse gas bubbles at the s.-l. interface.

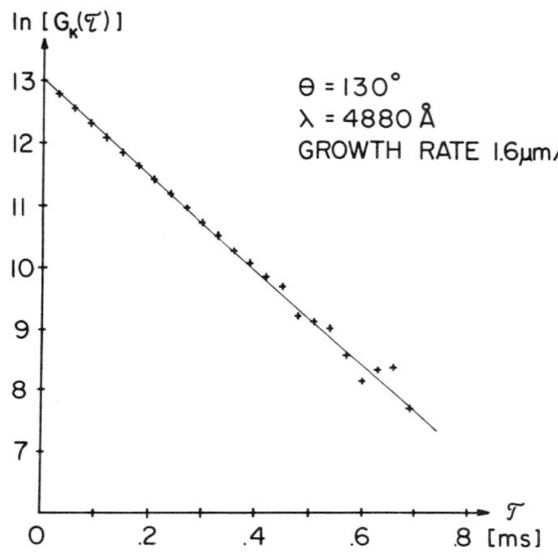

Fig.2 Semi-logarith-
mic plot of a inten-
sity autocorrelation
function [2]. It is
mandatory to show that
the correlation func-
tion can be fitted by
a single exponential
over a decay of 2 or-
ders of magnitude at
least. Light scattered
by dust particles does
not lead to such a
correlation function.

The nucleation and stability of gas bubbles

Cummins et al.[18] proposed the existence of gas bubbles which diffuse
freely in front of the s.-l. interface. It was assumed [18] that these bubbles are created by homogeneous nucleation. It was shown [15] that the probability for nucleation of a gas bubble in water contains a Boltzmann factor of $\exp(-5 \cdot 10^4)$. Thus the probability of homogeneous nucleation of gas
bubbles is very close to zero. Cummins et al. [24] give an estimation for
the Boltzmann factor for nucleation of gas bubbles in salol with $\exp(-2 \cdot 10^7)$.
There is general agreement that homogeneous nucleation of gas bubbles in
front of the s.-l. interface does not exist.

It was proposed [11,12,24,25], that once bubbles are formed by an unknown mechanism, these bubbles will float in front of the growing crystal.
The gas bubbles are assumed to have a radius corresponding to the hydrodynamic radius r as determined in the light scattering experiments (eq.5):
For the ice-water system $r \approx 25 \cdot 10^{-7}$ cm, for the s.-l. interface of cyclohexane $r \approx 250 \cdot 10^{-7}$ cm. The pressure inside of a gas bubble is given by the

external pressure p_e and a contribution due to the surface tension σ_{lv}. Following [24]

$$p = p_e + \frac{2\,\sigma_{lv}}{r} \tag{13}$$

Cummins et al. assume that the partial pressure of the gas outside of the bubble is the same as the gas pressure inside of the bubble. If the partial pressure outside of the bubble would be lower than the gas pressure in the bubble, the bubble would dissolve. The pressure inside of the bubble can be calculated. For the ice-water system σ_{lv}=75 dyn/cm, r=25·10^{-7} cm, $p_e \simeq p_{triple\ point}=p_t=6.10^{-3}$ bar. From eq.13 we obtain

$$p(\text{bubble in water}) \simeq 6 \cdot 10^7 \text{ dyn/cm}^2 = 60 \text{ bar} \tag{14}$$

In the following section we calculate the partial pressure at the s.-l. interface. Gases as well as other impurities with a concentration C_m in the melt far away from the interface are segregated at the surfaces of growing crystals, if they are less soluble in the solid than in the liquid phase. Segregation can be characterized by the segregation coefficient at the interface:

$$k^* = C_i/C_s \tag{15}$$

where C_i is the concentration of impurities in the melt at the interface and C_s is the concentration of impurities in the crystal. Typical values of k^* are $10^{-1} > k^* > 10^{-2}$ [24]. For our numerical estimations we use $k^*=10^{-2}$ in this paragraph. This is "best case" for the bubbles model according to Cummins et al. [24]. The segregated impurities are piled up in front of the crystal and may be removed by convection. It is very difficult to prevent convection at normal gravity conditions. Therefore the melt is stirred by convection and the concentration of impurities is given by the well known expression derived by Burton, Prim and Slichter [26]:

$$C_i = \frac{C_m}{k^* + (1-k^*)\,\exp(-v_k \delta/D_\ell)} \tag{16}$$

Here v_k is the growth velocity of the crystal, D_ℓ is the diffusion constant of the impurity and δ is the thickness of the unstirred diffusion layer in front of the growing crystal. $\delta \simeq 0.1$ cm if the melt is stirred very slowly by thermal convection. Inserting approximate numbers: $D_\ell=10^{-5}$ cm^2/s, $k^*=10^{-2}$, $v_k=10^{-4}$ cm/s, we obtain $C_i \simeq 3C_m$. The total gas pressure in the growth vessel is close to the triple point pressure p_t. Thus the partial pressure of any gas at the s.-l. interface will be of the order of magnitude p_t (p_t<<60bar). If there is no convection in the growth vessel then the concentration of the impurity at the interface will increase by another factor 2 i.e. $C_i \simeq 6C_m$. Thus we have to conclude: if there is an (up to now unknown) nucleation mechanism to create bubbles these bubbles would dissolve in the melt.

The size selection of the bubbles

Although there is no indication of a nucleation mechanism for free moving gas bubbles, we will assume in the following that such a mechanism does exist. Although we showed that these bubbles would dissolve in the melt, we now assume that these bubbles persist in front of the growing crystal.

If there are gas bubbles then we agree with Cummins et al. that a mechanism has to exist which provides the selection of size of the bubbles. The model was: small bubbles dissolve or they are captured by the crystal,

and bigger bubbles move out of the region of the interface. Two forces are
in balance for the monodisperse bubbles observed in the experiment: buoyancy force F_b and viscous force F_v. These forces keep the bubbles at constant
distance from the interface:

$$F_b = \frac{4}{3} \pi r^3 g \qquad (17)$$

$$F_v = 4 \pi r v_k \eta \qquad (18)$$

Using $v_k=10^{-4}$ cm/s this force balance defines a bubble size $r \simeq 5 \cdot 10^{-4}$ cm.
This is much larger than any size ever observed in a light scattering experiment under discussion. This discrepancy was realized by Mazur and Keizer [27]. The model predicts that r increases with v_k. All experiments show
that the diffusion constant is independent of v_k [2,14]. The most important
pitfall of this model is that it anticipates that the melt is above the
crystal. We did not find any change in the light scattering phenomena when
we turned the experiment upside down [4,14]. This experimental fact shows
that the model does not describe the experiment.

In a new model a third force is introduced [24,25]. Now buoyancy is
assumed to be unimportant but a thermophoretic force F_t is assumed to balance F_v

$$F_t = 2 \pi r^2 \frac{d\sigma_{1v}}{dT} G \qquad (19)$$

σ_{1v} is the surface free energy of the melt-gas interface and G is the thermal gradient in front of the s.-l. interface. The thermophoretic force was
introduced by Young [28]. This force is due to Marangoni convection. If
buoyancy can be neglected, the size of the gas bubbles is given by:

$$r = 2 \eta \frac{1}{d\sigma_{sl}/dT} \frac{1}{G} v_k \qquad (20)$$

We have varied v_k by more than two orders of magnitude [2,14]. We varied
G from positive to negitve sign [4] we did not find any influence of G or
v_k on the linewidth of the scattered light. No other experiments are published which show such a dependence. In addition to that there are doubts
that the Marangoni effect exists in small bubbles [29]. Thus we have to
conclude that the gas bubbles model does not describe the experiment.

<u>The intensity of the scattered light</u>

The scattering diagram for identical spherical particles is given by

$$I(\Theta) = N \frac{|S|^2}{k^2 R^2} I_o V \qquad (21)$$

where S is the amplitude scattering matix element, N the number density of
particles, R the distance from the scattering volume V to the detector and
$k=2\pi/\lambda$. We find the Rayleigh ratio R by comparing eq.21 with eq.6

$$R = N \frac{|S|^2}{k^2} \qquad (22)$$

The concentration of particles can be determined from relative measurements

$$N = \frac{k^2}{|S|^2} R_{ref} \frac{i}{i_{ref}} \qquad (23)$$

We used cyclohexane as reference substance. The Rayleigh ratio $R_{v,v+h}$ is
at T_m : $R_{cyclohexane}$= 14.6·10^{-4}m^{-1}. The matrix element S can be calculated
from Mie's scattering theory in terms of the parameter $\alpha=2\pi r n_{med}/\lambda$. The
parameter α has been determined from the angular dependence of the scattered
intensity $\alpha\approx 4$ [14]. A 100 fold increase in the scattering power is obtained
if there are about 10 gas bubbles in one sampling volume, i.e. the volume
which is imaged at the photosensitive spot of the photo multiplier. For
details of the calibration procedure see ref.14.

The intensity autocorrelation function of the scattered light can be
fitted by a single exponential. In intensity autocorrelation spectroscopy
a single exponential is obtained only if there are many identical scatterers in the scattering volume. If only a few scatterers would be present in
the scattering volume, fluctuations in the number of scatterers would lead
to intensity fluctuations, and thus to a deviation of the correlation
function from the exponential. Thus the single exponentials which were
found in all experiments are the result of a scattering process involving
a large number of scatterers. The experimental fact that intensity measurements are compatibel with about 10 gas bubbles only leads again to a contradiction between the experiment and the gas bubbles mode Changes in
number or size would lead to additional contradictions with the experiment.

Two different phenomena

The experiments performed up to now may be divided into two categories
a) Experiments using substances which are studied under high purity [3,4,14]
b) Experiments using substances which were not highly purified (Bridgman
growth [10,13,25] or which were intentionally saturated with gas [6]. Mazur
and Keizer [27] suggested that two different phenomena are studied in the
experiments with H_2O (type a experiments) and in the experiments with
cyclohexanol (type b experiments). There is experimental evidence that indeed two different phenomena are studied. We focus again at two experimental
facts:
1) The autocorrelation function of the photoncounts is a single exponential
   (no slow component) s. fig.2
2) The linewidth of the scattered light is proportional to the square of
   the scattering vector.
The photon correlation function can be fitted over more than two orders of
magnitude by a single exponential in type a) experiments. An example for
a correlation function obtained in type b) experiments was published in
ref. 11 fig.1. The exponential is fitted over a decay of less than 20%. It
is obvious that it is not possible to decide from a decay of about 20 %
whether the correlation function can be fitted by one single exponential
or not. For succinonitrile the situation is similar, ref.13 fig.2. The
correlation function cannot be fitted by a single exponential. No data are
given for the scale of the correlation function plotted. More information
can be obtained from the angular dependence of the linewidth of the
scattered light. A proportionality $\Gamma \propto q^2$ is found for the type a) experiments see ref. [2,3,7,14]. The date published for cyclohexanol do not show
this proportionality. These data might be fitted as well by a constant independent of scattering vector. This behavior is typical for a collection
of dust particles.

Conclusions on the gas bubbles model

1.) It was assumed that gas bubbles are nucleated homogeneously.
   - It was shown that the probability for homogeneous nucleation is zero.

2.) A nucleation mechanism may be given
   It is assumed that gas bubbles are in "equilibrium" with the melt.
   - It was shown that gas bubbles would dissolve.
3.) Stable gas bubbles may be given.
   It is assumed that F_v and F_h provide a selection mechanism.
   - Experiments and numerical estimations show that gravity does not influ-
     ence the size of the bubbles.
4.) Stable gas bubbles may be given
   It is assumed in a second model that F_v and F_t provide a selection
   mechanism.
   - It was shown that F_t does not exist in the form assumed [29].
5.) Stable gas bubbles and a thermophoretic force may be given.
   It is assumed that $r \propto v_k/G$ (eq.20)
   - It was shown that the size of the bubbles is independent of v_k.
   - It was shown that the size of the bubbles is independent of G.
6.) Monodisperse gas bubbles may be given
   It is assumed that the properties of the scattered light can be described
   by light scattered at gas bubbles.
   - It was shown that Mie-scattering does not describe the experimental
     scattering diagram, neither the absolute intensity, not the angular
     dependence of the intensity.

CONCLUSIONS

The combination of dynamic and static light scattering allows the determination of the inhomogeneities which scatter the light at the s.-l. interface of growing crystals. Two models proposed to explain the phenomenon
were tested thorougly. The enhanced scattering in pure systems cannot be
due to small gas bubbles. The second model assumes a layer of a mesophase
at the surface of growing crystals. This model is compatible with the experimental data. Fluctuations of the index of refraction scatter the light.

These fluctuations can be described by a Debye-Bueche pair correlation
function. From the enhancement of the scattered intensity we deduced a mean
amplitude $\Delta n \approx 10^{-4}$ and a typical correlation length of $\xi \approx 0.05$ µm of the
fluctuations in the interface layer of cyclohexane.

The resemblance to critical opalescence was tentatively used to describe the nature of the fluctuations. It was shown that the theoretical and
experimental results disagree. Thus fluctuations of the index of refraction
cannot be interpreted as thermodynamic temperature fluctuations in contrast to those observed in a liquid at the critical point. The mesophase
layer is thought to consist of small weakly bound aggregates of molecules.
In this case fluctuations of the index of refraction are not due to temperature fluctuations but to the formation and decay of intermolecular
structures.

Enhanced Rayleigh scattering has now been observed in various systems.
Cyclohexane, a plastic crystal without a permanent dipole moment is the
simplest substance which has been studied up to now. The occurence of enhanced Rayleigh scattering at the s.-l. interface of cyclohexane seems to
indicate that the mesophase layer is a general phenomenon in crystal growth.

ACKNOWLEDGEMENTS

    We thank Prof. H.-R. Ott for his support. This work was supported by
the Swiss National Science Foundation.

REFERENCES

1) J.H. Bilgram and R. Steininger, J. Cryst. Growth, $\underline{99}$,30(1990)
2) H. Güttinger, J.H. Bilgram and W. Känzig, J. Phys. Chem.Solids
   $\underline{40}$,55(1979)
3) P. Böni, J.H. Bilgram and W. Känzig, Phys. Rev. $\underline{A28}$, 2953(1983)
4) P.U. Halter, J.H. Bilgram and W. Känzig, J.Chem. Phys. $\underline{89}$,2622(1988)
5) R.A. Brown, J. Keizer, U. Steiger and Y.Yeh, J.Phys. Chem. $\underline{87}$,4135(1983)
6) J.P. Vasenka and Y. Yeh. Phys. Rev. $\underline{A38}$,5310(1988)
7) B. Zysset, P. Böni and J.H. Bilgram, Helv. Phys. Acta $\underline{54}$,265(1981)
8) U. Dürig, J.H. Bilgram and W. Känzig, Phys. Rev. $\underline{A30}$,946(1984)
9) O.N. Mesquita and H.Z. Cummins, Physico-Chem. Hydrodyn. $\underline{5}$,389(1984)
10) O.N. Mesquita, D.G. Neal, M. Copic and H.Z. Cummins, Phys. Rev. $\underline{B29}$,
    2846(1984)
11) G. Livescu, M.R. Srinivasan, H. Chou, O.N.Mesquita and H.Z. Cummins,
    Phys. Rev. $\underline{A36}$,2293(1987)
12) O.N. Mesquita, L.O. Ladeira, I. Gontijo, A.G. Oliveira and G.A, Barbosa,
    Phys. Rev. $\underline{B38}$,1550(1988)
13) J.M. Laherrère, H. Savary, R. Mellet and J.C. Tolédano, Phys. Rev. $\underline{A41}$,
    1142(1990)
14) R. Steininger and J.H. Bilgram, Helv. Phys. Acta $\underline{62}$,215(1989)
    R. Steininger and J.H. Bilgram, J. Cryst. Growth $\underline{99}$,98(1990)
    R. Steininger and J.H. Bilgram, J. Cryst. Growth
15) J.H. Bilgram, Phys. Reports $\underline{153}$,1(1987)
16) D. Cunningham, R.J. Davey, K.J. Roberts, J.N. Sherwood and T. Shripathi,
    J. Cryst. Growth $\underline{99}$,1065(1990)
17) E. Burke, J.Q. Broughton and G.H. Gilmer, J. Chem. Phys. $\underline{89}$,1030(1988)
18) H.Z. Cummins, G. Livescu, Henry Chou and M.R. Srinivasan, Solid State
    Com. $\underline{60}$,857(1986)
19) M. Kerker: The Scattering of Light and other Electromagnetic Radiation,
    Academic Press, London, (1969), p.459
20) P. Debye and A.M. Bueche, J.Appl. Phys. $\underline{20}$,518(1949)
21) M.E. Fisher, J.Math. Phys. $\underline{5}$,944(1964)
22) K. Kawasaki, Ann. Phys.(N.Y.) $\underline{61}$,1(1970)
23) H.L. Swinney, D.L. Henry and H.Z. Cummins, J.Physique $\underline{33}$,1,C1-C8(1972)
24) H.Z. Cummins and L.M. Williams: Dynamic light scattering at the non
    equilibrium crystal melt interface, in:"Light scattering by liquid
    surfaces, ed.: D. Langevin, Surfactant Sci. Ser. Marcel Dekker
25) L.M. Williams. M.R. Srinivasan and H.Z. Cummins, Phys.Rev. Lett. $\underline{64}$,
    1526(1990)
26) J.A. Burton, R.C. Prim, W.P. Slichter, J.Chem. Phys.$\underline{21}$,1987(1953)
27) P. Mazur and J.Keizer, Phys Rev. $\underline{A38}$,5267(1988)
28) N.O. Young, J.S. Goldstein and M.J.Block, J. Fluid Mech. $\underline{6}$,350(1959)
29) D. Neuhaus and B. Feuerbacher, Proc. 6th Europ. Symp. on Material
    Sci, under Microgravity Conditions, Bordeaux, 1986, ESA SP-256,
    (1987) p. 241

SIDEBRANCHING OF XENON DENDRITES

E. Hürlimann and J.H. Bilgram

Laboratory of Solid State Physics
ETH
CH-8093 Zürich

ABSTRACT

The sidebranches of xenon dendrites growing into supercooled melt have been investigated. The sidebranches grow on
top of the fins of the dendrite. First measurements of the
spatial evolution of the sidebranches have been performed.
The maximal amplitude w of the sidebranches has been measured
as a function of their position l_p at the fin. It has been
found $w \sim l_p^{1.72}$ independent of the supercooling. No decrease
of the amplitude for large l_p has been observed.

INTRODUCTION

Dendrites growing into a supercooled melt or into a
supersaturated solution have a very complex, time dependent
shape (fig.1). During the last years, great experimental and
theoretical efforts were made to understand the dendritic
crystallization. At present no theory is available which
describes the dendritic growth completely. Most of the models
describe the tips of the dendrites only. Many experiments
were performed to study the tips of dendrites: For several
substances the tip growth velocity, the tip radius and the
mean initial sidebranch spacing as a function of the supercooling were determined and compared with theoretical predictions [1-4]. There are only a few experiments, where the
behavior of the whole dendrite was investigated. One example
is the experiment where the volume of xenon dendrites as a
function of the overall length of the dendrite has been
measured [2,5]. Also little is known about the sidebranching:
Some experimental and theoretical work in this field are
found in the next paragraph.

In our experiment we are studying the sidebranching and
coarsening of free growing xenon dendrites. We are interested
in the spatial evolution of the sidebranches for a certain
time of the dendritic growth. For various supercoolings we
measure the amplitude of the sidebranches as a function of

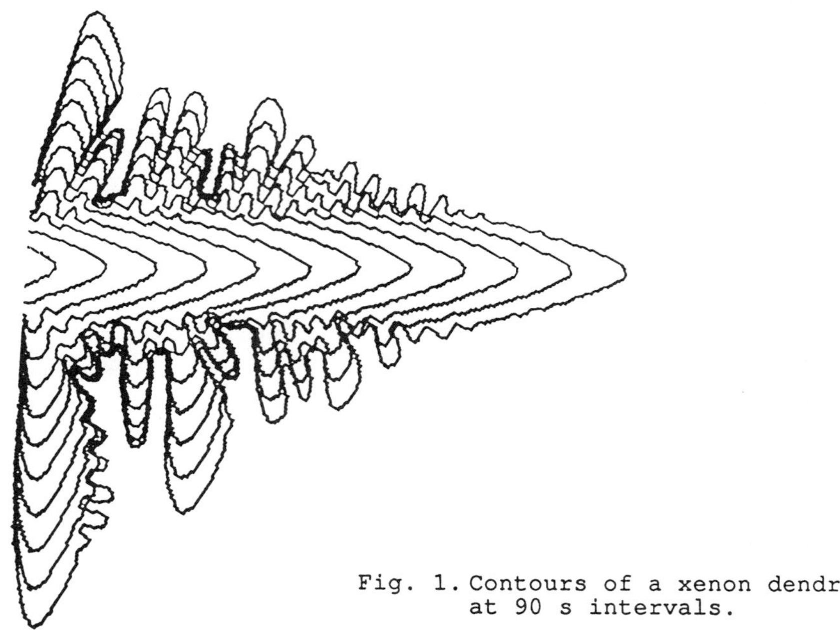

Fig. 1. Contours of a xenon dendrite
at 90 s intervals.

the distance from the tip. An analysis of the spatial frequency is performed in order to find out whether the initial
structure of the sidebranches can be characterized by one
unique wavelength and the subsequent coarsening by period
multiplication. First results of the amplitude measurements
and some qualitative observations of the dendritic shape are
presented in this paper.

SOME PREVIOUS STUDIES

Experimental results might be compared with two theoretical models. The first one proposes that sidebranches result
from the selective amplification of noise [6]. The initial
perturbation of the interface near the tip propagates away
from the tip and grows approximately exponentially with increasing distance from the tip. Perturbations with fixed frequency decay at large distances behind the tip, whereas wave
packets with low frequency components continue to grow at
arbitrarily large distances from the tip. In this case the
amplitude increases with $r^{3/8} \exp(\alpha \sqrt{r})$, where r means the
radial length of the dendrite. The structure is nonperiodic,
the characteristic wavelength grows slowly and the frequency
distribution becomes sharper with increasing r. Another consequence of this model is that the sidebranches on opposite
sides of the dendrite are not correlated. The second model
proposes a sidebranching which is driven by tip oscillations
[7]. A periodic structure and a correlation between sidebranches is expected. The amplitude of the sidebranches grows
exponentially, like $\exp(\alpha r)$, somewhat faster than in the
noise model.

80

At the moment there are still few experimental results available. Two types of experiments on dendritic sidebranching were performed:
a) Free growing dendrites were investigated. The typical sizes of the natural evolution of the sidebranches like the fre-quency, the amplitude and coarsening were measured [8].
b) The sidebranches are forced by an external mechanism: In one experiment by a heat pulse [9] and in another by a oscillating flow of the melt [10].
The results of these three experiments support the noise amplification model. But for a crucial test of the theories more experimental data are necessary. We decided to perform an experiment of the first type. But in contrast to the above mentioned experiments, which were performed with solute systems, we are studying an one component system.

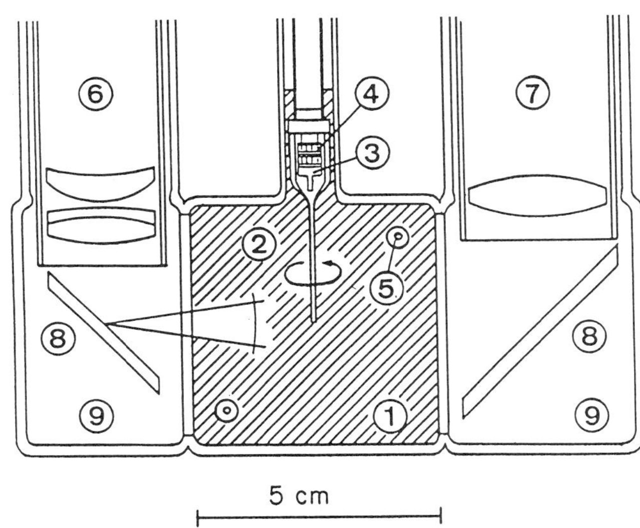

Fig. 2. Growth vessel: (1) liquid xenon; (2) rotatable capillary; (3) upper end of the capillary, where nucleation occurs; (4) Peltier element; (5) temperature sensor; (6) periscope; (7) illumination system; (8) mirror; (9) helium atmosphere.

EXPERIMENTAL SETUP

The growth vessel (fig. 2) in the cryostat (fig. 3) is filled with about 80 ccm pure, liquid xenon and cooled down to some mK below the triplepoint temperature of about 161 K. The temperature of the supercooled liquid can be stabilized within less than $\pm 10^{-4}$ K. For details on the temperature stabilization and the gas handling see [1]. Crystal growth is initiated by the capillary injection technique. At the upper end of the capillary a seed is nucleated by cooling with a Peltier element. This seed grows down inside the capillary into the melt and after a transient state stationary dendritic growth is observed. The dimensions of the dendrite are small compared to the dimensions of the growth vessel. There are no confinement effects [2,11].

The dendrites are imaged by a self-built periscope at the chip of a CCD-camera. This periscope has a F-number of about 1.7, which allows to work with a low intensity illumination. To improve the contrast, we use monochromatic light. These highly contrasted pictures are digitized with a framegrabber into 512x512 pixels and the dimensions of the sidebranches are determined by an image processing system. The resolution of about 8 μm is limited by the total number of pixels of the framegrabber as a result of a compromise between the size of the object plane and the magnification of the pictures.

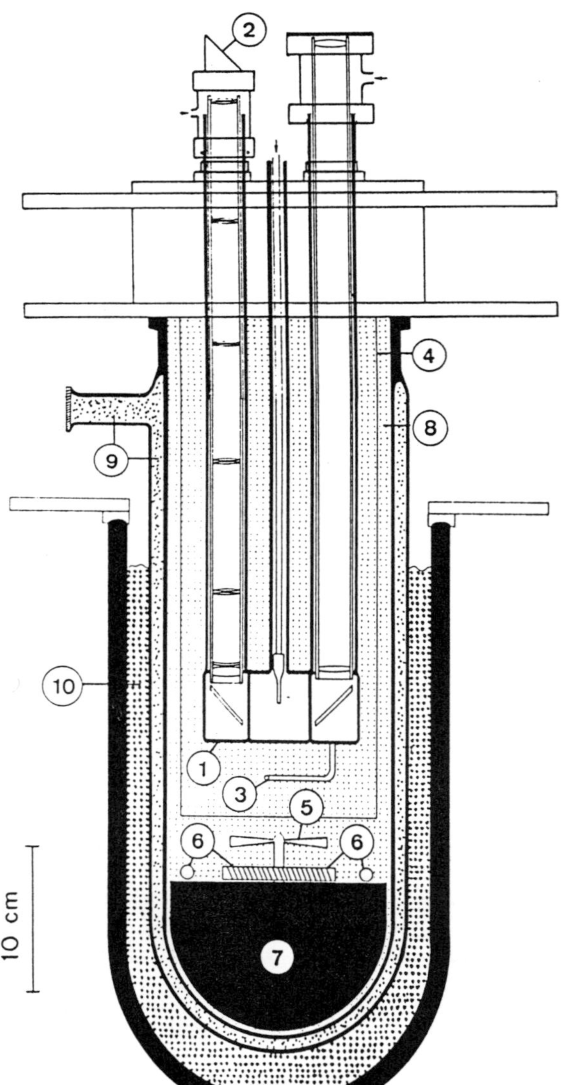

Fig. 3. Cryostat.
(1) growth vessel
(2) prism
(3) temperature sensor
(4) tube for a laminar flow of the thermostating liquid
(5) stirrer
(6) heater
(7) block of metal to damp vibrations of the stirrer
(8) thermostating liquid: isopentane
(9) adjustable vacuum to control cooling power
(10) liquid nitrogen

EXPERIMENTAL RESULTS

The xenon dendrites show, like all dendrites with a fcc
structure, a fourfold symmetry. The shape is not a paraboloid
of revolution. Four fins develop along the dendrite starting
very close to the tip. The sidebranches grow on the top of
these fins. The growth direction is normal to the back of the
fins. The sidebranches follow the steepest thermal gradient
and not the <100> direction. The first sidebranch appears
some tip radii behind the tip. In the tip frame of reference,
they propagate with the tip growth velocity along the
dendrite. This means they remain stationary in the laboratory
frame (fig.1). Initially they grow very regular. On average,
coarsening effects begin after about five sidebranches.

Fig. 4 shows a typical dendrite grown at $\Delta T = 150$ mK.
The shape of the fins is fitted by the parabola P which is

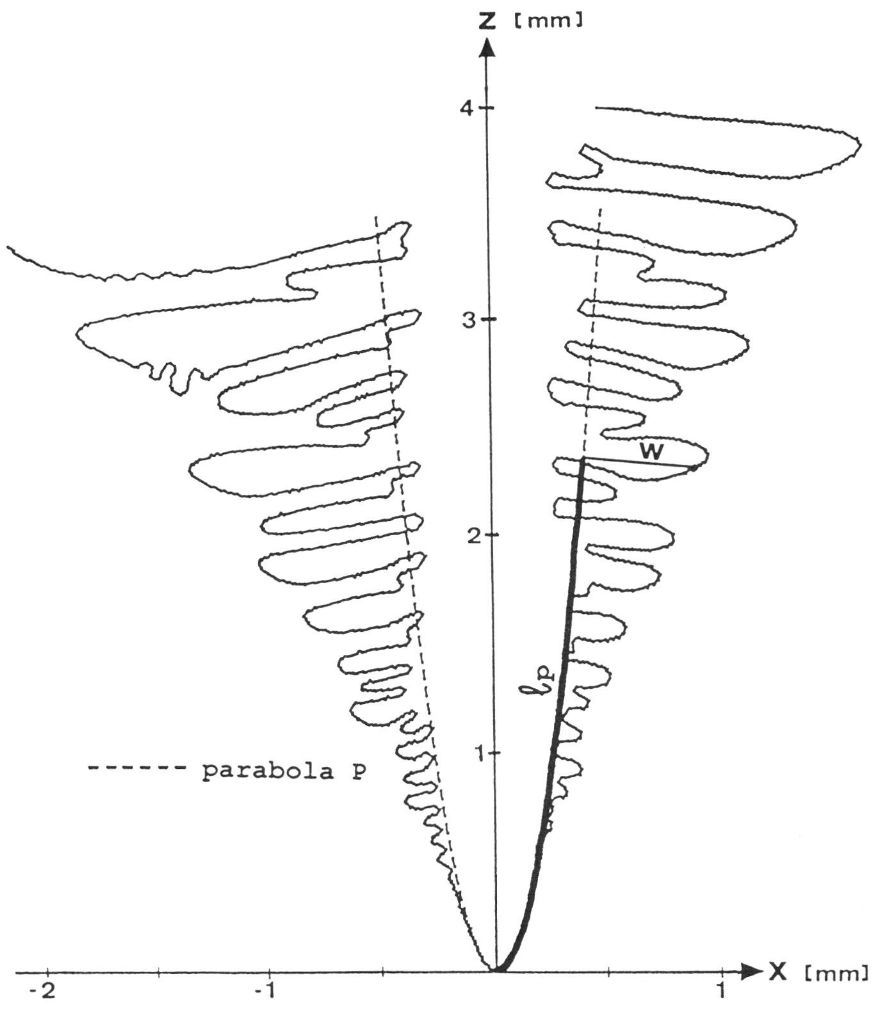

Fig. 4. Contour of a xenon dendrite.

drawn to fit the shape of the dendrite tip. We observe that
the fit is especially good as long as no coarsening effects
occur. The amplitude of the sidebranches, i.e. the difference
between the contour of the dendrite and the parabola, is
designated by w. The maximal amplitude of the sidebranches
vs. the parabola length l_p is plotted in fig. 5. Dendrites of
supercoolings 60 mK $\leq \Delta T \leq$ 150 mK are included. Dendrites
which grow at different supercoolings have different length
scales. Thus, we use the sidebranch spacing S [1] as first
attempt to scale the lengths w and l_p. The spread of the
plotted points may result from inaccuracies of the parabola
fit and of the length scaling. Nevertheless, a clear tendency
can be seen: In the log-log plot the amplitude of the sidebranches w increases linear with the parabola length l_p, independent from the supercooling. The line which is drawn to
guide the eye corresponds to $(w/S) \sim (l_p/S)^{1.72}$. According to
the predictions of the noise model for a fixed frequency
perturbation it is assumed that the amplitude decays at large
distances from the tip. We do not find such a decrease of the
amplitude w for large l_p. We have never seen such a behavior
during our experiments with rare gas dendrites. A quantitative comparison of our data with the theoretical predictions
is difficult at the moment. More data are necessary.

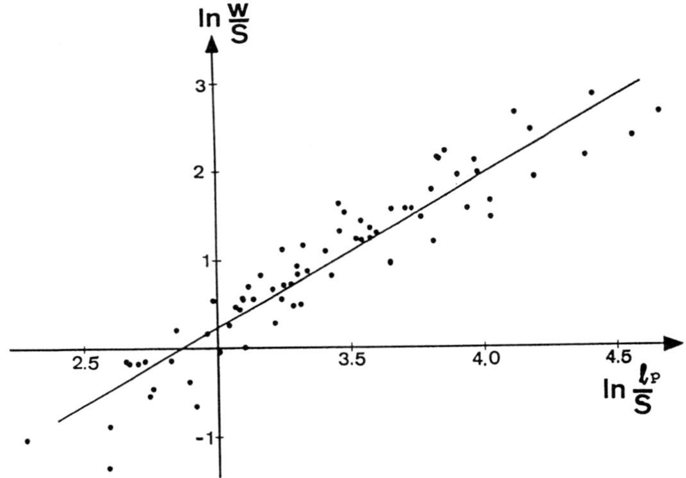

Fig. 5. Amplitude w of the sidebranches vs. the length l_p of the
parabola scaled with the sidebranch spacing S.

CONCLUSION

The shape of the dendrites is not a paraboloid of
revolution. The sidebranches appear some tip radii behind the
tip and propagate with the tip growth velocity along the
dendrite. A relationship between the amplitude of the sidebranches and the parabola length can be described by a power
law. A decrease of the amplitude at large distances from the
tip is not observed. More data are necessary for a comparison
of the experimental results with theoretical models.

ACKNOWLEDGEMENTS

We thank Prof. H.-R. Ott for his support. This work was supported by the Swiss National Science Foundation.

REFERENCES

1. J.H. Bilgram, M.Firmann and E.Hürlimann, J. Cryst. Growth
   96, 175 (1989)
2. J.H. Bilgram and E.Hürlimann in this conference
3. J.S. Langer, Lectures in the theory of pattern formation,
   in: Chance and Matter, Proc. Les Houches Summer School,
   Session XLVI, 1986, Eds. J. Souletie, J. Vannimenus and R.
   Stora, North-Holland, Amsterdam, 1987
4. Other papers of this conferences
5. J.H. Bilgram and E. Hürlimann, Proc. VIIth European Symp.
   on materials and fluid sciences in microgravity, ESA SP-
   295,173(1990)
6. J.S. Langer, Phys. Rev. A 36, 3350 (1987)
   M.N. Barber, A. Barbieri and J.S. Langer, Phys. Rev. A 36,
   3340 (1987)
7. O. Martin and N. Goldenfeld, Phys. Rev. A 35, 1382 (1987)
8. A. Dougherty, P.D. Kaplan and L.P.Golub, Phys. Rev. Lett.
   58, 1652 (1987)
9. X.W. Qian and H.Z. Cummins, Phys. Rev. Lett. 64, 3038
   (1990)
10.Ph. Bouissou, A. Chiffaudel, b. Perrin, P. Tabeling,
   Europhys. Lett. 13, 89 (1990)
11.R. Trivedi and J.T. Mason, Met. Trans. in press

EXPERIMENTAL DETERMINATION OF RAPID DENDRITE
GROWTH VELOCITIES IN LARGELY UNDERCOOLED METALS

D.M. Herlach and K. Eckler

Institut für Raumsimulation, DLR

P.O.B. 906058 D-5000 Köln 90, F.R. Germany

INTRODUCTION

Rapid solidification is a well established method for the preparation of metals and alloys in metastable states, giving access to a range of novel materials properties [1]. An important crystal-growth mechanism in these techniques is dendritic growth, which provides the most effective means of heat and solute rejection during solidification. Dendritic growth has attracted recent attention both from the theoretical [2 − 5] and the experimental side [6 − 9].

Theoretical models of dendritic growth rely on equilibrium thermodynamics and assume a local equilibrium at the solid-liquid interface. For low growth velocities these assumptions have been verified experimentally [6]. Analysis of rapidly solidified products points to the possibility of a breakdown of the local equilibrium at the interface [10]. As these conclusions are derived post mortem, after considerable modification of the material by aging processes, it is desirable to obtain direct evidence during the primary solidification stage. This turns out to be exceedingly difficult in techniques relying on rapid heat extraction, such as melt spinning, splat cooling, and atomization.

An alternative way to achieve high solidification velocities is to prevent nucleation in a bulk melt and thus undercool it by a substantial amount. The high driving force for crystallization accumulated in this way leads to a rapid solidification process whose solidification speed can be observed quantitatively.

The present paper reports on direct measurements of the solidification velocity in metals undercooled using an electromagnetic levitation technique. Solidification was initiated actively at a choosen undercooling temperature and in a well defined geometry. Measurements have been performed on pure Ni as well as on Ni-base alloys. The observed dendritic growth velocities show deviations from the predicted behaviour, which can be related to a relaxation of local interface equilibrium. This concerns both a kinetic undercooling of the interface and deviations from the chemical equilibrium at the solidification front. The latter can be described by a change in the partition coefficient with increasing growth velocity. For even larger growth rates, a change of the velocity dependence on undercooling is found. This occurs at a growth velocity where the morphology of the solidified product changes from coarse to fine grained.

EXPERIMENTAL DETAILS

Measurements were performed on pure Ni metal and Cu-Ni, Ni-B and Ni-Si alloys. Samples were prepared from constitutents of purity better than 99.99% by remelting in an arc furnace into spheres of about 6mm diam.. Undercooling conditions were established in an electromagnetic levitation apparatus described elsewhere [11] under clean environment conditions. The temperature of the sample was monitored in absolute terms by a two-colour pyrometer with an accuracy of +/- 3 K at a sampling rate of 100 Hz. Additionally, the image of the levitated molten drop was focused onto two silicon photodiodes [*Fig.1a*], which allowed a relative temperature measurement during the rapid recalescence phase with a time resolution of 1 μs [12]. In a typical observation cycle, the melt was undercooled to a predetermined temperature and crystallization was initiated by a solid needle at a well defined point in a plane normal to the viewing direction. Solidification proceeded radially from the trigger point, as confirmed by micrographs and indicated schematically in Fig. 1b. The corresponding recalescence signal as recorded by the two sensor elements is shown in Fig. 1c. Considering the path length difference s, the velocity of the thermal front can be derived independently for the two detectors, and turns out to be constant throughout the sample. The decay length of the thermal front is small compared to the sample dimensions as confirmed by the sharp transition between the lower and upper diode output [*Fig.1c*], so the observed signal can be related directly to the dendrite tip velocity.

RESULTS AND DISCUSSION

Results of a measurement of the solidification velocity as a function of undercooling are shown in Fig. 2 for a sample of pure Ni. Considerable velocities up to 70 m/s are observed at the highest undercooling values $\Delta T = 324$ K. At ΔT less than a critical limit, $\Delta T^* = 175$ K, the data follow a power law $V \propto \Delta T^\beta$ with $\beta = 3$, whereas for undercooling values larger than ΔT^* a linear dependence is found. Such an apparent break in the growth conditions has been observed earlier by Walker [13].

The recalescence phase takes place under near adiabatic conditions. During this stage of crystallization the growth velocity is limited by the heat transfer (in case of the pure metal), rejected from the advancing interface. The undercooled melt acts as a heat sink. Under these circumstances the interface is always unstable, leading to dendrite growth. The thermal gradient ahead of the interface drives the propagation of the solidification front into the undercooled melt. Following present theories of free dendritic growth [2], the total bath undercooling ΔT may be split into two contributions:

$$\Delta T = \Delta T_t + \Delta T_r \tag{1}$$

with ΔT_t the thermal undercooling and ΔT_r the Gibbs-Thomson depression of the melting temperature due to the curvature of the dendrite tip. The thermal undercooling ΔT_t [14] and the Gibb's Thomson undercooling ΔT_r can be calculated according to:

$$\Delta T_t = \frac{\Delta H_f}{C_p^l} P_t \exp(P_t) E_1(P_t) \equiv \frac{\Delta H_f}{C_p^l} Iv(P_t) \tag{1a}$$

$$\Delta T_r = 2 \frac{\Gamma}{R} \tag{1b}$$

with: ΔH_f: the heat of crystallization, C_p^l: the specific heat of the undercooled liquid, $P_t = (VR)/(2a)$: the thermal Peclet number, V: the dendrite growth velocity, R: the dendrite tip radius, $a = K/(\rho C_p^l)$: the thermal diffusivity, ρ: the mass density of the liquid, K: the thermal conductivity of the liquid, E_1: the first exponential integral function, Iv: the Ivantsov function, $\Gamma = \sigma/\Delta S_f$: the Gibbs-Thomson coefficient, σ: the interface energy, and ΔS_f: the entropy of fusion.

Eq. (1) gives the correlation between the undercooling and the product $V \cdot R$ in terms of the thermal Peclet number. Taking into account the marginal stability criterion [15], the general theory of free dendritic growth yields [2, 16]:

$$R = \frac{\Gamma/\sigma^*}{P_t \cdot \dfrac{\Delta H_f}{C_p^l} \cdot (1 - n)} \tag{2}$$

with:

$$n = \frac{1}{\sqrt{1 + \dfrac{1}{\sigma^* P_t^2}}}$$

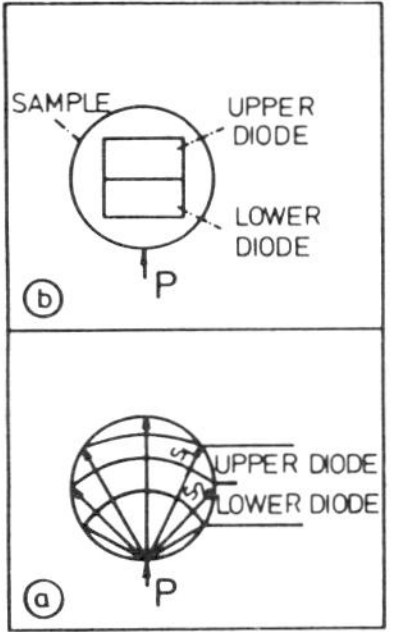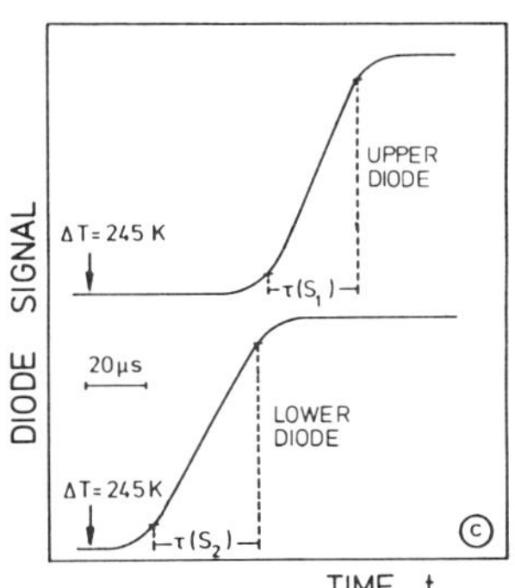

Figure 1. Principle of the measurements of solidfication rates in levitated undercooled drops. (a) An image of the sample is focused onto two fast-responding photodiodes. (b) Solidification is initiated intentionally at point P, from where it proceeds radially and is registered by the two diodes consecutively. (c) Typical output trace of the two photodiodes, shifted vertically for clarity.

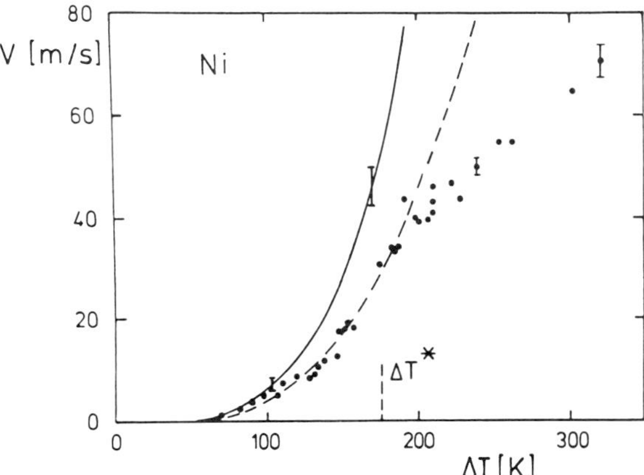

Figure 2. Growth velocity as a function of undercooling for primary solidification of pure Ni (Ref. 9). Dots, measured values; solid line, theoretical predictions according to Ref. 2 (error bar marks uncertainties in material parameters used); dashed line, theory including nonequilibrium at the interface by introducing atomic attachement kinetics.

σ^* denotes a stability constant, which can be approximated by $\sigma^* = 1/(4\pi^2)$ [15]. Eq. (2) together with Eq. (1) allows a unique calculation of the tip velocity as a function of undercooling.

Using characteristic data for Ni as given in Table 1, the dependence of the solidification rate V on the undercooling ΔT has been calculated (cf. solid line in Fig. 2). The theory leads obviously to an overestimation of the solidification velocity. However, a much better agreement, at least for undercoolings $\Delta T < \Delta T^* = 175$ K is obtained if the assumption of local equilibrium at the interface is relaxed by adding a kinetic term to Eq. (1), given by [17,18]:

$$\Delta T_k = \frac{V}{\mu} \tag{3}$$

μ is the interfacial kinetic coefficient. The dashed line in Fig. 2 is obtained when using a value of $\mu = 1.6$ m/(sK). This compares well to $\mu = 1.4$ m/(sK) derived from the collision limited growth model [17], $\mu = (\Delta H_f V_o)/(k_B T_E^2)$ with a sound velocity $V_o \approx 2000$ m/s. The calculation is in reasonable agreement with the value obtained by fitting the experimental data. The results obtained on pure Ni clearly reveal that the solidification rate of the undercooled melt is governed both by the release of the heat of fusion and the kinetics of atomic attachement at the interface.

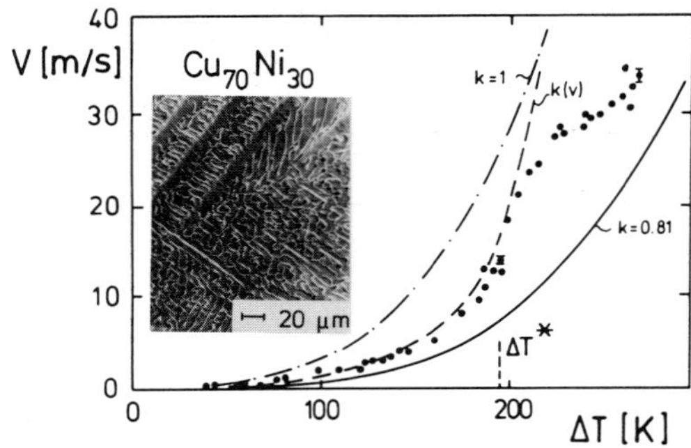

Figure 3. Growth velocity as a function of undercooling for primary solidification of $Cu_{70}Ni_{30}$ alloy (Ref. 9). Dots, measured values; solid line, theoretical prediction according to Ref. 2, using the equilibrium partition coefficient $k_E = 0.81$, dash-dotted line, same calculation for k = 1 (partitionless solidification), dashed line, theory including solute trapping by using a velocity-dependent partition coefficient (Ref. 19).

In particular, investigations on alloy systems are of interest with respect to the formation of metastable materials such as supersaturated alloys. Fig. 3 gives results of measurements on the alloy $Cu_{70}Ni_{30}$. Qualitatively, the same behaviour in $V(\Delta T)$ is observed, however, with solidification velocities smaller than for the pure metal. The temperature dependence of the growth velocity changes at a critical undercooling $\Delta T^* = 195$ K from a power law to a linear correlation. In contrast to the pure metal, this change is accompanied by a sudden rise of the growth velocity. Concerning the growth morphology, the insert of Fig. 3 shows, as an example, the dendritic microstructure of a $Cu_{70}Ni_{30}$ sample undercooled by an amount of $\Delta T = 190$ K prior to solidification.

The description of dendrite growth in alloys requires the consideration of constitutional effects both in the undercooling and in the marginal stability analysis. The constitutional undercooling ΔT_c can be evaluated to be [2]:

$$\Delta T_c = mc_o\left[\frac{1}{1-(1-k_E)/v(P_c)} - 1\right] \tag{4}$$

with: m: the slope of the liquidus, c_o: the initial alloy composition, k_E: the equilibrium partition coefficient, $P_c = V^*R/(2D)$: the chemical Peclet number, and D: the interdiffusion coefficient.

Similarly, the constitutional effect can be taken into account in the stabilty analysis resulting in an expression for the denrite tip radius of alloys [2, 16]:

$$R = \frac{\Gamma/\sigma^*}{\dfrac{P_t\Delta H_f}{C_p^l}(1-n) - \dfrac{2P_c mc_o(1-k_E)}{1-(1-k_E)/v(P_c)}(1+g)} \tag{5}$$

with:

$$g = \frac{2k_E}{1 - 2k_E - \sqrt{1 + 1/(\sigma^*P_c^2)}}$$

In order to calculate the growth velocities for the alloy system according to Eq. (1), but taking into account additionally the constitutional undercooling ΔT_c according to Eq. (4), kinetic effects are included by using the same kinetic coefficient μ as obtained for pure Ni. For the determination of ΔT_c the equilibrium partition coefficient $k_E = 0.81$ is used. The results of these calculations are given by the full line in Fig. 3. Obviously, there is an overestimation of the reduction in growth velocity due to solutal effects. On the other hand, assuming completely partitionless solidification, i.e. $k_E = 1$ (cf. dash-dotted line in Fig. 3), the calculated solidification rates are larger than the experimentally determined values over the whole temperature range investigated. The observation that the experimental data range between the extreme limits of $k_E = 0.81$ and $k_E = 1$, respectively, suggests that the growth velocities are too high to establish diffusional equilibrium, but not yet fast enough for completely partitionless solidification to occur. An agreement between experiment and theory is obtained when the partition coefficient k (V) is supposed to be velocity dependent as predicted by Aziz [19]

$$k(V) = \frac{k_E + (a_o/D)V}{1 + (a_o/D)V} \tag{6}$$

where a_o is a length scale in the order of the interatomic distance. As illustrated by the dashed line in Fig. 3, using a value of $a_o = 3 \cdot 10^{-10}$ m leads to an excellent description of the data even in the temperature range near ΔT^*. The sudden rise of V in the vicinity of ΔT^* with k (V) approaching unity suggests that at such undercooling the both thermally and chemically controlled growth changes to pure thermally governed growth with the consequence of almost partitionless solidification.

The alloy $Cu_{70}Ni_{30}$ shows a relatively high equilibrium partition coefficient $k_E = 0.81$. Deviations from the chemical equilibrium at the solid-liquid interface during rapid solidification should be more pronounced in alloys with a smaller equilibrium partition coefficient. This is demonstrated for dilute Ni-B alloys in the concentration range $c < 2$ at%. Fig. 4 shows results of measurements of the solidification velocity as a function of undercooling for two Ni-B alloys with a boron concentration of $c\approx0.7$ at% and $c\approx1.0$ at%, respectively. At moderate undercoolings, the dendrite growth velocity is essentially reduced in comparison with the pure metal Ni or the Cu-Ni alloy, indicating a much more pronounced influence of constitutional undercooling. This is due to the much lower partition coefficient $k_E\ll1$ of these alloys. However, when the undercooling approaches a value of $\Delta T = 200K$ (in case of $Ni_{99.3}B_{0.7}$) or $\Delta T = 220K$ (in case of $Ni_{99}B_1$) the growth speed rises rapidly. An analysis of these results within dendrite growth theory using the characteristic data given in Table 1 gives evidence that this sharp transition is caused by a change from mostly diffusion limited growth to purely thermally governed growth. These conditions correspond to a rapid increase of the partition coefficient from $k(V)\approx0$ to to $k(V) \to 1$ with the consequence of partitionless solidification at high dendrite growth velocities.

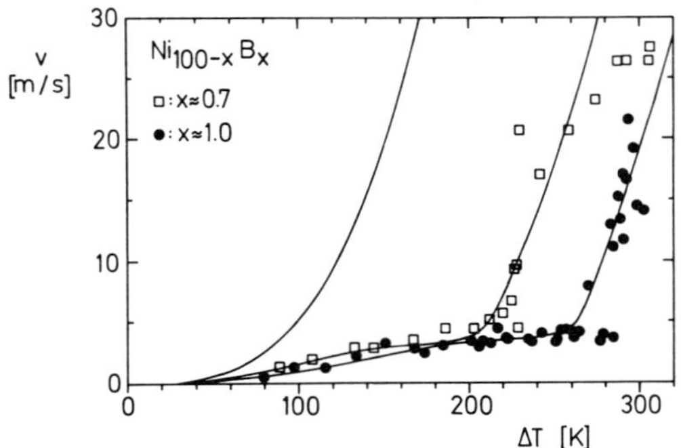

Figure 4. Growth velocity as a function of undercooling for primary solidification of dilute Ni-B alloys. Dots and squares, measured values; solid lines, theoretical prediction according to Ref. 3, using a velocity-dependent partition coefficient (Ref. 19). The steep rise of solidification velocity at large undercoolings marks a transition of mainly diffusion limited growth to pure thermally governed growth where k approaches unity.

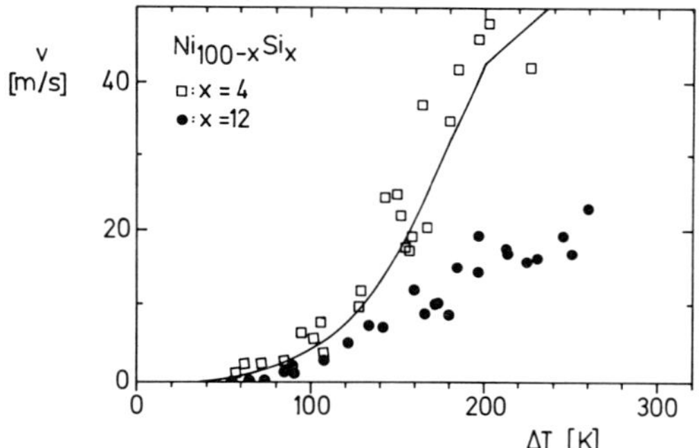

Figure 5. Growth velocity as a function of undercooling for primary solidification of dilute Ni-Si alloys. Dots and squares, measured values; solid line, the data of Ni (cf. Fig. 2) for comparison. In contrast to the dilute Ni-B alloys (Fig. 4) the velocity reducing effect of constitutional under-cooling is compensated by the velocity enhancement effect due to the sharpening of the alloy dendrites.

The effect of solute in dendrite growth is twofold: The constitutional undercooling leads to a reduction of the solidification rate. On the other hand, the addition of solute to a pure metal changes the conditions of morphological stability. The requirement of additional solute redistribution besides heat redistribution sharpens the dendrites leading to their faster propagation. In the Ni-B alloys, the first effect dominates probably due to the low mobility of boron in Ni-melts. Again, the experimental results are well described within dendrite growth theory when regarding a kinetic undercooling and the characteristic data for the system as given in Table 1.

Fig. 5 shows an example, in which the two countercurrent effects of solute addition almost equalize each other. This diagram gives the dendrite growth velocity as a function of undercooling for two dilute Ni-Si samples with silicon concentrations of c = 4 and 12 at%. One can see that in contrast to the Ni-B alloy systems the solidification velocity of the lower concentrated Ni-Si alloy is throughout comparable or even larger in comparison with pure Ni. Thus, the velocity reduction due to the constitutional undercooling is compensated by the velocity enhancement due to the sharpening of the alloy dendrites.

Table 1.   Material parameters as used in the calculation of dendritic growth.

Parameter	Dim.	Ni	$Cu_{70}Ni_{30}$	Ni - B
Heat of fusion	J/kg	$2.9 \cdot 10^5$	$2.3 \cdot 10^5$	$2.9 \cdot 10^5$
Specific heat	J/kg K	735	576	735
Interface energy	J/m^2	0.464	0.374	0.464
Thermal diffusivity	m^2/s	$6.5 \cdot 10^{-6}$	$3.0 \cdot 10^{-6}$	$1.0 \cdot 10^{-5}$
Liquidus slope	K/at%		-4.38	-14.3
Partition coefficient			0.81	$8 \cdot 10^{-6}$
Diffusion coefficient	m^2/s		$6.0 \cdot 10^{-9}$	$2.4 \cdot 10^{-9}$

SUMMARY

The discussion of the results given here is restricted for all samples on the undercooling range $\Delta T < \Delta T^*$ where ΔT^* is a critical undercooling at which deviations occur from the predicted behaviour. The values for the critical undercooling depend on the sample system investigated. In general, at undercoolings larger than ΔT^* a linear dependence of the dendrite growth velocity on the undercooling is found. The transition at ΔT^* to a weaker temperature dependence of V in the range $\Delta T > \Delta T^*$ hints at a suddenly occuring kinetic process which hinders a faster increase of V with ΔT. Very recently, it has been shown that the critical undercooling coincides with the critical undercooling for the onset of a drastic grain refinement process [20]. The origin of this grain refinement is possibly a fragmentation of primary formed dendrites when a critical velocity near the atomic diffusive speed is reached. Besides these observations, no detailed interpretation is presently available for the growth behaviour beyond ΔT^*. However, recent work indicates that the consideration of anisotropy effects in the interface energy may change the temperature dependence of the growth rate [21]. Also viscosity effects may have some influence on the morphological stability of dendrite growth in deeply undercooled melts [22].

ACKNOWLEDGEMENTS The present work contains partly results of previously published papers concerning the measurement technique of dendrite growth velocities (Ref. 12) and the analysis of growth behaviour of Ni as well as Cu-Ni (Ref. 9), in which E. Schleip and R. Willnecker, respectively, were essentially involved. The measurements on Ni-B and Ni-Si were performed together with R.F. Cochrane. The authors want to thank him also for valuable discussions and B. Feuerbacher for continous support. The authors also express their thanks to H. Mühlmeyer for the preparation of the samples.

REFERENCES

1. See, e.g., Proceedings Sixth International Conference on Rapidly Quenched Metals, Montreal, 1987, edited by R.W. Cochrane and J.O. Strom-Olsen [Mater. Sci. Eng. **97** (1988)].

2. J. Lipton, W. Kurz, and R. Trivedi, Acta Metall. **35** , 957 (1987); **35** , 965 (1987).

3. W.J. Boettinger, S.R. Coriell, and R. Trivedi, in Rapid Solidification Processing: Principles and Technology IV, edited by R. Mehrabian and P.A. Parrish (Claitor's Publishing Division, 1988), p. 13.

4. E. Ben-Jacob, N. Goldenfeld, J.S. Langer, and G. Schön. Phys. Rev. Lett. **51** , 1930 (1983).

5. D.A. Kessler and H. Levine, Phys. Rev. Lett. **57** , 3069 (1986).

6. M.E. Glicksman, R.J. Schäfer, and J.D. Ayers, Metall. Trans. **A&7** , 1747 (1976).

7. H. Honjo, S. Ohta, and Y. Sawada, Phys. Rev. Lett. **55** , 841 (1985).

8. A. Chou and H.Z. Cummins, Phys. Rev. Lett. **61** , 173 (1988).

9. R. Willnecker, D.M. Herlach, and B. Feuerbacher, Phys. Rev. Lett. **62** , 2707 (1990).

10. M.J. Aziz and C.W. White, Phys. Rev. Lett. **57** , 2675 (1986).

11. D.M. Herlach, R. Willnecker, and F. Gillessen, in Proceedings of the fourth European Symposium on Material Science under Microgravity (European Space Agency Report No. ESA SP-222, 1984), p. 399.

12. E. Schleip, R. Willnecker, D.M. Herlach, and G.P. Görler. Mater. Sci. Eng. **98** , 39 (1988).

13. J.L. Walker, Principles of Solidification, edited by B. Chalmers (J. Wiley & Sons Inc., New York, 1964), p. 114.

14. G.P. Ivantsov, Dokl. Akad. Nauk. SSSR **58** , 567 (1947).

15. J.S. Langer and H. Müller-Krumbhaar, Acta Metall. **26** , 1681 (1978).

16. R. Trivedi and W. Kurz, Acta Metall. **34** , 1663 (1986).

17. D. Turnbull, Metall. Trans. A **12** , 693 (1981).

18. S.R. Coriell and D. Turnbull, Acta Metall. **30** , 2315 (1982).

19. M. Aziz, J. Appl. Phys. **53** , 1158 (1982).

20. R. Willnecker, D.M. Herlach, and B. Feuerbacher, Appl. Phys. Lett. **56**, 324 (1990).

21. M. Ben-Amar, Phys. Rev. A **41**, 2080 (1990).

22. B. Caroli, C. Caroli, B. Roulet, and G. Fairre, J. Crystal Growth **94** , 253 (1989).

DENDRITES GROWING IN THE PRESENCE OF AN EXTERNAL FLOW :

THE CASES OF PVA AND NH4Br

P. Bouissou, B. Perrin[†], V. Emsellem, P. Tabeling

Laboratoire de Physique Statistique de l'ENS
24, rue Lhomond, 75231 Paris (France)
and
Laboratoire de Physique de la Matière Condensée de l'ENS
24 rue Lhomond, 75231 Paris (France)

INTRODUCTION

The problem of dendritic shape selection in crystal growth processes has been the subject of many theoretical and experimental studies these last years. Theoretical predictions[1] for the values of growth constant, $C = 4\ \rho^2 V/Dd_0$ are now available (here ρ, V are the tip radius and the growth velocity of the dendrite, and D, d_0 are the diffusion constant and the capillary length). Until now, only a few measurements of the growth constant have been done. Experiments performed with SCN [2] show good agreement between theory and experiment, whereas those carried out with NH_4Br [3] indicate agreement only within $\pm$ 25 %. Other measurements of C have also been performed on Pivalic Acid, indicating values much larger than those predicted by the theory ; actually they need to be completed by a reliable estimate of the crystal anisotropy [4]. Recently, dendrites of Ammonium Bromide growing in a supersaturated gel have been studied. The values of the growth constant are found significantly below the theoretical prediction.Until now, there is therefore no convincing agreement between theory and experiment concerning the stationary state of dendrites.

It seems that imposing a controlled external hydrodynamic flow should help to understand this problem. Indeed, one may notice that many perturbative effects are expected to play a role in the selection process, such as the existence of a flow around the crystal - due to natural convection or mass transfer -. But more fundamentally, perturbing a system in a controlled way may also allow to getting information on the unperturbed state.

Surprisingly very little is known about the influence of an external flow on dendritic growth. Recent theoretical studies[5] have been performed with zero surface tension and with a potential or Oseen-type flow around the crystal. Exact solutions have been found. The problem of the selection has further been considered by Bouissou and Pelcé[6]. Using a heuristic argument, they obtain an expression relating the growth constant to the external flow and the growth velocity. Ben Amar and Pomeau[7] have perfomed a general analysis of the problem, by using order of magnitude arguments. On the experimental side, work has been performed with an uncontrolled flow, essentially due to natural convection[8]. In such conditions, some qualitative effects - such as the amplification of the side-branching - have been observed.

The purpose of this paper is to report quantitative results concerning dendritic growth in the presence of a controlled external flow, and to compare with the existing theories ; experiments are carried out using solutions of pivalic acid/ethanol and ammonium bromide/water, corresponding hence to two different values of the anisotropy.Since the results obtained with pivalic acid were already published[9], we shall briefly summarize them.

Growth and Form, Edited by M. Ben Amar *et al.*
Plenum Press, New York, 1991

EXPERIMENTAL SET-UP

The system in which the crystals grow is essentially the same as the one used in a previous work[8]. The quartz cell is a parallelepiped of dimensions 45 x 5 x 1 mm. The Stokes flow is forced by gravitational forces. The cell is entirely immersed in a circulation of thermally regulated water maintained at a temperature T_1; the corresponding long and short term temperature fluctuations are about 20 mK. On the other hand, the tubes forming the hydraulic circuit are in thermal contact with another circulation of thermally regulated water ; such tubes can therefore be held at a temperature T_2 slightly above T_1. Most of the experiments are done with an overheating $\Delta T = T_2\text{-}T_1 \approx 0.2°C$, which is sufficient to avoid cristallization in the tubes and small enough to leave the growth process unperturbed by any temperature gradient effect.

The dendrites are observed with a microscope which allows magnification up to 750. The images are sent to a video camera and digitized into 512 x 512 pixels. The resulting overall image separation ranges from 0.6 to 0.9 mm/pixel.

The imposed upstream velocity of the flow can be measured directly by using either residual dust particules present in the fluid, or polystyren balls, 1 μm in diameter, added to the system. The tank capacity is large enough to impose a very large decay time for the flow ($\approx$ 3 hours) in comparison with the typical duration of the experiment ($\approx$ 20 min.).

RESULTS OBTAINED WITH PIVALIC ACID SOLUTIONS

For the case of solutions of Ethanol/Pivalic Acid, we have found the following results :
In the presence of a flow, the shape of the interface remains parabolic near the tip of the dendrite.
- The relation $\rho^2 V = cst$ - which is valid for dendritic growth without flow - remains true in the presence of an external flow.
- The constant value of $\rho^2 V$ does not depend on the transverse component of the flow and increases linearly with the longitudinal component $U_{//}$ of the flow.
 - An experimental law is obtained :

$$C = \frac{4\rho^2 V}{d_0 D} = \frac{8}{\sigma}(1+\chi\frac{U_{//}d_0}{D}) \qquad (1)$$

in which σ and χ are two numerical constants, estimated to 0.025 and 5300 respectively (assuming $D = 2\ 10^{-6} cm^2/s$ for the diffusion coefficient - which is only a rough estimate).

RESULTS OBTAINED WITH AMMONIUM BROMIDE SOLUTIONS

We present herein preliminary results obtained with ammonium bromide solutions.The qualitative features are similar to those of Pivalic acid. In particular, in the presence of the flow, the tip remains parabolic ; restricting the study to the influence of the longitudinal component, we have plotted in Fig. 1 the quantity $\rho^2 V$ as a function of $U_{//}$, for a saturation temperature $T_s = $ 19.5 °C. We find that within the experimental error, $\rho^2 V$ does not depend on $U_{//}$, throughout the range of variations of $U_{//}$, i.e, between 2 and 180 μm/s.Let us note that since the typical values of crystal growth velocities are 5 μm/s, this corresponds to ratios $U_{//}/ V$ lying typically between .5 and 40.
It is interesting to note that this result turns out to be consistent with the experimental law (1).

DISCUSSION

As mentionned in the introduction, two theoretical studies of dendrites growing in the presence of an external flow have been performed recently. M. Ben Amar and Y Pomeau[7] reveal the existence of several regimes. One of them, which should be relevant for our experiments,

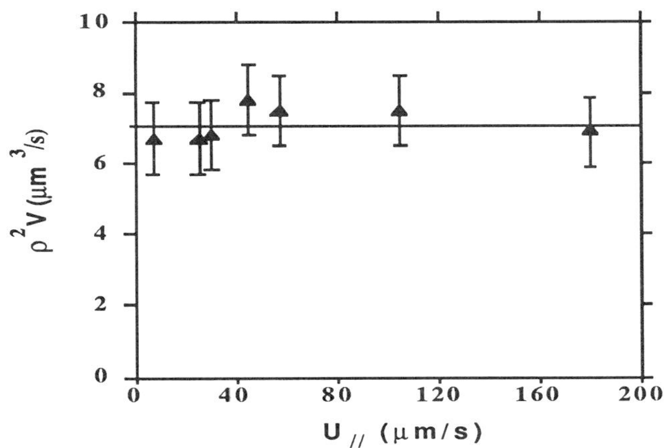

Figure 1. Dependance of product $\rho^2 V$ with the longitudinal component of the external flow for ammonium bromide dendrites.

corresponds to the case where we have :

$$\frac{U}{V} >> \left(\frac{v}{D}\right)^{\frac{1}{3}}$$

where v is the kinematic viscosity ; such a condition is satisfied at the largest values of the external flow velocity. In such a regime, M. Ben Amar and Y. Pomeau[7] predict (i) the existence of a growth constant, (ii) whose value is the same as that obtained when the fluid is at rest. Prediction (i) is consistent with the experiment performed on pivalic acid ; concerning now prediction (ii), there is agreement with the experiment on ammonium bromide, but disagreement is observed for the case of pivalic acid.

The second study, due to P. Bouissou and P. Pelcé[6] is an analytical approach to the problem.According to this theory, the growth constant C is determined by the following expression :

$$C = C_0\left[1 + \upsilon\left(\frac{d_0 U}{\beta^{\frac{3}{4}}\rho V}\right)^{\frac{11}{14}}\right]$$

in which C_0 is the value of C without flow, β is the cristal anisotropy, and υ is a term which depends logarithmically on the Reynolds number of the flow.It is interesting to note that, according to this theory,the growth constant depends on V when a flow is present.

In order to compare with the experiment, one can estimate a value for υ consistent with experimental law (1); we find the following crude estimates for the two experiments : $\upsilon \approx 200$ for PVA acid, and $\upsilon < 40$ for the case of ammonium bromide.This discrepancy remains to be interpreted.

CONCLUSION

The present experimental studies are in some sense, preliminary ; it would be interesting to increase the dynamical range of variation of the parameters, in order to be in position to test the existing predictions, and to measure the concentration field around the cristal. Beyond the experimental aspects of this problem, it is a striking fact that, even in the simplest configuration, there is still no definite description of the effect of a flow on the growth of dendrites.

ACKNOWLEDGMENTS

We thank P. Pelcé, M. Benamar for many helpful discussions related to this experiment.This study has been supported by CNES.

REFERENCES

1. P. Pelcé, Y. Pomeau, Stud. Appl. Math. 74, 245 (1986); M. Ben Amar, Y. Pomeau, Europhys. Lett, 2,307 (1986); D. A. Kessler, H. Levine, Phys. Rev. Lett., 57,24 (1986).
2. S.C. Huang, M.E. Glicksman, Acta Metall. 29, 717 (1981) ; A. Dougherty, J. P. Gollub Phys. RevA (to be published).
3. D.A. Saville, J.P. Beaghton, Phys. RevA, 37, 3423 (1988); S.K. Davh, W.N. Gill, J. Heat Transfer 27, 1345 (1984);
4. Glicksman, M.E., Singh, N.B., in "Rapidly Solidified Powder Aluminium Alloys", ASTM STP 890, M.E. Fine and E. A. Starke, Jr, Eds., 1986, 44-61.
5. M. Ben Amar, P. Bouissou, P. Pelcé, J. Crystal Growth 92, 97 (1988)
6. P. Bouissou, P. Pelcé, Phys RevA40 (1989).
7. M. Ben Amar, Y. Pomeau, C.R. Acad. Sci. Paris 308 (II), 907 (1989).
8. Ph. Bouissou, B. Perrin, P. Tabeling, *Phys. Rev. A*, **40**, 509,(1989).
9. M.E. Glicksman, S.R. Coriell, G.B. Mc Fadden, Ann Rev. Fluid. Mech, 307-335 (1986).

DENDRITIC GROWTH OF NH$_4$Cl: NUCLEATION, BOUNDARY CONDITIONS, PERIODIC AND APERIODIC MORPHOLOGIES

S. G. Lipson, Eli Raz and Shimon Kostianovski

Physics Dept.
Technion, Israel Institute of Technology
32000 Haifa, Israel

When we look at a collection of crystals in a museum, a souvenier store or a laboratory, we are always amazed by the enormous variety of beautiful shapes that nature creates. Even a single material can show different morphologies when it grows in various environments. Were the crystal in thermodynamic equilibrium its shape would be uniquely defined (up to a scale factor) at a given temperature, in a way which is in principle completely understood (although no single example has ever been calculated from first principles and compared to experimental data!). The basic theorem in the field was enunciated by Wulff (1901), developed in the early days by Landau (1950) and Herring (1952), and recently illustrated by work on h.c.p. ^{4}He (see, for example, Lipson, 1987). However, the richness of natural crystal shapes owes nothing to Wulff's theorem. Despite the millions of years that have passed by since their birth, most crystals found in nature reflect their growth processes and little, if any, of their characteristics have been modified by the passage of time. Today, we are trying to understand the much more complicated non-equilibrium processes involved in the relationship between growth dynamics and environmental conditions which dictate the morphology of the resulting crystal.

In the work which we shall describe here we are trying to disentangle the effects of the environment, a "soup" described by its local thermodynamical parameters, from the microscopical processes occuring at the interface, in the first atomic layers of the growing crystal itself. Our hope is to explain in physical terms why some crystals grow facetted and some dendritic, what are the basic characteristics of dendrites and how they are related to the properties of the "soup" and the crystal. Eventually, all this knowledge should be viewed as part of the general question of pattern formation in non-linear systems, of which crystal-growth is only one spectacular example.

Experiments on Ammonium Chloride

We chose one particular material for work so far, ammonium chloride. This has been used by several researchers, and there is already a considerable bank of data about it. The crystals are cubic, with the CsCl structure, and have no water of crystallization. They grow very anisotropically, $< 111 >$ being dominant in

Growth and Form, Edited by M. Ben Amar *et al.*
Plenum Press, New York, 1991

the range $5-15^\circ$C where we have worked although $<100>$ and $<110>$ growth occurs at other temperatures. By adding about .02% $CuSO_4$ to the solutions the $<111>$ growth can be made to dominate at all temperatures, and many of the experiments were made with this addition. The very high anisotropy of growth and its temperature dependence suggested to us that this crystal is growing below its surface roughening temperature*, a distinctive feature which of course may not apply to other crystals which have been investigated. Maurer et al (1989) have made the same suggestion for NH_4Br.

We observe the crystals microscopically. Our original aim was to relate quantitatively the crystal growth rate to the concentration profile around it, in the hope of distinguishing between various theoretical models claiming to describe this situation (see reviews by Langer, 1987, and Ben Jacob and Garik, 1990). The results of the research suggested that the method could be used to give information on the growth process at the liquid-crystal interface, and we are at present following up this study with more precise work to find out how much we can learn in detail about the interface kinetics.

On the way, we have also discovered several other things. The first is the great effect of some types of impurities on the crystal kinetics, apparently known to crystal growers, and which we have been able to use to advantage. Secondly, we have found how to grow ammonium chloride "flakes" showing the same beautiful high symmetry, but with detail which varies from example to example, as is seen in real winter snowflakes. Thirdly we have discovered a highly periodic form of dendritic growth which can be introduced by suitable boundary conditions and which has appparently not been previously reported.

When a crystal grows from water solution, both heat of crystallization and excess water are ejected at the interface. For the growth to continue, both of these must be dissipated through the surrounding fluid, processes which are generally considered as diffusive but which in practice may involve convection too. We grow the crystals under conditions which minimize temperature gradients in the fluid while affecting the concentration gradients very little. This is done by using a thin, quasi-two -dimensional horizontal cell (x, y plane) of which the upper and lower parallel boundaries can be considered as isothermal. The boundaries are separated by distances between 25 and 120μm. We investigate the concentration field around the growing crystal by means of optical interference microscopy, which measures, throughout the field of view, the optical thickness (integrated refractive index) of the medium in the $z-$direction. The refractive index is a function of both concentration and temperature, so that it is important to estimate how small the temperature field really is. The experimental cell (fig. 1) is cooled by a thermoelectric cooling device which allows convenient

---

* At T_R the surface of an equilibrium crystal undergoes a phase transition from atomically smooth at $T < T_R$ to atomically rough at $T > T_R$. The smooth surface corresponds macroscopically to a facet, and has thermal fluctuations of not more than one or two atoms. Above T_R, the equilibrium surface undergoes thermal fluctuations which are logarithmically infinite in amplitude; more exactly, their amplitude is of order $\ln(n)$ atoms, when the crystal dimension is n atoms. The rough surface is not faceted, and from the point of view of the crystal growth it is highly defective, and grows quickly.

control and stabilization. A drop of solution (unsaturated, at a relatively high temperature) is placed in the cell, the interferometer is adjusted with the uniform fluid sample to give a convenient set of straight-line fringes crossing the field of view, and the sample is then cooled to the working temperature. Tapping the cell lightly induces nucleation of a crystal. When it enters the field of view, the interference fringes distort, showing the establishment of a concentration field around the growing tip. The growth is recorded on video-tape, together with a meter showing the temperature continuously. Two examples of interferograms are shown in figs 2-3; a suitable picture for analysis shows a single tip which progresses at constant velocity across the field of view.

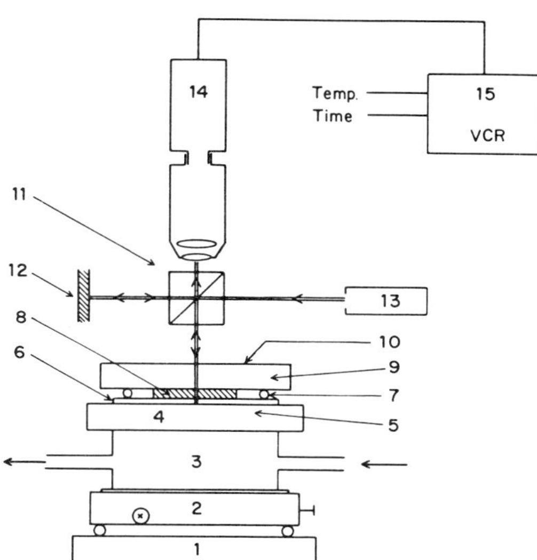

Fig. 1. Experimental system:
1-Microscope stage, 2-Tilting table,
3-Water-cooled chamber,
4-Copper cell base, 5-thermocouple,
6-Thin mirror, 7- Spacers, 8-Sample
solution, 9-Microscope slide,
10-Flowing nitrogen, 11-Interference
microscope, 12-Adjustable reference
mirror, 13-Sodium Lamp, 14-TV
camera, 15-VCR.

(From Raz et al, 1989)

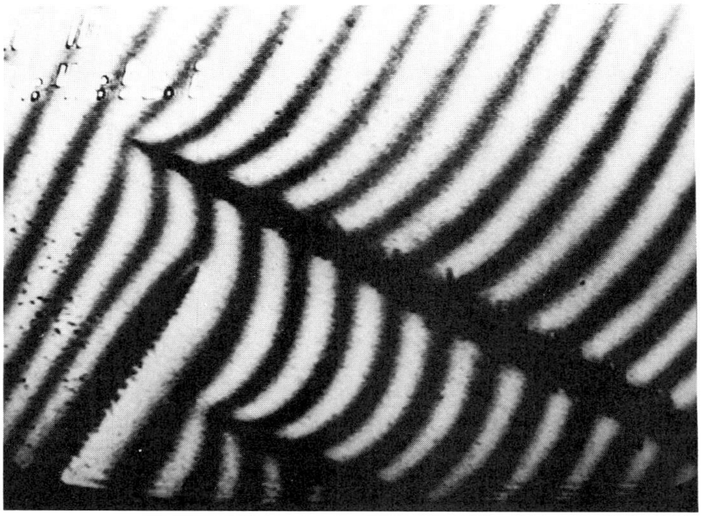

Fig. 2 Interferogram of a growing crystal, with a small number of fringes in the field of view.

Data Analysis

Quantitative data is extracted from the pictures by image analysis. Our initial experiments were analyzed by hand, but of course we have got the computer to work for us as fast as our research budget permitted. The method we use now (Takeda et al, 1982; Bore et al, 1986) is interesting as an example of image processing and we shall only outline it here. The interferogram is made in monochromatic light ($\lambda = 5640\text{Å}$) and arranged to have, in the uniform fluid, about 30 straight fringes across the field of view. It is best to use quasi-monochromatic light for imaging, but then it is necessary to adjust the interferometer rather precisely to get this number of fringes with uniform contrast. Their direction is unimportant. We use a frame-grabber to digitize a suitable image (fig. 3) using 512×512 pixels. Then we calculate its 2-D Fourier transform (fig. 4), which consists of a centro- symmetrical zero-order "patch" at the origin of the Fourier plane and two first-order "patches" symmetrically about it (sometimes weak second-order patches appear too, but they are the result of non-linearities in the video-recording and can be ignored). The first stage is to select the zero-order patch by defining a window around it. This is then re-transformed to produce an image in which the fringes have been eliminated, and which is used to define the shape of the dendrite boundary. A second selection in the Fourier plane selects one of the first-order patches only, and we then shift this to the origin of the Fourier plane and re-transform. The number of fringes in the field of view is only important insomuch as it must be sufficient for the zero and first order patches to be clearly separable in the Fourier plane. The *phase* of the resultant image is what interests us (fig. 5); it represents directly the concentration map of the fluid (on condition that the effect of temperature can be ignored). Phase variations occurring within the crystal boundaries (it is, after all, also transparent) are ignored and eliminated using the crystal boundary derived from the zero-order transform, as shown in the figure. An improvement to the technique subtracts the phases derived from the fringe pattern obtained from the uniform fluid (actually one or two frames before the crystal enters the field of view) which is technically equivalent to using holographic interferometry to remove all instrumental phase variations. This technique has been used in fig. 5.

Interpretation: a Nucleation Hypothesis

Inspection of the concentration map around the growing tip is instructive since it give a direct feel for the size of the "region of influence" of a growing crystal, the diffusion length $l_d = 2D/v$, where v is the velocity of the growth front. At the tip, where v is largest, the diffusion region is most restricted in extent. Neighbouring sidebranches are usually in one-another's diffusion fields (fig. 2). However, another thing is evident from the map: the concentration on the surface of the growing crystal is not uniform (i.e. a contour of equal concentration does not follow the crystal profile). This suggests that theories of dendrite growth which assume local equilibrium at the interface do not apply to this example. It is important to remark here that what is measured is the optical thickness, which measures the *mean* concentration through the cell and therefore samples concentrations at various distances from the growing front. However we can estimate the effect of this by studying an analytical form of the diffusion field around a growing paraboloidal tip assuming equilibrium at the interface

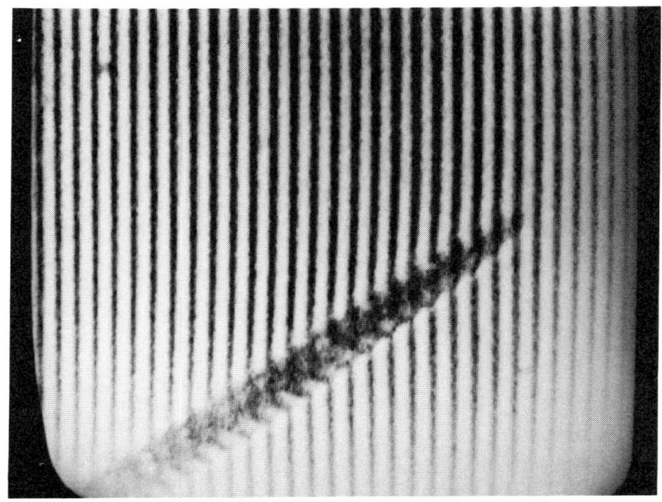

Fig. 3 Interferogram of a single stem growing in a 28μm cell at 7°C.

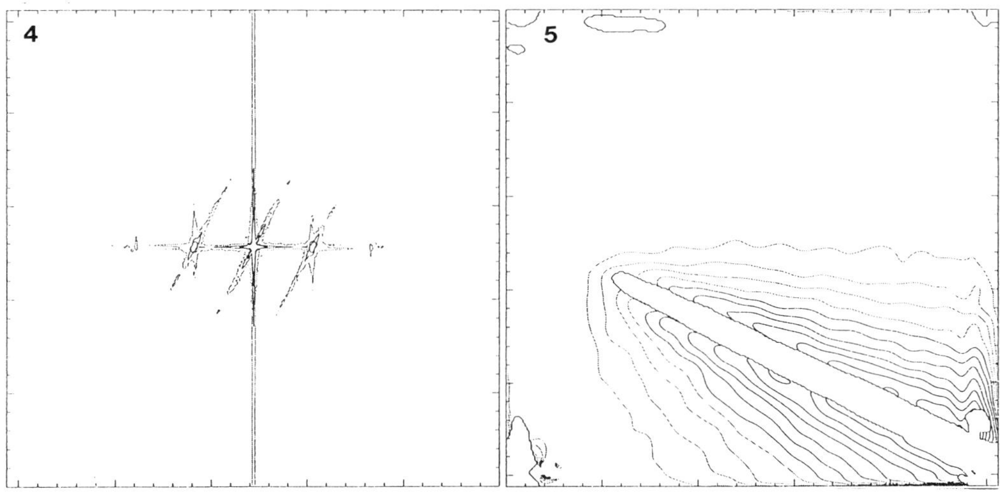

Fig. 4 Fourier transform (amplitude contours) of the interferogram of fig. 3. Notice the weak second orders, which are ignored in the data-processing.

Fig. 5 Phase contours of the interferogram in fig. 3, after removal of background effects, as explained in the text. These contours can be considered as loci of constant concentration.

(McFadden and Coriell, 1986). We calculated what the optical phase map would look like after integrating through the thickness, and showed that this explains only a small fraction of the observed effect (Raz et al, 1989).

We have made the above observations quantitative by relating the local rate of growth to the degree of supercooling on the interface. The former was determined by measuring the interface position on successsive photographs. The latter is determined from the interferograms, after correction for the averaging effect by using the analytical model. The results which have been analyzed give a consistent picture (fig. 6) which indicates a very non-linear relationship, approximating to a "super- cooling barrier". In other words, we found that v is very slow until a certain critical value of the interface supercooling, at which it suddenly incrases. Other experiments, carried out by slowly increasing the degree of cooling while the crystal is growing, show a sharp increase in v at a given supercooling.

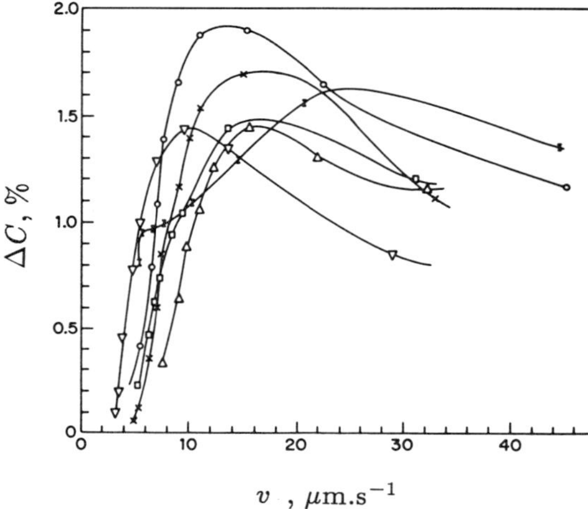

Fig. 6
Dependence of the super-saturation ΔC at the surface on the growth velocity, v , for several crystals.

(From Raz et al, 1989)

In our opinion, crystal growth by homogeneous nucleation can explain all the observations of growth of this type of crystal. This is a hypothesis, which must yet be proved. Other growth mechanisms could be relevant, and we hope that further experiments will make the distinction clearer. The case for homogeneous nucleation is as follows.

We assume that the relevant facet ($< 111 >$) of NH_4Cl is below its thermodynamic roughening temperature, T_R. If so, the crystal should grow faceted at infinitesimal rates, which it does. Now growth on a facet can take place around defects such as dislocations, or by homogeneous nucleation around statistically occuring islands. We tend to discount defect growth because we see, under certain circumstances, very highly ordered and periodic dendrites, and it seems unlikely that defects would occur just in the right places and at the right times to create them. On the other hand homogeneous nucleation is statistical in origin and from the macroscopic point of view is completely deterministic. It leads to two results which, qualitatively, agree with the experiments. The first is a non-linear relationship between the growth rate and the supersaturation. This takes the form (see, for example, Weeks and Gilmer, 1977):

$$v = \exp(-K/\Delta\mu),$$

where $\Delta\mu$ is the difference in chemical potential between the NH_4Cl in the crystal and the solution, proportional to the degree of supersaturation. This function is plotted in fig. 7, in which the initial part is also shown magnified. From the practical point of view, there is a sudden onset of growth at $\Delta\mu \simeq K$; we call this a "nucleation barrier". Another feature of fig. 7 is the saturation of v at large $\Delta\mu$.

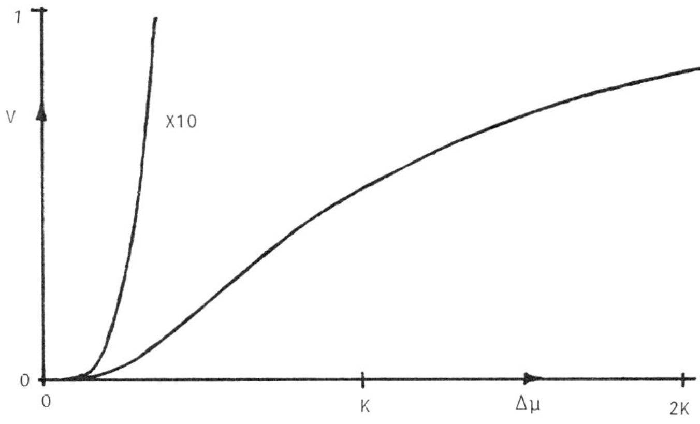

Fig. 7 The function $V = \exp(-K/\Delta\mu)$

Dynamic Roughening

Another aspect of facet growth is the possibility of dynamic roughening. It was suggested by Van Saarloos and Gilmer (1986) and has been observed in ^{3}He crystal growth by Rolley et al (1989) and by Maurer et al (1989) in NH_4Br dendrites. This mechanism enhances the growth rate of a facet at large $\Delta\mu$ and also makes the crystal grow rounded rather than faceted. But, more importantly in our opinion, it can introduce hysteresis into the $v(\Delta\mu)$ relationship since, when a crystal has become dynamically rough because of a large $\Delta\mu$, it can stay rough as $\Delta\mu$ falls. The hysteresis results in crystal growth like a relaxation oscillator, and can introduce a certain quasi- periodicity into the mophology. Thus the hypothesis of homogeneous nucleation together with dynamic roughening can explain, qualitatively, some of the major aspects of dendritic morphologies. Some simple simulations , which are today being improved, have shown dendritic structures to arise under these conditions.

Ammonium Chloride Snowflakes

When an ammonium chloride crystal grows from a stationary fluid, its external form has growth instabilities and shows no macroscopic symmetry. However, if the fluid is moving, it seems that large scale instabilities can be suppressed, and crystals which grow while falling through through a supersaturated solution develop highly symmetrical external morphologies. Experimentally, we prepare a solution in a glass tube and let the crystals grow while the tube rotates about its

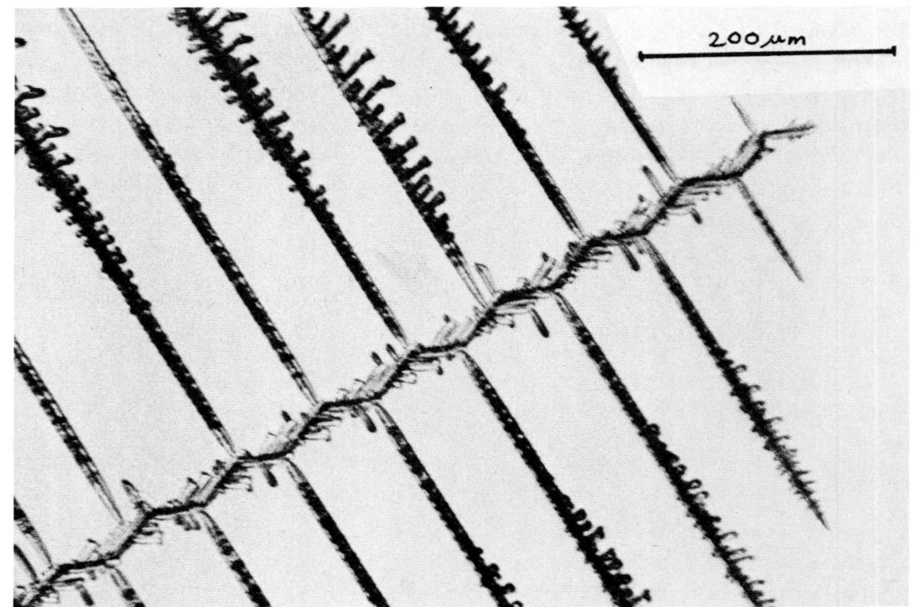

Fig. 8 An example of periodic dendrite structure

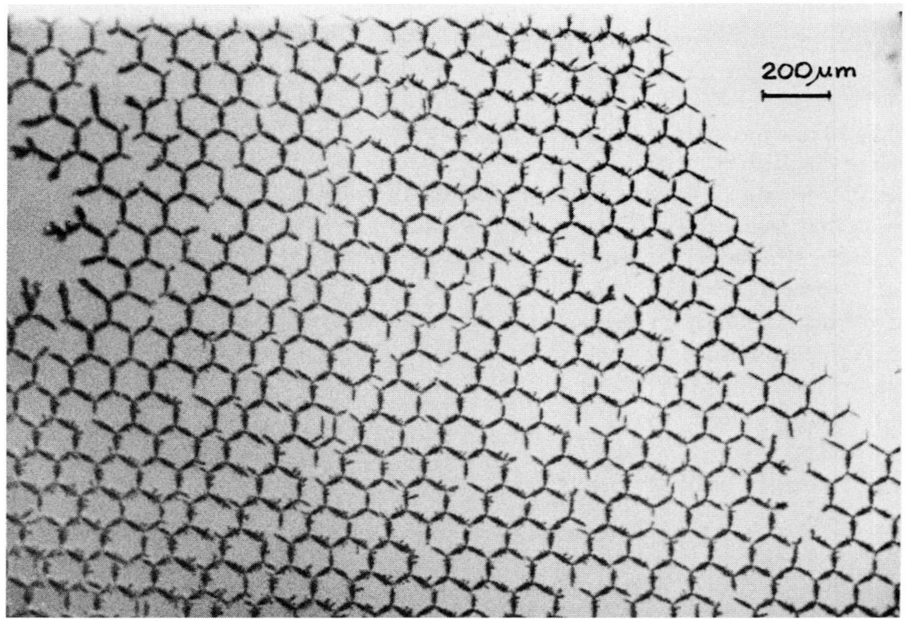

Fig. 9 An example of periodic dendrite structure

axis at a few revolutions per second. Since the crystals grow preferentially along the $< 111 >$ axes, the external shape becomes cubic, looking like sugar cubes with the growth axes as diagonals. Each cube is filled with a branched structure. This demonstrates that convection of the medium is important in determining the external symmetry of dendritic crystals, presumably because it replaces diffusion as the dominant macroscopic transport mechanism. We are still working on getting good photographs of this phenomenon.

Periodic Dendrites

An important aspect of the theory of dendritic structures is the appearance of two length scales, the capillary length and the diffusion length, $2D/v$. The latter relates to macroscopic dimensions of the dendrite such as sidebranch spacing. In our experiments a third length scale was introduced- the cell thickness. This also has a general influence, and its size is commensurate with the diffusion length so that the growth mode depends on which of the two is larger.

Consider the boundary conditions at the upper and lower cell boundaries, which are isotherms, and across which no solute diffuses. It is easy to see that, if diffusion dominates the crystal growth, the boundary conditions are equivalent to an image constuction (as in electrostatics) and the crystal "sees itself as if in a hall of parallel mirrors". If the diffusion length is greater than the cell thickness (low velocity), these parallel branches interfere with one-another's growth by mutual overlap of diffusion fields. But of course, no instability can arise. When the velocity is high, the images do not interfere with one another, and essentially free growth of the dendrites is observed, except when they approach within a diffusion length of the "mirrors". This results in some very striking periodic dendritic structures (figs 8 to 10). These structures show very long range order and we consider (Raz et al, 1990) that they give further credence to the homogeneous nucleation hypothesis.

Our argument is that the branching of the periodic structures occurs at a well-defined distance (in fact very close to) the cell boundaries, otherwise long range order would be lost. The nucleation mechanism is contolled by the environment alone (i.e. does not invoke noise or the presence of defects) and so this precision arises. Many of the structures in fact involve two branches nucleating simultaneously, in equivalent crystal axes. The actual structure observed in an experiment depends on the orientation of the initial seed, which is random; typical examples include zig-zags with periodic branches and hexagonal (honeycomb) cellular networks (figs 9 and 10).

Acknowledgements

We have greatly benefitted from discussions with J. S. Langer, Eshel Ben Jacob and Erez Ribak. This work was supported by the U.S.-Israel Binational Science Foundation and the Fund for Basic Research administered by the Israel Academy for Arts and Sciences. The technical assistance of Mr. S. Hoida is gratefully appreciated.

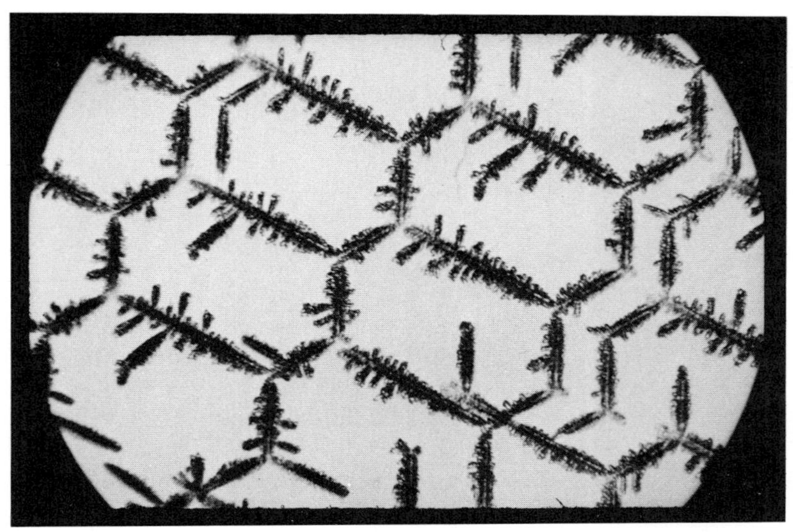

Fig. 10 An example of a periodic dendrite structure.

References

Ben-Jacob, E., and Garik, P., Nature 343, 523 (1990)

Bone, D. J., Bachor, H.-A. and Sandeman, J., Appl. Opt., 25, 1653 (1986)

Herring, C., in "Structure and Properties of Solid Surfaces", ed. R. Gomer and C. S. Smith, U. Chicago Press, Chicago (1952)

Landau, L. D., (1950) translated in "Collected Works", p.540, ed. D. Ter Haar, Pergamon, Oxford (1965)

Langer, J. S. in "Chance and Matter", Les Houches Summer School (ed. J. Souletie, J. Vannimenus and R. Stora) (1987)

Lipson, S. G., Contemp. Phys., 28, 117 (1987)

McFadden, G. B., and Coriell, S. R., J. Cryst. Growth 74, 507 (1986)

Maurer, J., Bouissou, P., Perrin, B., and Tabeling, P., Europhys. Lett., 8, 67, (1989)

Raz, E., Lipson, S. G., and Polturak, E., Phys. Rev., A40, 1088 (1989)

Raz, E., Lipson, S. G., and Ben-Jacob, E., submitted.

Rolley, E., Balibar, S., Gallet, F., Garner, F. and Guthmann, C., Europhys. Lett. 8, 523. (1989)

Takeda, M., Hideki, I. and Kobayashi, S., J. Opt. Soc. Am., 72, 156 (1982)

Van Saarloos, W., and Gilmer, G. H., Phys. Rev., B33, 4927 (1986)

Weeks, J. D., and Gilmer, G. H., Adv. Chem. Phys. 40, 157 (1979)

Wulff, G., Z. Crist. Mater., 34, 449 (1901)

ANISOTROPIC ORDERED PHASES IN LANGMUIR MONOLAYERS

Pierre Muller and François Gallet

Laboratoire de Physique Statistique de l'ENS
24, rue Lhomond
75231 Paris Cedex 05
France

INTRODUCTION

Langmuir films are two-dimensional monolayers made of amphiphilic molecules spread at the free surface of water. They are known to undergo various transitions between phases of increasing surface density : gaseous, liquid (expanded or condensed) and solid[1-7]. By using fluorescent molecules, it is now possible to visualize the different phases of these films[2-7], and to make a quantitative analysis of their growth and equilibrium properties. In this paper we present new results concerning the solid anisotropic phase of NBD-stearic acid films. First, we describe the experimental setup, allowing simultaneous observations under a microscope and measurements of the surface pressure Π. The solid phase appears as bright elongated needle-like domains, coexisting with the fluid phase, and corresponding to a long range orientational order of the molecular dipoles. We have studied the growth kinetics of the needles, and we have established that the longitudinal velocity of the needle is proportional to the departure from equilibrium $\Delta\Pi$. We have also shown that the number of nucleated needles is proportional to $\exp(-A/\Delta\Pi)$, verifying that the nucleation is homogeneous. We deduce from these experiments an order of magnitude for the energy per unit length of solid/liquid interface $\lambda \approx 1.5\ 10^{-6}$ erg/cm. Finally, we present a simple model for the equilibrium shape of the needles. Supposing that the permanent molecular dipoles are ordered like in a bidimensional ferroelectric, one calculates their electrostatic interaction energy By minimizing the contribution of the domain boundary to this energy, one finds needle-like equilibrium shapes. The predictions and the preliminary experimental results are in agreement.

Growth and Form, Edited by M. Ben Amar *et al.*
Plenum Press, New York, 1991

EXPERIMENTAL SET UP AND FIRST OBSERVATIONS

The molecule chosen to make films is a stearic acid labelled with a fluorescent NBD (NitroBenzoxaDiazol) group on the twelth carbon. Its structure is depicted on figure 1a. The NBD group carries in its plane a large permanent electrostatic dipole evaluated to 2.45 e.Å[8]. This fluorescent group absorbs visible light at 480 nm, and reemit at 520 nm. The absorption dipole is orthogonal to the NBD plane. The molecule, bought from Molecular Probes, is used as the single component of the film. Its purity is claimed to be higher than 99%, and no impurity was detected by a chromatography check. Our home-made Langmuir

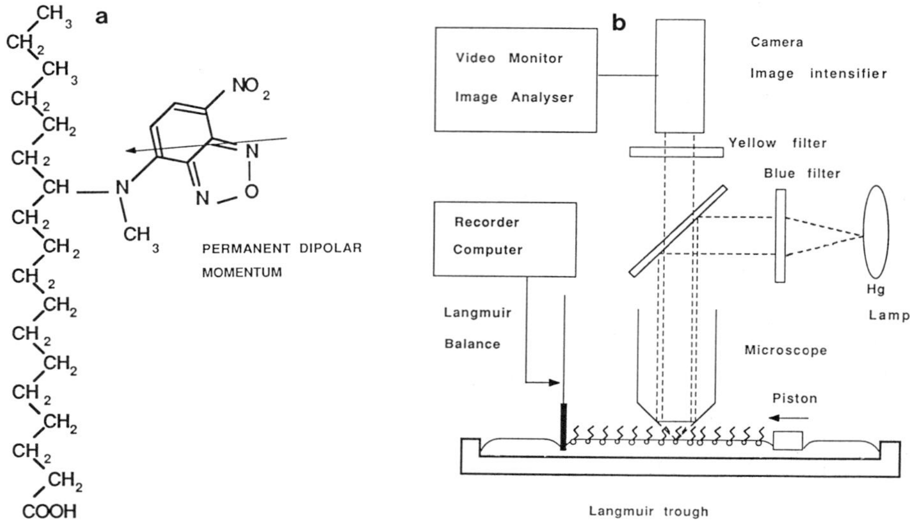

Figure 1. a - structure of the NBD-stearic acid molecule. The fluorescent NBD group is plane and carries a permanent dipolar momentum. The absoption momentum is orthogonal to this plane.

b - Scheme of the Langmuir trough allowing simultaneous observations and measurements of the surface pressure.

trough (figure 1b) is equiped with a Langmuir balance, and is small enough to be set under the objective of an optical microscope (Reichert Polyvar Met). A high sensitivity camera (Hamamatsu C 2400-11), including an image intensifier, is mounted on the microscope, so that we can simultaneously measure the surface pressure Π of the film and record its magnified image on a videotape. The trough is temperature regulated to about 0.05°C. The accuracy on the pressure measurement is 0.1dyne/cm, with a residual drift of about 0.2 dyne/cm per hour. The whole set up is confined inside a box cleared from dusts and organic impurities.

Earlier surface isotherms (Π as a function of the area A per molecule) have shown the existence of a quasi-plateau for $A_0 = 32 < A < A_c = 84$ Å^2/molecule, $\Pi \approx \Pi_c = 9$ dynes/cm (at 20°C)[7]. This quasi-plateau corresponds to a first order transition between two phases of respective density A_0 and A_c. Indeed, at equilibrium and for $A_0 < A < A_c$, one observes under microscope the coexistence of bright elongated domains with a dark fluid phase (figure 2). The length of the bright needles may be as long as a few millimiters. The aspect ratio is roughly the same for all the needles grown in the same conditions, and may vary from around 5 to 100, depending on these conditions. Some of them present random branching, but the generic shape is a single straight stripe with sharp ends. They are solid objects, since they can remain bent under a permanent mechanical stress without any visible plastic deformation, before breaking into pieces. The difference in contrast between the solid and the fluid phase is due to the quenching of the fluorescence in the fluid : there, the chains are probably lying on the surface, and the NBD cycles are in contact with water.

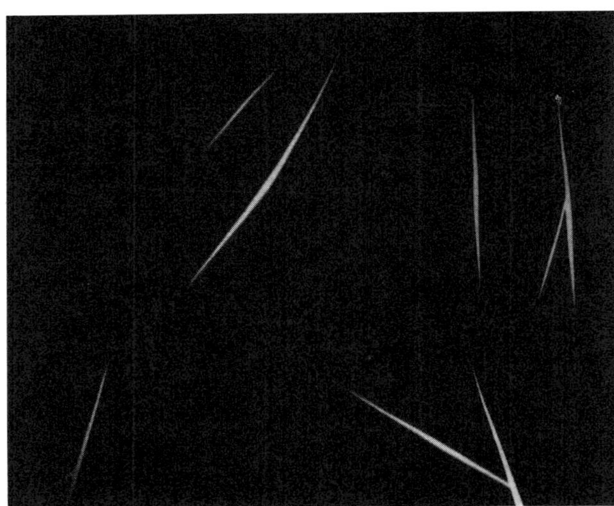

Figure 2. Photograph of bright solid domains coexisting with the dark fluid phase at T=20°C. The area per molecule is about 80Å^2. The picture size is 750 x 900 μ^2.

Observations between crossed polarizers have shown that the solid phase has a long range orientational order[7]. Indeed, when the film is illuminated at 480nm under normal incidence, the light polarization perpendicular to the main axis of a needle is partially absorbed, while the polarization parallel to it is not affected. The consequence is a rotation of the resultant polarization towards the principal axis. This means that the absoption dipoles (and thus the NBD planes) are aligned, perpendicular (respectively parallel) to the needle axis. In the solid phase, the molecules have at least an orientational order, but it is not clear at this stage of the experiment whether there is also a positional order, like in a bidimensional crystal.

GROWTH EXPERIMENTS

We decided to perform a quantitative study of the growth of the needles in order to get new information about the structure of the ordered/disordered interface. The experimental procedure is the following : starting with a film at equilibrium in the liquid state, we compress it quickly up to a surface pressure $\Pi = \Pi_0$ about 1dyne/cm above the equilibrium pressure Π_c. Then we measure the free relaxation of Π towards Π_c while the growth of the needles is simultaneously recorded. The typical duration of a run is quite long compared to other equivalent systems : the relaxation time for the pressure is about half an hour, meaning that the growth is particularly slow. In any case, this time is much larger than the mechanical equilibrium relaxation time, of the order of one second. Thus, during growth, the pressure is homogeneous over the whole film.

A crude analysis of the data indicates that Π-Π_c decays roughly exponentially with time, and so does the velocity of the tip dL/dt, with the same time constant. In order to establish an exact proportionality relation between these two quantities, we decided to plot the measured length L of the needles versus the integral $\int (\Pi$-$\Pi_c)$ dt. As a matter of fact, L can be measured with a much better accuracy than dL/dt, since the uncertainty over L is only a few microns over several hundred microns, and since dL/dt would have to be calculated by finite differences, with a bad accuracy. On the opposite, $\int (\Pi$-$\Pi_c)$ dt can be precisely calculated by integrating the continuous recording of Π. A typical set of data is plotted on figure 3. The lengths $L_i - L_i^0$ of twelve needles growing simultaneously at T=15°C are reported together. They all grow at the same velocity, although their lengths L_i^0

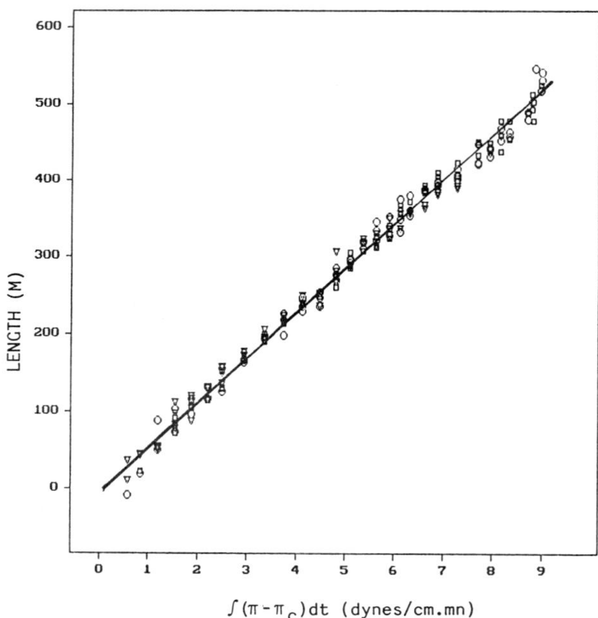

Figure 3 . plot of the lengths of twelve needles growing simultaneously at 15°C, versus the integral of the overpressure $\int (\Pi - \Pi_c)$ dt. The linear interpolation corresponds to the proportionality equation 1.

at time t = 0 are not the same. This is a first indication that the growth is not controlled by the diffusion of impurities : if it was the case, the larger needles, rejecting more impurities in front of their tips, would grow more slowly than the smaller ones. A linear fit is also drawn on figure 3, illustrating quite well the proportionality law :

$$\frac{dL}{dt} = K \, (\Pi - \Pi_c) \qquad (1)$$

The parameters of the fit are the initial lengths L_i^0 and the equilibrium pressure Π_c. The growth coefficient K is found equal to 54 ± 5 $\mu.cm/mn.dyne$ at 15°C. At T=20°C, one finds K'=58 ± 5 $\mu.cm/mn.dyne$, but we do not think that the difference between these two values is significant enough to be attributed to the temperature variation. Further experiments are in progress to determine the exact dependance of K with T.

Again, the proportionality law (1) is incompatible with a diffusion limited mechanism of growth. This experiment is thus directly sensitive to the kinetics of the interface, and more precisely to the attachment rate of the molecules at the tip of the needles. The relation (1) is similar to the one established for rough interfaces.

Concerning the width W of the needles, we could not yet determine the growth rate with enough accuracy. But it appears that, in this direction, the velocity is not proportional to $\Delta\Pi$: dW/dt decays with time much more rapidly than $\Delta\Pi$. The growth mechanism could then be different, reflecting the internal anisotropy of the ordered phase.

NUCLEATION RATE AND INTERFACE ENERGY

During a continuous compression of the film, starting from the fluid phase, one notices that, at the transition, there is a bump above Π_c in the recorded isotherm, and a delay in the nucleation of the needles. This bump is clearly visible (more than 0.5 dyne/cm) even if the compression is slow, i. e. around $2\text{\AA}^2$/molecule/mn. This phenomenon indicates that the energy barrier for the nucleation is large enough to be measured. More precisely, if λ is the energy per unit length of solid/fluid interface, supposed isotropic for simplicity, it is well known from the theory of homogeneous nucleation in two dimension that the radius of the critical circular seed is $r^* = \lambda/\{(A_c/A_0 - 1)\Delta\Pi\}$, and its energy is $E^*=\pi\lambda r^*$[9]. In this case, the number dN/dt of needles nucleated per unit time, under a supersaturation $\Delta\Pi$, at temperature T, is given by :

$$\frac{dN}{dt} = f(\Delta\Pi) \exp \left(- \frac{\pi\lambda^2}{(\frac{A_c}{A_o} - 1)\,\Delta\Pi\,kT} \right) \qquad (2)$$

The calculation of the prefactor f is not a trivial problem. In a simple-minded picture, f must be proportional to the number of nucleation sites. Since the size of each site is of order r^{*2}, this implies that $f(\Delta\Pi)$ is proportional to $1/r^{*2}$, and thus to $(\Delta\Pi)^2$. We admit this simple dependance in the following.

It is possible to compare the number of nucleated needles under a given supersaturation to the prediction of eq. 2. The experiment is performed as for the growth rate measurements : the pressure Π is rapidly increased up to Π_0, and then let relax towards Π_c, while one measures the final number of generated needles N per unit area. First, one notices that N drastically varies (from 1 to several thousands per mm^2) when the initial overpressure Π_0-Π_c is only changed from 2 to 4 dynes/cm. Secondly, most of the needles are nucleated in the first few seconds of the relaxation of Π towards Π_c, while Π remains close to Π_0. This indicates that the nucleation rate may effectively vary exponentially with $\Delta\Pi=\Pi-\Pi_c$. In the first stage of the relaxation, the variations of Π with time are approximately given by $\Pi(t) = \Pi_0(1-t/\tau)$. Equation (2) can thus be integrated to first order in $(\Pi_0-\Pi)/(\Pi_0-\Pi_c)$:

$$N = B\tau\,(\Pi_0-\Pi_c)^3 \exp\left(-\frac{\pi\lambda^2}{(\frac{A_c}{A_o}-1)\,(\Pi_0-\Pi_c)\,kT} \right) \tag{3}$$

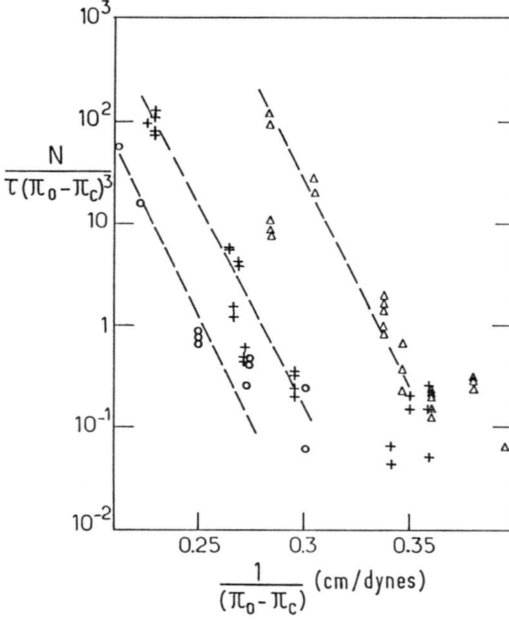

Figure 4. plot of the total number N of nucleated needles, as a function of the inverse of the applied overpressure $\Delta\Pi$. Within some prefactor, lnN is a linear function of $1/\Delta\Pi$. This result is consistent with equation 3, meaning that the nucleation is homogeneous.

B is a constant independant of Π_0-Π_c. A plot of the experimental variations of $\ln(N\tau^{-1}(\Pi_0-\Pi_c)^{-3})$ versus $1/(\Pi_0-\Pi_c)$ is represented on figure 4, for three different experimental runs, at three different temperatures (15, 19.5 and 20°C). Although there is some dispersion on the experimental points, a linear interpolation between the data can be drawn for each run. This means that the nucleation process is effectively homogeneous : for example, nucleation on surface impurities would not lead to such an exponential behaviour. Moreover, the common slope of these linear interpolations is related to the interface energy λ. Following equation 3, one finds $\lambda \approx 1.5\ 10^{-6}$ erg/cm. Of course, this value has to be considered as an order of magnitude, since λ has been supposed isotropic for simplicity, which is obviously not realistic for a needle. It is important to notice that, opposite to our expectation, the three experimental curves on figure 4 are not superimposed. This effect does not seem correlated to the temperature variations, and has no satisfactory explanation at the moment. Further measurements and refinements in the calculation of the nucleation rate, especially of the prefactor of the exponential, are needed to clarify this point. However, these preliminary results show for the first time that a direct measurement of the line energy between two phases can be achieved. This method can in principle be extended to other Langmuir monolayers.

CALCULATION OF THE EQUILIBRIUM SHAPE

A theoretical model for the shape of the ordered domains has been extensively developed in reference (10). We describe here the basic hypothesis of this model and its main results, which are in good agreement with our observations.

In the most general case, the equilibrium shape of an ordered domain minimizes the total energy U associated to its boundary, including the electrostatic interaction energy. In particular, if an anisotropic energy per unit length can be defined, this shape is given by the usual Wulff construction. In the case of NBD-stearic acid, it has been shown that all the dipoles in the solid phase are oriented in the same direction, along the axis of the needle. Their order should be either ferroelectric, or antiferroelectric. The following calculation of the electrostatic energy and of the resultant domain shape is made in the ferroelectric hypothesis, assuming that the dipoles are located above the water surface, and that the dipolar density P per unit area has a component $P_{//}$ (resp. $P_\perp$) along the film plane (resp. orthogonal to it). On the other hand, in the disordered phase, the horizontal dipolar density averages to zero, and P reduces to a vertical component $P'_\perp$. Let us first take into account the screening of a single dipole $p = p_{//} + p_\perp$ by the water. It can be shown[10] that, due to the difference in dielectric constant and to Debye-Hückel screening effect, p has an image p' $= -p_{//} + p_\perp$ located symmetrically below the water interface. If $p = p_{//}$, $p+p' = 0$, and the resultant interaction of two distant dipoles is then a quadrupolar short range interaction. On the contrary, if $p = p_\perp$, $p+p' = 2p_\perp$ and the interaction remains dipolar at large distance. The calculation of the electrostatic energy has then to be done separately for the two component of the dipole density.

In a first step, if $P_\perp = P'_\perp = 0$, the boundary energy U for a straight ordered/disordered interface is found proportional to its length. Thus, one can define and calculate an anisotropic line energy (per unit length) λ :

$$\lambda = \lambda_0 + \lambda_{//} \cos^2 \Phi \qquad (4)$$

In this expression Φ is the local orientation of the boundary (angle between its normal vector and $P_{//}$), $\lambda_{//} = \alpha \, P_{//}^2 / 4\pi\varepsilon_0$ (where α is a numerical constant of order unity), and λ_0 is a line energy accounting for local non-dipolar interactions (Van der Waals interactions for instance). λ_0 is supposed isotropic for simplicity. According to the Wulff construction, the equilibrium shape corresponding to equation 4 is elongated, and even presents sharp ends like a needle if $\lambda_{//} > \lambda_0$, i. e. if the horizontal dipole is large enough. In this case, the aspect ratio R (length L over width W of the needle) is equal to $2(\lambda_{//}/\lambda_0)^{1/2}$. This is consistent with our observations, all the more as the ratio $\lambda_{//}/\lambda_0$ is probably larger than one for NBD-stearic acid : taking $p = p_{//} \sim 2.45$ e.Å and the intermolecular distance a ~ 5Å, one finds $\lambda_{//}/a \sim 13$ kT, while λ_0 is expected to be comparable to kT.

Now, in the general case $P_\perp$, $P'_\perp \neq 0$, the interactions between vertical dipoles are not short range and one can no longer define an energy per unit length of boundary. It is neither possible to calculate and minimize the electrostatic energy for an arbitrary domain shape. Supposing that the above needle shape is preserved when $P_\perp$, $P'_\perp$ are small compared to $P_{//}$, one can do a variational calculation of the aspect ratio. After developping the energy in the limit $\lambda_{//} \gg \lambda_0 \gg \lambda_\perp = (P_\perp-P'_\perp)^2 / 4\pi\varepsilon_0$ (equivalent to R»1), one derives a relation between W and R = L/W at equilibrium[10] :

$$\ln \frac{W(R)}{W_\infty} = \frac{2}{R^2-2} \left(\ln \frac{R^2}{4} + 1 - \frac{\lambda_{//}}{\lambda_\perp} \right) \tag{5}$$

with
$$W_\infty = \frac{a}{2} \exp \left(3 + \frac{\lambda_0}{2\lambda_\perp} \right) \tag{6}$$

The two main features contained in equations 5 and 6 are the following : first, the width and length of a needle at equilibrium are no longer proportional ; a long needle is more elongated than a short one. Second, the width W reaches a finite asymptotic value W_∞ at large L. This result was previously found by Moy and McConnell[11], and also by Andelman et al.[12] for rectangular domains. The principle of their calculation was identical, but they did not properly account for the horizontal dipolar density $P_{//}$, and thus could not find sharp elongated needle shapes.

We already mentioned that the observed shapes are compatible with these predictions. As a further argument, we notice that the measured width of the needles never exceeds a fixed value between 20 and 50 μ, provided they have grown without interacting each other. A quantitative comparison between prediction (5) and measurements has even been made in a particular case : the variations of width versus length for three different needles are reported on figure 5, during their simultaneous slow growth at 20°C. There is a good agreement with the best theoretical fit corresponding to eq. 5, drawn with the following values of the parameters : $\lambda_{//} / \lambda_\perp = 200$, $\lambda_0 / \lambda_\perp = 17$ ($W_\infty = 25\mu$). Of course, one can argue that this comparison has little meaning, since the data are not taken at equilibrium. However we think that, because the growth is quite slow, the shapes are not drastically different from the equilibrium ones. At least the above values of the parameters should be considered as orders of magnitude. They show that the hypothesis of a ferroelectric order is reasonnable, and that in this case the dipole moment of each molecule should be slightly tilted but close to horizontal.

More generally, the model described here, taking carefully into account the horizontal dipolar density in a Langmuir film, allows to calculate exactly the shape of ferroelectric ordered domains, at least when this density is large. Its results should apply to a wide family of amphiphilic molecules.

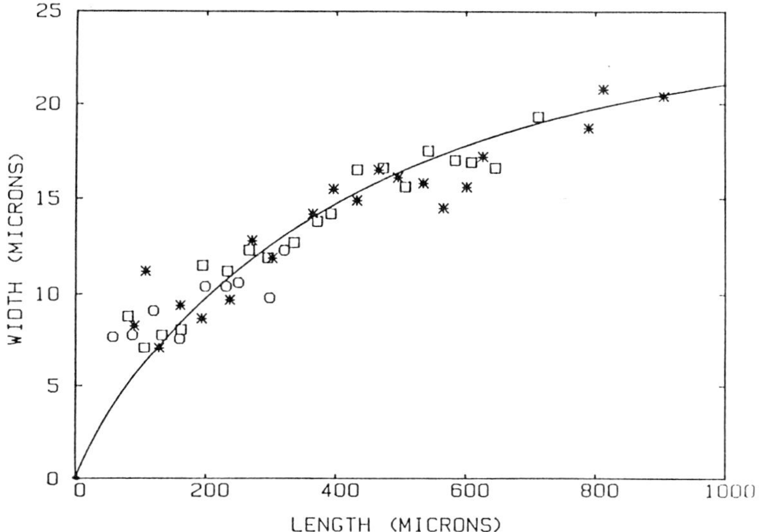

Figure 5. experimental variations of the width versus length of three needles slowly growing simultaneously at T=20°C. The theoretical prediction (equation 5 and 6) has been drawn for the following values of the parameters : $\lambda_{//} / \lambda_{\perp}$ = 200, $\lambda_0 / \lambda_{\perp}$ = 17 (W_∞ = 25μ). This comparison supposes that the shape of the needles always remains close to their equilibrium shape.

CONCLUSION

NBD-stearic acid Langmuir films present a solid ordered phase with an outstanding anisotropy, showing up in its optical properties and in its growth and equilibrium shapes. This anisotropy is due to the long range orientational order of the electric dipoles inside the "needle" phase. Growth studies indicate that the growth velocity, reflecting this anisotropy, is determined by the kinetics at the solid/liquid interface. The interface energy can be derived from measurements of the homogeneous nucleation rate. In the hypothesis of a ferroelectric structure, a simple model at equilibrium accounts for the observed morphology. Further experiments are needed to look for a long range positional order of the molecules, like in a bidimensional crystal.

REFERENCES

1 - G. L. Gaines, insoluble monolayers at liquid-gas interface (Wiley, New-York) 1966

2 - A. Miller, W. Knoll, H. Möhwald, Phys. Rev. Lett. 56, 2633 (1986)

   - M. Lösche, H. Möhwald, J. Colloid and Interface Sc. 131, 56 (1989) and ref. therein

3 - R. M. Weis, H. M. McConnell, J. Phys. Chem. 89, 4453 (1985)

   - V. T. Moy, D. J. Keller, H. E. Gaub, H. M. McConnell, J. Phys. Chem. 90, 3198
        (1986)

   - D. J. Keller, J. P. Korb, H. M. McConnell, J. Phys. Chem. 91, 6417 (1987)

4 - B. Moore, Ch. M. Knobler, D. Broseta, F. Rondelez, J. Chem. Soc., Far. Trans. 82,
        1753 (1986)

   - F. Rondelez, K.A. Suresh, Physics of amphiphilic layers, Eds J. Meunier, D.Langevin
        and N. Boccara (Springer) 1987.

5 - B. Berge, A. J. Simon and A. Libchaber, Phys. Rev. A 41, 6893 (1990)

6 - M. Seul, M. J. Sammon, Phys. Rev. Lett. 64, 1903 (1990).

7 - H. Bercegol, F. Gallet, D. Langevin, J. Meunier, J. Phys. France 50, 2277 (1989)

8 - G. Chambaud, private communication.

9 - See for example J.D. Weeks and G. H. Gilmer, adv. in Chem. Phys. 40, 157 (1979)

10 - P. Muller and F. Gallet, to be published in J. Chem. Phys, March 91

11 - H. McConnell and V. T. Moy, J. Phys. Chem. 92, 4520 (1988)

   - V. T. Moy, D. J. Keller and H. McConnell, J. Phys. Chem. 92, 5233 (1988)

12 - D. Andelman, F. Brochard, J. F. Joanny, J. Chem. Phys. 86, 3673 (1987)

DIRECTIONAL SOLIDIFICATION OF A BINARY MIXTURE

A NUMERICAL ANALYSIS OF DENDRITIC AND CELLULAR GROWTH OF A PURE MATERIAL INVESTIGATING THE TRANSITION FROM 'ARRAY' TO 'ISOLATED' GROWTH

J.D. Hunt

Department of Materials
University of Oxford
Oxford OX1 3PH, U.K.

ABSTRACT

A time dependent numerical model has been used to analyse the growth of dendrites into a supercooled liquid bath of a pure material. Both dendritic plates and dendrites having axial symmetry have been treated. For a finite surface energy, discrete solutions rather than a continuum of solutions were obtained for steady state dendritic growth at a fixed undercooling. The problem was considered in the limit of a zero specific heat so that the results could be compared with those obtained for the similar Hele – Shaw problem for which analytical solutions are available. It was found that there was an excellent agreement with the earlier work. The steady state results are plotted to show the relationship between the Hele – Shaw and dendrite problem. The transition between an array and isolated dendrites was examined. It was found that this two parameter problem could be reduced to a single line in the limit of a small thermal diffusion layer and a large dendrite spacing (an isolated dendrite) and also when the thermal layer was larger than the dendrite spacing (an array cell or dendrite).

Two types of solution, cell – like or dendritic, were found. The steady state dendritic results agreed quantitatively with the prediction of marginal stability but not with the physical basis of the model. It has previously been suggested that these steady shapes were not possible unless the surface energy is anisotropic. Possible reasons for the different behaviour found in the present work are discussed. The steady state shapes were tested for stability using time dependent growth.

INTRODUCTION

Recently a numerical model has been developed[1] to describe time dependent growth of solute dendrites or cells in a moving temperature gradient. As a test of the model, it was used to describe steady state growth at very large spacings in a zero temperature gradient. Under these conditions, previous work[2,3,4] suggested that near parabolic steady state shapes should be found for both zero and non – zero surface energies. In the numerical work[1] it was found that almost parabolic steady state interface shapes were developed. The behaviour was however quite different for zero and non – zero surface energies. For a given bath undercooling, solutions were obtained at any tip radius for a zero surface energy; whereas, solutions were obtained at one tip radius for a non zero surface energy. This observation is contrary to the marginal stability[5,6] approach to dendritic growth where it was assumed that marginal stability selects one of many steady state shapes. The observation is also contrary in detail to other more recent approaches[7–9] where it is suggested that no steady state solutions to the problem exist unless the surface energy is anisotropic. The discrete solution obtained in the numerical analysis[1] agreed quantitatively with the predictions of marginal stability condition even though the structures were not marginally stable.

The present paper extends the numerical work to the Hele – Shaw problem (and its three

Growth and Form, Edited by M. Ben Amar *et al.*
Plenum Press, New York, 1991

dimensional analogue) and shows how these results may be related to the three dimensional heat flow cells. Although treatments of the Hele – Shaw problem consider a rather different dendrite like situation it is equivalent to dendritic growth in a pure material which has no specific heat. Extensive work has been done on the Hele – Shaw problem, initially for a zero surface energy (Saffman and Taylor[10]), then including surface energy (McLean and Saffman[11], Vanden – Broeck[12] and Romero[13]). Solutions exist for both zero and non zero surface energies so that a comparison may be made with the present purely numerical calculations. Only heat flow in the liquid is considered so that a direct comparison can be made with the solutions to the Hele – Shaw problem; this restriction will be removed in future work. The numerical model has the advantage that very little change need be made to the program code to alter from two or three dimensions or from the Hele – Shaw problem to the heat flow problem.

One difficulty in discussing the Hele – Shaw problem and the heat flow dendrite problem is that different terminology and symbols are traditionally used. In the present work the cell or dendrite spacing 'a' should be distinguished from the cell width 'λa'. These are shown on fig. 1. The full problem to be solved including specific heat is as follows; the symbols are given in table 1. The heat flow equation in the liquid

$$K\nabla^2 T = c \frac{\delta T}{\delta t} \tag{1}$$

is solved for a shape where at the interface

$$v_{nl}L = -K \frac{\delta T}{\delta n} \tag{2}$$

The temperature at the interface is given by

$$T_I = -\Gamma(1/R_1 + 1/R_2) \tag{3}$$

(The freezing temperature of a flat surface is taken to be zero.)
The temperature ahead of the dendrite as x approaches infinity is given by

$$T = T_o \tag{4}$$

T_o is negative, provides the 'driving force' for the dendrite growth and can vary from zero to $-L/c$. The steady state fraction solid, f_s, will depend on the amount of this undercooling or supercooling. That is from global heat conservation

$$f_s = -c T_o/L \tag{5}$$

The temperature gradient in the y direction at y=0 and y=a is zero (see fig. 1). In the present work, the temperature gradient in the cell groove at x=0 is approximated to zero. This approximation was examined and improved when an analytic solution could be used to give greater accuracy.

The numerical model has been presented previously[1], so only essential detail will be included here. The cell or dendrite is assumed either to be a symmetric plate (two dimensional) or to have axial symmetry[1] (three dimensional). The heat diffusion equation is solved by considering the accumulation of heat in a control volume. The control volume box walls are orthogonal and are arranged so that the solid – liquid interface goes through diagonal corners of an interface box. This arrangement is shown schematically in fig.1. Each interface box corner is assumed to move with the dendrite tip velocity together with an additional velocity normal to the local interface position.

The accumulation of heat in a control volume in time δt is given by

$$(VT - V_{(i-1)}T_{(i-1)}) c = A_n (v_n cT_n + K \left[\frac{\delta T}{\delta y}\right]_n) \delta t - A_s (v_s cT_s + K \left[\frac{\delta T}{\delta y}\right]_s) \delta t$$

$$+ A_e (v_e cT_e + K \left[\frac{\delta T}{\delta x}\right]_e) \delta t - A_w (v_w cT_w + K \left[\frac{\delta T}{\delta x}\right]_w) \delta t \tag{6}$$

For the Hele – Shaw problem and its three dimensional analogue the specific heat c is put to zero. The present model is fully implicit so that the values other than $V_{(i-1)}T_{(i-1)}$ in this equation refer to values at the end of the time step, i, and are evaluated at the relevant walls.

For a non – zero surface energy, the undercooling equation is applied at each interface box corner by interpolating or extrapolating the temperatures at the interface box centres. The principal radii of curvature are obtained using a cubic spline interpolation. If there are n boxes and m interface points there are thus n composition equations and m + 1 undercooling equations. The compositions and normal shift of the box corners, other than the tip, are considered to be the variables. The final variable is the tip velocity.

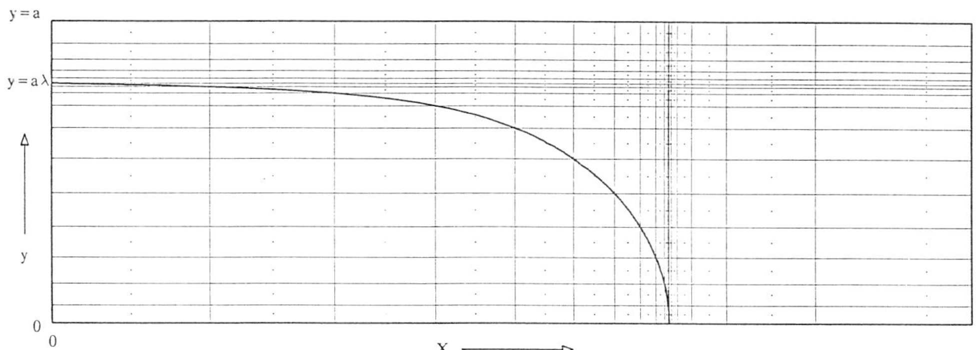

Figure 1 A schematic cell shape showing the nodes, the box wall and the definition x,y.a and λ.

As in the previous work[1] a slightly different procedure must be used when the surface energy is zero. The reason is that the $m+1$ undercooling equations, in the absence of surface energy, become linear combinations of only m different equations. One of the equations can be replaced by an equation that fixes the position of one point. This was done by fixing the box corner at $x=0$ (fig. 1). By changing the position of this point the family of solutions for a given set of growth conditions are obtained. (The undercooling equation is still satisfied at all box corners.)

The non$-$linear set of equations are solved using a Newton method. Since the numerical model is fully implicit, any time step could be used without mathematical instability. A convenient time step was found by experience to be that which allowed the tip to move about one tenth of the tip radius in one step. When solutions were found to be stable this step was further reduced to one hundredth of a tip radius. Steady state is effectively obtained by allowing the time step to become infinite. The solutions were iterated until the error in the temperature obtained from the flux equations and the error in temperature from the undercooling equations was typically less than $T_o \times 10^{-7}$. Solutions were obtained after three or four iterations for steady state and zero or one iteration for a finite time step where the previous changes could be used to predict the expected change. The number of points was kept to a minimum and was usually twenty five along the axis, sixteen in the radial direction and with eleven points on the interface. One of the reasons for the work was to find out whether reasonable results could be obtained with this number of points. To ensure that the correct density of points in important areas, the cell size and arrangement of points was periodically adjusted. The adjustment technique is described in the appendix A.

HELE$-$SHAW TWO DIMENSIONAL PROBLEM (Laplace 2D)

Experiments on a Hele$-$Shaw cell consider fluid flow between two plates. The problem is fully described in references 10,11. In the heat flow equivalent of the problem for a zero specific heat equations (1), (2) and (3) become

$$\nabla^2 T = 0 \tag{7}$$

As before at the interface

$$v_{nl}L = -K \frac{\delta T}{\delta n} \tag{8}$$

and the temperature at the interface is given by

$$T_I = -\Gamma/R \tag{9}$$

Instead of imposing a temperature at infinity (ie. the driving force) a uniform gradient is imposed as x becomes large.

$$\frac{\delta T}{\delta x} = -G_o \tag{10}$$

This gradient can also be thought of as imposing a planar interface velocity u_o where

$$G_o K = u_o L \tag{11}$$

At steady state the product of dendrite tip velocity and the cell width will equal this velocity

$$u_o = v_t \lambda \tag{12}$$

To maintain consistency with the earlier work it is convenient to non$-$dimensionalise distances as $x = ax'$, velocities as $v = u_o v'$, and temperatures as $T = G_o aT'$, giving instead of (7), (8), and (9)

$$\nabla^2 T' = 0 \tag{13}$$

$$v'_{nl} = - \frac{\delta T'}{\delta n'} \tag{14}$$

$$T'_I = - \Gamma/(G_o \, a^2 \, R') = - \Gamma_1/R' \tag{15}$$

where $\Gamma_1 = \Gamma/(G_o \, a^2)$. As x' becomes large

$$\frac{\delta T'}{\delta x'} = - 1 \quad \text{and} \quad u'_o = 1 = v'_t \lambda \tag{16}$$

The temperature gradient in the x direction at $x'=0$ was initially taken to be zero; it was later changed to the value given by the Saffman Taylor[10] solution, see below.

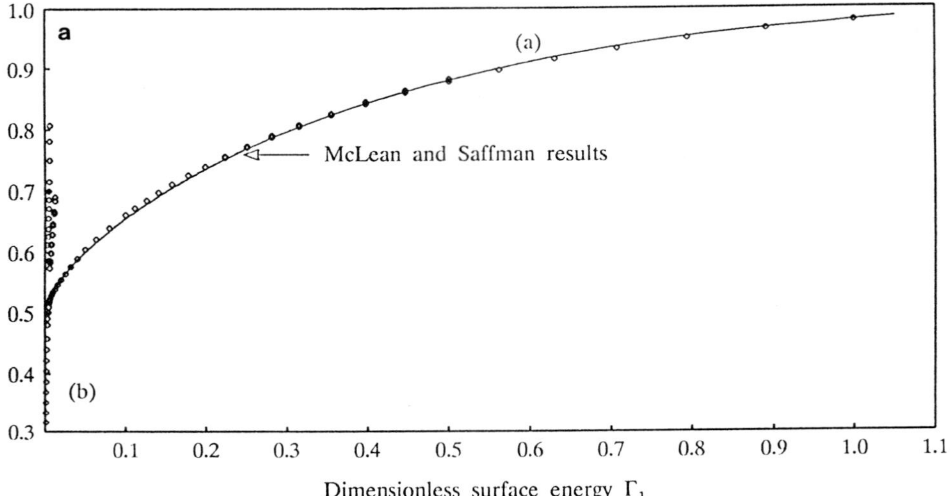

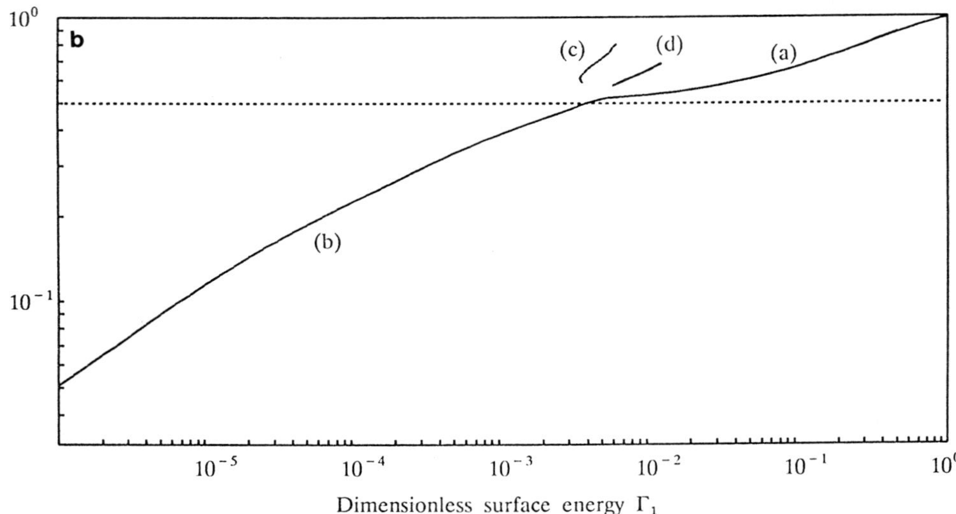

Figure 2. Plots of the calculated cell width, λ, against surface energy, Γ_1.
(a) The filled line shows results taken from McLean and Saffman[11]. The points are the present work.
(b) A log plot of the calculated results showing the very low velocity behaviour.

These equations are identical in form to those used to discuss the Hele – Shaw problem. The results may be expressed in terms of the single parameter, Γ_1. Following McLean and Saffman[11] the cell width can be calculated as a function of this dimensionless surface energy. The cell width is the solid volume fraction and from equation 16 is an inverse velocity.

In the earlier work for a zero surface energy, Saffman and Taylor[10] obtained an analytic expression for the interface shapes. They found that a complete set of solutions existed going from a fraction solid of 0 to 1. Their interface shape is given by:

$$ x_1{'} = \frac{(1 - \lambda)}{\pi} \ln \frac{1}{2} \left[1 + \cos \frac{\pi \, y'}{\lambda} \right] \tag{17} $$

Almost identical results were obtained in the present numerical work. The error in the interface position normal to the interface was usually less than 1×10^{-3}. The interface shapes obtained by the two methods could not be seen to be different on a plot.

The Saffman Taylor[10] solution allowed an analytic expression

$$ \frac{\delta T'}{\delta x'} = \pi(1 - \lambda)T'/2 \tag{18} $$

to be used for the temperature gradient in the x direction for x=0 (or large negative values of x in the Saffman and Taylor[10] analysis). Using this expression rather than the zero temperature gradient made practically no difference.

Even though a continuous set of solutions exist for a zero surface energy, McClean and Saffman[11] found that a unique volume fraction solid existed for a given dimensionless surface energy. Later Vanden – Broek[12] suggested that a set of discrete solutions exists. An important feature of the work is that steady state solutions were not found for volume fractions less than 0.5.

In the present numerical work unique steady state solutions were obtained; these are plotted as fraction solid as a function of dimensionless surface energy in fig. 2a and b. The results should be compared with the filled line giving the results of McLean and Saffman[11]. Their line was obtained from

Dimensionless cell width λ

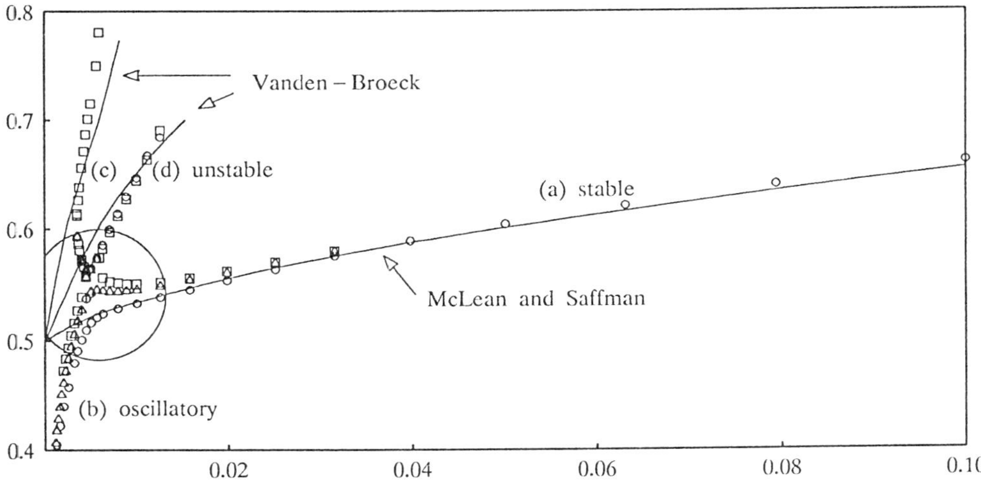

Dimensionless surface energy Γ_1

Figure 3. A plot of the calculated cell width, λ, for low surface energies, Γ_1, (points). The filled lines shows results taken from reference 11 and 12. Branches 'c' and 'd' are unstable, branch 'a' stable and branch 'b' oscillatory. Laplace 2D.

125

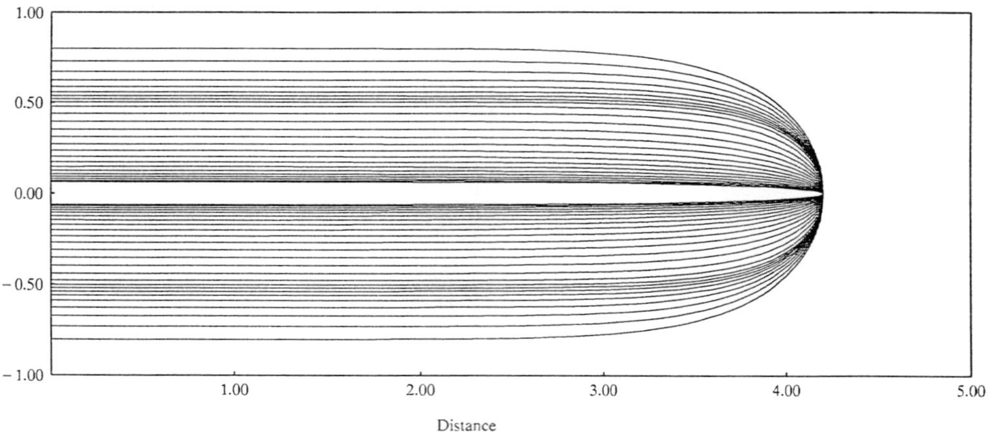

Figure 4. Calculated interface shapes varying from $0.3 - 5.0 \times 10^{-5}$ in dimensionless surface energy.

their fig. 4 and table 1 (Several values in their table have been neglected since they do not agree with their fig. 4.). The surface energy values must be multiplied by $(1 - \lambda)^2/(\pi^2\lambda)$ to allow for the different method of non – dimensionalisation. Again there is excellent agreement.

At low surface energies other sets of solutions were obtained. The different branches were found really by chance. Normally shapes would be found by scanning the dimensionless surface energy using a previous steady state shape as a starting point. For too large a change the calculation blows up or very occasionally moves to another branch. The small surface energy values are plotted on fig. 3. The different branches are labelled 'a','b','c','d' on fig. 3 (the branches 'a','b','c','d' and 'e' on figs. 2,3,6,8,9,11 and 12 refer to solutions of the same type). It was difficult to get solutions near other steady state solutions. In the area marked by a circle the results were particularly sensitive to any slight change, even changing the way the points were arranged in space or the length of cell being modelled produced

Stability parameter σ^*

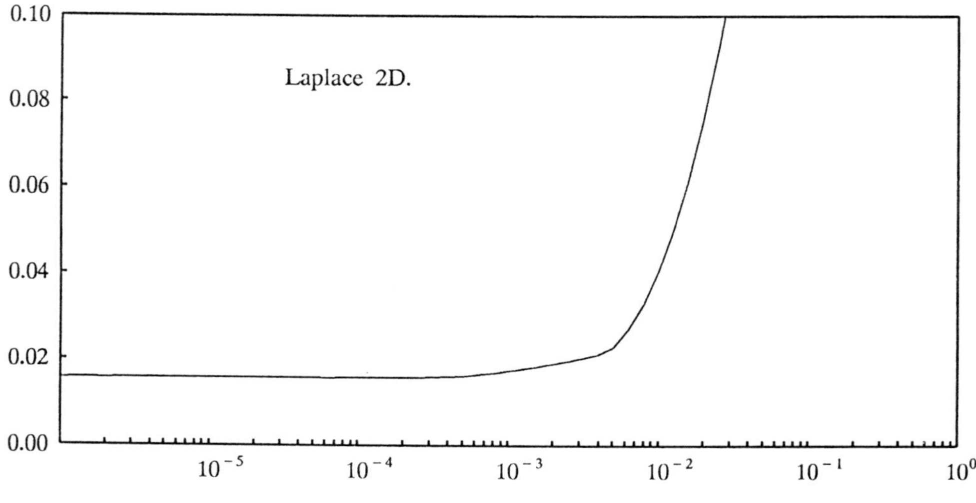

Figure 5. A plot of the stability parameter defined as $\sigma^* = \Gamma_{*1}/(v'_1 R'^2)$ plotted against the dimensionless surface energy, Γ_1 for branches 'a' and 'b'. Showing that σ^* reaches an almost constant value at small surface energies.

changes in the results. Outside the circle changes in numerical procedure made little difference (changes in modelled length produced small changes in the position of branches 'b','c' and 'd'. The branches were, however, always present). In the region of the circle the results should really be disregarded since the numerical method is clearly at the limit of its validity; elsewhere much more confidence can be placed in the results.

The shapes of the cell or dendrite for branches 'a' and 'b' (see fig. 3) are shown in fig. 4. The figure shows a geometric increment in the surface energy so that the sudden change in the rate of cell width change at 0.5 is readily apparent. Solutions were easily obtained for arbitrarily low surface energies for branch 'b'. These are plotted in fig. 2b down to 10^{-6}. Suggestions that dendrite like solutions exist for cell widths below 0.5 have been made by a number of authors[14-16]. These dendrite like solutions are clearly similar to those calculated in the previous work on solute dendrites. In the previous work it was found that the results agreed remarkably well with the predictions of marginal stability. The marginal stability condition in this case should be of the form

$$\sigma^* = \Gamma_\sigma K/(v_t\, L\, R^2) = \Gamma'/(v_t'R^{2'}) \tag{19}$$

The experimental results are shown in fig. 5. For branches 'a' and 'b' it can be seen that an almost constant value of 0.015 ± 0.002 is obtained once thin dendrite like shapes are formed.

The solutions on branches 'c' and 'd' can be compared with those of Vanden − Broeck[12] (filled lines on fig 3. The agreement in this case is not so good; a possible explanation could be lack of accuracy in this region of the plot although other suggestions will be discussed later. The interface shape for branches 'a','c' and 'd' for an identical cell width are shown in fig. 6. They are compared with the zero surface energy shape, the dotted lines. For branch 'a' the dotted line and the filled line do not cross, for 'd' they cross once and for 'c' twice.

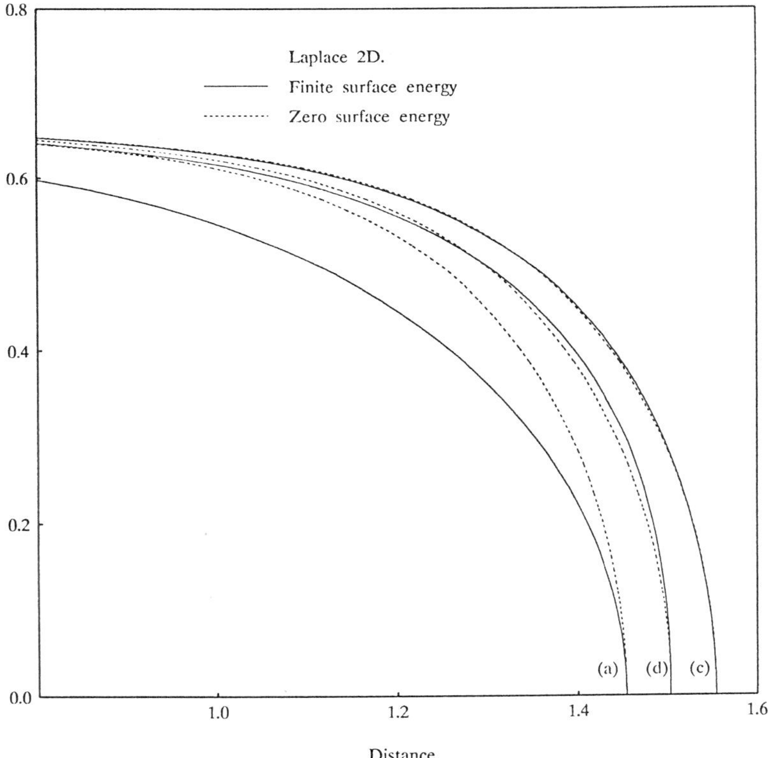

Figure 6. A comparison of the tip shapes for a fixed $\lambda = 0.66$ for branches 'a','c' and 'd' (filled lines). The dotted lines show the zero surface energy shape. For branch 'a' the dotted line and the filled line do not cross, for 'd' they cross once and for 'c' twice.

The steady state solutions were tested for stability by running the program with small time steps. Typically a time step which would give a movement of about 0.1 of the tip radius was used, ie. $\delta t = 0.1 \, R'/v'_t$. It was found that, provided the time step was less than about $0.25 \, R'/v'_t$, the results changed little with length of time step. For time steps much larger than this, steady state solutions were produced. For an unstable steady state solution numerical noise was sufficient to lead to instability, however this took time to develop. A quicker more reliable test was to suddenly change the surface

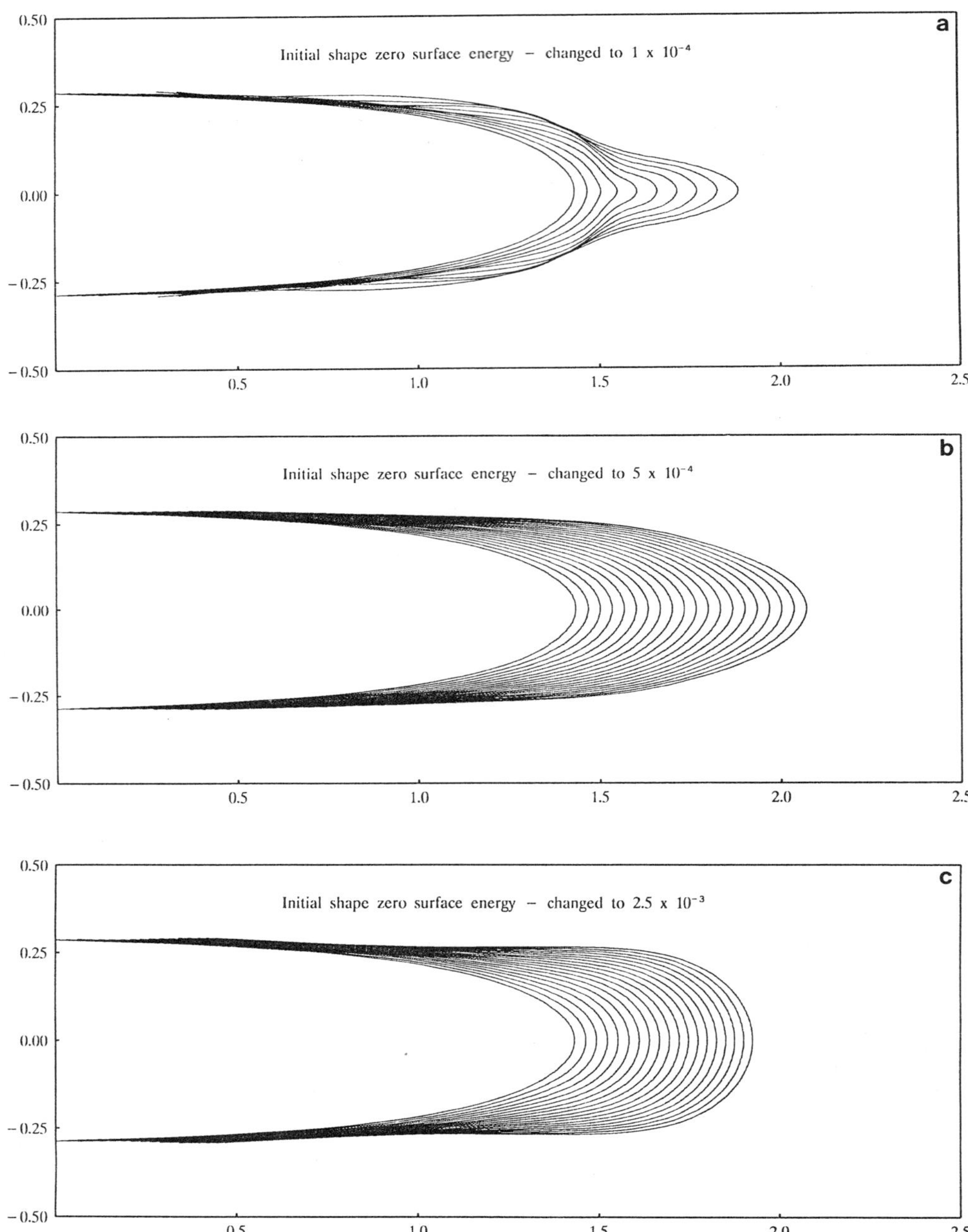

Figure 7. Plots of interface shape as a function of time showing how quickly the new tip shape is selected. a) A change in surface energy, Γ_1 from zero to 1×10^{-4}. b) A change in surface energy, Γ_1 from zero to 5×10^{-4}. c) A change in surface energy, Γ_1 from zero to 2.5×10^{-3}.

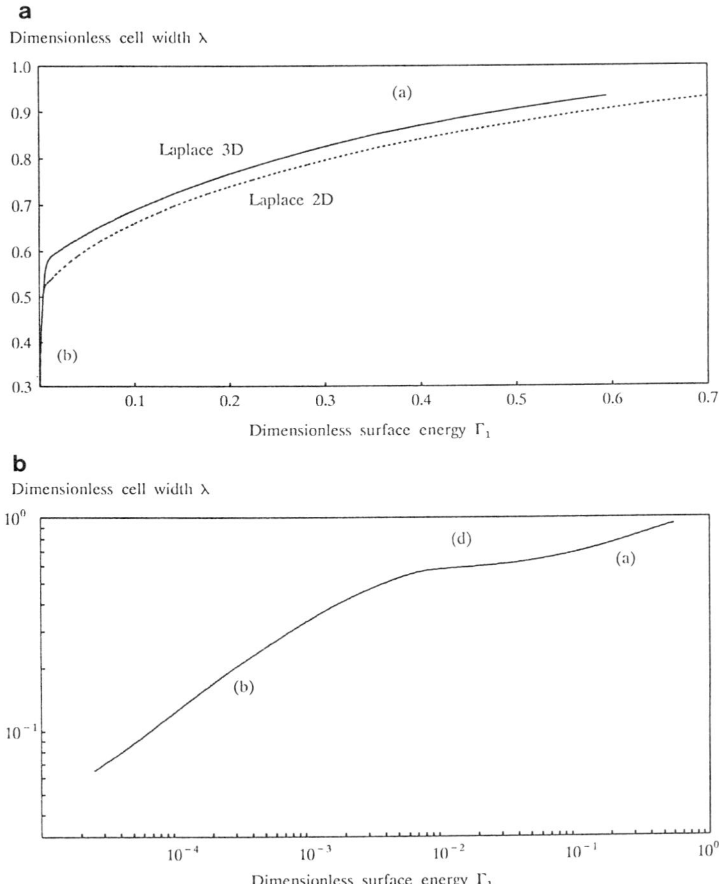

Figure 8. Plots of the calculated cell width, B, against surface energy, Γ_1.
(a) The filled line shows calculated results for Laplace 3D, the dotted line Laplace 2D.
(b) A log plot of the calculated results showing the very low velocity behaviour. Laplace 3D

energy by about 0.1%. Branches 'c' and 'd' were found to be unstable, that is, the solutions moved continuously away from the steady state. Eventually if the movement was towards branch 'a' a steady state solution on branch 'a' could be formed. More often bumps grew on the cell, becoming larger until they could not be handled by the numerical technique. On branch 'a' the solutions were stable, that is after a change the solution rapidly settled down to a steady state solution with no more than one crossing of the steady state velocity. On branch 'b' the solutions oscillated about the steady state velocity. The amplitude of the oscillation changed very little with time, either slowly increasing or decreasing. The period of the oscillation did not depend on the length of the time step, provided the time step was small enough. More detail on the oscillatory solutions will be presented later. The rapid approach towards a steady state solution for example on branch 'b' is shown in fig 7 a b and c. Here in each case the starting shape is the zero surface energy shape and the surface energy is then set to a value which would give a smaller (fig 7a), identical (fig 7b) and larger (fig 7c) cell width. Both the examples in fig. 7a and c blew up soon after the last shape was calculated.

HELE – SHAW THREE DIMENSIONAL PROBLEM. (Laplace 3D)

The three dimensional analogue of the Hele – Shaw problem does not appear to have been treated in as much detail as that in two dimensions . The problem is identical to that presented in the previous section and was non dimensionalised in the same way. The only difference is that instead of a plate, a cell having axial symmetry is considered. This means that equation (15) becomes

$$T'_I = - \Gamma'(1/R'_1 + 1/R'_2) \tag{20}$$

and equations (12), (16) become

$$u_o = v_t \lambda^2 = v_t f_s \quad \text{or} \quad u'_o = 1 = v'_t \lambda^2 = v'_t f_s \tag{21}$$

In this case since no analytic solution is available for the zero surface energy problem, the temperature gradient in the x direction at $x'=0$ is assumed to be zero. However this was shown in the previous section to be a negligible error. The zero surface energy shapes and finite surface energy shapes will be presented in a separate paper. The variation in cell width (obtained from $\lambda = 1/(v_t)^{0.5}$) is plotted against surface energy in fig. 8a and b. It can be seen that the results are very similar to those of the two dimensional case, the dotted line on fig 8a. The only difference is that the branch 'a' appears to go to a limiting value of about 0.56; this could be $(1/\pi)^{0.5}$. As before branch 'a' is stable, branch 'b' oscillatory and branch 'd' unstable.

THREE DIMENSIONAL HEAT FLOW DENDRITES. (Fick 3D)

Equations (1), (2) and (3) may be non – dimensionalised by writing $x = ax'$, $T = (L/c)\, T'$, $t = (a^2 c/K)\, t'$ or $t = (a^2/D_{th})\, t'$ and thus $v = (D_{th}/a)\, v'_2$ giving

$$\nabla^2 T' = \frac{\delta T'}{\delta t'} \tag{22}$$

is solved for a shape where at the interface

$$v'_{n2} = - \frac{\delta T'}{\delta n'} \tag{23}$$

The temperature at the interface is given by

$$T_I = - \Gamma_2(1/R'_1 + 1/R'_2) \tag{24}$$

where $\Gamma_2 = \Gamma c/La$

The temperature ahead of the dendrite as x approaches infinity is given by

$$T_o' = cT_o/L \tag{25}$$

At steady state using equation (5) T_o' equals the fraction solid or

$$T_o' = -f_s \tag{26}$$

The problem in this case needs two parameters. The isolated heat flow dendrite can be expressed in terms of a single parameter; two parameters are needed in the present work because of the finite cell spacing, 'a'. The additional parameter when compared to the previous sections arises because of the additional term in equation (22) as compared with (13). At steady state it is convenient to fix Γ_2 and scan the undercooling (that is fraction solid) calculating the tip velocity.

As in the previous work[1] and in the previous sections, discrete steady state solutions were obtained for a fixed undercooling and finite surface energy. (This is in contrast to the continuous set of solutions assumed to be present on a modified Ivantsov plot[2-6]). Again, as before, a continuous set of solutions was possible when the surface energy was zero. A plot of undercooling against tip velocity, v'_{2t} for different values of the surface energy term Γ_2 is shown in fig. 9a. At high velocities long thin dendrites are formed (branches 'b' and 'e' fig. 9a). At low velocities cell like structures are formed (branch 'a' fig 9a).

One feature to note is that the minimum on each plot occurs when $D_{th}/v_t \simeq a$, that is when the diffusion field just reaches the cell wall near the tip. The change in slope at low velocities arises when dendritic rather than cell like structures begin to form with increasing velocity. This aspect will be discussed in more detail later. At the lowest undercoolings it becomes increasingly difficult to model a length sufficient to approximate the gradient at $x'=0$ as equal to zero and simultaneously have enough points near the tip when the tip radius is getting ever sharper. This is the reason that the minimum has not been drawn for $\Gamma'_2 = 10^{-6}$.

The first problem is to combine the results of the previous sections with those in this section. In the McLean and Saffman[11] and the present Laplace 2D and 3D analyses, the volume fraction solid is obtained as the result of the calculation whereas in the present section it is imposed initially by equation (26) as soon as the undercooling is set. In the present Laplace 2D and 3D analyses the dimensionless surface energy was scanned (horizontal axis) and the velocity calculated; this was then inverted to give the fraction solid (vertical axis). In the heat flow analysis the undercooling, that is fraction solid, was scanned (vertical axis) and the velocity calculated (horizontal axis).

To draw the two types of solution together it would seem logical to start with the heat flow

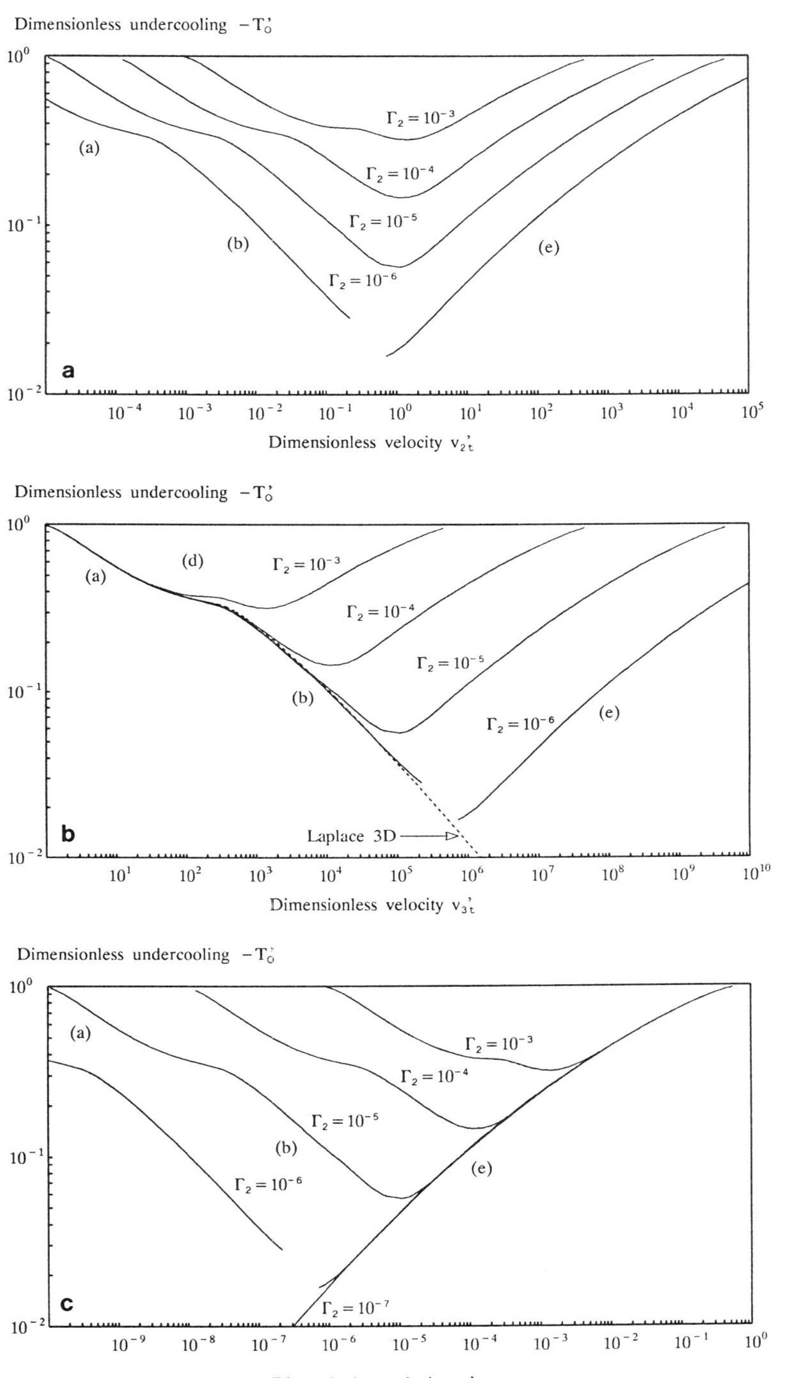

Figure 9. Plots of the dimensionless undercooling $(-T_0)$ against dimensionless velocities for different surface energies. a) The basic plot with velocity defined as $v_2' = av/D_{th}$. Fick 3D. b) The velocity defined as $v_3' = v_2'/\Gamma_2$. Showing the coincidence of all the low velocity Fick 3D results and the Laplace 3D results. c) The velocity defined as $v_4' = v_2'\Gamma_2$. Showing the coincidence of all the high velocity or isolated dendrite results. Fick 3D.

131

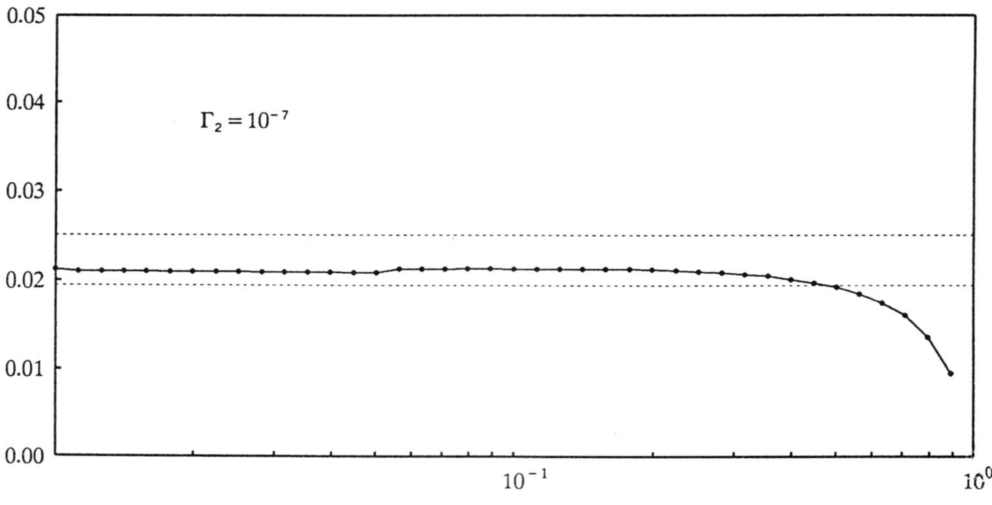

Figure 10. A plot of the stability parameter defined as $\sigma^* = \Gamma_2/(v_2' R'^2)$ plotted against the dimensionless surface energy, Γ_2 for branch 'e'. Showing that σ^* has an almost constant value for an isolated dendrite.

formulation since this should become the McLean and Saffman formulation in the limit of a zero specific heat. To show the effect of the specific heat approaching zero it is an advantage to have two axes which do not depend directly on the specific heat. A common feature of fig. 8 and fig. 9a is that the vertical axis may be regarded as the fraction solid and therefore may be considered to be independent of the specific heat (in fig. 8 fraction solid $f_s = \lambda^2$ and in fig. 9a the undercooling is $-T_o = f_s$ equation 26). For the other axis a dimensionless velocity is needed which does not contain the specific heat. A suitable velocity is

$$v_{3t}' = v_{2t}'/\Gamma_2 = a^2 L\, v_t/(K\Gamma) \tag{27}$$

The results of fig 9a are replotted using this velocity axis in fig. 9b. It can be seen that very conveniently the low velocity results collapse onto one line; deviation occurs when $D_{th}/v_t < a$. The dimensionless surface energy used in the sections treating the Hele–Shaw problem equation (15) may be expressed using equations (11) and (21) as

$$\Gamma_1 = K\Gamma/(a^2 L\, v_t\, f_s) \tag{28}$$

Inverting this expression and dividing by f_s gives the new velocity v_{3t}'. Using this relationship the points on fig. 8a and b are also plotted on fig 9b as the dotted line (since fig. 8b is a log plot the inversion reflects the points across unit velocity). It can be seen that the previous results coincide precisely with the results for the lines calculated in this section (provided $D_{th}/v_t \gg a$). The specific heat only appears in Γ_2 thus the term may be considered to be a modified specific heat and as it approaches zero the solutions becomes the solution to the three dimensional Hele–Shaw problem.

The low velocity terms on fig. 9b (branch 'a') are the McLean and Saffman cell like solutions (but for the 3D problem) and as before these appear for a volume fraction greater than about $1/\pi \simeq 0.32$ or $(0.56)^2$. The higher velocity solutions to the right of the change in slope (branch b fig. 9b) are the dendrite like shapes which were present at low surface energies on the McLean and Saffman plot branch b fig. 8. It is clear that the solutions which cannot be approximated in any form by an analysis of the Hele–Shaw problem are those to the right of the minimum, branches 'e'.

The solutions to the right of the minimum undercooling are those when the spacing 'a' is so large that the tip is not influenced significantly by the walls. These solutions may also be collapsed onto a single line using similar arguments. If a new dimensionless velocity is defined as

$$v_{4t}' = v_{2t}'\, \Gamma_2 = \Gamma c^2 v_t/(KL) \tag{29}$$

This velocity does not contain the cell spacing and so the parameter Γ_2 may be regarded as an inverse cell spacing. A plot of the results using this velocity is shown in fig. 9c. It can be seen that all the plots of fig 9c coincide provided $D_{th}/v_t \ll a$. The isolated dendrite is thus just the limiting solution for an infinite spacing and can grow at any arbitrarily low undercooling. (An alternative method of obtaining the same plot is to non $-$ dimensionalise distances as $x = (\Gamma c/L)x'$, temperature as $T = (L/c)T'$ and time as $t = (\Gamma c/L)^2 \, t'/D_{th}$ then use $a' = (L/\Gamma c)a$ and the undercooling as parameters. For a large enough a' the solution does not depend on a').

In the previous work on solute dendrites it was found that the steady state growth occurred very near the value of the parameter σ^* predicted by marginal stability. For this case in dimensionless units

$$\sigma^* \;=\; \Gamma_2 \, / \, (R'^2 \, v'_{2t}) \tag{30}$$

This parameter has been predicted to be about $1/(4\pi^2) = 0.025$; other work suggests between $0.0195 - 0.0253$. A plot from fig 9c for the line $\Gamma_2 = 10^{-7}$ is shown in fig 10. The value of σ^* is remarkably constant except at very large dimensionless undercoolings. The actual value does depend to a certain extent on the length of the cell being modelled and to a lesser extent on the way the points are arranged on the interface. The variation, for small undercoolings, was of the same order as the variation above $(0.019 - 0.026)$.

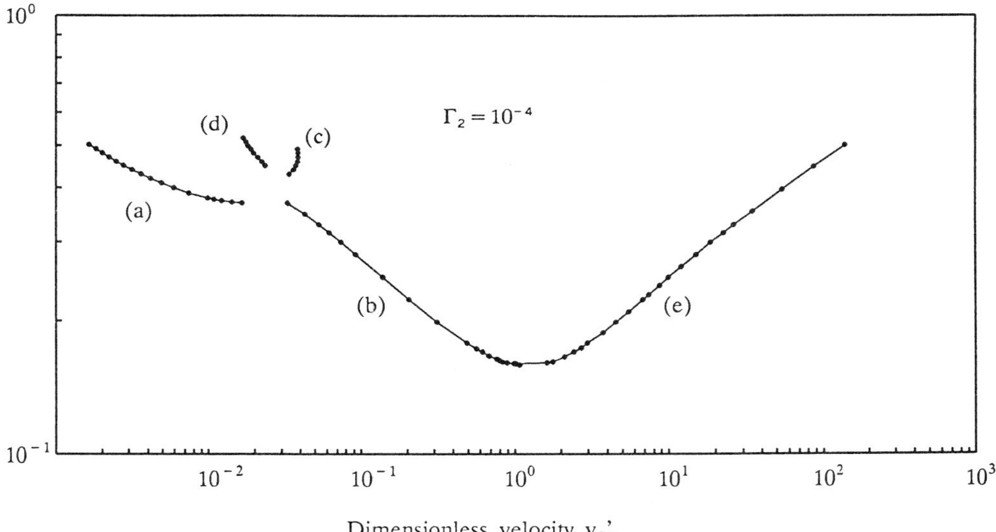

Figure 11. Plots of the dimensionless undercooling $(-T_o)$ against dimensionless velocity, v'_2. Showing the diferent branches a,b,c,d and e for one surface energy. Points show calculated values. Fick 3D.

When the direct analogy between the solutions to the Hele $-$ Shaw problem and that of the conduction equation was fully appreciated, attempts were made to find the other branches which had been produced on the McLean and Saffman plots. It was found that similar solutions were present, although these were much more difficult to obtain. Even very small changes in a scanned parameter led to the solution blowing up. No solutions at all were obtained near the break. The different branches 'a','b','c','d' and 'e' are shown on fig. 11. From this figure it appears as if the branches may cross; branch 'a' may continue as 'd' and branch 'b' as 'c'. A similar conclusion could be reached from figs. 2,3 and 8. This would mean that branches 'c' and 'd' in fig. 3 are rather different from the Vanden $-$ Broeck[12] solutions.

The stability parameter σ^* for the different branches is plotted on fig 12. It is apparent from this plot that dendritic type growth occurs near the marginal stability condition on branches 'b' and 'e', that is on either side of the minimum in the undercooling. They are thus both dendritic. The value deviates from the expected values for high undercoolings and strong interaction with the neighbouring dendrites. Branch 'a' behaves in a very different fashion to 'b' and 'e'. Again branch 'c' could be a continuation of 'b' and branch 'd' a continuation of branch 'a'.

As for the Laplace solutions, the Fick steady state solutions were tested for stability by reducing the time step to typically $\delta t = 0.1R'/v'_{2t}$. A check that the time step was small enough was made by running for some time with a value 0.1 of the initial value, if the behaviour was different the time step was reduced until similar results were obtained for a magnitude change in step size. The high velocity solutions, those to the right of the minimum, were tested at dimensionless undercoolings of 0.01, 0.03, 0.1 and 0.5. As before the surface energy was changed by 0.1%. A typical result is shown in fig. 13. Ten interface shapes were calculated between each marked interface position. In the lower plot of fig. 13, the tip velocity is plotted against tip position. It can be seen that the tip velocity oscillates about the steady state condition, gradually gaining in amplitude until the bumps on the interface are too large to be handled by the present model. The wavelength of the oscillation on the plot always appeared to be between 8 to 9 times the tip radius.

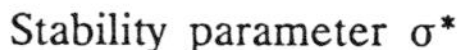

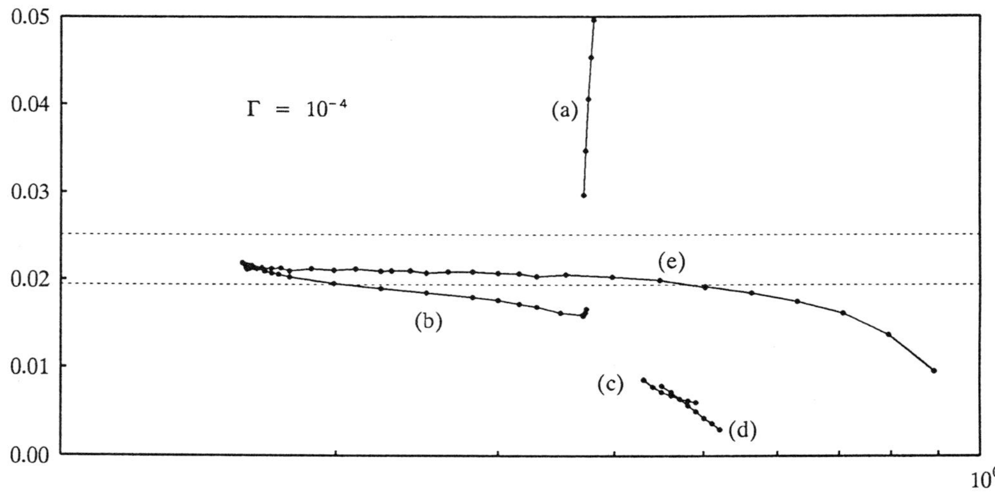

Figure 12. Plots of the stability parameter defined as $\sigma^* = \Gamma_2/(v'_2 R'^2)$ plotted against the dimensionless surface energy, Γ_2 for branches a,b,c,d and e. Showing that σ^* has an almost constant value for the dendritic solutions b and e. Fick 3D.

For the solutions to the left of the minimum, similar results were obtained for undercoolings above and below 0.32. After a perturbation in surface energy the solution crossed the steady state value once, then gradually moved away from the steady state value. It appears that in this case it is reaction of the shape with the long range diffusion field that leads to the instability. This would account for the difference between the stability of the Laplace and Fick solutions. Near the minimum on the undercooling plot, an oscillatory behaviour (similar to that occurring at high velocities) was superimposed on the gradual drift.

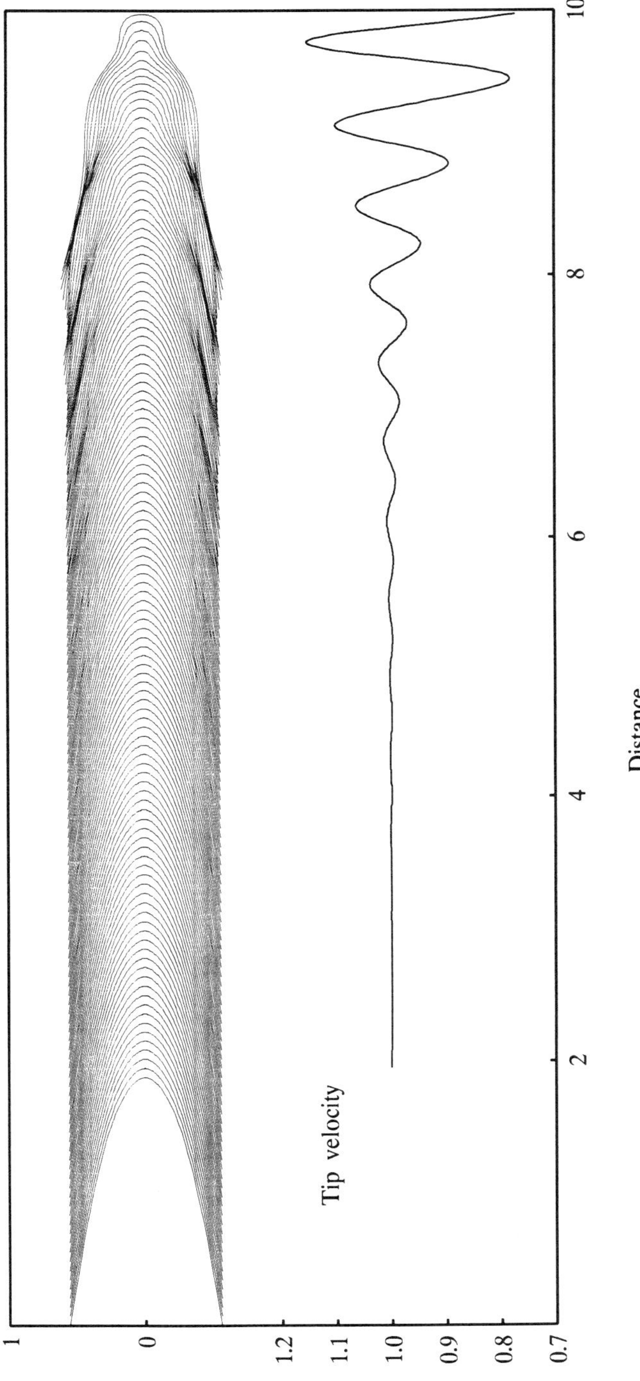

Figure 13. Plots of interface shape as a function of time showing a shape every ten iterations. The tip velocity plotted against position of the tip varies in a sinusoidal manner as is shown by the lower part of the diagram. Undercooling 0.1, $\Gamma_2 = 1 \times 10^{-6}$, branch 'e'. Fick 3D.

DISCUSSION

The present work has shown that the numerical analysis is sufficiently sensitive to obtain the zero surface energy shapes considered by Saffman and Taylor. As in the earlier work these occurred for any fraction solid. For finite surface energies discrete fractions solid were obtained and an excellent agreement was obtained with the work of McLean and Saffman. The work showed that heat flow solution produced as the specific heat approached zero were identical to those produced by considering the three dimensional analogue of the Hele – Shaw problem. In both problems the limiting volume fraction of the cell like or McLean and Saffman type solution occurred at a volume fraction of about $1/\pi$ in the three dimensional problem rather than the 0.5 found in the two dimensional problem. The cell like solution was found to be stable for the Hele – Shaw problem but unstable for the heat flow problem. This probably occurred because of an interaction of the growing cell with the exponential temperature field ahead of the tip in the heat flow problem.

As in the previous numerical work discrete steady state dendrite like solutions were obtained at high velocities. These grew very near the marginal stability condition except when the dimensionless undercooling was very high or when there was considerable interaction of the tip with the walls. When tested for stability none of the heat flow solutions were found to be truly stable. All the branches to the left of the minimum undercooling moved away from their discrete steady state solution and would probably have eventually reached the high velocity branch 'e'. This branch, although very nearly stable, oscillated about the steady state condition. It is to be expected that a model which could handle more pronounced arms or considered truly three dimensional growth (rather than axial symmetry) would lead to an oscillatory steady state where the amplitude of oscillation in the tip velocity remained constant and led to the repeated formation of dendrite arms. The present work indicates that, for a given bath undercooling, there is a minimum dendrite spacing for which a solution is possible. In practice the cell or dendrite spacing 'a' will be determined by the interaction of the dendrite with neighbouring dendrites.

There has been considerable discussion about whether steady state dendrite like solutions do or do not exist. Various workers[7–9] using different techniques have suggested that smooth steady state dendrite solutions are only possible when the surface energy is anisotropic. These treatments predict a large change in the growth conditions with extent of anisotropy for systems in which the anisotropy is very small. This variation does not appear to occur in practice[17].

Clearly, steady state dendrite like solutions have been found in the present numerical work. One conclusion might be that the earlier work which suggested that there are no solutions for isotropic surface energy is simply incorrect and that the present results show that discrete solutions can occur. The alternative conclusion would be that the present results are an artefact of the numerical method. Clearly it is not possible at this stage categorically to state which of these two conclusions is correct.

Although the conclusion which is reached may have mathematical significance, it is probably of little importance for the description of dendritic growth. In the earlier work the belief that steady state isotropic surface energy solutions were not possible led to the conclusion that dendrites grow because the surface energy must always be slightly anisotropic. An alternative approach would be to question whether the presence or absence of steady state growth over the complete dendrite was necessary. There is really no experimental evidence for truly steady state dendritic growth. After an initially smooth region dendrite arms are formed some distance behind the tip. Smooth shapes are only formed when solute cells are grown in an imposed moving temperature gradient and are not formed when growth is into a supercooled bath, the condition considered in the present work. Various authors have argued that the dendrite tip might be considered to be almost steady state and attempts have been made to detect fluctuations in the tip velocity due to (or leading to) branch formation. It is generally agreed that experimentally there is very little variation in tip velocity. Thus it is concluded that there is a region near the tip which is almost at steady state. On a macroscopic scale steady state exists, that is arms form and form again in a regular manner, but there is no evidence for smooth steady state shapes.

If it is assumed true that steady state isotropic surface energy solutions are not possible, the most likely reason for the present numerical results is that a dendrite tip having a length of only about twenty seven times the tip radius has been modelled. It is possible that approximately steady state shapes can be obtained when the solution is restricted to the vicinity of the tip and that it is, in fact, the relaxation far from the tip that allows a steady state solution to be found. The length modelled in the present work was chosen to be similar in size to that experimentally observed to be almost steady state. The analysis may thus describe the dendritic growth process well, precisely because only a restricted length of the dendrite is considered.

In practice it seems probable that the velocity of a dendrite tip oscillates very slightly (the variation may be too small to be seen experimentally) and that these oscillations lead to branching further down the dendrite. In the absence of a treatment which considers the formation and growth of the branches it seems reasonable to suggest that the steady state tip velocity obtained by considering either the complete dendrite or just the tip would give the average velocity with a sufficient degree of accuracy. Clearly this behaviour occurs in the numerical experiments shown in fig. 14.

Although cell – like, as opposed to dendritic structures, are not found when a pure material grows into a supercooled liquid, cells with a smooth completely steady state interface can be produced when an alloy is grown at a constant velocity by withdrawing a specimen from a furnace at a constant rate. This is precisely the problem for which the present numerical method was originally written. Although more work needs to be done, it seems possible that these cells are McLean and Saffman type solutions (branch 'a'). One difference, however, is that the volume fraction solid for the solute cell will not go to some limiting value but instead continually increases with decreasing temperature. It is possible, however, that there is a volume fraction solid near the tip (one cell spacing behind the tip) which cannot be less than $1/\pi$.

SUMMARY

It is concluded that the present numerical method is sufficiently accurate to reproduce the Saffman and Taylor and McLean and Saffman solutions of the Hele – Shaw problem. This conclusions allows more confidence to be placed in the other results. Solutions were obtained for the three dimensional Hele – Shaw problem and the dendritic heat flow problem. It was shown that the solution to the Hele – Shaw problem was identical to the limiting case of a zero specific heat. The limiting volume fraction in three dimensions for the McLean and Saffman type solutions was found to be about $1/\pi$ (as opposed to 0.5 for a plate). Despite the fact that the heat flow problem is a two parameter problem, the results could be plotted to give a single line when either the thermal diffusion interacted strongly with the cell walls (constrained cell or dendrite) or when the walls were so far away that there was little interaction (the isolated dendrite).

The dendrite like steady state solutions grew very near the value predicted by marginal stability even though only a discrete steady state solution was found. The solutions thus agree with much of the experimental work. The marginal stability condition should be regarded as a method for approximating the conditions that exist at steady state rather than a procedure for selection one of a continuum of steady state shapes. The dendrite tip at high velocities was found to be unstable and oscillated about the steady state condition with a wavelength on the plot of about $8-9$ times the tip radius. The low velocity shapes were found to be unstable but moved away from the steady state condition.

Basically two types of interface shape or solution were found. These were cell like shapes (McLean and Saffman solutions or branches 'a') where the fraction solid was greater than about $1/\pi$ (0.5 for the plate) and where the tip radius was similar to if not greater than the cell spacing. The second shape was dendritic (branches 'b' and 'e') where the tip was almost parabolic. The two shapes can be distinguished by the agreement or lack of agreement with the marginal stability condition. Visually it is very easy to separate shapes into the two different types.

Since the present work is purely numeric, although steady state dendrite like shapes were found for isotropic surface energy, it cannot be categorically concluded that infinitely long steady state shapes must be possible. If indeed solutions are mathematically not possible, the most probable reason for the present results is that only the region near the tip is treated, rather than a complete dendrite. The present work models a region similar in size to that seen to be approximately steady state experimentally. Since dendrites do not grow at steady state the present work probably approximates the growth of a dendrite whether smooth steady state shapes are or are not possible over the complete dendrite.

TABLE 1

a	Half repeat distance, cell or dendrite spacing shown on fig.1.
A_n, A_s, A_e, A_w	Areas of the north, south, east and west control volume walls.
c	Specific heat per unit volume.
D_{th}	Thermal diffusivity $D_{th} = K/c$
G_o	Temperature gradient as x becomes large fig.1 for Hele – Shaw problem.
f_s	Fraction solid.
K	Conductivity per unit area.
L	Latent heat per unit volume.
n, n'	The normal and dimensionless normal to the interface, n = an'.
R, R_1, R_2	Principal radii of curvature.
R', R_1', R_2'	Dimensionless radii of curvature non – dimensionlised as for x'.
ΔS	Entropy of melting per unit volume.
t, t'	Time and dimensionless time, t' is defined in each section.
$\delta t, \delta t'$	Time interval and dimensionless time interval.
T, T'	Temperature and dimensionless temperature, T' is defined in each section.
T_I, T_I'	Dimensional and dimensionless interface temperature, T_I' is defined as for T'.
T_n, T_s, T_e, T_w	The temperature at the north, south, east, west control volume walls.
T_o, T_o'	Temperature and dimensionless temperature as x goes to infinity (ie. the negative of the bath undercooling).
u_o, u_o'	Dimensional and dimensionless planar interface velocity $u_o' = 1$.
v, v', v_2'	Velocity and dimensionless velocities defined as $v' = v/u_o$ and $v_2' = (a/D_{th})\,v$.
v_n, v_s, v_e, v_w	Velocity of the north, south, east, west control volume walls.
v_{nl}, v_{nl}', v_{n2}'	The velocity normal to the interface, dimensionless normal velocities defined as for v' and v_2'.
v_t, v_t', v_{2t}'	Velocity of tip, dimensionless tip velocity defined as v' and v_2'.
v_{3t}', v_{4t}'	Dimensionless velocity defined as $v_{3t}' = v_{2t}'/\Gamma_2$ and $v_{4t}' = v_{2t}'\,\Gamma_2$.
V, V"	Volume of control volume time i and i – 1.
x, x'	Distance and dimensionless distance (see fig. 1), x = ax'.
x_1'	Dimensionless distance measured from tip, non – dimensionalised as x'.
y, y'	Distance and dimensionless distance (see fig. 1), y = ay'.
४	Surface energy.
Γ	Gibbs Thompson coefficient, $\Gamma = ४/\Delta S$, the surface energy term.
Γ_1, Γ_2	Dimensionless surface energy $\Gamma_1 = \Gamma/(G_o\,a^2)$ and $\Gamma_2 = \Gamma c/(La)$
λ	Cell width defined in fig. 1.

APPENDIX A

The cell or dendrite is assumed to have mirror or axial symmetry about the x direction (fig. 1). The control volume walls are orthogonal and the points of the control volume are at the centre of mass. The walls are arranged so that the solid – liquid interface goes through diagonal corners of an interface box. A schematic view of the points and box walls is shown in fig. 1. Each interface box corner is assumed to move with the tip velocity together with an additional velocity normal to the local interface position. This ensures that the tip stays in the same position within the total volume being modelled.

The depression in temperature due to curvature was obtained by getting the principle radii of curvature[1] from a cubic spline fitted through the box corners and from the angle the interface makes with the x axis[18]. Instabilities in this interpolation at high interface gradients were avoided by tilting the cell 45 degrees. The interface points were arranged to give a fixed angular difference (typically 0.07 radians) between the tip and the first point moving back along the cell. The remaining points were arranged on the interface using a geometric increment on the previous distance working from the tip backwards. The geometric increment was calculated to give the total length of the cell when all the points had been added. A similar technique was used to position the box walls moving out in the y direction from the last interface point but here the geometric increment was based on the width of the last interface box. The points ahead of the tip were calculated by doubling the size for each successive box. The modelled length was usually adjusted to be the smaller of either twenty seven times the tip radius or two times the cell or dendrite spacing 'a'. This procedure ensured that the approximately the same relative amount of cell or dendrite was modelled even when the tip radii varied over many orders of magnitude. The maximum length of twice the cell spacing was necessary to ensure the modelled length was not excessive when the tip radius was similar in size to the spacing.

138

ACKNOWLEDGEMENTS

The author would like to aknowledge Dr. J.R.Ockendon and Prof. P.Pelce for the suggestion of treating the Hele – Shaw problem. He would also like to aknowledge numerous constructive discussions of the work with Dr. J.R.Ockendon, Dr. S.D.Howison and C.Huntingford in the Centre for Industrial and Applied Mathematics, Oxford.

REFERENCES

1) J.D.Hunt, Acta Metall. $\underline{38}$, No 3, 411, (1990)
2) D.E.Temkin, Dokl. Akad. Nauk S.S.S.R., $\underline{132}$, 1307, (1960)
3) R.Trivedi, Acta Metall., $\underline{18}$, 287, (1970)
4) G.E.Nash and M.E.Glicksman, Acta Metall, $\underline{22}$, 1283, (1974)
5) J.S.Langer and H.Müller – Krumbhaar, Acta Metall. $\underline{26}$, 1681, 1689, 1697, (1978)
6) S.C.Huang and M.E.Glicksman, Acta Metall. $\underline{29}$, 701, (1981)
7) D.A.Kessler, J.Koplik and H.Levine, Phys. Rev, $\underline{A\ 33}$, 3352, (1986)
8) D.I.Meiron, Phys.Rev, $\underline{A\ 33}$, 2704, (1986)
9) P.Pelce and D.Bensimon, Nuclear Phys. B, Proc. Suppl. $\underline{2}$, 259, (1987)
10) P.G.Saffman and G.Taylor, Proc. Royal, Soc. $\underline{A\ 245}$, 312, (1958)
11) J.W.McLean and P.G.Saffman, J. Fluid Mech. $\underline{102}$, 455, (1981)
12) J – M Vanden – Broeck, Phys. Fluids, $\underline{26}$, 2033, (1983)
13) L.Romero, Ph.D. thesis, California Institute of Technology, (1982)
14) A.A.Lacey, S.D.Howison, J.R.Ockendon and P.Wilmott, to be published in Q. J. Mech. Appl. Math. (1990)
15) A.R.Kopf – Sill and G.M.Homsy, Phys. Fluids. $\underline{30}$, 2607, (1987)
16) Y.Couder, N.Gerard and M.Rabaud, Phys. Rev. $\underline{A34}$, 5175, (1986)
17) R.Trivedi and J.T.Mason, to be published Met. Trans. (1990)
18) K.A.Jackson and J.D.Hunt, Trans. Met. Soc. A.I.M.E. $\underline{236}$, 246, (1966)

CELL SHAPES AND WAVELENGTH SELECTION IN DIRECTIONAL SOLIDIFICATION

P. Kurowski, S. de Cheveigne, C. Guthmann

Groupe de Physique des Solides
Université Paris VII
2 place Jussieu, 75251 Paris CEDEX, France

INTRODUCTION

In directional solidification the sample, a binary alloy in the present case, initially liquid, is pulled at a given velocity V in a direction parallel to an imposed temperature gradient G set up around its melting temperature so as to solidify it progressively (fig. 1).

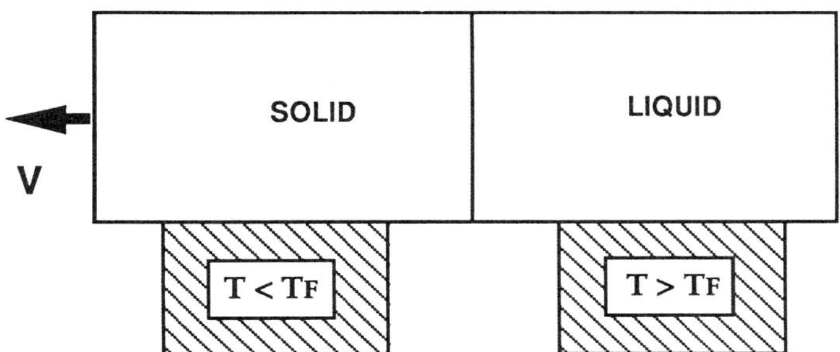

Fig. 1. Shematic view of directional solidification.

At low velocities the front is planar and the concentration of impurities in the solid is homogenous. Above a critical velocity V_C which depends on G and C_O (where C_O is the solute concentration of the mixture) the solid becomes inhomogenous and the interface presents a quasi-periodic cellular deformation on a scale of the order of 50-100 microns. The amplitude of the cells increases with increasing pulling speed.

At high velocities - V of the order of 10 V_C - the cells eventually develop side-branches and are then called dendrites.

Solidification is a first-order phase transition. So, as the solid grows, there will be production of latent heat and release (or absorption) of solute at the solid-liquid interface because of the different equilibrium concentrations of impurities in the two phases. But heat diffusion is instantenous with respect to solute diffusion (the heat diffusion coefficient is much greater than the solute diffusion coefficient). The dynamics of solidification is therefore controlled by the transport of solute.

Growth and Form, Edited by M. Ben Amar *et al.*
Plenum Press, New York, 1991

Mullins and Sekerka[1] analysed the physical mechanism of the instability that results from the competition between the destabilizing effect of the solute diffusion and the stabilizing effects of the temperature gradient and of the interfacial tension. In the limit of low velocities and for materials with equal heat conductivities in the solid and in the liquid, the bifurcation from a planar to a cellular front takes place at a critical velocity:

$$V_c = \frac{DGK}{m(K-1)C_o}$$

where D is the solute diffusion coefficient in the liquid, m the liquidus slope and K the partition coefficient of the alloy.

From an experimental point of view, we are confronted with the following questions common to all the systems far from equilibrium in which a periodic pattern develops :
- How can one describe the state of the system i.e the stationary shapes of the cells?
- To what extent is this structure periodic? Is the period well defined? How it is selected?

In the present article, we first examine cellular shapes in CBr_4-Br_2 and show that they follow a scaling law from V_c up to about 8-9V_c : the shapes depend on the relative velocity $\varepsilon = (V-V_c)/V_c$ and not separately on absolute velocity V or temperature gradient G. We then consider widths of individual cells along a given interface and their dispersion from threshold up to the appearance of dendrites.

EXPERIMENTAL

The experimental set-up is essentially similar to that suggested by Jackson and Hunt[2] and has been described elsewhere[3]. The experiments take place under a microscope and are recorded on videotape. To allow a more quantitative analysis, these videopictures are digitalized and cell contours are extracted and analysed.

For our solidification experiments we use CBr_4 with 0.12% impurities essentially Br_2. This material is a plastic and transparent crystal - which allows us to follow the dynamics of solidification - and presents a solid-liquid interface that is rough on the atomic scale as do many metals. The kinetics are then expected to be rapid and the interface to be locally at thermodynamic equilibrium. We also use thin samples, approximately 50 micrometers thick which exclude convection in the liquid phase.

CELL SHAPES

As the pulling speed is increased above threshold (for a given temperature gradient) the cells become deeper and more pointed and their average width decreases as $V^{-0.4\pm0.1}$. They nevertheless retain a more or less rounded shape, which H.Muller-Krumbhaar[4] has caracterized as "romanesque". These cells exist up to 10 times the threshold V_c.

The most obvious parameters with which to characterize cell shapes are the cell width λ and the tip radius ρ. The ratio of the two, the reduced tip curvature λ/ρ measures the "pointedness" of the cell.

Along a given front, one can observe an important dispersion in cell widths as will be discussed below. Nevertheless λ/ρ is constant to within less than 10% whereas the tip radius increases as cell width increases (fig. 2).

If now we compare experiments with different thresholds V_c (i.e with different temperature gradients) we find that the shapes are very different for equal velocities, but they are identical for equal relative velocities ε (fig. 3). We thus exhibit the following scaling law : cell shapes of the rounded "romanesque" type are superposable for a given value of ε when reduced in the same scale. They are determined by the unique parameter ε and not by V or G taken separately.

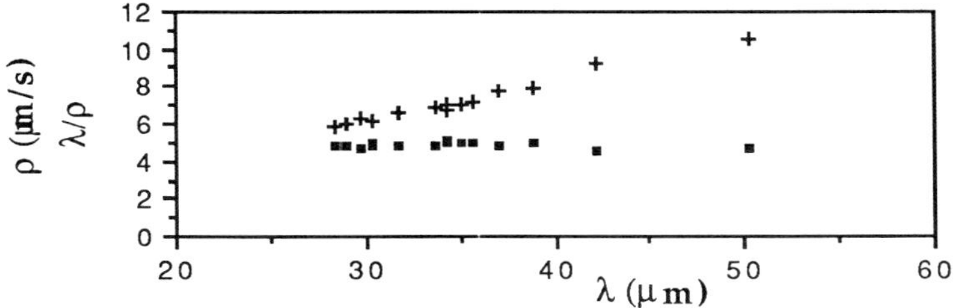

Fig. 2. Tip radius ρ (crosses) and reduced curvature λ/ρ (squares) versus cell width λ (G = 104 K/ cm, V = 30 μm/ s, ε = 5.5)

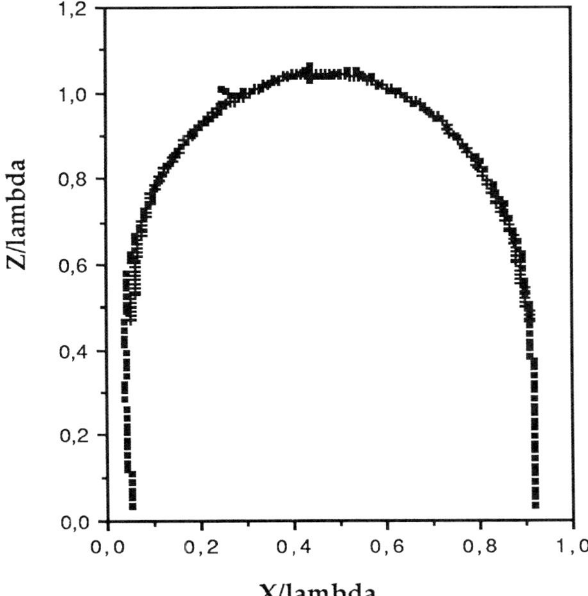

Fig. 3. Superposition of two cells of different widths taken from different experiments carried out at the same relative velocity ε, and reduced to unit width. Squares : G = 100 K/ cm, V = 49 μm/ s, ε = 5.5, λ/ρ = 3.6; crosses : G = 46 K/ cm, V = 21 μm/ s, λ = 36 μm, ε = 5.5, λ/ρ = 3.6.

WAVELENGTH DISPERSION

In the regime of low velocities - the cellular regime which corresponds to $\varepsilon \lesssim 8$ - cell widths are little dispersed : $d\lambda/\bar{\lambda}$ (where $\bar{\lambda}$ is the mean wavelength) is of the order of 10 % (fig.4). The experimental sources of fluctuations in cell widths are of the same order. These are due essentially to irregularities in pulling speed and to side and grain boundery effects. Because of this experimental "noise" we cannot conclude if there is, strictly speaking wavelength selection; but in any case the band of cell widths for a given velocity is very narrow compared

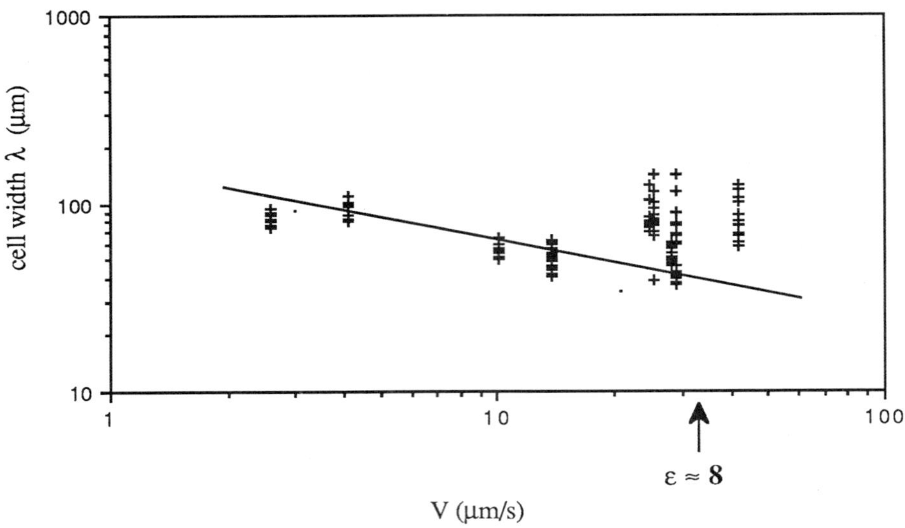

Fig. 4. Cell width λ versus pulling velocity V for a temperature gradient G $\approx$ 50 K/ cm, V_C = 2.7 μm/ s. The straight line represents previous results[3] concerning the average wavelength in the low velocity regime.

to the width of the Mullins and Sekerka neutral curve calculated by a linear analysis[1] or even to the Eckaus limit (calculated for a normal bifurcation[5]).

These studies have been carried out in a stationary state, when the front does not move anymore. But when we increase the pulling speed, the system goes through transients of a diffusive type before reaching this steady state. The mechanism of cell width adjustment during these transients are of three types in the cellular regime. One can observe :
- Disappearance of cells by pinching off that leads to a local increase in the wavelength.
- Appearance of cells by tip-splitting : two new cells are created and locally the wavelength decreases.
- A mechanism of a phase diffusion type[6] : when the interface is locally dilated after a death of a cell for example, one observes a progressive relaxation of cell widths to a periodic basic structure that takes place during times typically of the order of one minute.

For high velocities - $\varepsilon \gtrsim 8$ - we observe a spectacular increase in the cell width

dispersion (fig. 4). One can notice that the points are spread above the $V^{-0.4}$ curve : anomalously wide cells are appearing. On a given front cell widths can vary by a factor of 3 or 4. This announces the appearance of dendrites : that is why we call this velocity regime the predendritic regime. It therefore becomes meaningless to compute an average value of cell widths; one loses too much information on the physical mechanism behind the appearence of side-branches.

When we carry out experiments for different values of control parameters, we find that the increase in the dispersion takes place at different velocities but at about the same relative velocity ε of the order of 8.

The increase in the dispersion for $\varepsilon \gtrsim 8$ is due to the blocking of wavelength adjustement mechanisms other than pinching off of cells. When a cell disappears, its two neighbours become wider and do not adjust anymore.

The dispersion in cell widths is also associated with a dispersion in cell shapes. One can observe along a given front in a steady state coexistence of narrower "romanesque" cells, of cells with a "gothic" profile, wider and more pointed (Fig. 5). These cells are no longer homothetic. And finally the largest cells begin to develop side-branches.

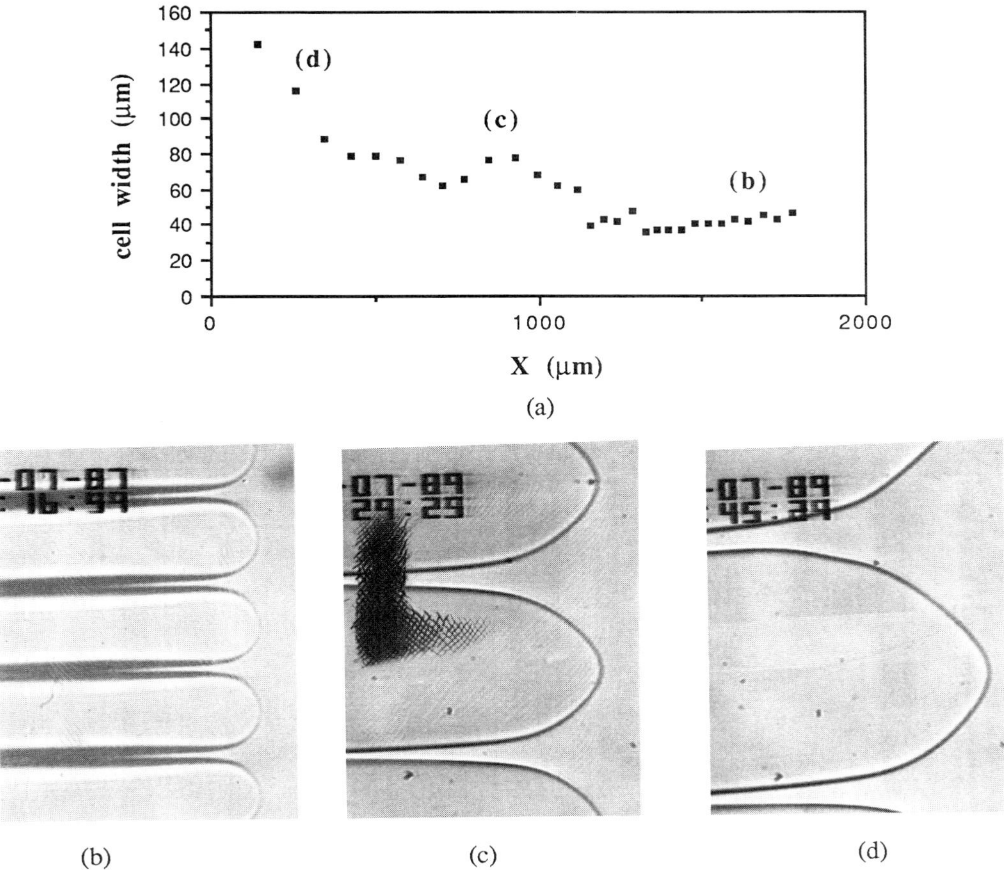

Fig. 5. Cell widths versus their position along the solidification front (a) and photographs of cells at the points indicated on the curve (b, c, d). Note the premices of sidebranches on (b).

FROM CELLS TO DENDRITES

The concept of a well defined "cell to dendrite transition" frequently used in the litterature is not appropriate here : we do not observe a global transition from a cellular front to a dendritic front. The appearance of side-branches is a local phenomenon:
- The first side-branches always appear on one or two adjacent cells.
- On a given front only the widest cells begin to emit side-branches.

To measure the diffusive interactions between cells, one defines the Peclet number $Pe = \lambda/l_d$ where $l_d = D/V$ is the diffusion length. When Pe is smaller than 1, the diffusion length is much greater than λ and the cells are not independant. On the contrary, when Pe is of the order of 1, one expects the dendrites to be (more or less) independant. We have calculated the Peclet number for the narrowest dendrites and for the widest cells along a given front for different experiments. These results are reported Fig.6 as a function of the pulling speed. These values of Peclet number increase with V and the minimum Peclet number is found equal to 2-3, therefore greater than 1. This experimental observation is confirmed by recent numerical simulations computed by Classen and al[7], introducing kinetics and kinetic anisotropy into the equations of solidification. They observe the appearance of side-branches for minimum Peclet

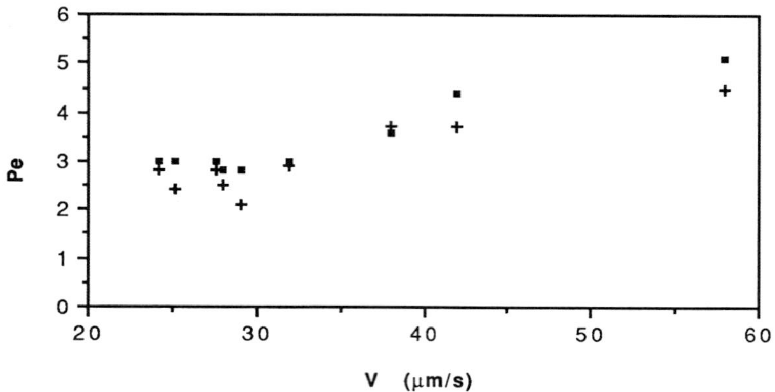

Fig. 6. Peclet numbers for the widest cells (squares) and for the narrowest dendrites (crosses) reported versus pulling speed.

number in the order of 3-4. This tends to suggest that kinetics should be taken into account in our experiments of directional solidification with CBr4-Br2.

CONCLUSION

We find two different velocities regimes in CBr4-Br2. The cellular regime, observed up to relative velocities ε of approximately 8 produces rounded cells, of well-defined widths, the shape of which only depends on ε and not on V or G taken separately. When ε increases, the dispersion in cell widths increases enormously and is associated with a dispersion in cell shapes. Some of the cells become more and more pointed and consequently tip-splitting no longer takes place. Abnormally wide cells appear and when their width exceeds 2 or 3 times the diffusion length they develop side-branches. The appearance of dendrites is therefore a local phenomenon in CBr4-Br2 and not a global transition.

BIBLIOGRAPHY

1) W.W.Mullins and R.F.Sekerka, J. Appl. Phys., 35, 444 (1964).
2) K.A.Jackson and Hunt, Trans. Met. Soc. AIME , 236, 1929 (1966).
3) S. de Cheveigné, C. Guthmann and M.M. Lebrun, J. de Phys., 47, 2095 (1986).
4) H. Müller-Krumbhaar, private communication.
5) K. Brattkus and C. Misbah, Phys. Rev. Lett., 64, 1935 (1990).
6) P. Kurowski, Thesis, Université Paris VII (1990).
7) A. Classen, C. Misbah, H. Müller-Krumbhaar, Y. Saïto, preprint (1990).

DEEP CELLS IN

DIRECTIONAL SOLIDIFICATION

Alain Karma

Pierre Pelcé

Dept. Physics
Northeastern University
Boston, MA 02115, U.S.A.

Laboratoire de Recherche en Combustion
Université de Provence, St Jerome
13397 Marseille Cedex 13, France

INTRODUCTION

The deep cells regime in directional solidification of binary mixtures[1] is currently under great investigation. The basic problems posed by the corresponding pattern are the following:

For given control parameters, i.e. the pulling velocity V, the temperature gradient G and the concentration of the solute in the liquid far away from the crystal c_∞, what are:

-The steady state shapes of the cells,

-The wavelength Λ of the pattern,

-The instabilities of the pattern, as for instance the cell to dendrite transition.

Significant progress towards an understanding of these questions was performed recently with the study of the small Péclet number limit[2-6]. The Péclet number of the cells is defined as the ratio between the wavelength of the cell array Λ and the diffusive length of the concentration field $l = 2D / V$, i.e. $Pe \approx \Lambda V / D$, where D is the coefficient of diffusion of the solute. A basic difficulty of this approach is that this number is not directly defined from the initial control parameters because of the initial indetermination of the wavelength and thus it is not certain that the patterns obtained in this limit will be observed in experiments. However, an encouraging experimental observation is that steady and stable cell arrays have been observed at $Pe \approx .1$.

The simplification which occurs in this limit is that the concentration field around the cells varies on two different scales[2]. Close to the cells on the scale Λ, the concentration field satisfies the Laplace equation and conformal mapping methods of two dimensional media can be used. Far from the cells the concentration decreases in the direction normal to the envelope of the cell tips on the scale of the diffusive length. This separation of scales allows us to consider the envelope of the cell tips as an effective liquid-solid interface on which boundary conditions for the concentration field are applied.The periodic cell array appears as an effective planar interface whose position in the temperature gradient can be determined. Instabilities of the deep cell array can be studied in a similar way as the Mullins-Sekerka instability of the planar interface.

Thus, the program which is a summary of the studies[4,5] is the following : One determines first the boundary conditions that must be applied on the effective interface.Then, one determines the steady states and studies their stability. Then, one discusses the results found in relation to the experimental observations and presents a list of problems that remain to be solved in order to arrive to a complete understanding of the deep cell regime.

Growth and Form, Edited by M. Ben Amar *et al.*
Plenum Press, New York, 1991

THE EFFECTIVE LIQUID-SOLID INTERFACE

In directional solidification of a binary liquid mixture , the motion of the solidification front is determined by the diffusion of solute in the melt together with a set of boundary conditions on the interface.The solute concentration c satisfies the usual diffusion equation

$$\frac{\partial c}{\partial t} = D \Delta c \qquad (1)$$

in the liquid and diffusion in the solid is neglected. Temperature and solute concentration at the interface are related to the local interfacial curvature by the Gibbs-Thomson relation

$$T = T_0 + mc - \frac{\sigma T_0}{QR} \qquad (2)$$

Here, D is the coefficient of solute diffusivity in the liquid , m the slope of the liquidus line in the binary phase diagram, T_0 the crystallisation temperature of the pure melt, R the local radius of curvature, Q the latent heat released per unit volume and σ the liquid-solid surface tension.In the slow growth regime the latent heat release at the interface can be neglected. Consequently, the temperature in the sample is simply $T = T_p + Gz$, where $T_p = T_0 + m\,c_\infty$ / K is the temperature of the planar interface. G is the temperature gradient imposed on the sample, K the partition coefficient and c_∞ the bulk mixture composition.

In addition , solute conservation at the interface requires

$$-D \vec{\nabla} c . \vec{n} = c\,(1 - K)\,\vec{v}.\vec{n} \qquad (3)$$

Consider a periodic cellular interface and a contour drawn along the cell shape as it is shown on Fig.1, delimiting an area of typical size Λ. As the Peclet number is small , the diffusive term in eqn. (1) dominates sothat $\Delta c = 0$ in this region. The cell shape obtained after application of the boundary conditions (2) and (3) is a Saffman-Taylor shape, i.e. an elongated shape of relative size λ. Far from the tip, on a distance much larger than Λ, the Laplace equation breaks down and the shape satisfies the Scheil equation, i.e. a balance between longitudinal advection and transverse diffusion. The corresponding shape is a long link almost parallel to the direction of propagation which prolongates the sides of the finger up to the walls.

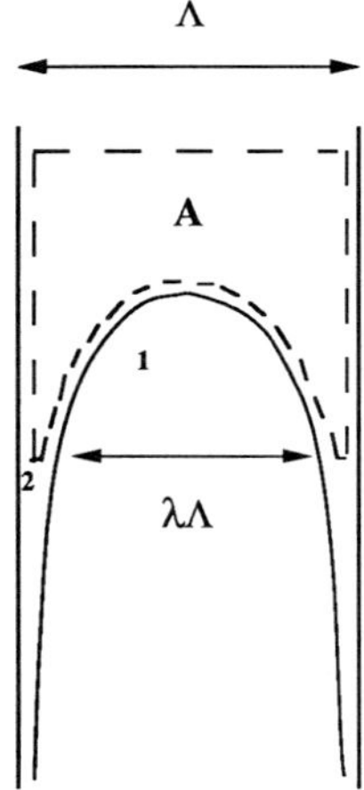

Fig.1. Contour of integration

After integration of the Laplace equation on the area A , one obtains the concentration flux released ahead of the cells:

$$-D\frac{dc}{dz} = \lambda V c_0 (1 - K) - (1 - \lambda)\frac{DG}{m} \qquad (4)$$

Here, the first term of the r.h.s. of eqn. (4) corresponds to the flux released by the portion (1) of the contour (Fig.1) and the second term to the portion (2) (in this region the concentration gradient is almost uniform since the shape is nearly parallel to the wall). λ is a decreasing function of the effective Saffman-Taylor control parameter C defined as

$$C = \frac{\Lambda^2 V}{D d_0}(1 - (1 - K)\frac{z_0}{l_T} - \frac{1}{2v}) \qquad (5)$$

where $l_T = | m | c_\infty (1 - K) / KG$ is the thermal length, $v = Vl_T / 2D$ the dimensionless control parameter , $d_0 = \sigma T_0 / | m | Q$ the capillary length , K the partition coefficient and z_0 the position of the cell tips. In the small Peclet number limit, the relative variation of the concentration around the tip is small compared to the supersaturation. Furthermore cells are assumed to have a large wavelength compared to the Mullins-Sekerka length sothat curvature effects on the value of the tip concentration can be neglected. It follows that the concentration of the effective interface can be approximated to

$$c_0 = \frac{c_\infty}{K} + z_0\frac{G}{m} \qquad (6)$$

Consider now a wrinkled periodic cellular interface and that the number of cells is conserved (no tip-splitting nor cell death). As the total length of the effective interface increases cells need to increase or decrease their size depending on the local geometric quantities of the effective interface with a net increase of their mean size. Consider first the case with zero temperature gradient, i.e. when growth proceeds isotropically. The simplest configuration is the case of a circular effective germ of radius R (Fig.2).

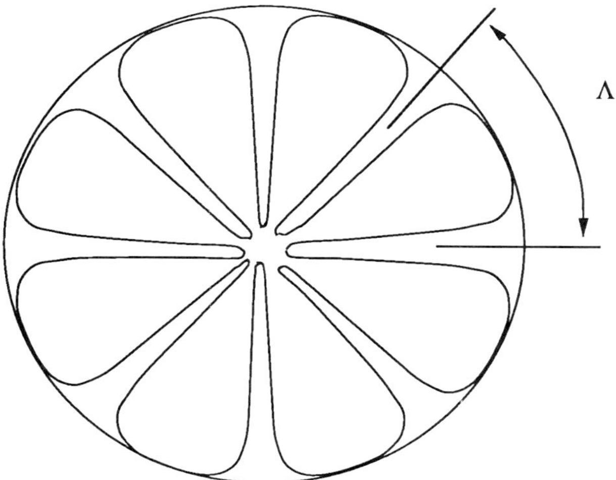

Fig.2. Sketch of a circular germ with an isotropic microstructure

For reasons of symmetry, cells must grow perpendicularly to the effective interface so that $(d\Lambda / dt) / \Lambda = (dR / dt) / R$. In order to generalize this relation in a covariant way, one obtains:

$$\frac{1}{\Lambda} \frac{d\Lambda}{dt} = \vec{v}.\vec{n} \, \frac{1}{R} \qquad (7)$$

Here, the cell tips are assumed to move in the normal direction of the effective interface with the normal velocity v.n. When the cells grow in a temperature gradient , there is a privileged direction in the system and other scalars quantities must be used to complete relation (7). Furthermore cell tips may move with a tangential velocity, which in this case remains to be determined. In the following, we assume that the temperature gradient is sufficiently weak that relation (7) remains valid. Similar relation was already used in the context of eutectic growth[6], where the temperature gradient effects can usually be neglected. Thus, eqn.(1) used with eqns. (4) to (7) describe the motion of the effective liquid-solid interface when the Peclet number of the cells is small.

THE STEADY ARRAY OF CELLS

To obtain the steady-state solutions of this set of equations we first rewrite eqn.(1) in a frame moving with the interface at velocity V :

$$D \frac{d^2 c}{dz^2} + V \frac{dc}{dz} = 0 \qquad (8)$$

The steady-state concentration profile which solves eqn. (8) subject to the boundary condition at the interface eqns (6) is then simply :

$$\frac{(c - c_\infty)}{c_\infty \frac{(1 - K)}{K}} = (1 - \frac{z_0}{l_T}) \exp (- \frac{V}{D} (z - z_0)) \qquad (9)$$

Substituting eqn. (9) into the flux conservation relation eqn. (4) we obtain:

$$\Delta = \frac{K \lambda + \dfrac{(1 - \lambda)}{2 v}}{1 - (1 - K) \lambda} \qquad (10)$$

where we have defined the dimensionless tip undercooling :

$$\Delta = 1 - \frac{z_0}{l_T} \qquad (11)$$

The steady-state solutions of our model are then completely determined by eqn. (10) together with the relation between the finger width λ and the cell spacing Λ with

$$C = \frac{K (1 - \dfrac{1}{2 v})}{1 - (1 - K) \lambda} \frac{\Lambda^2 U}{d_0 D} \qquad (12)$$

obtained by combining eqn. (5) and (10).

For a material with an isotropic surface tension the function $\lambda = f (C)$ is a decreasing function of C which approaches asymptotically the value 1/2 for large values of C and has been calculated numerically[7]. We use here an interpolation formula for f (C) :

$$\lambda = f(C) = \frac{(C - \tilde{C}) + \tilde{A}}{2(C - \tilde{C}) + \tilde{A}} \qquad (13)$$

with $C = 2.87$ and $A = 50.8$ which reproduces within a few percents their numerical data for values of λ ranging between 1.0 and 0.6. The numerical value of C which we use was determined by Dombre and Hakim[3] who analysed theSaffman-Taylor problem in the limit λ close to unity and corresponds to the minimum allowed value of C for steady-state shapes. In Fig.3 we show a plot of the dimensionless tip undercooling Δ (C) obtained by combining eqns. (10) and (12) for three values of v . Since at fixed velocity V the cell spacing Λ is uniquely determined by the control parameter C via eqn. (12) , Δ (C) represents a one parameter family of steady-state shapes with varying cell spacing and tip undercooling. Note that the tip undercooling varies very little with Λ for sufficiently large values of Λ (or equivalently sufficiently large values of C) and approaches the limiting value :

$$\tilde{\Delta} = \frac{K\left(1 + \dfrac{1}{2v}\right)}{1 + K} \qquad (14)$$

as λ goes to 1/2. As v increases Δ decreases. Both the relative flatness of the curve Δ (C) at sufficiently large cell spacings and the decrease of Δ with increasing velocity have been observed previously by Hunt and McCartney[8] who determined numerically the shape of deep cells for small values of the Peclet number.

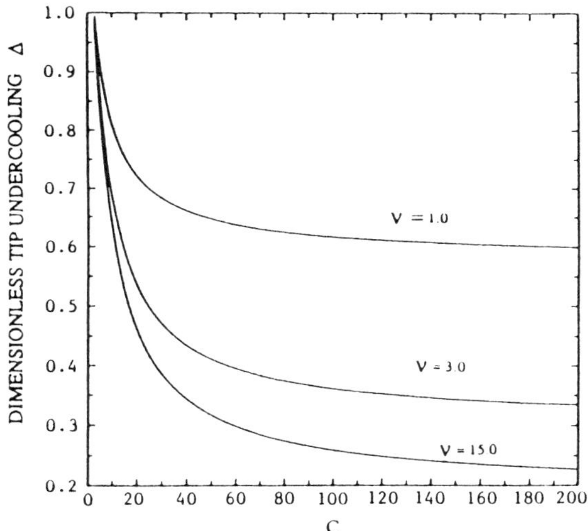

Fig.3. Dimensionless undercooling Δ (C) for $K = .2$ and three values of v

On Fig.4 we draw two cellular arrays corresponding to small and large wavelength. The first one has a λ close to 1 and the positions of the cell tips are close to the planar interface position. The second one has a λ close to 1/2 and the positions of the cell tips are largely displaced towards the liquid.

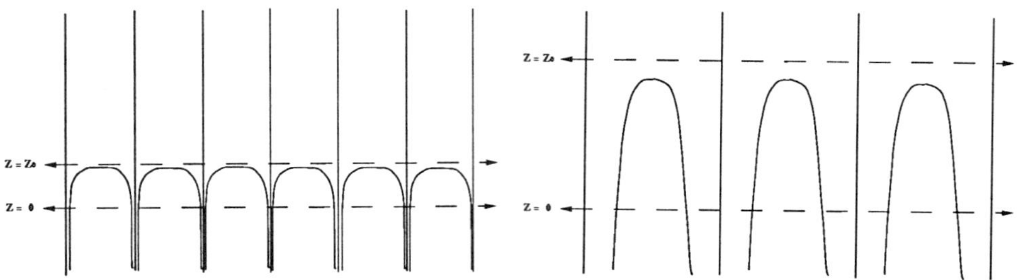

Fig.4. Cellular arrays corresponding to small (left) and large (right) wavelength.

LINEAR STABILITY ANALYSIS

Eigenvalue equation

To derive the eigenvalue equation for the linear stability problem, we linearize the equations (1) and (4-7) around the steady state solutions defined by eqn. (9) to (13) and restrict our attention to perturbations of wave number k and growth rate ω of the form

$$\delta u = A \exp (\omega t + iky + qz) \ , \ Re\,(q) < 0 \qquad (15)$$

$$\delta z_0 = B \exp(\omega t + iky) \quad , \quad \delta \Lambda = E \exp(\omega t + iky) \qquad (16)$$

After some calculations[4,5], one obtains the following eigenvalue equation for the dimensionless growth rate $\Omega = \omega\, 2D / V^2$:

$$(\Delta - \frac{1}{2v}) \sqrt{ 1 + 2\Omega + k^2 l^2 } = a\,\Omega + b + c\,\frac{k^2 l^2}{\Omega} \qquad (17)$$

where, a, b and c are two quantities depending on the steady state shapes.
Consider first the case of the planar interface, formally obtained when $\lambda = 1$. In this limit one obtains the eigenvalue equation

$$(1 - \frac{1}{2v}) \sqrt{ 1 + 2\Omega + k^2 l^2 } = \Omega + 1 - \frac{1}{2v} + \frac{K}{v} \qquad (18)$$

whose solution is the Mullins-Sekerka growth rate in which surface tension effects have been neglected. At small wave number, this mode is damped with a non zero imaginary part due to the restabilizing effect of the temperature gradient. At large wave number it is unstable when v is larger than 1/2 due to the destabilizing effect of the diffusion field. The growth rate is then proportional to kl . As can be seen from eqns (17) and (18) a similarity exists between the eigenvalue equations corresponding to planar and effective interfaces.

<u>Modes of the effective liquid-solid interface</u>

<u>The oscillatory mode</u>

Consider first the k=0 mode , i.e. the effective interface remains planar. It is found that for large enough velocity the real part of Ω becomes positive with a finite imaginary part, this corresponding to an oscillatory instability of the cell tips (Fig.5). Above the threshold of this instability, the width and the vertical tip position of all cells oscillate in phase coherently.

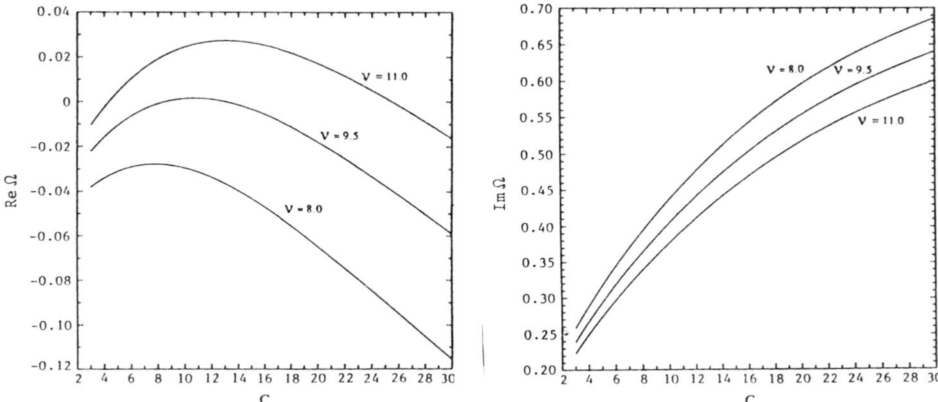

Fig.5. Real (left) and Imaginary (right) part of Ω vs C for K = .2 and three values of v.

The physical origin of the oscillatory instability can be understood more clearly by examining two important limiting cases of the eigenvalue equation (17).

The planar interface.

This limit is formally obtained when $\lambda = 1$.When $1/v$ is small, the growth rate is simply

$$\Omega = -\frac{1}{v}(1 + 2K)\pm i\sqrt{\frac{2K}{v}} \qquad (19)$$

Thus, in this limit, the interface motion undergoes damped oscillations.The physical origin of the damping can be explained as follows: First, consider a perturbation where the interface is sligthly displaced towards the high temperature region.The effective undercooling sligthly decreases and the released concentration flux is reduced.As a result, the interface velocity decreases and the interface comes back to the low temperature region.

Crystal growth in a capillary tube

When there is no temperature gradient ($1/v = 0$),the system is equivalent to the growth of a crystal in a capillary tube.The crystal is assimilated to one of the cells of the array and the distance between the walls to the primary spacing Λ .In this limit, the eigenvalue equation reduces to a form derived previously by Pelcé[9] . It has a nonzero solution equal to

$$\Omega = - \frac{C \, f(C) \frac{df(C)}{dC}}{\left(\frac{d}{dC}(C \, f(C))\right)^2} \qquad (20)$$

Since f(C) is a decreasing function of C, the growth rate is real and positive and leads to an instability of the stationary cell.To understand this effect consider a perturbation which increases sligthly the relative finger width.More solute will be rejected and the effective undercooling will be reduced.As a result, the crystal will slow down and become even thicker, since a slower motion corresponds to a larger relative width (Fig.6).

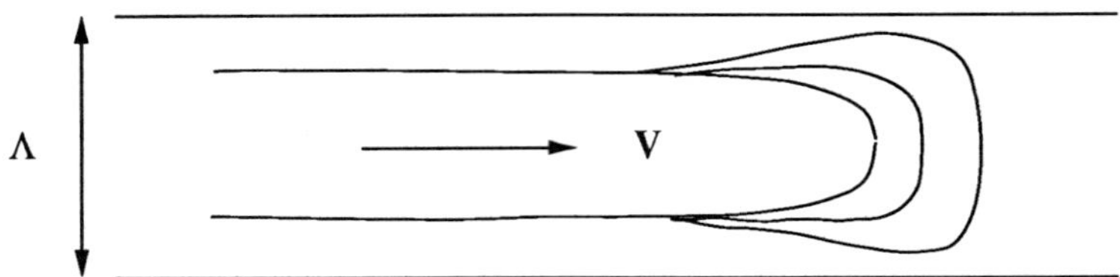

Fig.6. Illustration of the instability of the growth in a capillary tube at small Péclet number.

When both effects are considered, as it is the case in the complete growth rate equation one expects that for some values of the control parameter, an oscillatory instability will appear.Consider a sufficiently large value of $1/v$, i.e a strong temperature gradient.The first effect (planar interface) will dominate and the interface will be damped in an oscillatory way. If the temperature gradient is decreased, the second effect (crystal in the capillary tube) will appear and the real part of the growth rate will become positive, still with a finite imaginary part. The competition between these two effects will therefore lead to an oscillatory instability.

<u>The long wavelength mode</u>

The second mode is analogous to the Eckhaus mode found for instance in small amplitude periodic convective structures. At small wave number it is of the diffusive form:

$$\Omega = - 2v \, C \, \frac{df}{dC} \, \frac{K\left(1 - \frac{1}{2v}\right)}{(1 - (1 - K)\lambda)(1 - (1 - K)(\lambda + C\frac{df}{dC}))} (kl)^2 \qquad (21)$$

The coefficient of proportionality -Cdf/dC is always positive in the small Peclet number limit considered here. The mode is therefore always unstable in this limit. To understand the nature of the instability consider a bulge in the effective interface towards the liquid.The local wavelength is larger at the tip of the buldge where it is stretched than in the other parts of the interface. Since larger spacings correspond to smaller tip undercooling the cells at the tip of the bulge will move up towards warmer liquid region and consequently the amplitude of the bulge will increase. This instability eventually leads to the elimination of cells and an increase in the average wavelength of the array.

In the small kl limit the time scale on which long-wavelength perturbations of the array grow (or decay diffusively when $df/dC > 0$) is proportional to $(kl)^2$ and therefore much longer than the diffusive time scale on which the composition field readjusts itself spatially over the scale of the diffusion length l. Consequently, on the time scale where long-wavelength perturbations of the array vary significantly the composition field $c(z)$ can be assumed to relax instantaneously to its steady-state profile given by eqn.9 with z_0 replaced by $z_0(y)$. The dynamics of long-wavelength perturbations on a scale much larger than l is then governed solely by the geometrical streching of the local cell spacing (eqn.7) and the steady-state relation between the local height of the interface (i.e.the local tip undercooling) and the cell spacing: $z_0(\Lambda)=l_T[1 - \Delta(\Lambda)]$. Substituting this relation for $z_0(\Lambda)$ in eqn.7 we obtain the growth rate of the long wavelength mode as:

$$\Omega = -\nu\,\Lambda\,\frac{d\Delta}{d\Lambda_0}\,(kl)^2 \qquad\qquad (22)$$

The quantity $d\Delta/d\Lambda_0$ in eqn.22 can be evaluated directly by using the steady-state eqn.10 together with eqn.12. Then, one recovers at once eqn.22 as expected.

One immediate consequence of eqn.22 is that the point of minimum undercooling for which $d\Delta/d\Lambda_0 = 0$ at $\Lambda_0 = \Lambda_m$ corresponds to the point of marginal stability of the array. A cellular array with $\Lambda_0 < \Lambda_m$ ($\Lambda_0 > \Lambda_m$) is unstable (stable) with respect to long-wavelength perturbations of the array. This same correspondance between marginal stability and minimum undercooling was found previously by Langer for the case of thin lamellar eutectic growth[6]. As in the small Péclet number limit there is no minimum undercooling, even if at large C, $d\Delta/d\Lambda_0$ is very small, the conclusion is that the deep cells arrays are always unstable.

POSSIBLE ROUTES TO UNDERSTAND THE DEEP CELL ARRAYS

Thus, the conclusion of the studies performed in the small Péclet number limit is for the moment negative: The steady deep cell arrays determined in this limit are always unstable. So what can we do now?

First, one believes that there are effectively stable steady-state arrays in the small Péclet number regime. In this case there is something wrong in the previous analysis. Relation (7), which determines how the local wavelength is changed with the curvature of the effective interface , is valid for small temperature gradient and with the assumption that the cell tips move in the normal direction to the effective interface. It is necessary to incorporate the effects of the temperature gradient in eqn. (7) , determine the possible tangential velocity of the cell tips, and see how the criterium marginal stability-minimum undercooling is modified. An approach in this direction was adopted recently by Caroli et al.[10] in the eutectic problem. They found that in the large gradient limit , the eutectic array is completely stable, showing indirectly that relation (7) needs to be modified when temperature gradient effects are important.

Secondly , it is possible that $Pe \approx .1$ is not a so small Péclet number and that effects of Péclet number are important for the stability of the deep cells array. In that case, the previous analysis may help to understand the transitory dynamics which leads to the final steady state: Above threshold of instability of the planar interface,   cellular arrays of small wavelength and thus small Péclet number (close to the Mullins-Sekerka wavelength) are first formed. They are unstable because of the longwavelength mode and cells are eliminated. Thus the Péclet number of the array increases and the steady state can become stable only for a Péclet number of order unity.

In order to understand these steady states, the first problem is to understand completely the growth of a crystal in a capillary at zero temperature gradient (Brener et al[11], Molho et al.[12]) and study the transition between the Saffman-Taylor shapes at small Péclet number and the dendritic branch at Péclet number larger than 1. When the temperature gradient is taken into account, VanSaarloos and Weeks[13] exhibit shapes at $Pe \approx 1$ and partition coefficient close to zero in disagreement with the experiments[14]. They suggest that this discrepancy is largely due to the presence of two branches of solutions in the problem of the growth in a capillary.

Another important question which has not yet definitive answer is the problem of the cell-to dendrite transition. Is the oscillatory mode found in this study responsible for this transition? Here too there are two possibilities:

-Steady states cells are stable at small Péclet number and cell to dendrite transition can occur via this oscillatory mode, i.e. a mode where the sidebranches are coherent from one cell to another. In this case the discrepancy with the numerical results of Kessler and Levine[15] may be relied to the disparity of the models considered.

-Steady states are stable only at Peclet number larger than one and cell-to dendrite transition may be induced by noise[14] as it the case for a large class of curved interfaces[16].

Acknowledgments: This research is supported by a collaborative research grant NATO and a contract CNES-PIRMAT.

REFERENCES

1. M.A.ESHELMAN,V.SEETHARAMAN and R.TRIVEDI,(1988) Acta Metall. 36, 1165.
2. P.PELCE and A.PUMIR,(1985) J.Cryst.Growth 73, 337.
3. T.DOMBRE and V.HAKIM,(1987) Phys.Rev.A36, 2811.
4. A.KARMA and P.PELCE,(1989) Phys.Rev.A 39, 4162 and Europhys.Lett. 9, 713.
5. A.KARMA and P.PELCE,(1990) Phys.Rev.A 41, 6741.
6. J.S.LANGER,(1980) Phys.Rev.Lett. 44, 1023.
7. J.W. McLEAN and P.G. SAFFMAN,(1981) J.Fluid.Mech. 102, 445.
8. J.D. HUNT and D.G. McCARTNEY,(1987) Acta Metall. 35, 89.
9. P.PELCE,(1988) Europhys.Lett. 7, 453.
10. B.CAROLI,C.CAROLI and B.ROULET,(1990) Lamellar Eutectic Growth at Large Thermal Gradient, Linear stability, Preprint.
11. E.A.BRENER, M.GEILIKMAN and D.E.TEMKIN,(1988) Sov.Phys.JETP 67, 565.
12. P.MOLHO, A.J.SIMON and A.LIBCHABER,(1990) Phys.Rev.A42, 904.
13. J.D.WEEKS and W.VAN SAARLOOS,(1989) Phys.Rev.A39, 2772.
14. P.KUROWSKI, C.GUTHMANN and S.DE CHEVEIGNE,(1990) On the so-called Cellular-Dendritic transition: Shapes and Wavelength Selection in Directional Solidification, Preprint
15. D.A.KESSLER and H.LEVINE,(1990) Phys.Rev.A41, 3197.
16. P.PELCE,(1988) Dynamics of curved fronts, Academic Press.

CELLULAR PROFILES IN DIRECTIONAL SOLIDIFICATION: IS THE SAFFMAN-TAYLOR BRANCH OF SOLUTIONS THE PHYSICALLY RELEVANT ONE?

Wim van Saarloos* and John D. Weeks**

AT&T Bell Laboratories
Murray Hill, New Jersey 07974
USA

ABSTRACT

We summarize the main results and implications of our work on the calculation of cellular shapes in directional solidification using the asymptotic matching method introduced by Dombre and Hakim. For cells with narrow grooves, the finite Péclet number corrections to the cellular profiles that reduce to the Saffman-Taylor solutions for Péclet number $p \to 0$, turn out to be small. We argue that there are several discrepancies between the behavior of these Saffman-Taylor like cells and those observed in experiments as well as numerical studies that suggest that this branch of solutions is not always the physically relevant one for directional solidification.

INTRODUCTION

In the last few years, most analytic approaches[1-5] aimed at understanding the problem of directional solidification (DS) have been based on the observation (originally due to Pelcé and Pumir[1]) that the directional solidification equations for the two-dimensional one-sided model can be reduced to those for the Saffman-Taylor (ST) problem in the small Péclet number limit $p \to 0$. With this mapping one can therefore easily obtain explicit results[3,4] for a continuous branch of cellular shapes with deep grooves.

For cells with narrow grooves, Dombre and Hakim[2] showed that one can exploit the smallness of the groove width for the calculation of cell parameters using asymptotic matching methods. For the limit $p \to 0$ they studied, their results reduce to those derived directly from the mapping onto the ST problem. Motivated by the observation that experimental cells often have rather narrow grooves, we have extended[6,7] their treatment of cells with narrow grooves to the case of small partition coefficient k ($k \leq 0.15$, say) but p of order unity. This is the range relevant for many experiments. Our results reduce smoothly to those of Dombre and Hakim[2] for $p \to 0$, where the

---

* Present address: Instituut-Lorentz, University of Leiden, P.O. Box 9506, 2300 RA Leiden, The Netherlands

** Present Address: Institute for Physical Science and Technology, University of Maryland, College Park, MD 20742, U.S.A.

Growth and Form, Edited by M. Ben Amar *et al.*
Plenum Press, New York, 1991

restriction to small k is then not required, and we will therefor refer to this particular branch of solutions we calculate as the Saffman-Taylor-like branch.

These results for cells with narrow grooves are disappointing in that they do not compare well with typical experimental results. Moreover, they also show discrepancies with a number of numerical studies of the same two-dimensional one-sided model. They therefore raise the question whether this branch of solutions is the experimentally relevant one, and also whether this is in branch of solutions studied in most numerical approaches.[8-10]

The essential problem is easy to understand qualitatively for $p \to 0$. It is well-known that the appropriate dimensionless surface tension parameter σ of experimental cells usually turns out to be rather small, of the order of 10^{-2}, and that cells typically have rather *narrow* grooves. This behavior is incompatible with that of the ST branch of solutions, for which small surface tension cells correspond to *wide* groove solutions: in the absence of surface tension anisotropy these solutions have grooves whose width not too far from the tip is roughly half of the pattern wavelength λ (See, e.g., Refs 3 and 4).

This discrepancy is just one of the ways in which ST type of cellular solutions appear to behave significantly different from experimental cells as well as from most cellular solutions studied numerically for the same model. We have recently given an extensive discussion of these issues,[7] and therefore confine ourselves here to highlighting some of our main points, in particular those that touch on other topics addressed at the summerschool.

SUMMARY OF RESULTS AND IMPLICATIONS

In this section, we briefly summarize the main results and conclusions from our analysis of cellular profiles with narrow grooves using matched asymptotic expansions.[7] We then illustrate these points in some more detail in the subsequent sections. Throughout this paper, we will denote the relative groove width of cells near their tip by ε, since this is the small parameter in the matching method. Note that this quantity is usually referred to as $1-\lambda$ in the ST literature. However, we prefer to use λ for the wavelength.

1. For systems with small k, we can calculate the properties of a particular continuous branch of steady state solutions describing cellular profiles with deep narrow grooves. The Péclet number dependence of these solutions is weak, and for $p \to 0$ they reduce to the Saffman-Taylor-like solutions found earlier by Dombre and Hakim[2]. The variation in wavelength of this branch is parametrized in the matching theory by the relative groove width ε near the tip, or, more convenient to experiment, by a dimensionless parameter ζ_t giving the tip position (see the next section and Eq. (3) below).

2. A fundamental assumption self-consistently satisfied by these solutions is the existence of a single matching region controlled by the small parameter ε where deep narrow grooves can be matched to a particular class of finite amplitude cellular solutions describing the fingertip region. These finite amplitude solutions have wavelength lying *outside* the planar instability band and evolve from the (infinitesimal amplitude) neutrally stable modes forming the short wavelength side of the planar instability band (the left hand solid curve in Fig. 1; see below). These solutions have a regular expansion in powers of k. The fingerlike solutions with narrow grooves (small ε) resulting from the matching also have wavelengths lying outside the planar instability band and have tips that do not move up nearly as much as one expects[11] on the basis of a simple extension of

the classical constitutional supercooling argument (the Local Constitutional Supercooling criterion or LCS – see the next section).

3. Most numerical studies, using the same two-dimensional model we consider here, find narrow-grooved fingerlike patterns, but with wavelength lying *inside* the planar instability band. These solutions are found well above threshold and evolve from small amplitude cells with wavelengths lying inside the planar instability band, presumably related for small k to the cells studied by Sivashinsky[12] and others.[13,14] The latter bifurcate off from the *center* of the planar instability band (indicated by the square in Fig. 1), and do not have a regular k expansion. The contrast with the matching solutions discussed in 1) is evident.

4. Most experimental patterns also have narrow grooves, but with wavelength lying inside the planar instability band. In the few cases where experimental data on the tip position is available, small values of the parameter ζ_t are found, in qualitative agreement with our approximate stability criterion[11] (LCS). This implies that the tips of experimentally selected patterns have moved up substantially in the temperature gradient relative to the planar position.

5. In contrast, the matching solutions for small ε have tip positions that remain relatively close to the planar position. More detailed experimental measurements of the tip position would be extremely useful to determine the accuracy of these stability ideas.

6. The analysis of Brener et al.[15] for crystal growth in a channel, as well as arguments by Ungar and Brown[8], again suggest the existence of a branch of steady state solutions in DS other than the ST branch. The analytic mapping of the $p \to 0$ DS equations to the ST problem breaks down near the LCS line $\zeta_t \approx 0$. Thus we argue that the matching solutions lie on a different, and physically irrelevant, branch of steady state solutions for DS.

COMPARISON OF PREDICTIONS FOR GROOVE WIDTH, WAVELENGTH AND TIP POSITION

In this section, we illustrate the points 1, 2, 4 and 5 made above in more detail.

Let us first make the comparison of our results with experimental observations more quantitative with the aid of the data of de Cheveigné et al.[16] shown in Fig. 1. In this figure, the wavelength of perturbations about the planar interface that are neutrally stable (i.e. neither grow nor decay) is indicated by a solid line. Curves for two different temperature gradients are shown. The minimum threshold velocity V_c at which the planar interface first goes unstable is denoted by the square symbol at the bottom of the curves. Perturbations about the planar interface with wavelengths in the wide band between these two solid lines are linearly unstable; we therefore refer to this region as the planar instability band. The selected cells for a given pulling velocity in the experiments of de Cheveigné et al.[16] have wavelengths that lie within this planar instability band in the narrow dashed region whose wavelength is a factor $3 \sim 5$ larger than the neutral stability value. This is what is typically observed.

A quantitative measure of the importance of surface tension effects is the parameter σ, defined as

$$\sigma \equiv \left(\frac{\lambda_s^0}{\pi\lambda} \right)^2 , \tag{1}$$

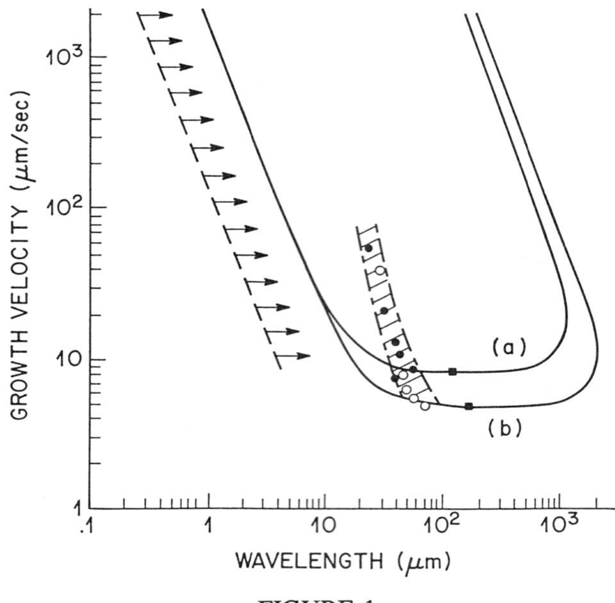

FIGURE 1

Plot of the growth velocity versus wavelength for the experiments of de Cheveigné et al.[16]. The solid line marks the neutral stability wavelength as a function of the growth velocity for two values of the thermal gradient G. Experiments for (a) $G = 120\,°C/cm$, full circles and (b) $G = 70\,°C/cm$, open circles lie in the narrow shaded band. The dashed line indicates the minimum deep cell wavelength calculated in this paper, and the arrows illustrate that the matched asymptotic expansion employed in this paper is an expansion towards larger wavelength.

where

$$\lambda_s^0 = 2\pi \left[\frac{\nu d_0 \ell_D}{\nu - 1} \right]^{1/2} \tag{2}$$

is the $k \to 0$ limit of the smallest neutral stability wavelength λ_s. Here d_0 is the usual capillary length, $\ell_D = D / V$ the diffusion length and $\nu = \ell_T / \ell_D$ the ratio of the thermal length ℓ_T and the diffusion length. (Throughout this paper, we will use the notation of Ref. 7) Since $k \approx 0.16$ for de Cheveigné's data[16] and since the k dependence of λ_s is rather weak for such small values of k, we may ignore the difference between λ_s and λ_s^0 and estimate $\lambda_s^0 / \lambda = 0.2$~$0.25$ for the data of Fig. 1. Thus $\sigma \approx 0.006$ in this experiment. On the other hand, as the arrows in Fig. 1 indicate, the narrow groove solutions that we can calculate perturbatively using matched asymptotic expansion methods have $\lambda_s^0 / \lambda = 2$~$3$, so that for these $\sigma = O(1)$.

A quantity that is explicitly predicted from the mapping onto the ST problem as well as from our matched asymptotic expansion approach for finite p, is the tip position z_t of the cells. We take the convention that the cells grow in the $+z$ direction and that the planar position is at $z = 0$. According to our calculations, the p-dependence of the ratio z_t / ℓ_T of cells on the ST branch is quite weak, and we will therefore approximate $z_t(p) / \ell_T$ by $z_t(p \to 0) / \ell_T$ as predicted by the ST analogy. The results are plotted in Fig. 2 as a function of v for several values of the relative groove with ε. As Fig. 2 illustrates, the larger the groove width ε, the more the cells are predicted to move up for fixed v. The data points in Fig. 2 refer to experimental measurements of Esaka and Kurz[17] on succinonitrile-acetone mixtures with $k = 0.1$. Clearly, the experimental cells have moved up in the temperature gradient more than any of the ST solutions predicts they should in the absence of crystalline anisotropy. Moreover, the solutions of the ST branch that come closest to the experimental data points are those corresponding to *wide* grooves ($\varepsilon \to 1/2$). Therefore, both Fig. 1 and Fig. 2 give evidence that the solutions on the ST branch have much wider grooves than one commonly observes.

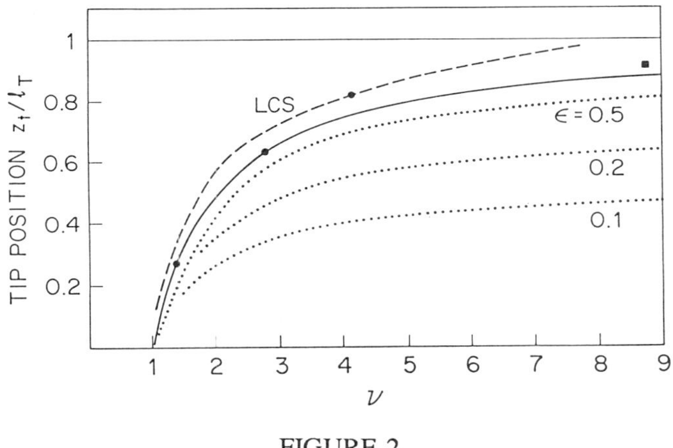

FIGURE 2

Plot of the tip position z_t/ℓ_T versus v. The dashed line corresponds to the LCS line given by $\zeta_t = 0$ or $z_t/\ell_T = (1 - 1/v)/(1 - k)$ for $k = 0.1$. The full line gives the $k \to 0$ limit of the LCS line, $z_t/\ell_T = 1 - 1/v$, which is a relation first derived by Brody and Flemings (See Ref. 7). The dotted lines are the expression for $z_t(p)/\ell_T$ as calculated in Ref. 3 and 4. The solid dot at $v = 4.2$ and the square dot at $v = 8.7$ represent two data points obtained and published by Esaka and Kurz,[17] while the data points for $v = 1.3$ and $v = 4.2$ refer to measurements by the same authors reported by Billia et al.[17]

The dashed line in Fig. 2 is labeled LCS. This stands for Local Constitutional Supercooling, an approximate stability and selection criterion for cells that we have advanced recently.[11] This criterion is a straightforward extension of the classical constitutional supercooling theory for the stability of a plane, whose basic idea is that in the steady state we can apply the concept of local thermodynamic equilibrium to the *melt just in front* of the moving planar interface. The interface is supposed to remain stable as long as the impurity concentration in the melt just in front lies in the stable single phase region as determined from the phase diagram. As is well-known, this idea predicts that a planar interface remains stable as long as $v \leq 1$. For small values of k, this criterion is quite accurate.

These constitutional supercooling arguments seem nearly as plausible when applied *locally* to the melt in front of the tips of certain (non-planar) patterns, provided that the curvature corrections are small (i.e. $\sigma \ll 1$), and that k is reasonably small, e.g., $k \leq 0.2$. In contrast to the plane, whose position in the gradient is fixed through conservation, cells can regain stability by moving up in the temperature gradient. A simple analysis based on this idea leads to the LCS prediction that cells restabilize at a position z_t such that

$$\zeta_t \equiv (v-1) - (1-k)z_t/\ell_D \leq 0 \ . \tag{3}$$

Equation (3) confirms that the tip position z_t of stable patterns with $v > 1$ must move *up* in the cell, relative to the planar position at $z = 0$. It is natural to conjecture that the operating point in real experiments should be close to the one that requires the least forward motion, i.e., the region where $\zeta_t \approx 0$. We refer to this as the LCS criterion.[11] The dashed line in Fig. 2 is the line where $\zeta_t = 0$. The fact that the experimental data points lie rather close to this line suggests that the LCS criterion is rather accurate.

As discussed in more detail in Ref. 7, at the line $\zeta_t = 0$ the mathematical mapping onto the ST problem breaks down. Close to this line, the temperature dependence of the miscibility gap can not be ignored and an analysis in terms of a single matching region (as is assumed in the mapping) no longer suffices. Thus, the LCS line roughly marks the region where one could (even for small p) expect to find cellular shapes whose appearance is distinctly different from the characteristic ST cell shape.

COMPARISON WITH SCENARIOS FOR MULTIPLE BRANCHES OF SOLUTIONS

In this section, we discuss our results in the light of other work on bifurcations and multiplicity of steady state cellular patterns (points 3 and 6 above). For a detailed discussion of the differences and similarities with the cell shapes calculate numerically by several groups we refer to Ref. 7.

In the work of Dombre and Hakim[2] and our extension thereof, cellular shapes with narrow grooves are calculated by matching an extension of the Scheil equation for the grooves to profiles that resemble finite amplitude solutions. However, the finite amplitude solutions that are used in this procedure are most probably physically irrelevant, since they lie *outside* the planar instability band. Moreover, these solutions on the ST branch are found to exist for *any* value of v above the threshold v_c, rather than a finite distance above threshold.[7] We believe that other possible branches of cells with grooves could be thought of as arising from joining up grooves to finite amplitude cells that lie within the planar instability band, even though the idea of a single matching region need not necessarily hold true. Evidence for the latter type of finite amplitude cells is given by the analytical work of Langer,[13] Sivashinsky[12] and Kurtze[14] as well as by the numerical work of Ungar and Brown.[8] Note in this regard that in contrast to the solutions our matching method finds, the cellular profiles discussed by

Sivashinsky[12] and Kurtze[14] for small k bifurcate off from the *center* of the planer instability band (indicated by the square in Fig. 1) and do *not* have a regular k expansion. Presumably some members of this latter finite amplitude branch of solutions develop deep grooves as v is increased. Indeed, to our knowledge most numerical calculations[8–10] find deep cells by evolving continuously from small amplitude cells near the center of the planar instability band (near the square in Fig. 1).

Our conclusion that there is possibly another branch of steady state cellular solutions which is the physically relevant one for DS appear to be in line with the arguments of Brener et al.[15] discussed at this summerschool. These authors studied crystal growth in a channel, which corresponds to the $\ell_T \to \infty$ limit of DS. Not surprisingly, for

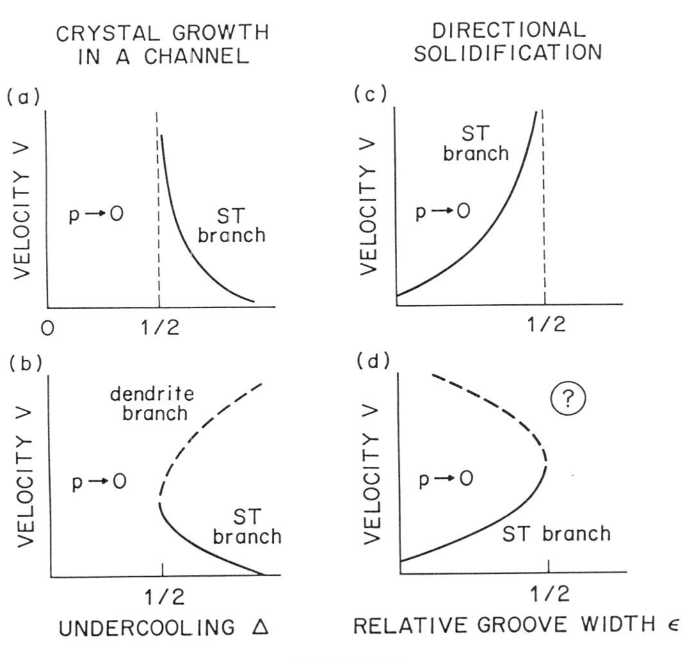

FIGURE 3

(a),(b) Summary of the bifurcation structure for crystal growth in a channel as argued by Brener et al.[15] (a) If the cellular profiles are modeled by the ST expression for the shape, one finds a ST branch for $\Delta > 1/2$. The velocity of these solutions diverges as $\Delta \to 1/2$. (b) Brener et al.[15] argue that if finite Péclet number corrections to the shape are taken into account, there is a bifurcation point for Δ close to 1/2. From this point, both the ST branch and another "dendritic" branch bifurcate. Solutions on the ST branch are unstable, and those on the dendritic branch stable. (c), (d) Possible bifurcation diagram for directional solidification for small p, based on the analogous conjecture for crystal growth in a channel. (c) The velocity V diverges as the relative groove width approaches 1/2. (d) In analogy with (b), it is possible that the ST branch merges with another cellular or dendrite branch near $\varepsilon = 1/2$ and that solutions on the ST branch are unstable.

small p, there is again a branch of Saffman-Taylor like solutions. From the mapping to the ST problem, it follows that these solutions exist only for dimensionless under-coolings $\Delta > 1/2$, and that the velocity is a *decreasing* function of Δ. This is illustrated in Fig. 3a. The decrease of V with increasing undercooling is, of course, counterintuitive, and both analytical[18] and numerical[19] work indicate that the solutions in this branch are *unstable*. Note also that on this ST branch, the velocity V becomes arbitrarily large as $\Delta \to 1/2$. For fixed channel width, this means that finite Péclet number corrections to the shape will become more and more important as $\Delta \to 1/2$. Motivated by this observation, Brener et al.[15] argue that in the absence of surface tension anisotropy, the ST branch ends at a bifurcation point near $\Delta = 1/2$. See Fig. 3b. Another "dendrite-like" branch bifurcates from this point as well, and it is this second branch of solutions that is believed to be stable[15,20].

Although the results of the simulations which Hunt discussed at the school[20] give some support for this scenario, we emphasize that the analysis of Brener et al.[15] is non-rigorous in that it is based on a physically motivated but ad-hoc ansatz for the finite Péclet number corrections to the zero surface tension shape. At present, the arguments of Brener et al.[15] should therefore be considered somewhat speculative. The experimental situation is not completely clear either. Molho et al.[21] observe that there is a low-velocity small-Péclet number regime where the crystal shapes fit ST solutions very well and where the relative cell width $1 - \varepsilon$ ($= \Delta$) is a decreasing function of V, as in Fig. 3a. On the other hand, these shapes may be weakly unstable as expected theoretically[18,19]. Indeed, at higher growth rates a crossover to a different high velocity regime is observed.

Returning now to the DS problem, numerical and analytical work has shown that multiple branches of solutions exist at finite Péclet numbers. We are not aware of any systematic study of the multiplicity of solutions for small p. However, the work of Brener et al.[15] illustrates that the behavior in this limit may be quite subtle.

Indeed, in analogy to crystal growth in a channel, let us consider the DS problem for a fixed set of experimental parameters. For *fixed wavelength*, the ST branch of solutions gives a relation between the growth velocity V and the relative groove width ε whose behavior is illustrated in Fig. 3c. As in Fig 3a, the velocity diverges as $\varepsilon \to 1/2$, where σ tends to zero. [For fixed λ, Eqs. (1) and (2) show that the vanishing of σ implies $V \to \infty$.] However, based on the similarity of crystal growth in a channel and the DS problem, we speculate that if the velocity dependence of the shape would be taken into account properly, one would likewise find that the ST branch merges with another cellular branch at a bifurcation point near $\varepsilon = 1/2$ and and that solutions on the ST branch are all unstable. Compare Figs. 3b and 3d. The small ε solutions are almost certainly unstable since they lie outside the planar stability band, and both the analogy to crystal growth in a channel and our stability argument (LCS) suggest that all members of this branch are unstable (Karma and Pelcé [22], on the other hand, believe that the ST branch is stable for small v, and that it exhibits an oscillatory instability as v increases). Note also that the postulated non-ST branch corresponds to much larger velocities at small ε than the ST branch. For fixed wavelength λ, (1) shows that this conjectures branch is likely to have small groove width ε *and* small σ − precisely the characteristics that distinguish experimental cells form those on the ST branch! We hope that future analysis of the ideas of Brener et al.[15] will establish whether this scenario is correct.

CONCLUSION

In conclusion, we believe that there are serious indications that the Saffman-Taylor branch of cellular solutions is not the relevant one for DS. This issue is an important assumption that this branch does in fact correspond to the one that one observes in experiments. The resolution of this question will have important implications for our understanding of DS as well as crystal growth in a channel.

REFERENCES

1. P. Pelcé and A. Pumir, J. Cryst. Growth 73 (1985) 337.

2. T. Dombre and V. Hakim, Phys. Rev. A36 (1987) 2811.

3. A. Karma and P. Pelcé, Phys. Rev. A39 (1989) 4162 and Europhys. Lett. 9 (1989) 713.

4. M. Mashaal, M. Ben-Amar and V. Hakim, Phys. Rev. A41 (1990) 4421.

5. B. Billia, H. Jamgotchian and L. Capella, J. Cryst. Growth 82 (1987) 747; 94 (1989) 987.

6. J. D. Weeks and W. van Saarloos, Phys. Rev. A39 (1989) 2772.

7. J. D. Weeks, W. van Saarloos and M. Grant, to appear in J. Cryst. Growth.

8. L. H. Ungar and R. A. Brown, Phys. Rev. B31 (1985) 5931; B29, 1367 (1984); B30 (1984) 3993; L. H. Ungar, M. J. Bennett and R. A. Brown, Phys. Rev. B31 (1985) 5923.

9. D. A. Kessler and H. Levine, Phys. Rev. A39 (1989) 3041; Phys. Rev. A39 (1989) 3208.

10. C. Misbah, H. Müller-Krumbhaar and co-workers (this conference).

11. J. D. Weeks and W. van Saarloos, Phys. Rev. A42, 5056 (1990).

12. G. I. Sivashinsky, Physica 8D (1983) 243; see also D. Wollkind and L. Segel, Phil. Trans. R. Soc. 268 (1970) 351.

13. J. S. Langer, Acta Metall. 25 (1977) 1121.

14. D. A. Kurtze, Phys. Rev. B37 (1988) 370.

15. E. A. Brener, M. Geilikman and D. E. Temkin, Zh. Eksp. Teor. Fiz. 94 (1988) 241 [Sov. Phys. JETP 67 (1988) 565.]

16. S. de Cheveigné, C. Guthmann and M. Lebrun, J. Phys. (France) 47 (1986) 2095. See also P. Kurowski (this conference).

17. W. Kurz and D. J. Fisher, Acta. Metall. 29 (1981) 11. See also B. Billia, H. Jamgotchian and R. Trivedi, J. Cryst. Growth, to appear.

18. P. Pelcé, Europhys. Lett. 7 (1988) 453.

19. See, e.g., D. A. Kessler and H. Levine, Mod. Phys. Lett. B2 (1988) 945 and references therein.

20. J. D. Hunt, this conference and "A numerical analysis of dendritic and cellular growth of a pure material investigating the transition from 'array' to 'isolated' growth" (preprint).

21. P. Molho, A. J. Simon and A. Libchaber, Phys. Rev. A42, 904 (1990).

22. A. Karma and P. Pelcé, Phys. Rev. 41 (1990) 6741, and this conference.

PATTERN FORMATION IN DIRECTIONAL SOLIDIFICATION

C. Misbah[a], H. Müller-Krumbhaar[b], Y. Saito[c] and D.E. Temkin[d]

[a] *GPS, associé au CNRS, Place Jussieu,Paris, France*
[b] *IFF des Forschungszentrums, Jülich, Germany*
[c] *Physics Department, Keio University, Yokohama, Japan*
[d] *I.P. Bardin Institut for Ferrous Metals, Moscow, USSR*

The current status of our understanding of pattern formation in directional solidification of binary alloys is reviewed. We focus, in particular, on the weakly nonlinear regime and on transition from cells to dendrites. Interface kinetics is included and seems to be decisive in some systems as impure CBr_4. We briefly discuss the wavelength selection dilemma. We say a few words on the transition to the restabilization of the planar front at high growth speeds. This restabilization is likely preceeded by a chaotic regime.

I. INTRODUCTION

Pattern formation in nonequilibrium systems constitutes a major branch of research both experimentally and theoretically. The crucial question that is addressed pertains to the understanding of why order may appear and which structures are selected among a large manifold of possibilities. Directional solidification has become an important component of this branch[1], not only because of its potential technological importance, but also because it constitutes an interesting free boundary pattern forming system on the fundamental level.

Already in 1953 Chalmers and Rutter [2] recognized that the microsegregation of impurities (which alters the physical properties of the material) found in the solid phase obtained by directionaly solidifying an impure liquid is nothig but a signature of a morphological instability of the solidification front. They have provided a stability criterion based on thermodynamical considerations. This is the so-called constitutional supercooling criterion. It states that if the actual interface impurities concentration falls into the unstable region for the two phases coexistence, the planar front becomes unstable. Mullins and Sekerka[3] were the first to perform systematic linear stability analyses and to point out the underlying kinetic nature of the process. In Segel's survey (1966) Kirkaldy[4] pointed out the similarity between the Rayleigh-Bénard problem and cellular patterns observed in directional solidification of alloys. It is by now well known that the dynamics of systems, that look extremely diverse, is described, close to the instability threshold, by a universal equation of Landau-Ginzburg type. This feature is similar to that encountered in phase transition phenomena. Indeed Landau's theory, or more advanced scaling theories, tell us that the macroscopic properties of seemingly quite different systems can be described near the critical point with an identical equation if one makes an appropriate scaling of some physical quantities. However whether the amplitude theory accurately describes or not the physical phenomenon that is observed in a given experiment strongly depends on the details of the underlying physical mechanisms. Indeed if a such theory was succesfull in describing some features in Rayleigh-Bénard system for example, it is far from doing so in directional solidification. The Main reason is that the two capital competing phenomena, which are the impurities diffusion and capillary forces, operate, in general, on much different scales (the diffusion length scale is of the order of a few microns, while the capillary forces act on an atomic scale). As a consequence the Mullins-Sekerka[3] spectrum for regression of fluctuations about a planar front is flat. This means that higher harmonics should be active even close to threshold, thus causing interface fluctuations to escape the lowest order amplitude description. Accordingly one should resort to other methods of investigation, even when dealing with structures close to threshold.

A first step towards the understanding of cellular shapes in directional solidification came from numerical simulations of steady solutions[5-7]. Such a procedure does not allow, however, to deal with the dendritic regime. Moreover the stationary calculations sometimes give solutions which are unreachable dynamically. Therefore a time-dependent treatment seems unavoidable. This method first developed in the context of free dendrites[8] was adapted to directional solidification[9]. It consists of a forward time-dependent calculation based on the quasi-stationary approximation. This approximation still correctly identifies the occurence of a stationary growing pattern if the bifurcation is not of Hopf type. Below we will outline the main development made in the domain of pattern formation in directional solidification. We send sometimes the reader to other refererences in this volume for some questions which are, somehow, tangential to this presentation.

II. THE GROWTH MODEL

The 'minimum version' of the front dynamics model in directional solidification of an impure solid is defined as follows. If $u = (C - C_\infty)/\Delta C$ (C is the impurities concentration in the liquid phase, C_∞ its value far ahead of the solidification front, and $\Delta C = (1 - k)C_\infty/k$ is the equilibrium miscibility gap, where k is the partition coefficient) designates the dimensionless concentration, the mass diffusion equation in the liquid phase reads

$$\nabla^2 u + 2\frac{\partial u}{\partial z} = \frac{\partial u}{\partial t}, \quad z > \zeta(x,t) \tag{1}$$

where lengths and time are reduced by $l = 2D/V$ and l^2/D respectively, with D the diffusion constant and V the pullig speed. We assume that mass diffusion is negligible in the solid phase since the corresponding diffusion constant is several orders of magnitude smaller than that in the liquid phase. At the liquid-solid interface $(z = \zeta(x,t))$ using a pill box type argument we obtain

$$\{1 + (u-1)(1-k)\}v_n = -\frac{\partial u}{\partial n}, \tag{2}$$

where v_n is the normal velocity. The temperatures in both phases are assumed to satisfy Laplace's equation. This is justified by the fact that heat diffuses in a much faster way that mass. Furthermore if one neglects the latent heat and assumes that heat diffusion takes place symmetrically in both phases (a situation encountered when the plates are highly conducting so that heat diffuses mainly trough them), then the temperature profile along the sample is simply fixed by the external thermal contacts and is decoupled from the concentration field. Now using the fact that the actual temperature is shifted from the melting temperature due (i) to impurities and (ii) to curvature (Gibbs-Thomson effect) we obtain the last condition

$$u = 1 - d\kappa - \frac{\zeta}{l_T} \tag{3}$$

where κ is the interface curvature and is counted to be positive for a convex solid, $d = \frac{\gamma T_m}{lLm_l\Delta C}$ is the dimensionless capillary length, with γ the surface tension, T_m the melting temperature of the pure substance, m_l the liquidus slope and L the latent heat of fusion per unit volume, $l_T = \frac{Gm_l}{l\Delta C}$ is the thermal length, where G is the applied thermal gradient. Note that d is an orientation-dependent quantity. For a four-fold cristalline anisotropy d takes the form

$$d = d_0\{1 - \alpha_4 cos(4\theta)\}, \tag{4}$$

where d_0 is the isotropic part, α_4 represents the strength of cristalline anisotropy and θ is the angle between the growth axis and the normal to the interface. Eqs.(1)-(3) with the condition $u(z = \infty) = 0$ completely describe dynamics of solidification. This set admits a planar front solution characterized by $u_0(z) = exp(-2z)$, the linear stability of which was performed by Mullins and Sekerka[3]. Several authors[10-14] have, since then, performed calculations in the weakly nonlinear regime based on standard amplitude expansions. As mentioned above the amplitude theory is inadequate. We should therefore have recourse to other methods. Recently Weeks *et al.*[15] (see also W. Van Saarloos in this volume) have discussed the validity of the asymptotic method performed for small partition coefficient, which is inspired by that developed by Dombre and Hakim[16] at small Péclet number P (here $P = \lambda/l$, λ being the periodicity of the pattern). They have shown that a such method may provide an accurate description only in a region of parameters that lies far beyond the experimental regime. Despite its own merit a such method is most likely not relevant for actual experiments, not talking about the stability of the pattern[17].

The first essential step towards the understanding of pattern formation in directional solidification is due to Ungar and Brown[5]. They have demonstrated the complex nature of the bifurcation diagram. In particular they have shown that the bifurcating state, which emerges at the threshold, ceases to exist slightly above the critical velocity V_c (saddle-node bifurcation) and merges with solutions at half the wavelength... The next step was accomplished thanks to a forward time-dependent calculation[9]. Besides the fact that it allows to follow the dynamics of the above mentioned scenarios, this method has, at least, two other merits. (i) It is capable, unlike steady-calculations[5−7], of dealing with the dendritic regime. (ii) It detects short wavelength instabilities(shorter than the basic periodicity) and permits (unlike a linear stability analysis) us to study the nonlinear development of the instability.

II. RESULTS

The numerical technique was described in details elswhere[18]. We will summarize below our findings. We have first reproduced the Mullins-Sekerka[3] instability within 1% of uncertainty.

a) MORPHOLGY OBTAINED CLOSE TO THRESHOLD

To study the weakly nonlinear dynamics we should distinguish between two cases.

(i) Supercritical bifurcation: As shown by Wollkind ans Segel[10] (see also Caroli *et al.*[13]) the bifurcation is supercritical for $k > 0.4..$ and subcritical instead. Very close above threshold $((V - V_c)/V_c = 1 - 5\%)$ the interface is smooth (Fig.1a) but it already departures from a sinusoidal morphology. This is a consequence of the flatness of the Mullis-Sekerka[3] spectrum. On further increase of V $((V - V_c)/V_c \sim 10 - 20\%)$ the interface develops moderately deep grooves.

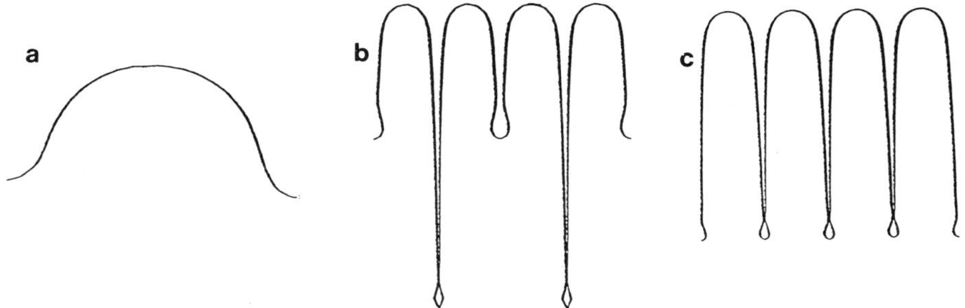

Fig.1 a: A typical morphology close to the threshold. b: A tip splitting instability; the tip deepens in the course of time. c: The interface reaches a stationary regime where the initial wavelength has halved.

At a critical velocity the mode with the bifurcating wavelength ceases to exist. The interface undergoes a tip splitting instability (Fig.1b). The tip deepens as time elapses until it reaches the depth of the primary groove. The periodicity has been halved (Fig.1c). This is the dynamical version of the saddle-node bifurcation discovered by Ungar and Brown[5].

(ii) Subcritical bifurcation: even below threshlod the interface may exhibit a strong modulation similar to that displayed in Fig.1c provided that the initial amplitude is large enough to overcome the metastability barrier. One remarkable point is that the the critical mode, that follows from the Mulins-Sekerka analysis, is unstable at the linear threshold against tip splitting modes. This result seems to be in agreement with various experiments[21]. Far from threshold the nature of the primary bifurcation is irrelevant.

b) *STRUCTURES FAR AWAY FROM THRESHOLD*

(i) Saffman-Taylor like morphology: This corresponds to situations where the Péclet number is small. The interface shape is similar[9], at first glance, to that obtained by Saffman-Taylor[19,20] (Fig.2a). Measuring the width of the finger λ_f at about one wavelength behind the tip, the predicted Saffman-Taylor [9] scaling ($\lambda_f = \Delta\lambda$, with Δ the actual tip undercooling) is recovered within a few percent. However the actual shape shows a significant departure from a Saffman-Taylor finger as soon as one moves away from the tip.

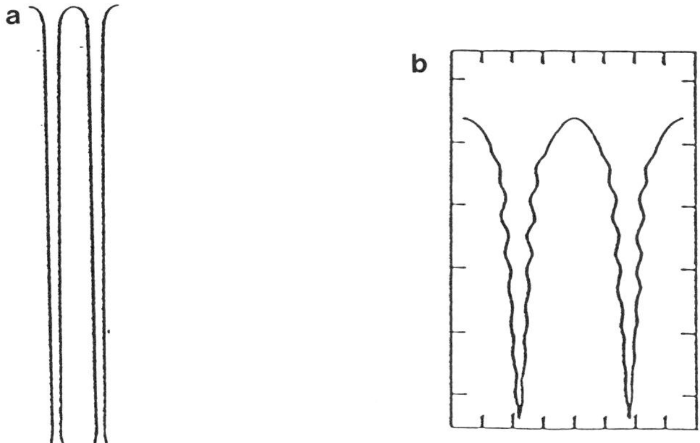

Fig2. a: The Saffman-Taylor like structure. b: The dendritic structure.

(ii) The dendritic regime: When P approaches unity, that is when the primary spacing becomes comparable to the diffusion length, dendritic growth becomes apparent[9] (Fig1.b). As in free growth[8] cristalline anisotropy is necessary to stabilize the tip against tip splitting modes. The scaling law for the tip radius agrees well with that obtained in free growth[9]. Not surprisingly the Ivantsov relation is however recovered only at sufficiently large velocities ($P > 3$). This relation represents indeed a global conservation law for a single dendrite, while selection of the actual tip radius depends basically only on details close to the tip. On further increase of V a tail instability takes place. This corresponds to the uprise of a new cell out of a sidebranch in the groove. The primary cell maintains its identity. This observation is quite in agreement with experiments[22], where the wavelength adjustment seems to occur via such 'hard' mode instabilities rather than by the much slower phase diffusion process.

III. KINETICS-CONTROLLED GROWTH

Recent experiments[23] on impure CBr_4 have shown that for $1 < P < 3$, where each cell operates quasi-independently in the diffusion field, the interface exhibits an 'angular' structure. Moreover the sidebranches are not visible (they become actually visible only for $P \geq 3$). These features strongly differ from those met in, say, 'the classical' dendrites (Fig.2b) described above. The question thus arises of whether this phenomenon is a signature of a fundamental ingredient that is missing in the usually called 'minimum version' of crystal growth. We have deferred untill now a discussion of the most important physical condition under which eq.(3) is derived. Indeed the most serious point that is emphasized by the derivation of the growth equations is that the chemical boundary condition (eq.3) is an equilibrium condition. Such a condition is *a priori* fulfilled for not too large growth speeds. It would seem at first sight that interface kinetics is relevant for rough parts only for speeds of the order of molecular speeds. In the absence of a complete microscopic description it is not obvious to have precise information. However the planar front recession[23] measured on impure CBr_4 seems to indicate that kinetics is important even at velocities of the order of a few $\mu m/s$. As a first step towards testing the importance of kinetics it is natural to assume a linear law à la Onsager. Then the modified Gibbs-Thomson condition takes the form

$$u = 1 - d\kappa - \frac{\zeta}{l_T} - B_{kin} v_n \qquad (5)$$

where B_{kin} is a phenomenological kinetic coefficient, which is, in general, anisotropic. A complete description of the kinetic effect was given elsewhere[18]. We focus here on the dendritic regime only. The magnitude of B_{kin} was taken even one order of magnitude smaller than that calculated from planar front recessions[23] on impure CBr_4.

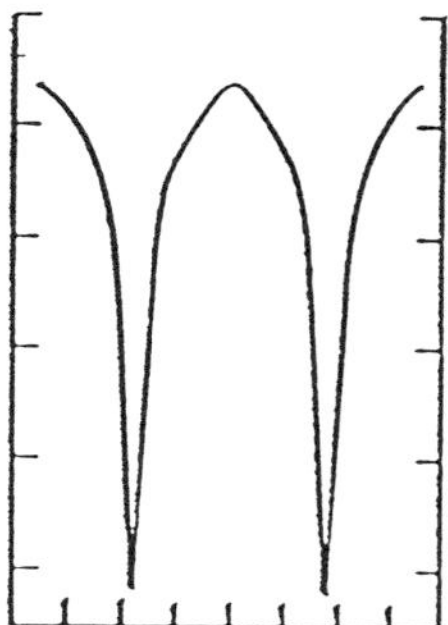

Fig.3 *Kinetics-controlled dendrite.*

For isotropic surface tension, kinetics anisotropy is found to stabilize the tip of dendrite. The front morphology obtained by simulation (Fig.3) is very similar to that observed[24] on impure CBr_4. In particular the front develops planar far sections and sidebranches are not visibile ($1 < P < 3$). We are naturally led to postulate that dendrites in CBr_4 are selected by kinetics anisotropy rather than by surface energy anisotropy.

IV. PHASE DIFFUSION AND WAVELENGTH SELECTION

Hitherto we have considered only situations where the system size is equal to one or two wavelengths. This means that we could detect only short wavelength instabilities (tip splitting, tail instability..). If the instabilities are of soft type, however, we would require to perform simulations on large sizes systems. This would imply a huge computing time. We should therefore resort to other methods. Since the basic pattern is periodic, it is invariant under a constant phase shift. This means that the translational mode is a neutral mode of the linearized equations about the basic state. It is therefore appealing to expect large scale phase modulations to be dangerous. The phase instability[25] leads to a restriction of the band of possible solutions. But this band is wide enough so that we can still discriminate in practice between two nonequivalent stable states. On the other hand, in most experiments the pattern wavelength is greatly restricted or even uniquely determined. Hence, the understanding of whether pattern selection is an intrinsic property of the systems themeseleves, or rather depends on experimental protocols is a question which still poses a formidable challenge to theoretical investigations. Usually boundary conditions simply cause a restriction of the allowed band[26]. Dynamical selection [27] based on marginal stability criterion was succesfull in model equations but its general application remains to be shown. Selection by soft boundaries or ramps[28] (that is a situation where the control parameter is slowly varied from above to below threshold) has know a certain success in hydrodynamics. The rotating solidification set up, not yet exploited experimentally, should lead to a unique selection[29]. We were able to perform a numerical study for small sizes systems (5 wavelengths) where a discrete set of stable solutions exist. Here also a perfect selection is obtained[30]. The wavelength adjustment is found to occur via a phase diffusion process. As the depth of the cells become larger and larger we observe a drastic slowing down of phase diffusion. We expect there the wavelength adjustment to occur via hard mode instabilities, as a tail instability in dendritic growth. We conclude that experimental investigations on selection by a ramp, in small systems as well as in large systems, should focus on the nematic crystal where the cells never develop deep grooves.

V. GENERIC INTERFACE STRUCTURES

At low speeds the Mullins-Sekerka instability intervenes, at a certain threshold, to destabilize the planar front. The interface assumes a cellular structure. On further increase of V cells bifurcate into dendrites. If the kinetic coefficient is not too large dendrites should be selected by surface tension anisotropy. As the velocity increases interface kinetics becomes of great importance and should become decisive. We expect a transiton to kinetics-controlled pattern. At larger $V's$, that is when diffusion length approaches the capillary length the planar front becomes again stable, even for a vanishing thermal gradient. Slightly below the critical velocity for the planar front restabilization the most dangerous mode is the one with a small wavenumber (in diffusion lenght unit). There the front dynamics is governed by a Kuramoto-Sivashinsky[31] equation, whith a small damping term due to the external thermal gradient. Based on the properties of a such equation[32] we expect that the interface, generically, should undergoe a chaotic motion before restabilization. We would like to conclude by saying that one major handicap, in our opinion, we are faced to is to have precise information on the kinetic effect. We hope that this work will constitute a call for carreful experiments with the aim to extract information on kinetics.

Due to the limited space of the manuscript, we did not devote a discussion to the possibility of non axisymmetric growth, which is one of the latest subject in patern formation. The discovery of this phenomenon is due to Simon, Bechoefer and Libchaber[33] during the growth of a nematic crystal. Since then other systems[34,35] have revealed similar phenomena. We send the reader to the talks of Y. Couder, W. Rappel and C. Misbah in this volume.

Most of this work was accomplished in the Forschungszentrum at Jülich (Germany) where C.M., Y.S. and D.E.T. were visiting and we would like to express our gratitude for hospitality and financial support.

REFERENCES

1. For a recent review see: H. Müller-Krumbhaar and W. Kurz, preprint 1990.
2. B. Chalmers and J.W. Rutter, *Canad. J. Phys.* **31**: 15(1953).
3 W.W. Mullins and R.F. Sekerka, *J. Appl. Phys.* **35**, 444 (1964).
4 J.S. Kirkaldy, in 'Non-equilibrium thermodynamics, variational techniques, ans stability', (ed. R. Donnely *et al.*) pp. 281-282. University Chicago Press.
5. L.H. Ungar and R.A. Brown, *Phys. Rev. B* **29**, 1367 (1984); **30**, 3993 (1984); **31**, 5923 (1985); **31**, 5931 (1985).
6. N. Ramprassad, M.J. Benett and R.A. Brown, *Phys. Rev. B* **38**, 583 (1988).
7. D. Kessler and H. Levine, *Phys. Rev. A* **39**, 3041(1989).
8. Y. Saito, G. Goldbeck-Wood and H. Müller-Krumbhaar, *Phys. Rev. A* **38**, 2148 (1988).
9. Y. Saito, C. Misbah and H. Müller-Krumbhaar, *Phys. Rev. Lett.* **63**, 2377 (1989).
10. D.J. Wollkind and L.A. Segel, *Philos. Trans. R. Soc. London* **51**, 268 (1970).
11. J.S. Langer and L.A. Turski, *Acta Met.* **24**, 1113 (1977).
12. J.S. Langer , *Acta Met.* **25**, 1121 (1977).
13. B. Caroli, C. Caroli and B. Roulet, *J. Phys.* **43**, 1767 (1982).
14. G. Dee and R. Mathur, *Phys. Rev. B* **27**, 7073 (1983).

15. J.D. Weeks, W. van Saarloos and M. Grant, preprint (1990).

16. T. Dombre and V. Hakim *Phys. Rev. A* **36**, 2811 (1987).

17. J.D. Weeks and W. van Saarloos, preprint (1990).

18. A. Classen, C. Misbah, H. Müller-Krumbhaar and Y. Saito, preprint(1990).

19. P.G. Saffman and G.I. Taylor, *Proc. R. Soc.* **A 245**, 312 (1958).

20. P. Pelcé and A. Pumir, J. Crys. Growth **73** 337 (1985).

21. S. de Cheveigné, C. Guthmann, P. Kurowski, E. Vicente and H. Biloni, *J. Cryst. Growth* **92**, 616 (1988).

22. H. Esaka and W. Kurz, *J. Crys. Growth* **72**, 578 (1985).

23. S. de Cheveigné, G. Faivre, C. Guthmann, P. Kurowski and J. Mergy (private communication).

24. P. Kurowski, *Thèse d'Université, Paris 7*, 1990; P. Kurowski, C. Guthmann and S. de Cheveigné (to be published).

25. K. Brattkus and C. Misbah *Phys. Rev. Lett.* **64**, 1935(1990).

26. M.C. Cross, P.G. Daniels, P.C. Hohenberg and E.D. Siggia, *J. Fluid. Mech.* **127**, 155 (1983), and references therein.

27. See for example W. van Saarloos, *Phys. Rev. A* **37**, 211 (1988) and references therein.

28. See for example H. Riecke, *Europhys. Lett.* **2**, 1 (1986) and references therein.

29. C. Misbah, *J. Phys.* **50**, 971 (1989).

30. C. Misbah, H. Müller-Krumbhaar and Y. Saito, *J. Cryst. Growth* **99**, 156 (1990).

31. C. Misbah, H. Müller-Krumbhaar and D.E. Temkin, preprint (1990).

32. See for example P. Manneville *in Popagation in systems far from equilibrium*, edited by J.E. Wesfreid, H.R. Brand, P. Manneville, G. Albinet and N. Boccara, Springer-Verlag Berlin (1988).

33. A.J. Simon, J. Bechoefer and A. Libchaber, *Phys. Rev. Lett.* **61**, 2574 (1988).

34. G. Faivre, S. de Cheveigné, C. Guthmann and P. Kurowski, *Europhys. Lett.* **9**, 779 (1989).

35. M. Rabaud, S. Michalland and Y. Couder, *Phys. Rev. Lett.* **64**, 184 (1990).

DIRECTIONAL GROWTH OF A FACETED SMECTIC B PLASTIC CRYSTAL

Patrick Oswald and Francisco Melo

Laboratoire de Physique
Ecole Normale Supérieure de Lyon
46 Allée d'Italie
69364 Lyon Cedex 07 France

1-INTRODUCTION

Directional solidification of metals, alloys and organic materials (including plastic crystals and liquid crystals) has been intensively studied over the past several years. Most of the experiments have been performed on non-faceting materials, for which the "solid-liquid" interface is atomically rough. It is well known that the front is unstable above a critical growth velocity and that the cellular bifurcation is either subcritical or supercritical depending upon the material. This instability results from a competition between the destabilizing effect of the diffusion field and the stabilizing effects of both the temperature gradient and the surface tension. The linear stability analysis of this problem was first carried out by Mullins and Sekerka[1] in 1964. This calculation is the fundamental key of our understanding of this instability. Later, Woolkind and Segel[2] and Caroli et al [3] did a weakly non-linear bifurcation analysis. They showed that the nature of the bifurcation depends on the value of the solute partition coefficient. This prediction has been well verified experimentally[4]. The current problem is to understand the non-linear evolution of deep-cell arrays and their dynamics. So far, there is no unified theory describing the wavelength selection as well as the cell to dendrite transition. Nevertheless, it is now established both experimentally[5] and theoretically[6] that there exists a band of allowed wavelengths and that through a still unknown dynamical process the system chooses a particular periodicity. By contrast, the cell to dendrite transition is poorly understood. A possible mechanism was proposed recently by Karma and Pelce[7]. They suggest that the cellular structure exhibits a linear instability to an oscillatory state corresponding to a periodic emission of sidebranches (Hopf bifurcation). This scenario has not yet been proved to exist experimentally. We shall return to this problem in Section 3. On the other hand, Kessler and Levine[8] have performed numerical simulations and claim that there is no evidence for this mechanism. Experimentally, the cell-to-dendrite transition takes place over a rather wide range of velocities[9]. In this intermediate regime, cells and dendrites of different wavelengths

Growth and Form, Edited by M. Ben Amar *et al.*
Plenum Press, New York, 1991

coexist[10]. Finally, Seetharaman *et al* [9] propose that this transition is due to a jump in the mean selected wavelength, but there is no theoretical or numerical evidence of this phenomenon[11].

More recently, faceted cellular growth has become of interest because of its importance to the understanding of the growth of thin film silicon single crystals produced by zone-melting recrystallisation[12]. In these experiments, the cell spacing selection is quite different from non-faceted cellular growth. For example, the wavelength λ increases as $V^{1/2}$ instead of decreasing as $V^{-1/2}$. Here, facet-growth kinetics dominate[13].

In this article, we review a recent experiment on a partly faceted interface between a smectic A and a smectic B phase of a liquid crystal. We have chosen for this study the compound 408 (butyloxybenzilidene octylaniline), which has a smectic A-smectic B transition at 49.9ºC. We recall that the smectic A is a lamellar phase with fluid layers whereas the smectic B is a plastic hexagonal close packed crystal with layers stacked ABAB...[14]. As we shall see in the following, the behavior in directional growth of the Sm.A-Sm.B front strongly depends upon the growth orientation with respect to the smectic layers. This is related to some unusual surface-tension properties of 408, which we shall describe in section 3. In particular, we shall show that there exist unstable orientations. Section 4 deals with the dynamical properties of the front in directional solidification. The cases corresponding to a forbidden orientation and to a facet will be separately described. In the next section, we briefly describe the sample preparation and the experimental set up and recall a few physical properties of this material.

2-THE EXPERIMENT

Our sample consists of two glass plates separated by two 15-μm-thick spacers. Into the gap between the two plates, we introduced 408. To obtain planar alignment of the molecules on the glass (smectic layers perpendicular to the glass plates), a 300-Å-thick layer of polyimide ZLI-2650 (Merck Corp.) was deposited on the inner surfaces. This layers was then rubbed in a single direction in order to orient the smectic layers perpendicular to the scratches. Experiments at fixed temperatures have been made in a hot stage controlled to about ±5 mK. Growth experiments have been performed in a directional solidification apparatus. Here, the sample straddles the space between two ovens whose temperatures are controlled to ±50mK. Temperature gradients are measured with a Cu-constantan thermocouple and the sample is pushed by a fine screw that is driven by a stepping motor. Observations are made via phase contrast microscopy.

A limitation on our experiment is a rapid degradation of 408 at temperatures larger than 50ºC. For this reason, we often change samples and we systematically measure the liquidus and the solidus temperatures before each run. We shall call ΔT the freezing range in the following. The partition coefficient k of impurity is approximately 0.56[15].

3-WULFF PLOT AND SURFACE TENSION MEASUREMENT[16]

It is well known that many physical properties of the B phase are anisotropic. This is primarily due to the big difference between the spacing of molecules in each layer ($\approx$5Å) and the spacing between layers ($\approx$ 30 Å). We are going to see that the surface properties are also anisotropic.

Fig 1 shows a smectic B monodomain in equilibrium with the smectic A phase. Although small, it took nearly one day to equilibrate, as both the surface tension γ and the impurity diffusion coefficient D are very small, as we shall see later. The equilibration time is inversely proportional to γD.

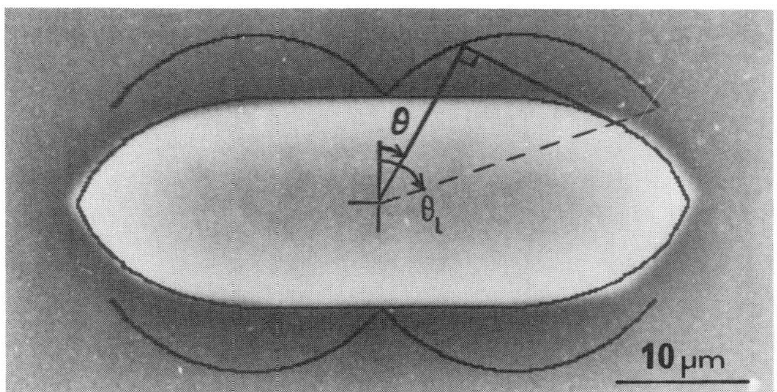

Fig. 1. Sm. B monodomain in equilibrium with the smectic A phase and the corresponding Wulff plot (from Ref. 16).

The main properties of this germ, in addition to its elongated shape, are the following:
-the interface is faceted parallel to the layers;
-the facet matches tangentially with the contiguous curved region;
-the faces perpendicular to the smectic layers (and those which are close to this orientation) do not appear in the equilibrium form, which has sharp edges.
The corresponding $\gamma(\theta)$ plot is given by the Wulff construction and is shown in Fig. 1. The cusps correspond to the facets ($\theta=0$) and the missing parts ot the γ-plot ($\theta>\theta_1$) to the missing orientations in the equilibrium shape. A detailed analysis of this plot, using a generalization of the terrace-ledge-kink model, is given in Ref. 16 along with a measurement of the surface energy of a facet γ_0 using the classical grain boundary method. We found $\gamma_0 \approx 0.1$ erg/cm^2. This small value results from the large layer thickness b ($\gamma_0 b^2 \approx k_B T$).
One of the most interesting properties of 408 is the existence of forbidden orientations in the equilibrium shape. It is possible to show that some of them are unstable ($\gamma + d^2\gamma/d\theta^2 < 0$) while the others are metastable. In our system and because of its symmetry, faces perpendicular to the layers are necessarily unstable. What happens if one tries to force the interface to lie along a forbidden orientation, for example by putting the sample in a temperature gradient parallel to the smectic layers? Herring[17] has shown that there is always a hill-and-valley structure of smaller energy (fig.2).

Fig.2. Hill and valley structure

The wavelength λ of this structure results from the competition between the gain in surface free energy, the loss in the volume free energy (the hatched regions in fig.2 correspond to, either superheated solid, or undercooled liquid) and the energy of the points $\gamma\xi$. ξ is a cutoff length that we expect to be of a molecular size. A detailed calculation gives[18]

$$\lambda = (48 \; l_c^2 \; \xi \; (\tan\theta_1)^2)^{1/3} \qquad (1)$$

where $l_c = (2 \; \gamma \; T^*/GL)^{1/2}$ is a capillary length. T^* is the melting temperature and L the latent heat. Note that λ scales like $G^{-1/3}$. In this formula everything is known except ξ. To estimate it, we performed the experiment (see fig.6a of &4) and we measured λ versus the temperature gradient. These data are shown in fig. 3. The fit with the theoretical law (1) is very good and gives $\xi \approx 700\text{Å}$.

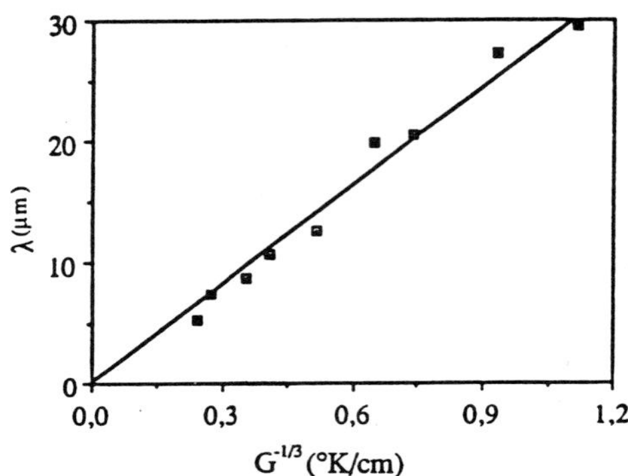

Fig.3. λ versus $G^{-1/3}$. The dots are experimental and the solid line is the best fit with eq. 1.

4-INTERFACE DYNAMICS IN DIRECTIONAL GROWTH[15,19,20]

Let us first discuss what happens in the general case when the layers make an angle with the temperature gradient. We exclude from this discussion the unstable orientations close to $\theta=\pi/2$ and the faceted case ($\theta=0$). The behavior of the front at increasing velocities is shown in fig.4. At rest and at small velocities (fig. 4a), the interface is smooth and stable. At a critical velocity V_c, a sinusoidal undulation occurs (fig. 2b) whose amplitude slowly increases until the cells reach their stationary shape. At this velocity, the cells are rounded (fig. 4c) (no unstable orientations). At larger velocity, $V>1.2V_c$, (fig. 4d) the cells develop facets and sharp tips that are a consequence of missing orientations. Careful observation of the undulation at onset shows that it moves slowly along the unperturbed interface in the direction of the layers. We measured its drift velocity V_d as a function of the angle θ and we found that it passes through a maximum close to 0.2V for $\theta\approx45°$. The critical velocity V_c also depends on the angle θ. These data are shown in fig.5.

To explain the θ-dependence of the critical velocity and the drift of the cells at onset, we have taken into account the diffusion anisotropy in the linear stability analysis of the

180

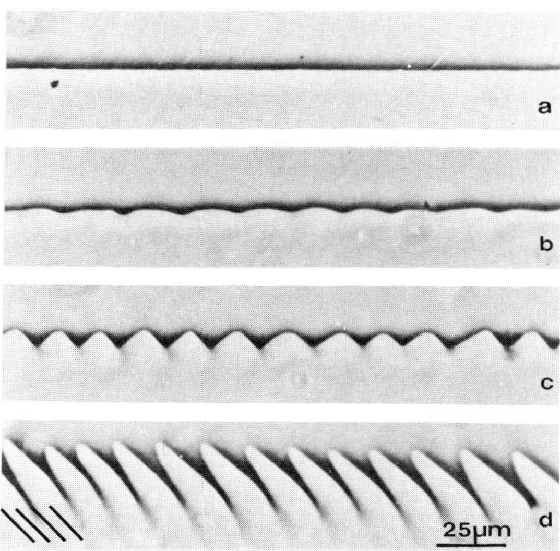

Fig.4. Destabilization of the front when the layers make an angle θ=45° with respect to the temperature gradient. The Sm.B is at the bottom and the Sm.A is at the top (the same in the other figures). (a) flat stationary interface (b),(c) V=0.8μm/s (d)V=0.95μm/s(from Ref. 20). Inset: schematic representation of the smectic layers.

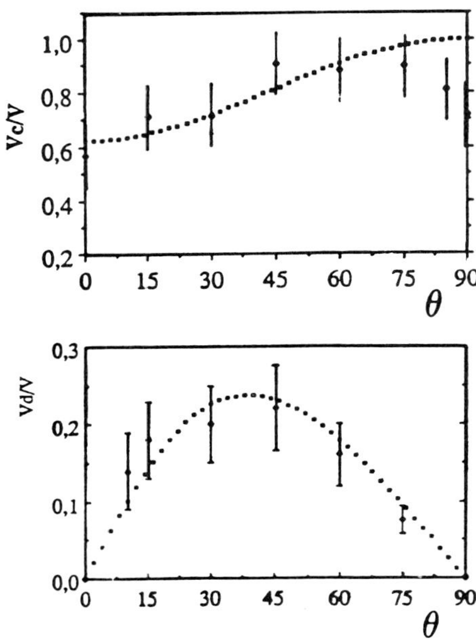

Fig.5. Critical velocity and drift velocity versus the angle θ. The solid lines have been calculated from eqs. (2) with $D_{\perp}=5\text{x}10^{-7}\text{cm}^2/\text{s}$ and $D_{//}=3\text{x}10^{-7}\text{cm}^2/\text{s}$.

planar front. The impurities indeed diffuse more easily in the
smectic A layers than perpendicular to them ($D_\perp > D_{//}$). The
calculation gives[20]

$$V_c/V= D_{eff}G/\Delta T \qquad \text{and} \qquad V_d/V= (D_\perp - D_{//}) \sin\theta \cos\theta / D_{eff} \qquad (2)$$

where $D_{eff}=D_{//}\cos^2\theta + D_\perp \sin^2\theta$ is an effective diffusion
coefficient. These two formulae account for our experimental
results (see fig. 5). Furthermore, we have shown that kinetic
effects are negligible in our system (see Ref. 20 for more
detail).

The case of an unstable orientation ($\theta = \pi/2$) is more
complicated because of Herring's instability. Here the layers are
parallel to the temperature gradient. The behavior of the front is
shown in fig.6.

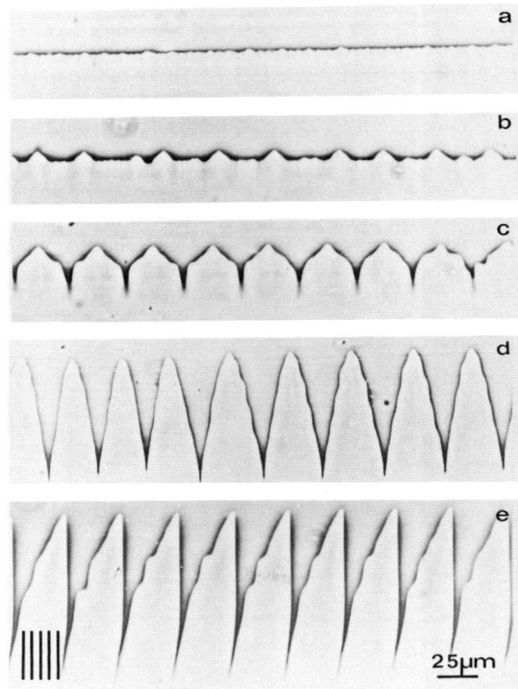

Fig.6. (a)Stationary interface: the hill and valley structure is clearly
visible (G=76 K/cm) (b),(c) V=0.38 μm/s, (d),(e) V=0.45 μm/s. Inset:
schematic representation of the smectic layers (from Ref. 20).

This sequence of pictures shows that the cellular bifurcation
is quite different from the preceding case and from what has been
currently observed in plastic crystals or metals. Here, there is a
critical velocity V_c above which isolated "triangles" of the
hill-and-valley structure strongly develop (fig.6a). These
"triangles" grow and join to form a periodic and stationary
cellular front. The resulting cells are "brace"-shaped and pointed
with an angular discontinuity at the tip (fig. 6b), showing that
orientations that are unstable at equilibrium do not appear during

slow growth. The cell amplitude increases rapidly with the
velocity. Above 1.2 V_C, the cells become unstable and lateral
undulations or sidebranches develop (fig. 6d). The fact that
sidebranching occurs very close to the instability threshold is
perhaps due to the tendency to facet. This tendency is clearly
responsible for the formation, after several hours, of asymmetric
cells. These cells are composed of a large, single facet and a
rough, unstable part (fig. 6e). Careful observation shows that the
sidebranches of both symmetrical and asymmetrical cells are
correlated across many cells (at least ten in our experiment).
This long-range correlation is compatible with the model of Karma
and Pelcé mentioned above[7]. Nevertheless, we have not yet been
able to prove whether there exists a collective oscillation of the
cell tips as predicted by this model. Let us emphasize that, in
our experiment, the Peclet number is very small (less than 0.1),
which is an essential assumption in the Karma-Pelcé model.

Finally, we describe the destabilization of a facet ($\theta=0$).
Here, the layers are perpendicular to the temperature gradient.
This case was recently studied theoretically (R. Bowley et al.
[21] and B. Caroli et al.[22]). They predict the existence of
crenellated solutions consisting of alternating hot and cold
facets connected by curved regions. In fact, this solution is
always unstable so that they suggest that there exists another
stationary solution consisting of partly faceted, large-amplitude
cells. Experimentally, we have never observed, for any velocity,
a periodic, stationary solution. On the contrary, we have observed
that the front destabilizes by generating macrosteps, which are
always unstable and drift along the facet (fig. 7).

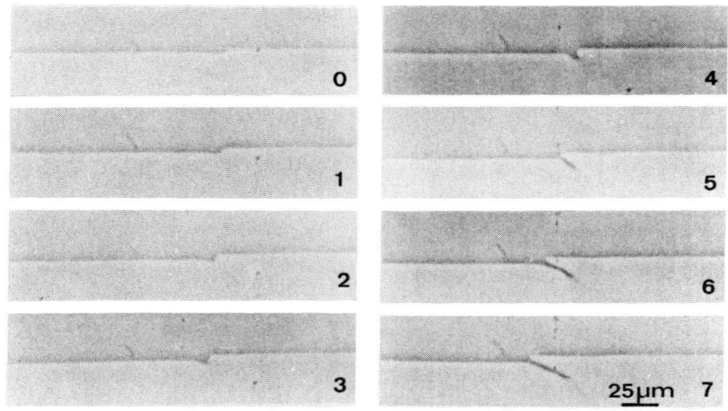

Fig. 7. Destabilization of a macrostep propagating along the macroscopic
interface (from Ref.15)

There are two types of macrosteps. Either they are rounded
and match tangentially the two facets, or they are cusp-like with
a sharp edge. In general, a rounded macrostep becomes
progressively sharp-edged when it propagates along the interface.
Its shape also depends on the nucleation process: when a bump
facets, it generates two rounded macrosteps; by contrast, a
sharp-edged macrostep immediately develops when a pit occurs in
the facet. The first process is common at low velocity whereas the
second one is only present at large velocity. These mechanisms are

described in more detail in Ref. 15. Finally, we have verified that the onset of instability is given by the constitutional undercooling criterion (eq.2), to within our experimental error (±30%). In particular, there is no measurable restabilization effect due to a possible kinetic anisotropy.

4-CONCLUSION

We have shown that the behavior of the Sm.A-Sm.B interface strongly depends upon its orientation with respect to the molecular layers. One of the major results is the existence of missing orientations in the equilibrium shape and the observation of the Herring instability. Another important result is that unstable orientations do not appear during growth, leading to an angular discontinuity at the cell tip. This tip boundary condition and the tendancy to facet parallel to the layers explain the nearly immediate development of sidebranching above the critical velocity. We have also seen coherent sidebranches over many cells which is possible evidence for the Hopf bifurcation predicted by Karma and Pelcé. Concerning the facet, we have never observed any stationary or periodic solution. On the other hand, we have observed the formation of macrosteps above the onset of instability, which drift and change form continually. Finally, we stress the crucial role played by diffusion anisotropy in explaining the drift of the cells. Note also that non-equilibrium effects (i.e. kinetics and associated anisotropy) do not seem to be very important, even in the faceted case.

ACKNOWLEDGMENTS

This work was partly supported by a contract with the Rhône-Alpes Région and by a C.N.R.S.-C.N.E.S. contract.

REFERENCES

1. W. W. Mullins, R. F. Sekerka, Stability of Planar Interface during Solidification of a Dilute Binary Alloy, J. Appl. Phys., 35:444 (1964).
2. D. J. Wollkind, L.A. Segel, A Nonlinear Stability Analysis of the Freezing of a Dilute Binary Alloy, Phil. Trans. Roy. Soc. London, 268:351 (1970).
3. B. Caroli, C. Caroli, B. Roulet, On the Emergence of One-dimensional Front Instabilies in Directional Solidification and Fusion of Binary Mixtures, J. Physique France, 43:1767 (1982).
4. S. de Cheveigné, C. Guthmann, P. Kurowski, E. Vicente, H. Biloni, Directional Solidification of Metallic Alloys: the Nature of the Bifurcation from Planar to Cellular Interface, J. Cryst. Growth, 92:616 (1988).
5. P. Kurowski, Etude Expérimentale d'un Système hors d'Equilibre Fortement Non-linéaire: Fronts cellulaires en solidification Directionnelle d'un Alliage Dilué (CBr$_4$), PhD Thesis, Paris (1990).
6. M. Ben Amar, B. Moussalam, Absence of Selection in Directional Solidification, Phys. Rev. Lett., 60:317 (1988).
7. A. Karma, P. Pelcé, Oscillatory Instability of Deep Cells in Directional Solidification, Phys. Rev. A, 39:4162 (1989).
8. D. A. Kessler, H. Levine, Linear Stability of Directional Solidification Cells, Phys. Rev. A, 41:3197 (1990).
9. M. A. Eshelman, V. Seetharaman, R. Trivedi, Cellular Spacing-I. Steady State Growth, Acta Metall., 36:1165 (1988).
   V. Seetharaman, M.A. Eshelman, R. Trivedi, Cellular spacing-II. Dynamical Studies, Acta Metall., 36:1175 (1988).
10. J. Bechhoefer, A. Libchaber, Testing Shape in Directional Solidification, Phys. Rev Lett., 35:1393 (1987).

11. Y. Saito, C. Misbah, H. Müller-Krumbhaar, Directional Solidification: Transition from Cells to Dendrites, *Phys. Rev. Lett.*, 63:2377 (1989).

12. L. Pfeiffer, S. Paine, G.H. Gilmer, W. van Saarloos, K. W. West, Pattern Formation Resulting from Faceted Growth in Zone-Melted Thin Films, *Phys. Rev. Lett.* 54:1944 (1985).

13. D. K. Shangguan, J. D. Hunt, Dynamical Study of the Pattern Formation of Faceted Cells, *J. Cryst. Growth*, 96:856 (1989).

14. J. Doucet, A.M. Levelut, X-ray Study of the Ordered Smectic Phases in some Benzylideneaniline, *J. Physique France*, 38:1163 (1977).

15. F. Melo, P. Oswald, Facet Destabilization and macrostep Dynamics at the Smectic-A Smectic B Interface, submitted to *J. Physique*.

16. P. Oswald, F.Melo, C. Germain, Smectic A-Smectic B Interface: Faceting and Surface Free Energy Measurement, *J. Physique France*, 50:3527 (1989).

17. C. Herring, Some Theorems on the Energies of Crystal Surfaces, *Phys. Rev.*, 82:87 (1951).

18. F. Melo, P. Oswald, Herring Instability and its Consequences on the Directional Growth of a Plastic Smectic B Phase, to be published in *Annales de Chimie*.

19. J. Bechhoefer, P. Oswald, A. Libchaber, C. Germain, Observation of Cellular and Dendritic Growth of Smectic A-Smectic B Interface, *Phys. Rev. A*, 37:1691 (1988).

20. F. Melo, P. Oswald, Destabilization of a Faceted Smectic A-Smectic B Interface, *Phys Rev . Lett.*, 64:1381 (1990).

21. R. Bowley, B. Caroli, C. Caroli, F. Graner, P. Nozières, B. Roulet, On Directional Solidification of a Faceted Crystal, *J. Physique France*, 50:1377 (1989).

22. B. Caroli, C. Caroli, B. Roulet, Directional Solidification of Faceted Crystal II. Phase Dynamics of Crenelatted Front Patterns, *J. Physique France*, 50:3075 (1989).

FACETTING OF WEAKLY ANISOTROPIC MATERIALS

Mokhtar ADDA BEDIA and Martine BEN AMAR

Laboratoire de Physique Statistique de l'Ecole Normale Supérieure
associé aux universités Paris 6 et Paris 7
24 Rue Lhomond, 75231, Paris Cedex 05, FRANCE

INTRODUCTION

In nature, most of the time, crystals show facets in specific directions which correspond to cristallographic symmetries. When perpendicular to the $\mathbf{n}$ direction, these facets are observed if the temperature is below the roughening temperature $T_r(\mathbf{n})$ characteristic of this direction $\mathbf{n}$. It is why, at ordinary room temperature, crystals show both rough and facetted parts. The existence of these facets is not restricted to equilibrium situations [1]. It is known for a long time, and it is used until now experimentally, that one way to get larger facets consists in imposing a slow growth to the crystal. The aim of our paper is to introduce a theoretical treatment of facetted crystal growth in situations far from equilibrium. This means that we take explicitly into account the diffusion in the liquid phase and, if necessary in the solid phase. We apply it to the needle crystal growth and the directional solidification.

This paper is organized as follows. Section I recalls how one can predict the equilibrium shape of crystals once the Wulff's plot is known and shows how to extend the theory of diffusion limited growth to facetted morphologies. In section II, we apply it to the free dendritic growth. Section III gives the first numerical solutions of facetted growing patterns: directional solidification in the low velocity regime.

I. FACETTED CRYSTAL SHAPES

I.a) EQUILIBRIUM SHAPES

At equilibrium the shape of a crystal is entirely fixed [2] by its surface tension $\gamma(\theta)$, if θ is the polar angle of the surface normal $\mathbf{n}$. As shown by Wulff, it corresponds to a minimum of the Gibbs free energy at total constant volume: $G = \int ds\, \gamma(\theta)$, where s and θ are the curvilinear coordinates of the interface.

At a point $\mathbf{r}$ of the interface which is locally orthogonal to $\mathbf{n}$, this minimization gives: $\gamma(\mathbf{n}) = \Delta P/2\; \mathbf{r}.\mathbf{n}$, with ΔP the solid-liquid pressure variation at the interface. It can be interpreted geometrically by the Wulff's construction (W-construction below): to obtain the crystal shape, one has to take the perpendicular to **n** (of polar angle θ) at the intersection of **n** with the Wulf-plot (W-plot which represents $\gamma(\theta)$). The shape is the <u>convex</u> envelope of the orthogonal lines to the **n** direction at a distance $\gamma(\mathbf{n})$. Moreover, the minimization of the thermodynamic potentiel fixes, in one hand, the local pressure variation ΔP in term of the local curvature Ω: $\Delta P = (\gamma(\theta) + d^2\gamma(\theta)/d\theta^2) \Omega$, where the quantity in parenthesis is called the surface stiffness. The equilibrium is possible if the stiffness is a positive quantity. In other hand, this minimization also gives a relation between ΔP and the liquid-solid chemical potentiel variation per unit volume of solid $\Delta\mu = \mu_l - \mu_s$. The combination of these two equalities yields to the Gibbs-Thomson law [3]:

$$\delta p \left(\frac{v_l}{v_s} - 1\right) - \frac{\delta T}{T_m} L - \Delta\mu = (\gamma(\theta) + d^2\gamma(\theta)/d\theta^2) \Omega \qquad (1)$$

where L is the latent heat per unit volume of solid, v_l (resp. v_s) is the specific volume of the liquid (resp. the solid) and δP (resp. δT) is the local variation of the liquid pressure (resp. the temperature) with respect to P_m (resp. T_m), the melting pressure (resp. temperature). The first term in the l.h.s of (1) is neglected because one often considers the pressure of the liquid constant and uniform. In general, there are two cases which are mostly considered: when one takes the case of a pure sample, the shape of the crystal is essentially due to the diffusion of latent heat, so what remains in the l.h.s of (1) is only the temperature field variation term. The second case is when the process includes the diffusion of impurities. In this case, both the second and the third term in the l.h.s of (1) are related to the effective concentration of impurities at the interface.

When the W-plot presents a conic point for some specific direction θ_0 as shown in Fig.1, the W-construction leads geometrically to a facet with a length $\pounds$ given by :

$$\Delta P \; \pounds = \gamma'(\theta_0{}^+) - \gamma'(\theta_0{}^-) \qquad (2)$$

once one assumes the stiffness positive on each side of the facet which implies tangential junctions of the rough parts with the facet. Moreover, for this direction, one can notice from relation (1) that the surface stiffness is not defined. Since the field variation ΔP cannot be singular for this direction, the curvature must vanish. The dominant term in (1) is $d^2\gamma(\theta)/d\theta^2$ and we recover Eqn.(2) by an average on the facet length. The geometrical W-construction and the Gibbs-Thomson relation are completely equivalent. The facet length is just given by the cusp in the W-plot. In the following, we will choose a simple form for the W-plot of a cubic crystal with four symmetric facets at $\theta_0 = \pm\pi/4$ and $\pm 3\pi/4$. For θ between 0 and $\pi/2$, we can write it as follows:

$$\gamma = \gamma_0 \; (1 + \delta \; (\sin|\theta - \theta_0| + \cos|\theta - \theta_0|)) \qquad (3)$$

δ is the cusp coefficient. This W-plot is shown in fig.1 and it gives for the facet length :

$$\pounds = 2 \gamma_0 \; \delta/\Delta P \qquad (4)$$

<u>I.b) NON-EQUILIBRIUM SHAPES</u>

For a theorist, it is very complicated to predict the shape of a growing crystal. There is no equivalent of a W-construction here, even if one restricts to steady-state situations, when

188

the crystal grows at constant velocity without modifying its shape. Even for the simplest W-plot of a completely isotropic material, the shape of a crystal in an infinite medium is not known analytically or geometrically. Moreover, it depends on the geometry of the experiment. The standard theory of crystal growth developed by Ivantsov [3] determines the crystal shape by the diffusion of latent heat (or impurities) in the liquid (and eventually in the solid). If one restricts to steady situations, it reads:

$$D_{s,l} \Delta\phi + v \, \partial\phi/\partial y = 0 \tag{5}$$

with $D_{s,l}$ the diffusion coefficient in the solid and liquid phases, v the rate along the y axis, the growing direction. For solidification of a pure liquid, the dynamic of growth is governed by the diffusion of latent heat L both in the solid and liquid phases. The dimensionless field ϕ is given in this case by:

$$\phi = \frac{c\,(T-T_m)}{L}$$

with c the heat capacity and T_m the melting temperature. A convenient model is the "symmetric" one [3] which assumes equality of all the thermodynamic coefficients of the two phases. When the solidification is the result of diffusion of impurity (in the case of an alloy), one usually neglects the diffusion in the solid phase (the one sided model with $D_s=0$) and the dimensionless impurity amount is :

$$\phi = \frac{c_l - c_\infty}{\Delta c}$$

where c_l is the impurity amount in the liquid phase and Δc is the miscibility gap as given by the phase diagram. In any case, the crystal interface is a source line which separates the liquid from the solid phases. This line of discontinuity gives the Stefan's condition :

$$\mathbf{v.n} = (D_s \nabla\phi_s - D_l \nabla\phi_l).\mathbf{n} \tag{6}$$

Capillary effects should not be forgotten, since they are responsible for the crystal shape at equilibrium. An important theoretical point is how to introduce them in a growth process? The theory of capillarity is well established for equilibrium situations but remains uncertain for non-equilibriun states. As usual [3] we will assume that the interface is in a quasi equilibrium situation and that we can apply the Gibbs-Thomson-Herring (GTH) [3,4] condition fixing the curvature of the surface in term of the diffusion field. If one wants to be more rigorous we can take into account the departure from equilibrium and adds the so-called "kinetic effects" which slow down the growth. For a pure sample, the GTH condition states the continuity of the chemical potential across the solid-melt boundary: $\Delta\mu = 0$. Including the kinetic effects means that $\Delta\mu$ and the growth rate have a one to one relationship. For rough interfaces, the attachment of the atoms at the interface is rather quick compared to a typical diffusion time except at high undercooling or supersaturation. So this discontinuity $\Delta\mu$ is very often neglected or reduced to a linear term $-\beta\mathbf{v.n}$ added to the GTH relation.

The GTH relation is valid in any smooth part of the crystal but has to be modified at $\theta=\theta_0$ where Ω vanishes. It can be modified according to Eqn. (2) but it is impossible to fix a constant ϕ on the facet. The profile of a facetted part is known so the solidification free-boundary problem is overdetermined when one fixes the value of ϕ for each point of the

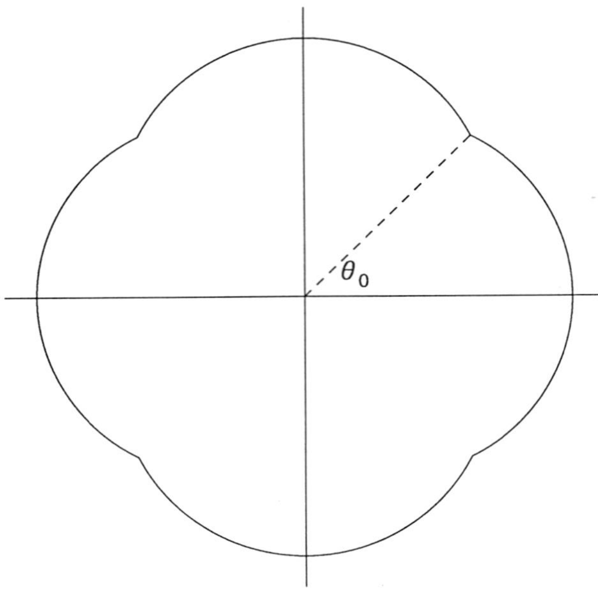

Figure 1. Wulff's plot for a cubic crystal: $\gamma / \gamma_0 = 1 + \delta \, (\sin | \theta - \theta_0 | + \cos | \theta - \theta_0 |)$, with $\theta_0 = \pi/4$ and $\delta = 0.5$

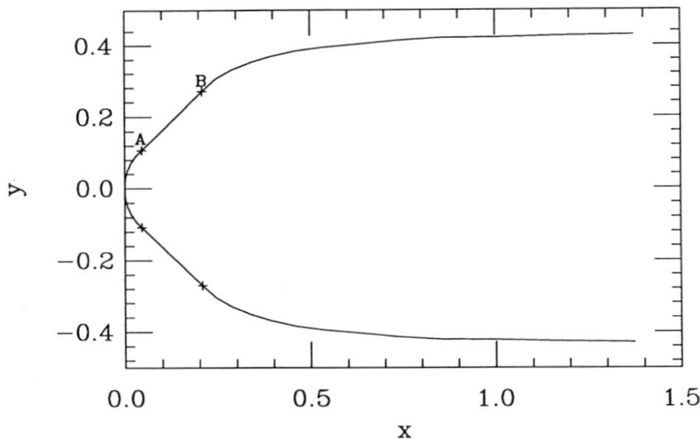

Figure 2. Numerical facetted cell, solution of the profile equation for $\sigma = 0.0033$, $P = 0.2$, $\delta = 0.5$ and $\nu = 1$. Definition of our coordinate frame and our profile partition A and B.

facet. One way to handle this is to replace the left-hand side of Eqn (2) by an integral over the facet length $\pounds$:

$$\int_{\pounds} ds \, \phi \; = <\phi> \pounds = - \; (\, d\gamma/d\theta|_+ - d\gamma/d\theta|_-) \tag{7}$$

and we recover a relation shown by Herring in the context of sintering. This relation may be interpreted physically, as saying that, if the growth is slow enough, fluctuations of the facet due to lateral motion of steps are fast enough to sample the average difference of chemical potential represented by the right hand side of (7) If we take the same W-plot as in (3), predicting now only two facets at $\theta_0 = \pm\pi/4$, we will deduce that the mean field on the facet is given by (4). The interface quasi-equilibrium hypothesis, when extended to facetting, neglects completely the various aspects of dynamical roughening transition [5]. In fact, in a dynamical process, the W-plot is not really cusped, it has only a pronounced dip which is therefore less and less pronounced as the velocity increases and as one is closer to the roughening temperature. In absence of a detailed theory of the dynamical roughening transition except when the growth of the facets is due to nucleations of bidimensional terraces, we will assume that the W-plot remains singular in growth situations, eventually with a dynamical cusp $\delta(v)$. This hypothesis is probably valid for experiments far below $T_r(\mathbf{n})$ at low undercooling. But in this regime of growth, it is generally observed that the facets inhibits the growth, compared to the rough parts. Two physical processes of facet growth have been studied: either the addition of steps around loops of dislocations or by nucleation of bidimensional germs. The two processes have different growth kinetics. Most of the time, far below the roughening temperature and in solutions, the first one predominates. The kinetic law: $\Delta\mu = F(v.n)$, for each of them, is well established when $\Delta\mu$ is homogeneous, which is clearly not the case here. One can wonder how to adapt this law when the field ϕ, so $\Delta\mu$, although unknown, varies from one point to another on the facet. The situation seems clearer in case of nucleation: the characteristic time of a germ creation is greater than the surface diffusion of steps. Since this time is governed by an Arrhenius law, the dynamics will be controlled by the extremum value of $\Delta\mu$. In case of growth by dislocations, we will assume that the one to one relation between $\Delta\mu$ and the normal velocity concerns only the average $<\Delta\mu>$ on the facet. So now Eqn (2) has to be transformed into:

$$\phi = (\gamma \; + d^2\gamma/d\theta^2) \, \Omega \; - \; F(v.n) \tag{8}$$

on the rough surfaces with F simply proportional to the normal velocity, which makes it irrelevant at low undercooling. On the facets, if we assume that the growth mechanism is by Franck and Read's sources [6], F(v.n) will be represented by various power laws :

$$F(\mathbf{v.n}) \approx \beta \; (v\cos\theta_0) \, ^\alpha \qquad \text{with } 1/2 \, \leq \alpha \leq 1 \qquad \text{and } \beta > 0 \tag{9}$$

at least above some rather small threshold value v_c. Eqn.(7) has to be changed into:

$$<\phi> \pounds = - \; (\, d\gamma/d\theta|_+ - d\gamma/d\theta|_-) \; - \beta \, \pounds \, (v\cos\theta_0)^\alpha \tag{10}$$

For any α, we see that the kinetic term does not change anything in the analysis. Its role is limited to increase the facet length $\pounds$ which can explain why large facets are observed at low velocities. It is important to note that averaged laws are necessary to solve mathematically the free boundary problem of diffusion. They are also the only way to fix the continuity of the temperature field and the chemical potential on both sides of the facet.

II. FACETTED NEEDLE CRYSTALS

During these last years, the theory of rough growing needle crystals has known important improvements. Both their shapes, their growth rates and their stability are now well explained. In the case of steady and invariant shapes, the classical theory has been developed [7] by Ivantsov and by Nash and Glicksman [7]. First, Ivantsov has shown that, at vanishing surface tension, the solution of the needle crystal problem gives a continuous set of parabolic shapes, such that only the velocity v times the tip radius of curvature ρ is function of the undercooling Δ imposed at infinity. At low undercooling, introducing surface tension adds a new relation between the velocity and the tip radius of the needle crystal. They are given in term of a nonlinear eigenvalue $C = v\,\rho^2/Dd_0$ [8], (d_0 is the capillary length: $d_0 = \gamma c T_m/L^2$, of order of the lattice spacing), such that:

$$v = (16\,/\pi^2 C)\,\Delta^4 \quad (D/d_0) \qquad\qquad \rho = C\,\pi\,\Delta^{-2}/4\,d_0 \qquad\qquad (11)$$

These relations are valid at low Δ, note that ρ is the radius of curvature of the asymptotic parabola and in the following we will choose 2ρ as length unit. So in principle, the determination of C will allow the selection mechanism. Since we focus on low undercoolings, we will restrict here to the Pelce-Pomeau [8] equation which is the limit of the Nash and Glicksman [7] one for low growth rates. This equation is an integro-differential one that gives both the dimensionless profile represented by $\zeta(x)$ and also the eigenvalue C. It reads:

$$\zeta''/(1+\zeta'^2)^{3/2} = -\,C/4\pi \int dt\,\mathrm{Log}\{((x-t)^2+(\zeta(x)-\zeta(t))^2)/((x-t)^2+(x^2-t^2)^2)\} \qquad (12)$$

where ζ' and ζ'' are respectively the first and second derivatives with respect to x. The numerical and analytical solutions of (12) show that for <u>isotropic</u> crystals, physical solutions <u>do not</u> exist for finite values of C. Nevertheless, in experiments the needle crystals show privileged directions of growth. This suggests to introduce also the effect of the anisotropy of surface tension [9a] which precisely will remove this indefiniteness. By taking into account this anisotropy, one finds [9a], at small growth rates, that there exists a discrete set of solutions C_n, so the quantity $\rho^2 v$, independent of the undercooling Δ depends only on the anisotropy coefficient ε. The stability analyis completes this study [10] by showing that among all this set of solutions $C_n(\varepsilon)$, the stable one corresponds to n=0, which agrees rather well whith experimental findings. In spite of this, the shape of the needle crystal remains parabolic in the tail regions. It differs only at the tip, on distance of order ρ, where the crystal looks more pointed, because of the capillary effect. The experimental findings are very limited by the difficulties of measuring the intrinsic parameters of the crystal, such as γ, Δ, D and ε. So in general, one limits oneself, in addition to draw the shape of the crystal, to measure the geometric parameter ρ and the velocity v. The only possible comparison with the theoretical results rests on the verification that $\rho^2 v$ versus v is a plateau, so is independent of the undercooling.

Experiments on growing needle crystals have shown [1] the existence of facetted ones, that is needle crystals with general parbolic shapes and facets close to the tip. A quantitative study of these objects was done by Maurer et al.[1] on solidification of NH_4Br in water below the roughening temperature of the (110) direction. In opposite to the rough case where $\rho^2 v$ is a constant, they found that this quantity versus v defines now two plateaus, according as v is greater or smaller than a certain critical velocity v_c. Between these two plateaus, one notes a transition zone of small extension in velocity which looks like more or less a jump. Moreover, they found that the facet length £ always scales as $v^{-1/2}$, at least in the first plateau of smaller velocities where facets appear unambiguously. It follows that £ is always proportional to ρ independently of v, so facets do not introduce a new length scale with respect to the case of nonfacetted crystals [9b]. One problem which remains questionable is about the origin of the transition between the two plateaus: is it a dynamical roughening transition or do facets exist in the second plateau too? Because the experiment is done far below the roughening transition temperature and the observed velocities are low, one may think that a roughening transition is not implicated. In addition of that, the plateaus confirm that the intrinsic parameters d_0 and the cusp δ (characteristics of the W-plot) are independent of the velocity which is not in favor of the dynamical roughening transition theory. In the following, we will propose an explanation connected to the kinetic effects behaviour. This argument supposes the existence of tiny facets in the second growth region, but this is not obviously the case experimentally.

When introducing facets, and following the approach exposed above, the most important consequence we find is that the scaling laws are unchanged, which implies in turn that the shape of the needle crystal should be always the same up to a dilation depending on the undercooling. The temperature field, $c(T-T_m)/L$, once replaced in (7) gives the mean temperature of the facet in dimensionless variables [9b]:

$$c (<T_f> - T_m) \, £/L = - 2 \, \delta \tag{13}$$

we recall that our length scale is 2ρ. Relation (13) shows that introducing facets does not introduce any new length scale, proving that the facet length scales as the tip radius ρ. This prohibits the disappearance of the facets as the growth velocity increases, at a temperature much lower than T_r as growing needle crystals should be homothetic to each other for a given W-plot and various undercoolings, even if they have facets. As a consequence, the facet length £ is scaled by $v^{-1/2}$, in agreement with the experimental results of [1] and checks the scaling laws for the rough needle crystal [8]. For a facetted needle crystal, the profile $\zeta(x)$ is obtained by solving equation (12) in the rough parts and supplementing it by the condition (13) on facets. This gives a new nonlinear eigenvalue C'. Since the quasi-equilibrium assumption implies that the ratio $£/\rho$ is Δ-independent, we conclude that even for a facetted needle crystal:

$$\rho^2 \, v = C' \tag{14}$$

The existence of two levels in experiment, instead of one, can be explained if we take into account the kinetic effects. As said above, the kinetic term F is simply proportional to the normal velocity on the rough surfaces, which makes it irrelevant at small Δ: from the scaling laws (11), the ordinary GTH term is of order Δ^2 while the kinetic term, when proportional to v, is of order Δ^4, so is negligible. This remark justifies the hypothesis of quasi-equilibrium used in D.L.G. For the facets, we will assume that the microscopic growth mechanism is by Franck and Read sources [6]. So for any $\alpha > 0.5$, the same argument as before, shows that the kinetic term is negligible, at least at the low undercooling limit. In this case (14) remains valid with the same eigenvalue C'. On the contrary, for v less then v_c ($\alpha=0.5$), both terms on the r.h.s of (10) have the same Δ dependence, so they contribute to the facet length in exactly the same way, although from (10) the kinetic term increases £, everything else being kept constant. Thus the plot $\rho^2 v$ versus v must show a new level, at any velocity greater than v_c, corresponding to a new nonlinear eigenvalue C" in (14). The low undercooling limit predicts an abrupt jump between the two levels but a transition zone of small extension (scaled with some positive power of Δ) seems more reasonnable.

The experimental result of the existence of plateaus suggests that the intrinsic parameters d_0 and δ are poorly dependent of the velocity. That is a great handicap for the roughening transition explanation which assumes a rather sharp dependence of these parameters to explain the disappearance of the facets. This is why we think that the approach of the influence of kinetic effects is more plausible.

III. FACETTING IN DIRECTIONAL SOLIDIFICATION

We present here both an analytical and a numerical treatment of facetted crystal growth [see also 9b] far from equilibrium, in the case of directional solidification in the cellular regime. Such a free boundary problem with volume diffusion in the liquid phase and specific interface laws has apparently never been solved before. In directional solidification, the shape is dominated by the volume diffusion of the impurities, which amount is characterized by $(c_l - c_\infty)/\Delta c$. The interface is stabilized by the exterior temperature gradient G and by the capillary effects.

The linear stability of the facetted crenellated front, for two orientations θ_0 of the facet, have been examined in [11a,b]. It has been found solutions of small extension beyond the usual Mullins-Sekerka [3] threshold: $v_{MS} = \dfrac{m\Delta c}{G} \dfrac{U_{MS}}{2D} \approx 0.5$, with D the diffusion coefficient in the liquid phase, m the absolute value of the slope of the solidus and U the pulling speed. We consider a cusp-like-cell pattern with two facets at $\theta_0 = \pm 45°$ of the velocity U. We compute infinite steady-state cells by solving the integro-differential equation which comes from the Green's function formulation of directional solidification [12]. The treatment of the rough parts of the cell is analogous to what has been explained in [12]. On the facet, because the shape of the crystal is known, we must look for the local supersaturation which gives the cusp coefficient δ by using the averaged GTH law as emphasized above. Our main results are the existence of such cells (see an example in Fig.2) for a surface tension parameter σ less than a maximum value $\sigma_{max}(\delta)$ (see Fig.3a), where σ is defined as:

$$\sigma = \frac{T_m \gamma_0}{G\alpha^2(2v-1)} \tag{15}$$

with $(2v-1)$ is the deviation from the Mullins-Sekerka threshold and α the pattern wavelength which is our length unit. At fixed cusp δ and Peclet number $P = \alpha U/D$, we have found a continuum of facetted cells labelled by σ. The facet length l increases with σ (Fig.4), and reaches its maximum value l_0 for $\sigma_{max}(\delta)$ (Fig.3b). For small Peclet number, note that σ_{max} varies almost linearly in P (Fig.3a) but l is quite independent of P as shown in Fig.3b and 4. We conclude that the facet length decreases with σ when the growth rate U increases. This can explain why the facets seem to disappear without any change of the W-plot. Even with facets, we found again known results concerning the Saffman-Taylor (S-T) instability and the directional solidification cellular regime. As a consequence, we proved that facetting does not introduce a new kind of selection mechanism: it does not restrict the space of solutions, rather it increases it since it introduces its own physical parameters such as δ.

To interpret our numerical results, at least in a limiting case, we have adapted an analytical treatment of directional solidification given in [13]. It relies on a viscous fingering analogy, valid at low Peclet numbers. This boundary layer analysis, splits the cell profile in two parts: the tip region for $\theta \leq \pi/2$ and the grooves for $\theta \approx \pi/2$. The cell tip has an extension of order one. the cell reaches its asymptotic behavior in the grooves, on a distance of order $(1-\lambda)^{-1/3}$ which is assumed to be small when compared to the characteristic length scale $1/P$ of the grooves, λ is the parameter of the equivalent S-T finger. The asymptotic matching procedure [13] proves that the surface tension fixes the parameter λ while the Peclet number is left arbitrary. Since the facets occur in part one ($\theta_0 << \pi/2$), our treatment of facetting modifies only the tip region and not the grooves so our analysis focuses mainly on the cell tip.

For the surface tension, we assume the same singular function as in (3). Due to the linear behavior of the impurity concentration field, valid to leading order in P, rough parts of the tip are deduced from the GTH law:

$$\omega_0 - \sigma \, \Omega_{int} = x_{int} \tag{16}$$

with Ω_{int} the positive curvature of the interface, σ is the dimensionless surface tension parameter which must be identified to the S-T parameter [13]. Hereafter, we will call part A the rough cell tip, part B the facet and part C the end of the cell tip before the beginning of the groove. From (16), we deduce the half profile of part A and C, the whole profile is obtained by symmetry about the x axis. In part A:

$$x(\theta) = (2\sigma)^{1/2} [(1+C_A)^{1/2} - (\cos\theta + C_A)^{1/2}]$$

$$y(\theta) = \sigma^{1/2} \Psi(\theta, C_A) \quad \text{with} \quad \Psi(\theta, C_A) = 2^{-1/2} \int_0^\theta d\theta' \, \frac{\cos\theta'}{(\cos\theta' + C_A)^{1/2}} \tag{17}$$

θ is less than θ_0 which defines the beginning of part B. For the moment, C_A is an unknown constant of integration. The profile in part B is a segment which begins at the point $A(X_A, Y_A)$ and stops at $B(X_B, Y_B)$ such that: $Y_A - Y_B = \cotan(\theta_0)(X_A - X_B)$. X_A and Y_A satisfies (17) for $\theta = \theta_0$. Finally, in part C of the cell:

$$x(\theta) = (2\sigma)^{1/2} [(\cos\theta_0 + C_C)^{1/2} - (\cos\theta + C_C)^{1/2}] + X_B$$
$$y(\theta) = \sigma^{1/2} [\Psi(\theta, C_C) - \Psi(\theta_0, C_C)] + Y_B$$

When applied both on the left and right hand sides of the facet, relation (16) gives the facet length l:

$$l = \sigma^{1/2} L = 2 \, \sigma^{1/2} [(\cos\theta_0 + C_A)^{1/2} - (\cos\theta_0 + C_C)^{1/2}] / \sin\theta_0$$

The condition that the tip fills nearly completely the cell ($\theta = \text{Arcos}(|C_C|) \approx \pi/2$) gives the surface tension coefficient:

$$\sigma^{-1} = 4 [\Psi(\theta_0, C_A) + \Psi(\text{Arcos}(|C_C|), C_C) - \Psi(\theta_0, C_C) + L\cos(\theta_0)]^2 \tag{18}$$

The matching of the profile to the grooves leads to the selection of λ by σ:

$$|C_C| \approx 2.17 \; (1-\lambda)^{2/3}/\sigma^{1/3}$$

In order to discuss this result, let us determine the facet length as a function of δ, the cusp coefficient. By performing the average of the GTH relation (16) along the facet, one recovers that:

$$(C_A - C_C)/\sin\theta_0 = 2\,\delta \tag{19}$$

After linearisation of (18) in the limit of vanishing $|C_C|$, we find the S-T parameter λ in term of the surface tension: $\sigma = \sigma_{max}(\delta)\;[1-\tau(\delta)\;(1-\lambda)^{2/3}]$. $\sigma_{max}(\delta)$ is the largest allowed surface tension for cusp-like-cells. λ identifies the S-T finger which fits perfectly the cell tip, at least at low Peclet numbers. The expressions of $\sigma_{max}(\delta)$ and $\tau(\delta)$ involve incomplete elliptic integrals and can be computed numerically. It is easy to check that $\tau(\delta)$ is always positive. In figures (3), we have plotted the theoretical maximum facet length and the maximum surface tension for different δ. Also, in theses figures numerical points, calculated with our code of directional solidification for various P values, are indicated. Note the nice agreement between the model predictions and the numerics and the weak effect of P. The most important result of our work is that with or without a cusp in surface tension, there exists an upper bound for the surface tension. Its value depends on the cusp amplitude. Its origin comes from the S-T picture of directional solidification, valid at low Peclet numbers. Note that, when $\lambda \neq 1$, at fixed δ, the facet length is reduced from its value at $\lambda=1$ (Fig.4). As expected, the maximum facet length is reached when both δ and σ take their largest allowed value. This model does not allow facets greater than 0.405 for $\theta_0=\pi/4$, even for $\delta=1$, although the geometry of the experiment allows a length of $2^{-1/2}$. Even for this particular value, the facet does not occupy the whole available space. To explain larger facets in directional solidification, one may put forward kinetic effects. Once added in (19), this kinetic supersaturation increases the effect of δ and the facet length.

This purely kinematic model where the surface tension is independent of the temperature contrary to any traditional dynamical roughening transition model, proves that the relative facet length decreases when the growth rate increases: first the dimensionless surface tension σ (Eqn.15) decreases when one moves away from the Mullins-Sekerka threshold; second, at low velocities, the kinetic "undercooling" is more efficient. This can explain the visual disappearing of experimental facets at temperature rather below the roughening transition temperature.

We have underlined that the growth at low undercooling is rather well explained by the diffusion limited growth theory. Unfortunately, even for completely rough crystal, this theory raises difficult mathematical problem although these last years, important improvements have been made concerning diffusion-controlled instabilities. We show that this theory can include facetted growing crystals from the melt. By scaling arguments we are able to explain experimental results on dendritic growth. Concerning directional solidification, we succeed in solving completely numerically the free boundary problem of solidification. This remains to be done for the needle crystal [14]. We show in particular that facetting does not modify the selection mechanism of this instability.

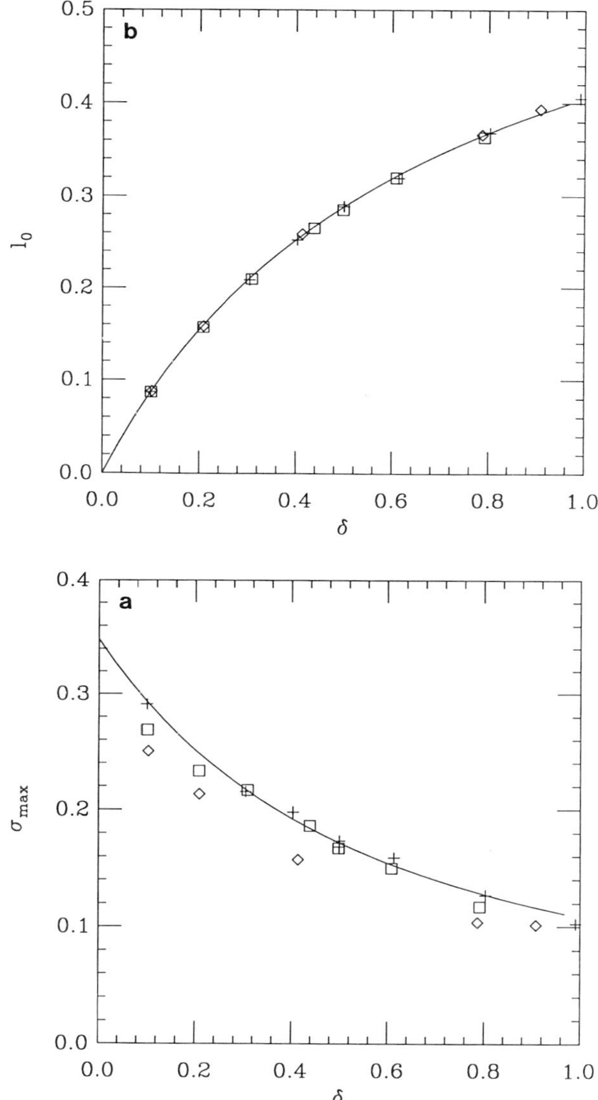

Figure 3. Maximum value of the surface tension parameter σ (Fig.3a) and the corresponding facet length l_0 (Fig.3b) versus the anisotropy coefficient δ.

___ Theoretical predictions for twice the value of the Mullins-Sekerka threshold and a coefficient of segregation equal to one .

Numerical points: + for P = 0.01, for P = 0.1, $\Diamond$ for P = 0.4

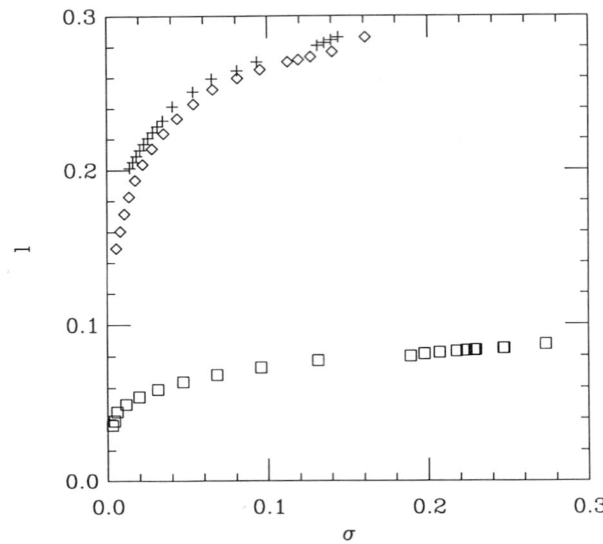

Figure 4. Facet length l versus σ for different values of P and δ
   + for P = 0.2 and δ = 0.5,   for P = 0.01 and δ = 0.1, ◊ for P = 0.01 and δ = 0.5

REFERENCES

[1] J. Maurer, B. Perrin, P. Bouissou and P. Tabeling, Europhys. Lett. **8**, 67 (1988). The occurence of facets at low velocities of growth seems to have been known for some time by metallurgists, see for instance K.A. Jackson and J.D. Hunt, Acta Metall. **13**, 1212 (1965) and references quoted therein. It has been observed too on Helium 3 (see ref.5); see also E. Raz, S.G. Lipson and E. Polturak, Phys. Rev. A **40**, 1088 (1989) and S.G Lipson, this proceeding

[2] L.D Landau, Statistical Physics (Pergamon Press, London, 1980)

[3] J. Langer, Rev. Mod. Phys. **52**, 1 (1980); J. Langer, "Chance and Matter", Ed. J. Souletie, J. Vannimenus, R. Stora (North Holland, Amsterdam, 1987), D. Kessler, J. Koplik and H. Levine, Adv. in Phys. **37**, 255 (1988); P. Pelcé, "Dynamics of curved fronts", Perspective in Physics (Ed. H Araki, A. Libchaber, G. Parisi, 1988); Y. Pomeau and M Ben Amar, "Dendritic growth and related Topics", (Solids Far from equilibrium, Beg Rohu Lectures, Ed. C Godreche, to be published)

[4] C. Herring, Phys. Rev. **82**, 87 (1951); C. Herring, "The physics of powder metallurgy" (Mc Graw-Hill, New-York, Ed.W.E. Kingston, 1951); M. Ben Amar and Y. Pomeau, Euro Phys. Lett. **6**, 609 (1988)

[5] P. Nozières Unpublished College de France Lecture, 1983/1984; F. Gallet, thèse de Doctorat d'état de l'université de Paris VI, 1986; E. Rolley, S. Balibar, F. Gallet, Euro. phys. Lett. **2(3)**, 247 (1986); P. Nozières and F. Gallet, Journ. de Phys. **48**, 353 (1987)

[6] W. K. Burton, N. Cabrera and F. C. Frank, Philos. Trans. R. Soc. London, Ser A **243**, 299 (1951); J. D. Weeks and G. H. Gilmer, Adv. In Chem. Phys. (Vol XL, Prigogine I. and Rice S.A. Ed. Wiley, New-York,1979)

[7] G. P. Ivantsov, Dokl. Akad. Nauk. SSR. **58**, 567 (1947); G. C. Nash and M. E. Glicksman, Acta. Metal. **22**, 1283 (1974)

[8] P. Pelcé and Y. Pomeau, Studies in Appl. Math. **74**, 245 (1986)

[9] a) M. Ben Amar and Y. Pomeau, Euro. Phys. Lett. **2**, 307 (1986); M. Ben Amar,

Physica **25**D, 155 (1987); b) M. Ben Amar and Y. Pomeau, Euro. Phys. Lett. **2**, 307 (1986), M. Adda Bedia and M. Ben Amar to be published

[10] D. Kessler, J. Koplik and H. Levine, Phys. Rev. A **33**, 3352 (1986)

[11] a) R. Bowley, B. Caroli, C. Caroli, F. Graner, P. Nozières and B. Roulet, Journ. de Phys. **50**, 1377 (1989); b) P. Pelcé unpublished

[12] M. Ben Amar and B. Moussallam, Phys. Rev. Lett. **60**, 317 (1988); M. Mashaal, M. Ben Amar and V. Hakim, Phys. Rev. A **41**, 4421 (1990)

[13] T. Dombre and V. Hakim, Phys. Rev. A **36**, 2811 (1987)

[14] M. Adda Bedia and M. Ben Amar in preparation

DIRECTIONAL GROWTH OF LAMELLAR EUTECTICS

K. Kassner[a] and C. Misbah[b]

[a] *IFF des Forschungszentrums, Jülich, Germany*
[b] *GPS, associé au CNRS, Place Jussieu, Paris, France*

We briefly review the problem of pattern formation in directional solidification of lamellar eutectics. We pay special attention to the recent development. In particular parity-breaking and similarity laws are discussed.

I. INTRODUCTION

When eutectics are submitted to directional solidification - that is by pulling the sample at a constant velocity V in an external thermal gradient G- the liquid-solid interface forms parallel lamellae of the two coexisting solid phases [1,2]. This problem, like the analogous directional solidification of binary alloys problem (see Misbah et *al.* in this volume) is of interest both from the point of view of its potential technological importance and as an example of pattern formation on the fundamental level.

The steady-state theory of eutectic growth has been described in a pioneering work by Jackson and Hunt[3]. It emerged from their analysis that steady-state growth exists within a wide range (infinite in principle) of wavelengths. Experiments, however, seem to indicate a unique spacing for given external constraints. Hence the question of whether wavelength selection is an intrinsic property of the system itself, or rather depends on experimental protocols, is a fascinating question, which still poses a formidable challenge to theoretical investgations. In a paper dealing with diffusive instabilities Langer[4] put forward the idea that the steady-state growth of a eutectic should occur at its point of marginal stability. He showed that this point happens to coincide with the minimum undercoling, which was often suspected to be the operating point. To arrive to a such result, however, use was made of the assumption that the solid-solid boundary is locally normal to the liquid-solid interface. The justification of a such hypothesis remained open. Relaxing the Langer[4] assumption, Caroli et *al*[5] have recently shown that, in the large thermal gradient limit, the steady solution at the minimum undercooling is not marginally stable (a pattern about the minimum undercooling is stable), thus indicating that the Langer[4] assumption is - at least for large thermal gradients- inappropriate. A full diffusive instability treatment[6], similar to that performed in directional solidification of binary alloys[7], is necessary to provide information at arbitrary thermal gradient.

Another line of development concerns parity-breaking transitions. Recent experiments on the $CBr_4 - C_2Cl_6$ transparent eutectic have revealed[8] the possibility for the lamellae to grow in a nonaxisymmetric manner. This corresponds to domains of tilted lamellae which move transversely to the growth front. In reality the observation of a such phenomenon can be traced back two decades at least, but only short comments were devoted to them[9] . It was also often argued that they owe to the underlying cristalline anisotropy. This is, as will be described below, not the case; they are completely intrinsic to the isotropic model. Similar phenomena have also been identified in different systems[10].

Below we will briefly describe the state of affairs in pattern formation in eutectics. We pay special attention to the recent development.

II. BASIC EQUATIONS

The basic equations of growth follow from the first and second laws of thermodynamics, applied on the local scale, combined with kinetic laws. Interface dissipation is most likely not a crucial ingredient and will not be accounted for here (of course this should be taken, in general, with certain caution; see Misbah et *al.* in this volume). The mass conservation law in the liquid phase reads

$$\frac{\partial u}{\partial t} = \nabla^2 u + 2\frac{\partial u}{\partial z}. \tag{1}$$

$u = (c - c_e)/\Delta c$, where c is the actual liquid concentration , c_e the eutectic concentration and $\Delta c = c_\beta - c_\alpha$ (Fig.1). In eq.1 lengths and time are reduced by $l = 2D/V$ and l^2/D respectively, where D is the diffusion constant and V the pulling speed. The concentration far ahead of the solidification front is maintained at a value c_∞, so that the condition on u at infinity reads $u(z = \infty) = (c_\infty - c_e)/\Delta c \equiv u_\infty$.

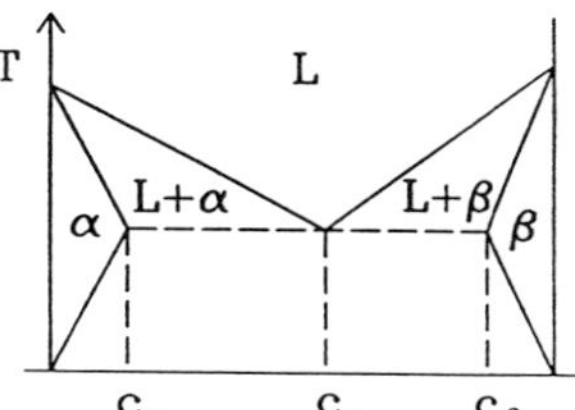

Fig.1. Eutectic phase diagram.

At the liquid-solid boundaries (liquid-α phase and liquid-β phase) the continuity equation and the Gibbs-Thomson condition take the form:

$$\frac{\partial u}{\partial n} = -(2 + \dot\zeta)\{(1 - k_\alpha)u + \delta\}n_z \tag{2a}$$

$$\frac{\partial u}{\partial n} = -(2 + \dot\zeta)\{(1 - k_\beta)u + \delta - 1\}n_z \tag{2b}$$

$$u = -\frac{\zeta}{l_T^\alpha} - d_0^\alpha \kappa \tag{3a}$$

$$u = \frac{\zeta}{l_T^\beta} + d_0^\beta \kappa \tag{3b}$$

where $\delta = (c_e - c_\alpha)/\Delta c$, k_i is the partition coefficient for the liquid-i solid ($i = \alpha, \beta$) coexistence, d_0^i and l_T^i are the usual (chemical) capillary and thermal lengths for the i phase. Finally κ is the surface curvature counted to be positive for a convex solid.

Eqs. (1)-(3) should be supplemented by the mechanical equilibrium conditions where the three phases intersect (triple point)

$$\vec{\gamma}_{\alpha\beta} + \vec{\gamma}_{\alpha L} + \vec{\gamma}_{\beta L} = 0, \tag{4}$$

where $\vec{\gamma}_{ij}$ designates the surface tension between phase i and phase j. If the magnitudes of surface tensions are known, then condition (4) determines uniquely the contanct angles θ_α and θ_β. If the pattern is tilted at an angle ϕ then the pinning angles at the two ends of a given lamella have different values. For example if θ_α and θ'_α denote these two angles (with, for example, $\theta_\alpha > \theta'_\alpha$), they are related by

$$\theta_\alpha = \theta'_\alpha + 2\phi \tag{5}$$

and a similar relation for θ_β and θ'_β. Relation (5) follows frow a simple geometrical argument. Eqs. (1)-(4) completely describe the dynamics of solidification of a lamellar eutectic.

III. STEADY-STATE SOLUTIONS AND STABILITY

The steady-state theory of axisymmetric lamellar eutectic growth has been described in an important paper by Jackson and Hunt[3]. Their idea is the replacement of conditions (2) and (3) by averages over individual lamellae, which are assumed, to first approximation, to be flat. They then compute the steady diffusion field everywhere in the liquid phase in terms of a Fourier series. The Fourier coefficients are computed by imposing the conservation equations (2a,b) at the (still-undetermined) average interface position. The final step consists in reporting the solution for u into eqs.(3). Then taking averages over each lamella the physical average front position ($< \tilde{\zeta} >$), which is a measure of the average undercooling Δ ($\Delta = -G < \tilde{\zeta} >$), is obtained as

$$\Delta = A/\lambda + BV\lambda, \tag{6}$$

where A and B are constant quantities which depend on material parameters only[3]. The first term expresses a capillary effect ($A \sim d_0$) and acts, as expected, at small λ where pinning at the triple point is important. At large λ the second term, which is proportional to the rate of phase transformation dominates. $\Delta(\lambda)$ takes on a minimum at

$$\lambda = \lambda_{min} = \sqrt{A/BV}. \tag{7}$$

Since $A \sim d_0$, λ_{min} is nothing but (up to a constant prefactor) the geometrical mean of the capillary and diffusion lengths. Eq.(6) indicates that there is a continuous family of possible wavelengths. In other words the present calculation is incapable of predicting the unique behaviour observed in experiments. It has often been argued that eutectic growth should occur at the minimum undercooling. If such is the case then the selected wavelength should scale, according to eq.(7), as $\lambda \sim V^{-1/2}$. This scaling was observed in various experiments[11]. However careful experimental investigations show that such a law holds only if the velocity is large enough (for more details see Sec. V). At small velocities, however, a strong deviation from that law is observed[11] (see Sec.V).

The study of soft instabilities of phase type was performed by Langer[4]. The analysis led to the conclusion that the state with the minimum undercooling is marginally stable (a phase fluctuation does not grow nor decay in the course of time). Langer[4] suggested that the system should operate at its marginal point. This speculation was likely motivated by the apparent success of the marginal stability criterion in predicting a velocity selection in dendritic growth. Despite the success of such a criterion in some model equations[13], its general application remains to be shown. We should mention in addition that the calculation of Caroli et al[5] performed in the large thermal gradient limit, without resorting to the Langer[4] assumption that the solid-solid line remains orthogonal to the liquid-solid interface, has shown that the state with the minimum undercooling is not marginal (all patterns about the minimum undercooling are stable). This indicates that the Langer assumption is not legitimate. The full study of phase dynamics at arbitrary thermal gradients is currently under investigation[6].

Datye and Langer[4] have shown that the interface undergoes an oscillatory instability for sufficiently large values of u_∞, which is a measure of departure from the eutectic concentration. This instability corresponds to a situation where two consecutive triple points oscillate at opposite phases perpendicularly to the growth direction.

IV. PARITY-BREAKING

Recent experiments[8] on the $CBr_4 - C_2Cl_6$ have shown that under some circumstances, a domain of tilted lamellae may appear. This domain may either spread out to contaminate adjacent axisymmetric domains, shrink, or even keep the same width. This depends both on the sign of the tilt angle and relative values of the wavelengths of the untilted domains that escort the tilted one (for example if the wavelengths on both sides of the tilted domain are equal, the width of the domain remains constant). As time elapses the tilted lamellae move at a constant speed transversely to the growth front. Similar phenomena have been observed in directional growth of a nematic crystal and in directional viscous fingering[10].

Coullet et al.[14] suggested that tilted domains are localized inclusions of a new antisymmetric state which emerges as a result of a secondary bifurcation of the symmetric state. One expects, generically, bifurcations to be accompanied with a loss of some symmetry (for example the Rayleigh-Bénard instability leads to the breaking of the translational symmetry of the quiescent fluid). In the present situation it is parity-breaking that accompanies the bifurcation. If one admits such an idea, one should expect that there exist solutions of the eutectic problem with periodic tilted lamellae filling the whole space.

To deal with this question we have first converted the basic equations (1)-(3) into a single closed integral equation for the interface profile $\zeta(x,t)$, by paying the price of nonlocality. The derivation of the front equation uses standard Green's function techniques[16]. Tilted lamellae move at a constant speed, $V_t = V\tan(\phi)$, transversely to the growth front. In the rest frame of the pattern the integral equation may be written as

$$\int d\Gamma g(\mathbf{r},\mathbf{r}')\frac{\partial w(\mathbf{r})}{\partial n} = \int d\Gamma w(\mathbf{r})\left\{\frac{\partial}{\partial n}g(\mathbf{r},\mathbf{r}') - \frac{1}{2\ell}(n_z + n_x\tan\phi)g(\mathbf{r},\mathbf{r}') - \frac{\delta(\mathbf{r}-\mathbf{r}')}{2}\right\},$$

$$(8)$$

where integration is performed along the interface (Γ designates the coordinate along the surface). g is the propagator associated with the stationary form of the diffusion equation(1), $w = u - u_\infty$ and ϕ is the tilt angle, unknown for the moment. Eq.(1) is discretized by means of a boundary element scheme[17]. Use is made of an intrinsic representation of the curve, $\theta(s)$, where θ is the angle between the growth axis and normal to the interface, and s is the curvilinear coordinate. We seek periodic solutions with periodicity λ. Choosing $N = N_\alpha + N_\beta$ discretization points over one period we have $N - 2$ angle variables. Because the front position along the growth axis is not arbitrary due to the presence of the external thermal gradient, we fix its position by, for example, the coordinates of one triple point (x_e, z_e), not known _a priori_. The number of unknowns is N. We impose the integral equation everywhere except at the triple points. There we require mechanical equilibrium instead (eq.4) (four conditions since the contact angle at the $\alpha - \beta$ intersection is different from that at the $\beta - \alpha$ one, because of asymmetry; see eq.(5).). One sees that the number of equations is also N.

It might therefore seem at first sight that our equations could be solved for any value of ϕ. This is however not the case. Indeed, giving a set of the N unknowns ($N-2$ $\theta's$ and x_e, z_e) that solve our equations does not necessarily imply, unless accidentally, that the two end points of the interval, ζ_1 and ζ_N, will lie at the same height. This would imply, in general, a jump of the interface at these end points. Hence, the requirement that our solutions be physical is that $\zeta_1(\phi) - \zeta_N(\phi) = 0$.

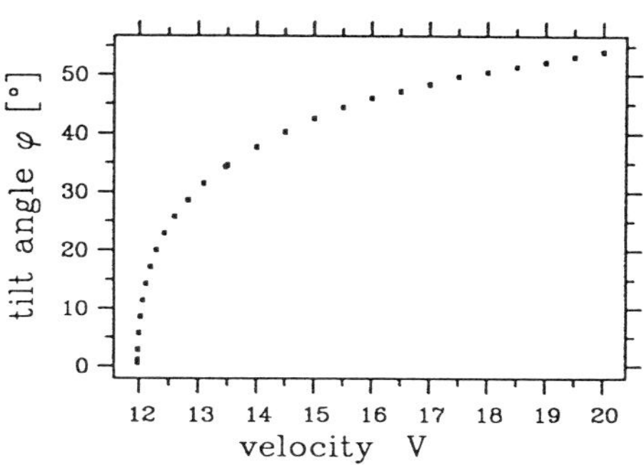

Fig.2. The tilt angle as a function of V.

This is a quantization condition which leads to the selection of a discrete set of possible ϕ values.

Focusing on the Faivre et *al.*[8] experimental set-up , we have shown that the fully isotropic model of eutectic growth supports tilted solutions with a well defined tilt angle. For a given wavelength tilted solutions exist[15] only above a velocity V_c which scales with λ as $V_c \sim \lambda^{-2}$. Fig.2 shows the dependence of the tilt angle on V for a fixed λ. One sees there that parity-breaking is a forward bifurcation of the basic symmetric state. Fig.3a displays an axisymmetric profile at $V = V_c$, while Fig.3b shows a front profile slightly above V_c. A remarkable point is that the tilt angle is as large as $20°$ very close above V_c. Computing the wavelength λ_{min} at $V = V_c$ for the axisymmetric state with the minimum undercooling, we find, for the set of parameters used so far, that $\lambda \simeq 2\lambda_{min}..$

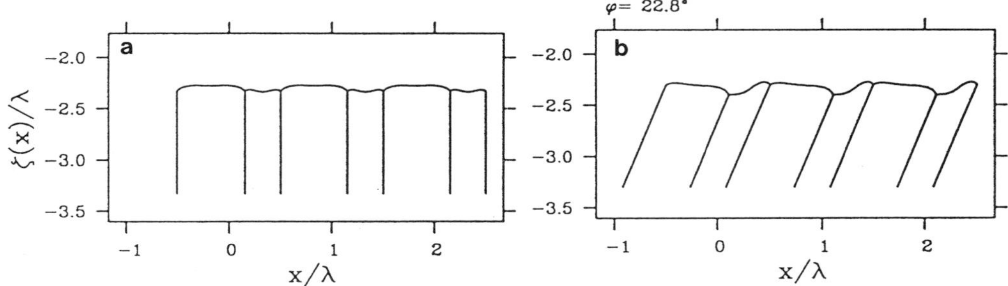

Fig.3a,b. Untilted and tilted growth respectively.

This agrees with experiments[8]. This result seems to be general for different systems undergoing a parity-breaking transition. A further point worthmentioning is that, for not too small V's (see Ref. 18), the tilt angle depends on $\lambda^2 V$ only. This result agrees with experimental investigations[8] where no noticeable variations of ϕ were detected, and where the quantity $\lambda^2 V \simeq constant$.

V. SIMILARITY LAWS

A common result reported by many experiments[11] is that the lamellar spacing scales as $\lambda \sim V^{-1/2}$. The same scaling holds at the minimum undercooling point in the Jackson and Hunt theory[3] (see eq. 7). Moreover the tilt angle for non axisymmetric growth[15] depends, approximately, on $\lambda^2 V$ only... These results seem to indicate that the quantity $\lambda^2 V$ is the only relevant parameter. Hence it is of great importance to understand this fact, and to specify its degree of validity.

The eutectic problem contains, for periodic structures, four independent lengths: d_0, l, λ and l_T. One of these lengths may serve as a length unit so that the number of independent dimensionless quantities is equal to three. Or more generally, we are at liberty to choose any three independent dimensionless combinations to characterize the problem. We take for example $\sigma \equiv d_0 l/\lambda^2$, $\chi \equiv l/l_T$ and $P \equiv \lambda/l$ (the Péclet number).

From dimensional considerations the physical quantites should depend on these three combinations only. To eventually reduce the number of degree of freedom, we should go into details of the governing equations and the order of magnitude of different contributions. In standard experiments the diffusion length is much bigger than the wavelength λ; that is $P \sim 10^{-2}$. We have taken advantage of this fact[18] to show that P scales out from the integral equation (8) so that the remaining parameters are χ and σ only. The small Péclet number limit is not uniform and care should be taken when performing it. Moreover the kernel of the integral equation diverges logarithmically. We have explained[18] how to circumvent this difficulty. The similarity equation derived in this limit indicates that the pattern is self similar under simultaneous stretching (or shrinkage) of λ and (G, V) by α and α^{-2} respectively, where α is a real number. Numerical integration of the original equation has confirmed our expectations: the physical quantities depend on σ and χ only. For example the tilt angle $\phi = \phi(\sigma, \chi)$. For small χ (which we call the large velocity domain) we find, as mentioned above, that the dependence on χ is so smooth that ϕ seems to depend on σ only. As the velocity decreases, however, that is when χ increases we observe a strong dependence on χ (see ref. 18 for more details).

An important consequence of the similarity is that the wavelength of the pattern should scale as

$$\lambda \sim V^{-1/2} f(\chi), \tag{9}$$

where f is a function of χ only. It appears from eq.(9) that $\lambda \sim V^{-1/2}$ only if $f \simeq constant$. We have computed f for the minimum undercooling point.

The condition of minimum undercooling has conventionally been assumed in the metallurgical literature to locate the operating point; and the resulting predictions seem to be in good agreement with experiments. Fig.4 shows the behaviour of f. At large $V's$ f is a smooth function and can hardly be distinguished from a constant in practice. At small V f starts decreasing significantly. This entails that $\lambda^2 V$ deviates from a constant. More precisely at small V, λ is smaller than what is predicted from the J-H theory.

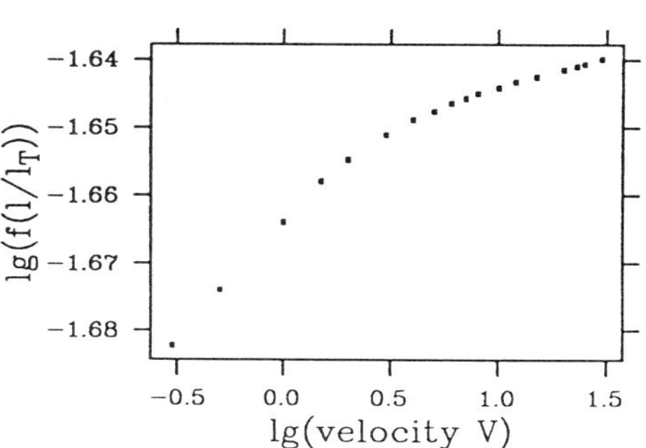

Fig.4. The behavior of the scaling function

Our results agree well with a large number of experiments[11]. Indeed careful experiments[11] have shown that the law $\lambda^2 \sim V^{-1}$ holds only at large enough growth velocities. At small velocities a deviation from that law is observed. That deviation has the same tendency as that reflected by the 'scaling' function (Fig.4).

A natural extension of this work consists in dealing with the stability of the pattern. Firstly the full study of diffusive instabilities performed in directional solidification of dilute mixtures[7] can be extended in principle to the present problem, but the mathematics require some sophistication[6]. The other type pertains to short wavelength instabilities. This will allow us to, alternatively, locate the point of parity-breaking. Moreover there is some experimental evidence in the literature that domains of sufficiently large wavelength preferentially exhibit an oscillatory instability. If such is the case then we expect that, generically, a secondary instability of the oscillatory state should lead to a chaotic regime. For more precise indications it seems to us imperative, as a first step, to perform the full stability analysis of the pattern. We hope to report on these aspects in the future.

Note : In an interesting paper Alain Karma[19] has studied eutectic growth using a random walk model. He found a 'tilting instability' for off-eutectic concentrations. Such an instability is most likely the dynamical manifestation of parity-breaking transitions.

REFERENCES

1. J.D. Hunt and K.A. Jackson , *Trans. Metall. Soc. AIME* **236**, 843 (1966).

2. V. Seetharaman and R. Trivedi, *Metall. Trans.* **19A.**, 2955 (1988).

3. K.A. Jackson and J.D. Hunt, *Trans. Mettal. Soc. AIME* **236**, 1129 (1966).

4. J.S. Langer, *Phys. Rev. Lett.* **44**, 1023 (1980); V. Datye and J.S. Langer, *Phys. Rev. B* **24**, 4155 (1981).

5. B. Caroli, C. Caroli and B. Roulet, *J. Phys.* **51**, 1865 (1990).

6. K. Kassner and C. Misbah (in progress).

7. K. Brattkus and C. Misbah, *Phys. Rev. Lett.* **64**, 1935 (1990).

8. G. Faivre, S. de Cheveigné, C. Guthmann, and P. Kurowski, *Europhys. Lett.* **9**, 779 (1989).

9. R. Racek, *Thèse d'Université Nancy I*, (1973); see also H.E. Cline, *Mat. Sci. Engr.* **65**, 93 (1984).

10. A.J. Simon, J. Bechhofer, and A. Libchaber, *Phys. Rev. Lett.* **61**, 2574 (1988); M. Rabaud, S. Michalland and Y. Couder, *Phys. Rev. Lett.* **64**, 184 (1990).

11. For a review see G. Lesoult, *Ann. Chim. Fr.* **5**, 154 (1980).

12. J.S. Langer and H. Müller-Krumbhaar, *Acta Metallurgica* **26**, 1681; 1689; 1697 (1978).

13. W. van Saarloos, *Phys. Rev. A* **37**, 211 (1988) and references therein.

14. P. Coullet, R.E. Goldstein and G.H. Gunaratne, *Phys. Rev. Lett.* **63**, 1954 (1989).

15. K. Kassner and C. Misbah, *Phys. Rev. Lett.* **65**, 1458 (1990); erratum *Phys. Rev. Lett.* **66**, 522 (1991).

16. B. Caroli, C. Caroli, B. Roulet and J.S. Langer, *Phys. Rev. A* **33**, 422 (1986).

17. Y. Saito, G. Goldbeck-Wood, and H. Müller-Krumbhaar, *Phys. Rev. A* **38**, 2148 (1988).

18. K. Kassner and C. Misbah, *Phys. Rev. Lett.* **66**, 445 (1991).

19. A. Karma, *Phys. Rev. Lett.* **59**, 71 (1987).

ARRAY DISORDER, PATTERN SELECTION AND LOCAL DENDRITIC TRANSITION
IN CELLULAR DIRECTIONAL SOLIDIFICATION

Bernard Billia, Haïk Jamgotchian and Henri Nguyen Thi

Laboratoire de Physique Cristalline (URA CNRS n° 797)
Université d'Aix-Marseille III, Faculté des Sciences de
St Jérôme, Case 151, 13397 Marseille Cedex 13, France

INTRODUCTION

Controlled solidification is one of the major techniques to produce engineering components since a close control of the microstructure, and thus of the properties of the material, is possible. Therefore, it is important to establish the precise correlation between the microstructure and the processing conditions. On the other hand, in recent years interesting similarities have been appreciated between various pattern-forming instabilities in different fields of science[1-3], which have led to significant theoretical advances in the understanding of the selection and stability of structures. Cells and dendrites, which result from morphological instabilities of the solidification front[4], belong to this family. Therefore, besides their long-time known technological importance, these non-linear patterns have now gained further attention as archetypes. The present paper concerns cellular patterns in directional solidification of a binary alloy, with an emphasis on the defects and disorder of the arrays. Indeed, in dissipative structures, disorder is linked to the basic problem of the wavenumber selection[5] and it even seems to play a decisive role in the transition to turbulence[6].

In contrast with other instabilities, e.g. Bénard-Marangoni cells and nematic-shear-flow patterns[7], the analysis of cellular arrays is at its very beginning. For the first time in directional solidification, we have recently carried out such an analysis[8] on Pb-30wt% Tl alloys grown upwards. A Wigner-Seitz construction[9] was performed which, besides an accurate determination of the primary spacing, enabled the statistical analysis of the defects and of the disorder of the arrays. The major results are: i) the basic defects are polygons with less, or more, than six sides, ii) the percentage of defects d_s is high so that the regular honeycomb, which would exist for a "2D-solid" array, is actually "melted" by the defects and iii) the correlation functions concomitantly are typical of only a short-range order.

First, the analysis of the disorder of 2D-cellular arrays will be improved by using a *weighted* Wigner-Seitz construction. Then, the selection of the primary spacing will be analyzed. Further informations will be obtained by using the *Minimal Spanning Tree* approach and the universal (m,σ)-diagram on which the standard deviation σ is plotted versus the average length m of the tree edges[10]. Finally, a theory will be sketched to estimate when, upon changes in primary spacing, unbalanced constitutional supercooling will locally build up in a 1D-cellular array, where it will drive *local dendritic transition*.

WIGNER-SEITZ APPROACH TO DISORDER OF 2D-CELLULAR ARRAYS

Weighted Wigner-Seitz Construction

In directional solidification of a binary alloy in a temperature gradient G, a cellular microstructure develops at the solid - liquid interface above a critical value of the growth

Fig. 1. Binary image. Pb-30 % Tl. G = 45°C/cm.
V = 1.1 cm/h. Image width = 3.2 mm.

velocity V^{11}. In a bulk sample, the cells self-organize into an array of columns which, by chemical etching on a transverse section, usually shows up as an imperfect 2D-array. With a CCD camera and a video digitizer, this array is acquired in 256 grey levels on a computer hard disk. A binary image is then obtained by appropriate filterings and scalings (Fig. 1). The black lines indicate the intercellullar grooves. The centers of gravity of the white areas are taken as the cell centers which is sufficient for a Wigner-Seitz construction of the array[8]. Such a construction fits well when the cells are of about the same size but it is no longer satisfactory when a large cell and a small cell are neighbours, the latter being enlarged at the expense of the former[11]. These findings have motivated the introduction of a *weighted* Wigner-Seitz construction.

The basic idea is to associate to each cell a weight related to the corresponding white surface on the binary image. The relevant weight is the inverse of the square root of that surface. The cell sides are then obtained one point after the other (Fig. 2) by using the relation

$$\frac{d_i}{d_j} = \sqrt{\frac{S_i}{S_j}} \tag{1}$$

where d_i (d_j) is the distance from the point under consideration to the center of cell i (j), whose surface is S_i (S_j). The physical meaning of the weight we have introduced becomes clear when each cell is replaced by a soft disk of identical surface. Indeed,

$$\frac{d_i}{d_j} = \sqrt{\frac{S_i}{S_j}} = \frac{\phi_i}{\phi_j} \tag{2}$$

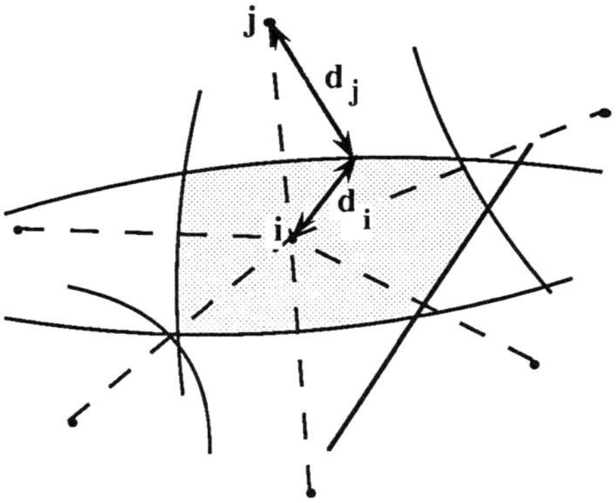

Fig. 2. Weighted Wigner-Seitz construction ($d_i / S_i^{1/2} = d_j / S_j^{1/2}$).

212

Fig. 3. Superposition of the 2D-cellular array obtained by a weighted
Wigner-Seitz construction on the binary image. Pb-30 % Tl.
G = 45 °C/cm. V = 1.1 cm/h. Image width = 3.2 mm.

where ϕ_i (ϕ_j) is the diameter of cell i (j) so that, in a picturesque way, we can say that the
cellular array is obtained by compaction of a set of soft disks, in order to suppress the
interstices[11].

Cell after cell, a 2D-array is constructed whose superposition on the binary image is
now excellent, even when the variation of the cell size locally is important (Fig. 3). As the
averaging effect, which is inherent to the classical Wigner - Seitz construction, is eliminated

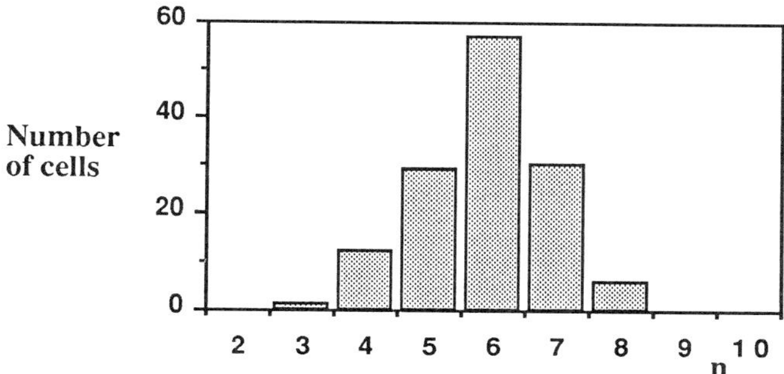

Fig. 4. Histogram of the n-sided cells. Pb-30 % Tl.
G = 45 °C/cm. V = 1.1 cm/h.

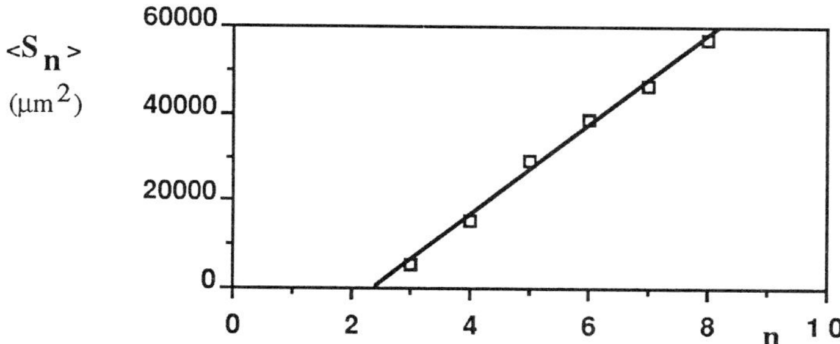

Fig. 5. Variation of the average area of an n-sided cell with the number n
of its sides. Pb-30 % Tl. G = 45 °C/cm. V = 1.1 cm/h.

by a proper weighing, it follows that the "liquid" nature of the cellular array is even more pronounced. Consequently, the histogram of the n-sided cells (Fig. 4) shows a significant increase of the number of defects, the 6-sided cells now being only 40 % of all. Besides, the percentages of the squares or octagons, previously of the order of 1 % [8], now is in between 5 and 10 %.

Figure 5, which gives the variation of the average cell surface with the number of sides, shows that Lewis's rule is satisfied by the 2D-cellular arrays observed in directional solidification. According to this rule, first proposed by Lewis for biological cells[12], the average area $<S_n>$ of an n-sided cell is a linear function of the number n of its sides. Since, it was established by Rivier and Lissowski[13] that Lewis's rule actually is a mathematical law which has to be verified by any space-filling structure, which leads to the identification of a parameter measuring the ageing of the array[9].

Pattern Selection

If the definition of the primary spacing λ is simple for a perfectly periodic array, with long range order, or for a random distribution of cells,

$$\lambda = \alpha <S>^{1/2} \tag{3}$$

where α is a coefficient which depends on the arrangement[14] and $<S>$ the average cell surface, there is no obvious choice as soon as the structure becomes disordered until it becomes completely random, as it is the case for the lead-thallium alloys which are under study. Then, the value of the parameter α is unknown. In such a situation of high disorder, it is far more convenient to characterize each cell by the diameter ϕ of the disk of equal area and to extract λ from the distribution of ϕ. The major advantage being to get rid off of any estimation of the parameter α, otherwise more or less affected by subjectivity.

From figure 6, which gives the histogram of the diameter ϕ, for a set of 440 cells, the distribution is symmetrical with a mean value $<\phi> = 130$ μm, which is identified with the primary spacing λ, and a standard deviation $\Delta\phi = 30$ μm, which measures the dispersion of the data. It follows from the mere presence of a peak that, by some mechanism, the spacing is well and truly selected in these experiments. In the cellular range, the most advanced theoretical studies[15] predict the existence of a band of steady states, whose width is difficult to appreciate, rather than the sharp selection of a unique spacing. Nevertheless, the question of the stability of these steady states is up to now unanswered. It should be emphasized that the shape of the distribution of ϕ strongly depends on the total number N of cells. For these lead-thallium alloys, a sound analysis of the selection of the primary spacing is possible only for N≥200. It cannot be excluded that this last remark may be valid for other binary systems.

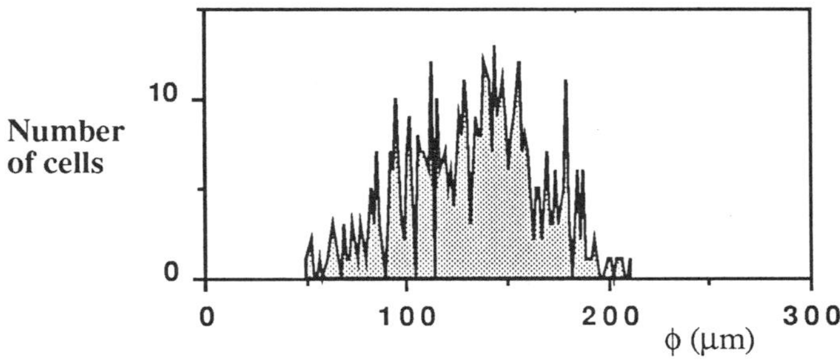

Fig. 6. Histogram of the diameter ϕ. Pb-30 % Tl. G = 45 °C/cm.
V = 0.75 cm/h.

MINIMAL SPANNING TREE APPROACH TO DISORDER

The Minimal Spanning Tree and the (m,σ)-Diagram

The Minimal Spanning Tree (MST) can be constructed for any distribution of points on a surface so that the organization of any 2D - array can be analyzed. As the basic

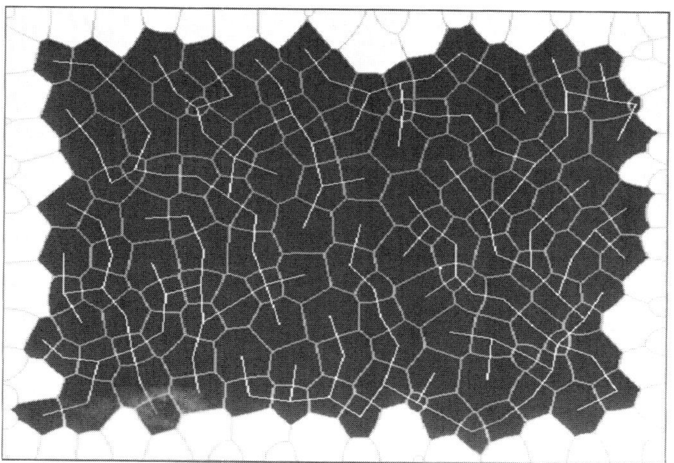

Fig. 7. Minimal Spanning Tree superimposed on the 2D-cellular array
obtained by a weighted Wigner-Seitz construction. Pb-30 % Tl.
G = 45 °C/cm. V = 1.1 cm/h. Image width = 3.2 mm.

definitions of graph theory can be easily found[10,16-18], let us just recall here that an edge-weighted linear graph is composed of a set of nodes (the cell centers for the case of cellular solidification) and a set of edges, an edge being defined by a pair of cell centers, with a weight assigned to each edge (the corresponding center to center distance). A Minimal Spanning Tree is a connected graph without any closed loop, which contains all the cell centers and for which the sum of the edge weigths is minimum. The most illustrative way to construct a MST, which yet leads the most time-consuming algorithm, is to start from any cell center and to add at each step the cell center which is closest to the current tree, together with the weight associated to that shortest distance (Fig. 7).

There are a few particular situations in which, for a given distribution of cell centers on a surface, the MST may not be unique and locally vary depending on the choice of the starting point for its construction. Nevertheless, all the possible MSTs do have the *same* edge-length histogram. Therefore, parameters deduced from the statistical analysis of that histogram can be used to characterize the arangement of the cell centers. The most informative parameters are the average edge length m^* and the standard deviation σ^*. Dussert and co-workers[10,18] have shown that it is convenient to normalize, roughly by the square root of the average cell surface $<S>$, m^* and σ^*

$$ m = \frac{m^*}{\sqrt{<S>}} \frac{N-1}{N} \qquad (4.a) $$

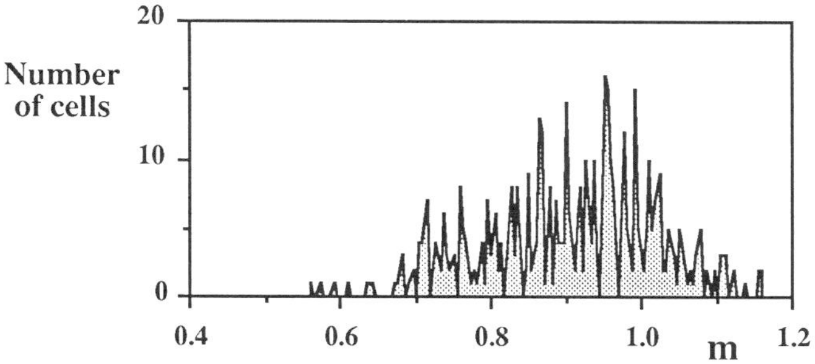

Fig. 8. Histogram of the normalized edge lengths m. Pb-30 % Tl.
G = 45 °C/cm. V = 0.75 cm/h.

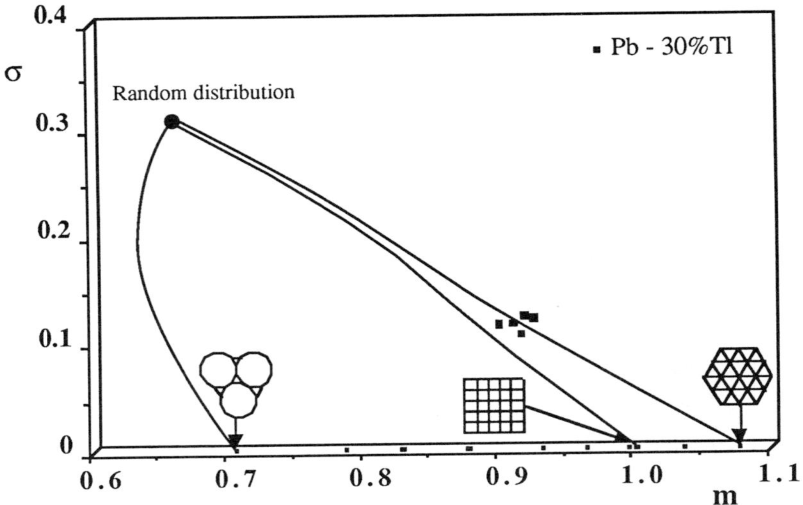

Fig. 9. The (m,σ)-diagram

$$\sigma = \frac{\sigma^{*}}{\sqrt{<S>}} \frac{N-1}{N} \qquad (4.b)$$

and to consider the histogram of the dimensionless edge lengths m (Fig. 8). Then, one is able to plot all distributions in the (m,σ)-plane, where they can be compared, e.g. to perfectly ordered mosaics (σ=0) or to the random distribution (Fig. 9). As any regular lattice may be progressively randomized by giving each point a new position deduced from its previous one using a Gaussian distribution of increasing standard deviation, Dussert et al. have computed the trajectories joining some regular mosaics (triangular, square, ...) to the random distribution[10,18]. It results that the points obtained for the Pb-30% Tl alloys precisely fall on such a trajectory. It is therefore possible to determine the underlying arrangement of these 2D-cellular arrays: the cell centers form a triangular array, as it is expected for hexagonal cells, whose disorder can be described by a Gaussian "noise" associated to the topological defects. It is remarkable that, despite the very high percentage of defects and liquid-type correlation functions, there is still a detectable manifestation of some underlying mosaic, which incidentally would make any attempt to a priori estimate of the coefficient α, equation (3), extremely hazardous.

Convergence problems. When determining m and σ, great care should be taken of the surface S of the sampling window. Indeed, these values must be known with at least three significant figures for a good positioning in the (m,σ)-plane. In order to check the convergence of m and σ, the total area can be reduced by successive halvings and the MSTs constructed at each stage. The most striking feature is that the smaller the sampling area the strongest the influence of the too long tree edges which are incorporated at the periphery. Indeed, as the choice for the connection to the current tree is restricted to the cells which are in full in the sampling window, more natural edges are impeded as the size is reduced, which obviously leads to an overestimate of the value of m, which increases as the surface S diminishes. It follows from figure 10 that the dispersion of the values is important for both m and σ when S is divided in 16 parts and continuously goes down as the sampling surface is enlarged. Nevertheless, the average values of these two parameters for a given division of S, $<m \ (S/2^{n})>$ and $<\sigma \ (S/2^{n})>$, have very different behaviours: the latter quantity is rougly constant whereas the former slowly converges to its asymptotic value, which is reached to a sufficient precision only for S/2, i.e. for a total number of cells N = 200 or thereabout. For a smaller number of cells, it is meaningless to use the (m,σ)-plane to analyze the disorder of 2D-cellular arrays.

216

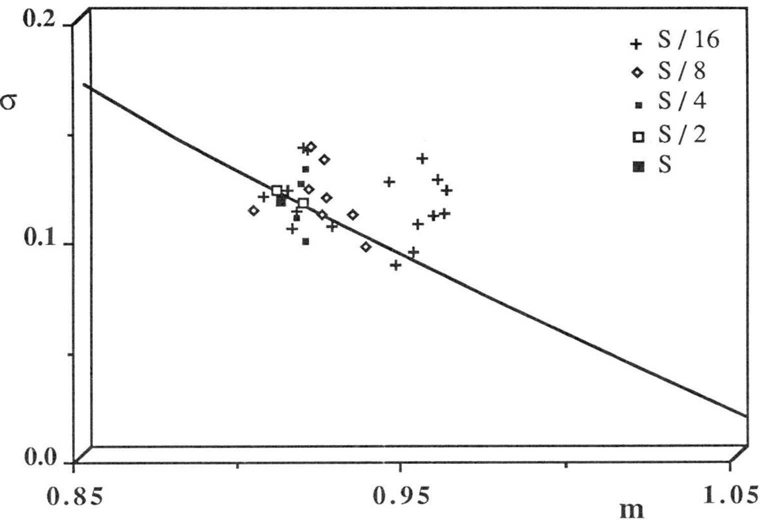

Fig. 10. Variation of m and σ with the size of the sampling window.
Pb-30 % Tl. G = 45 °C/cm. V = 0.75 cm/h. N=440 cells for S.

LOCAL DENDRITIC TRANSITION

Experimental Evidence

The Deep Cellular-Dendritic transition is a fundamental problem of pattern formation in nonlinear systems[4,19]. When the primary spacing is unique, two mechanisms have been proposed: a noise-driven fingering instability for which the generation of secondary arms would be random (see e.g. ref. 3), and an oscillatory instability of the cell tips which would result in the coherent emission of sidebranches[20]. These mechanisms still are under controversy. Actually, the dendritic transition occurs locally in standard experiments where the growth velocity is suddenly switched on after thermal stabilization, the largest cells developing sidebranches first. This *Local Dendritic Transition* is directly associated to the existence of a distribution of the primary spacing, i.e. to the array disorder. For the cellular solidification of succinonitrile-acetone alloys in a Hele-Shaw cell, the primary spacing first increases by the elimination of cells and it is in the course of this process that, in a cellular array, sidebranches may appear depending on the local enlargement of the *half* spacing, often when the evolution of the pattern has slowed down very much (Fig. 11).

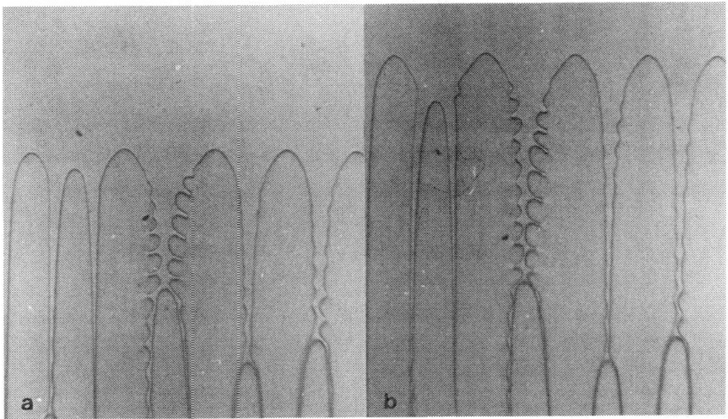

Fig. 11. Local Dendritic Transition. Succinonitrile-1.3 % acetone
(courtesy of W. Kurz). G = 72.3 °C/cm. V = 0.83 cm/h.
a: Time t - b: t + 200 s. Photograph width = 1.4 mm.

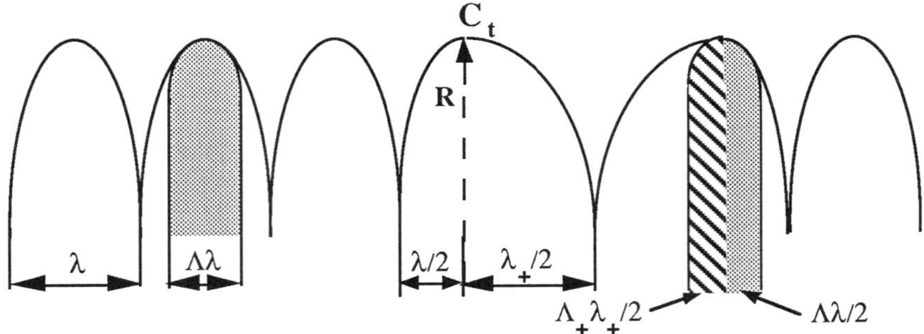

Fig. 12. Sketch of the changes induced in a 1D-cellular array by the local increase of the half primary spacing upon the elimination of a cell, as it is shown on figure 11.

Threshold of Local Dendritic Transition

Unbalanced constitutional supercooling. The local elimination of a cell in a 1D-cellular array is mimicked on figure 12. The major change is the increase of the half spacing ($\lambda/2 \rightarrow \lambda_+/2$). As only one side is concerned by the elimination of a neighbouring cell, the radius R and solute concentration at the tip C_t in a rough description are unaffected, from which it results that the constitutional supercooling at the tip, $\nu C_t^* - 1$ where ν is the level of morphological instability and C_t^* ($=KC_t/C_\infty$, with K the segregation coefficient and C_∞ the initial solute concentration) a nondimensional tip concentration[19], is also preserved. Within this frame, in the condition of microsolvability, where it balances constitutional supercooling, the capillary term undergoes the following variation

$$\nu C_t^* - 1 = \nu A F_\Lambda / K P_\lambda^2 \quad \rightarrow \quad \nu C_t^* - 1 = \nu A F_{\Lambda+} / K P_{\lambda+}^2 \ . \tag{5}$$

P_λ is the Peclet number, A the Sekerka parameter[19] and F_Λ a function, which depends on the relative cellular width $\Lambda = 2 / [(1 + 4\lambda/\pi R)^{1/2} + 1]$.

If the capillary term in equation 5 increases with the half spacing, the cell elimination induces an overcompensation of constitutional supercooling, which cannot favor any further morphological instability. Conversely, *unbalanced constitutional supercooling* occurs when the capillary term is decreased, i.e. some driving force is generated which might drive the formation of sidebranches and lead to local dendritic growth.

By using a Taylor expansion, the microsolvability condition reads

$$\nu C_t^* - 1 = \frac{\nu A F_\Lambda}{K P_\lambda^2}\left[1 + \frac{1}{F_\Lambda}\frac{dF_\Lambda}{d\Lambda}\Delta\Lambda - 2\frac{\Delta P_\lambda}{P_\lambda}\right] \ . \tag{6}$$

As the tip radius is unchanged, $\Delta P_\lambda/P_\lambda = \Delta(\lambda/R)/(\lambda/R)$ and thus can be expressed as a function of Λ by using the geometrical definition of the relative width. Finally, one gets for the condition for unbalanced constitutional supercooling upon a cell elimination ($\Delta\Lambda<0$)

$$\frac{1}{F_\Lambda}\frac{dF_\Lambda}{d\Lambda} - 2\frac{\Lambda - 2}{\Lambda(1 - \Lambda)} > 0 \ . \tag{7}$$

Criterion for local dendritic growth. For succinonitrile-acetone alloys, a logarithmic fit through the experimental data[21], in the diagram giving $(1 - \Lambda)/\Lambda$ versus F_Λ, leads to $F_\Lambda^{0.40}/32 = (1 - \Lambda)/\Lambda$ and $(dF_\Lambda/d\Lambda)/F_\Lambda = -2.5/[\Lambda(1 - \Lambda)]$. Therefore, for these alloys, the criterion for local dendritic solidification reads

$$(1.5 - 2\Lambda) / [\Lambda(1 - \Lambda)] \geq 0 \ . \tag{8}$$

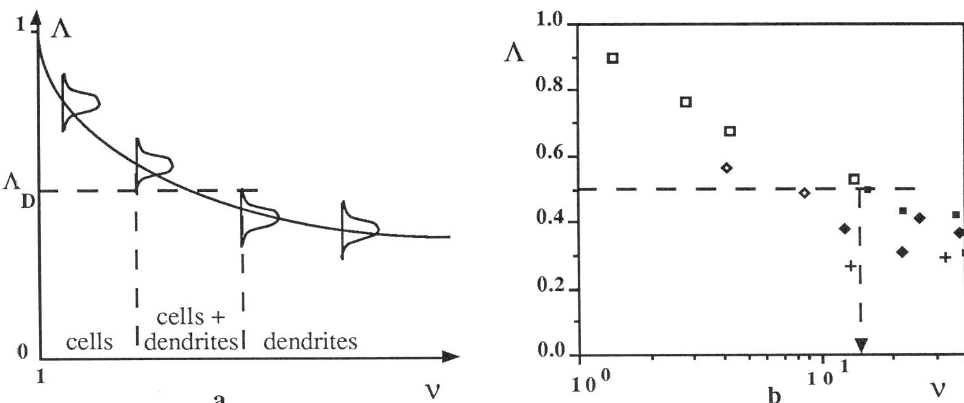

Fig. 13. Cell-dendrite transition in standard experiments exhibiting a distribution of the primary spacing. a: Sketch of the variation of the relative cellular width Λ with the level of morphological instability v, showing the fading of the dendritic threshold due to the presence of a spacing distribution. b: The cellular and dendritic regimes observed for succinonitrile-acetone alloys grown in a Hele-Shaw cell[22] (open symbols: cells, full symbols: dendrites).

As the denominator in relation 8 is always positive, unbalanced constitutional supercooling will drive the development of sidebranches in a cellular array everywhere the local relative width $\Lambda \le 0.75 \ (= \Lambda_D)$.

The most important consequence is that, in experimental situations characterized by a distribution of the primary spacing, the transition between the cellular and the dendritic regimes will be blurred as it is sketched on figure 13a, dendrites progressively appearing in places where the relative width is too small. Such a scenario is actually seen in experiments (Fig. 13b), the critical value Λ_D being yet about 0.5. The difference with the prediction (Λ_D=0.75) should not be considered as significant due to the assumptions introduced in the derivation of relation 8. It is likely that a more sophisticated theory would improve the fit, but we consider that a leading physical mechanism has been evidenced.

CONCLUSION

Disorder in 2D-cellular arrays has been analyzed by using -i) a weighted Wigner-Seitz construction and -ii) the Minimal Spanning Tree and the (m,σ)-diagram. Usually, interface patterns which are steady on the average are obtained soon in experiments but the characteristic time τ for the array to achieve the state of order corresponding to the imposed value of the level v of morphological instability is absolutely not known. If the time τ were short, this would mean that a high degree of disorder is intrinsic to the cellular arrays. From the present understanding of pattern-forming instabilites, the likeliness of such a scenario is weak.

For instance, a different behaviour is observed for Bénard-Marangoni convection[7]. The patterns are well organized near the onset and disorder progressively increases with instability. Nevertheless, it is well recognized that the percentage of defects is very sensitive to the experimental procedure which is used to establish the structures[23]. As our results are for "standard" directional solidification, in which the growth velocity V is suddenly switched on, it is still an open question whether an optimized procedure is feasible or not.

In other words, it would be worth to know the time dependences of the percentage of defects d_s and of m and σ. If there is some analogy with Bénard-Marangoni convection, it is likely that d_s will decrease slowly with time by the transient elimination of defects by pair annihilation and on the periphery, down to a limiting value that is characteristic of v, and that (m,σ) will move towards the underlying mosaic. Moreover, it would be precious to clarify the role played by the defects in the dynamics of the adjustement of the average primary spacing to a steady-state value. The fact that the cellular arrays are far from being perfect honeycombs might be a clue to the understanding of pattern selection.

ACKNOWLEDGEMENTS

The support of the Centre National d'Etudes Spatiales and the Centre National de la Recherche Scientifique is gratefully acknowledged. The authors are moreover indebted to W. Kurz, P. Pelcé and R. Trivedi for many stimulating discussions.

REFERENCES

1. J. S. Langer, in "Chance and Matter", Les Houches Proceedings, vol. 46, J. Souletie and J. Vanimenus Eds., North Holland, Amsterdam (1987) p.631
2. D. A. Kessler, J. Koplik and H. Levine, *Advan. Phys.* 37:255 (1988)
3. P. Pelcé," Dynamics of Curved Fronts", Academic Press, Boston (1988)
4. J. S. Langer, *Rev. Mod. Phys.* 52:1 (1980)
5. A. Pocheau and V. Croquette, *J. Phys. France* 45:489 (1984)
6. L. Gil, J. Lega and J.L. Meunier, *Phys. Rev. A* 41:1138 (1990)
7. R. Occelli, E. Guazzelli and J. Pantaloni, *J. Phys. France* 44:L567 (1983)
8. H. Nguyen Thi, B. Billia and L. Capella, *J. Phys. France* 51:625 (1990)
9. D. Weaire and N. Rivier, *Contemp. Phys.* 25:59 (1984)
10. C. Dussert, G. Rasigni, M. Rasigni, J. Palmari and A. Llebaria, *Phys. Rev. B* 34:3528 (1986)
11. H. Jamgotchian, B. Billia and H. Nguyen Thi, Ann. Chimie, to appear
12. E. T. Lewis, *Anat. Rec.*38:341 (1928)
13. N. Rivier and A. Lissowski, *J. Phys. A* 15:L143 (1982)
14. D. G. McCartney and J. D. Hunt, *Acta. Met.* 29:1851 (1981)
15. M. Ben Amar and B. Moussalam, *Phys. Rev. Letters* 60:317 (1988)
16. F. Harary, R. Z. Norman and D. Cartwright, "Structural Models", Wiley, New York (1965)
17. C. T. Zahn, *IEEE Trans. Comput. C* 20:66 (1971)
18. C. Dussert, Ph. D. Thesis, University of Aix-Marseille III (1988)
19. B. Billia, H. Jamgotchian and R. Trivedi, *J. Crystal Growth* (1990) to appear
20. A. Karma and P. Pelcé, *Phys. Rev. A* 39:4162 (1989)
21. B. Billia, H. Jamgotchian and R. Trivedi, in *Materials and Fluid Sciences in Microgravity*, ESA SP-295, ESA Publications, Noordwijk (1989) p.189
22. H. Esaka, Ph. D. Thesis, Ecole Polytechnique Fédérale de Lausanne (1986)
23. P. Bigazzi, S. Ciliberto and V. Croquette, *J. Phys. France* 51:611 (1990)

FLAME PROPAGATION

SOUNDS AND FLAMES

Paul Clavin

Laboratoire de Recherche en Combustion URA D 1117
CNRS/Université de Provence, S.252 St Jérôme
13397 Marseille Cedex 13, France

INTRODUCTION

As reported by Rayleigh[1], the interaction between acoustic and combustion is known
from more than two centuries. The sounds emitted by a jet of hydrogen burning in a pipe
were noticed by Higgins soon after the discovery of the gas. At the beginning of the XIXth
century, Faraday shows that such singing flames may be produced with other reacting gases.
Later, the reverse phenomenon was observed by Tyndall: a voice pitched at the note of the
tube causes the flame to quiver.

The excitation of acoustic vibrations by heat is related to a more general
phenomenon, the thermoacoustic instability which was first pointed out by Rijke in the
middle of the XIXth century: a metallic heated gauze streching across the lower part of a tube
open at both ends produce a sound of considerable power. Because of gas expansion, heat
release acts as a volume source which may easily excite acoustic vibrations. Rayleigh[1] gave a
general criterion for a spontaneous amplification of acoustic modes by any local heat source:
the local-energy addition or substraction must be positively correlated with pressure
fluctuations (Strehlow[2]). But this does not tell us what are the coupling mechanisms.
Different examples are presented in this volume. The theoretical analysis of Pelcé and
Rochewerger[3] concerns cellular flames with the objective of explaining the primary sound
emitted by a curved flame propagating in a tube. The work of Searby[4] concerns a parametric
instability of planar fronts for describing a secondary instability mechanism of flame
propagating in tubes breaking down into an incoherent self-turbulizing regime.

This paper is concerned with two aspects of the interaction between sound and
combustion, the generation of sound by flames and the acoustic feedback on combustion
leading to instabilities. The basic equations governing the interaction between sound and
combustion are recalled in the next section where the general background of the problem is
presented. The rest of the paper is in two parts:
- The first one presents a work[5] with Siggia on noise production by a premixed
turbulent flame radiating to a free field. The main concern here is to relate the frequency
content of combustion noise to the geometrical and statistical properties of the turbulent
flame front controlled by the turbulent flow.
- The second part is a brief survey of different coupling mechanisms between planar
flame structures and acoustic waves of a cavity which have been studied recently ranging
from premixed gaseous mixtures[6] to sprays[7] and to solid propellants[8].

Growth and Form, Edited by M. Ben Amar *et al.*
Plenum Press, New York, 1991

GOVERNING EQUATIONS

The equation for temperature in a reactive gaseous mixture is

$$\rho C_p \, DT/Dt = \nabla . (\lambda \nabla T) + \dot{Q} + Dp/Dt \qquad (1a)$$

where the heat capacity C_p of the mixture has been assumed to be constant for simplicity, p is the pressure and $\dot{Q}$ the rate of heat release per unit volume. D/Dt is the Lagrangian derivative. It is convenient to eliminate the temperature from eq.(1a) by using the ideal gas laws

$$p/\rho = RT = (C_v/C_p)c^2 \quad \text{with} \quad R = C_p - C_v \qquad (1b)$$

where c is the sound speed. Neglecting the heat conduction term in eq.(1a), one obtains:

$$\frac{Dp}{Dt} = c^2 \frac{D\rho}{Dt} + (\frac{C_p - C_v}{C_v})\dot{Q} \qquad (1c)$$

Two limiting cases of eq.(1c) are particularly instructive. When the heat release is neglected, eq.(1c) yields the ordinary relation for isentropic compressible fluids. In the quasi isobaric approximation where the left hand side of eq.(1c) is neglected, this equation expresses the heat release induced gas expension. This exhibits clearly how heat release in gases acts as a volume source and may trigger acoustics. In the notation

$$a = \bar{a} + a' \qquad (1d)$$

the linearized version of the continuity equation and the invicid momentum conservation equations together with (1c) yields:

$$\frac{\partial^2}{\partial t^2}p' - c^2\nabla^2 p' = (\frac{C_p - C_v}{C_v})\frac{\partial}{\partial t}\dot{Q} \qquad (1e)$$

The time derivative of the heat release rate may be considered as a forcing term in the acoustic wave equation. When the fluctuations in the rate of heat release can be expressed linearly only in terms of the local pressure fluctuations, eq.(1e) takes the form of a closed linear equation for the pressure field. This equation describes the acoustic combustion instabilities and exhibits clearly the Rayleigh criterion. It may be solved by a perturbative method when the forcing term is weak which is generally the case in combustion, as we will see later. At the leading order, the acoustic frequencies of the cavity are not perturbed by the combustion fluctuations. The growth (or damping) rate is small compared to the acoustic frequency and may be easily computed from eq.(1e) by a perturbative expansion in powers of the Mach number. When combustion has a destabilizing effect, damping terms by dissipative mechanisms must be introduced to determine the stability limits. In many cases the most efficient damping mechanisms are associated with boundary conditions, such as the viscous and thermal transfert to the walls in boundary layers or radiation of acoustic energy through exits (open ends, nozzle, ...).

In the case of flames, another simplification generally occurs when the flame front is localized in a region of space with a characteristic length scale much smaller than the acoustic wavelength. The pressure may be considered as uniform in the flame zone and the forcing term in eq.(1e) takes the simplified form of a δ-source in space. For premixed flames propagating in tubes, the acoustic field of the longitudinal modes is one-dimensionnal outside the boundary layers. The combustion zone is a thin layer separating two regions with different temperatures and sound speeds. The corresponding stability limits has been solved analytically by a multiscale analysis for both planar flame fronts[6] and cellular flames (see the paper by Pelcé and Rochewerger in this volume). The difference in the sound speed between the burned and unburned gases introduces only technical difficulties but no fundamental new mechanisms. The basic linear phenomena may be understood by considering the sound speed as constant in eq.(1e).

SOUND GENERATION BY TURBULENT PREMIXED FLAMES

Before going back to combustion instabilities, the sound generation by a turbulent flame is presented in this section. Let's consider a geometry in which a fire ball is burning in free suroundings, fed by jets of fresh premixed gas which also pump the velocity field and maintain the turbulence. The fluctuations in the rate of heat release is due to the time dependent surface area of the flame front which is wrinkled by the turbulent flow. No acoustic feedback is considered here. The average position of the flame is maintained stationary around the origin. The characteristic turbulence frequencies are u'/L where u' is the turbulence intensity and L the characteristic size of the gas jet. For a subsonic turbulent flow, $u'/c \equiv M \ll 1$, the acoustic wave length $\lambda = cL/u'$ is much larger than the dimension of the flame L. The δ-source approximation is valid in eq.(1e) and when the mass rate of fresh mixture in the jets is kept constant, the acoustic pressure fluctuation is given by the retarded Green function to give at sufficiently large distance (r>>L):

$$p' = \frac{1}{4\pi \, r \, c^2} \, (\frac{C_p - C_v}{C_v}) \int \frac{\partial}{\partial t} \dot{Q}(\mathbf{x_1}, t - r/c) \, d^3\mathbf{x_1} \tag{2a}$$

Then, the total intensity of the sound emitted by the flame, $I = 4\pi r^2 p'^2/\rho c$, is obtained directly from (2a) in a form first derived by Strahle[9]. In the corrugated flamelets regime defined by $v_K \ll u_L \ll u'$, where v_K is the fluctuating turbulent velocity at the viscous Kolmogorov scale and where u_L is the laminar flame speed, the flame thickness d_L is much smaller than all the lengthscales of the turbulent flow. In such cases, a flame front is properly defined at each time, with a local structure close to the 1-D laminar flame one. Then, the mass and energy fluxes across this flame front are $\rho_u u_L$ and $\rho_u u_L C_p (T_b - T_u)$ where the subscripts u and b refer to the unburned and burned mixture. The instantaneous total heat release is thus directly proportional to the, surface area of flame S,

$$\int \dot{Q}(\mathbf{x_1}, t) \, d^3\mathbf{x_1} = C_p (T_b - T_u) \, \rho_u u_L \, S(t) \tag{2b}$$

and (2a) yields,

$$I = \frac{1}{4\pi c} \, (\frac{1}{\rho_b} - \frac{1}{\rho_u})^2 \, \rho_b \, (\rho_u u_L)^2 \left\langle (\frac{dS}{dt})^2 \right\rangle \tag{2c}$$

where the brackets $\langle \ \rangle$ denote time average.Then, the power spectrum of the noise is proportional to the Fourier transform of the autocorrelation function of the time derivative of the total surface area of flame. The problem is now to relate this quantity to the characteristics of the turbulent jet flows.

The mean square fluctuation of the total flame surface $\langle (\delta S)^2 \rangle$ may be computed as follows.We mark off regular grids of spacing k_i^{-1} and consider all the families of such subgrids with meshes ranging from the Gibson scale $k_G^{-1} = (k_{i=n})^{-1}$ to the integral scale $L = (k_{i=0})^{-1}$, $k_{i+1} = bk_i$ with an integer $b \geq 2$. The Gibson scale corresponds to the length below which the turbulent flow is not able to further wrinkle the flame because the corresponding eddies will burn before they turn over[10]. One defines N_i as the number of blocks of the ith grid intersecting the surface of the flame at a given time. Lets consider one such block labelled i' ranging from 1 to N_i and introduce $\eta_{i+1}(i')$, the number of blocks of the $i+1^{st}$ grid intersecting the piece of flame surface located in the blocks i' , $1 < \eta_{i+1}(i') < b^3$. One has:

$$N_{i+1} = \Sigma_{i'=1}^{N_i} \eta_{i+1}(i') \tag{3a}$$

the η_{i+1} are a local measure of the increase of surface area at the $i+1^{st}$ scale. By assuming N_i and $\{\eta_{i+1}\}$ to be mutually independent, and by introducing ensemble averages (denoted by brackets), one has $\langle N_{i+1} \rangle = \langle N_i \rangle \langle \eta_i \rangle$ for the means and

$$\frac{\langle \delta N_n^2 \rangle}{\langle N_n \rangle^2} = \Sigma_{i=0}^{n-1} \frac{1}{\langle N_i \rangle} \frac{\langle \delta \eta_{i+1}^2 \rangle}{\langle \eta_{i+1} \rangle^2} \tag{3b}$$

for the mean square fluctuations. Equation (3b) becomes relevant to sound generation if we remember that $S \approx k_G^{-2} N_n$ and observe that the same formula holds if we replace

$$\langle \delta N_n^2 \rangle \quad \text{by} \quad \langle \dot{S}(t)\dot{S}(0) \rangle$$

$$\text{and} \quad \langle \delta \eta_i^2 \rangle \quad \text{by} \quad \langle \dot{\eta}_i(t)\dot{\eta}_i(0) \rangle$$

For t=0 we have for the frequency integrated acoustic power, in order of magnitude,

$$\langle \dot{S}^2 \rangle \approx \langle S \rangle^2 \Sigma_{i=0}^{n-1} \frac{\omega_i^2}{\langle N_i \rangle} \tag{3c}$$

where we have set $\omega_i^2 \equiv \langle \dot{\eta}_i^2 \rangle / \langle \eta_i^2 \rangle$, and ignored the numerical factor relating ω_i to ω_{i+1}.

We may reasonably assume that ω_i, as defined above, corresponds to the characterisitc frequency of the turbulent flow at the scale k_i^{-1}. Rough arguments[5] have also been given to show that combustion effects on fully developed turbulence are not expected to change the Kolmogorov scaling laws $\omega_i^2 \approx \varepsilon^{2/3} k_i^{4/3}$ where $\varepsilon = u'^3/L$ is the energy dissipation rate per mass. Consequently in the range between the Gibson scale and the integral scale, the flame front has a fractal structure, $\langle N_i \rangle \approx (Lk_i)^{D_f}$, characterized by a fractal dimension $D_f = 7/3$, as first proposed by Kerstein[11] and Peters[10] (see ref 5 for more details). Thus, the sum in (3c) is rapidly convergent in i, and the integrated power in a band of frequencies around ω_i comes essentially from the ith term in (3c). For each integral volume L^3, we have

$$\langle \dot{S}^2 \rangle \approx \langle S \rangle^2 \left(\frac{u'}{L}\right)^2 \Sigma_{i=0}^{n-1} (k_i L)^{4/3-D_f} \tag{3d}$$

The band integrated wave number spectrum in (3d) is converted into a frequency spectrum using $k_i \approx \omega_i^{3/2}$ and $D_f = 7/3$ yielding finally

$$dI(\omega) \approx \omega^{-5/2} d\omega \tag{3e}$$

The coefficient in (3e) is determined by making the integral down $\omega = u'/L$ agree with (2c) and (3d)

In conclusion, as first pointed by Strahle[9], the total sound intensity scales with the volume of the flame brush and is predominantly at low frequency. The relation, (2c), of the acoustic intensity to the surface fluctuations is valid within the corrugated flamelets regime without assumptions about the flow. When N_i and $\{\eta_{i+1}\}$ are mutually independent, eq.(3c) is true, even as an expression for the acoustic power per frequency band, provided one inserts the correct ω_i for the flow and the measured $\langle N_i \rangle$. Eq.(3e) is restricted to turbulent premixed flames in a fully developed turbulent flow when the Kolmogorov scalings hold. Finally it must be pointed out that the ratio of the radiated energy of the sound to the energy dissipated by the turbulence is, according to (2c) and (3e), of the order of magnitude of the Mach number M=u'/c. In this evaluation one has used $\langle S \rangle \approx L^2 (u'/u_L)^{3D-6} = L^2 (u'/u_L)$. This result has to be compared with M^5 which is obtained for the same turbulent flow in the non reactive case (without combustion) by the theory of Lighthill[12].

PRESSURE COUPLING IN SOLID PROPELLANT BURNING

Some basic mechanisms of thermoacoustic instabilities may be described within the framework of a simple model of a non-steady one-dimensional gas flow in a tube in which heat release is localized in a region much smaller than the acoustic wavelength. Axial oscillations in an end-burning solid rocket is the prototype of problems which are well represented by such a simple model (see fig.1). It is instructive to present first this case. The boundary condition at one end of the tube is associated with the burning rate of the solid propellant whose the burning surface stands perpendicular to the axis of the tube. Let $\hat{u}$ and $\hat{p}$ be the time Fourier transform of the disturbances of the axial gas velocity δu and of the pressure δp. The linear response of the burning rate to an harmonic pressure fluctuation at a frequency ω may be characterized by the nondimensional admittance function $Z(\omega)$:

$$\hat{u}_b = Z(\omega)\, \hat{p}/\rho c \tag{4a}$$

where ρ and c are the density and the sound speed in the burned gases. The burning velocity u_b is defined here as the normal velocity of burned gases relative to the burning surface (an equivalent flame speed u_L may be defined as $\rho_u u_L = \rho_b u_b$). Equation (4a) yields the boundary condition at the end of the tube delimited by the burning solid propellant denoted by the subscript 2 :

$$\hat{u}_2 = Z(\omega)\hat{p}_2 / \rho_2 c_2. \tag{4b}$$

The boundary condition at the other end of the tube is provided by the conditions at the entrance plane of a choked exhaust nozzle denoted by subscript o. In the linear regime one has:

$$\hat{u}_o = Y(\omega)\, \hat{p}_o / \rho_2 c_2. \tag{4c}$$

where $Y(\omega)$ is characteristic of the nozzle.

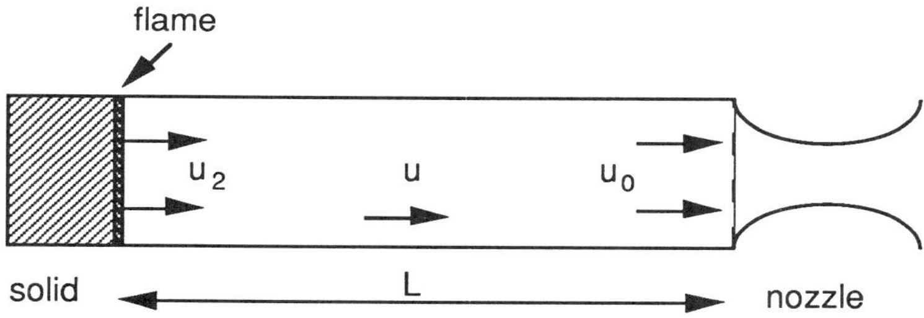

Fig 1. Sketch of an end burning solid rocket

Scalings in (4a-c) are defined in relation with the order of magnitude of pressure fluctuations in an acoustic wave, $\delta p \approx \rho c \delta u$, in such a way that $|Z(\omega)|$ characterizes the relative modification of the acoustic velocity field by the combustion, $|Z(\omega)| = O(\delta u_b/\delta u)$.

In the framework of a "pressure coupling" mechanism, the basic assumption is that the relative modifications of the burning rate and of the pressure are of the same order of magnitude, yielding:

$$\delta u_b/u_b = O(\delta p/p). \tag{5a}$$

This is the case when the dominant coupling mechanism is developped through the pressure dependence of the prefactor of the Arrhenius law contolling the combustion rate. Then, according to (4a) and $\delta p \approx \rho c \delta u$ with $p \approx \rho c^2$ ($\delta p/p = O(M\, \delta u/u)$ in an acoustic wave), the admittance function $Z(\omega)$ is proportional to the Mach number. For similar reasons, this is also true for $Y(\omega)$ (see Crocco and Cheng[13]):

$$Y(\omega) = M_2\, \mathcal{Y}(\omega)\ ,\qquad Z(\omega) = M_2\, \mathcal{Z}(\omega) \tag{5b}$$

with $M_2 \equiv u_b/c_2$ and

$$\mathcal{Y}(\omega) = O(1)\ ,\qquad \mathcal{Z}(\omega) = O(1)\quad \text{in the limit } M_2 \to 0. \tag{5c}$$

According (5a), the fluctuations of the burning velocity δu_b have an amplitude smaller by a factor Mach than the acoustic velocity fluctuations of the gases in the combustion chamber. Thus, the perturbation of the acoustic field by combustion is weak in subsonic flows and a perturbative method may be used to solve the combustion instability problem (see Clavin et al.[6] for a recent calculation). This problem will not be considered in details here, only orders of magnitude will be presented. The time derivative of the energy per unit of cross-sectional area, $\mathcal{E}$, stored in the acoustic modes of the tube is given by an expression similar to the case of a closed tube with positions of the two end walls fluctuating with the velocities δu_2 and δu_0:

$$d\mathcal{E}/dt = \delta p_2 \delta u_2 - \delta p_0 \delta u_0. \tag{6a}$$

The second term in the r.h.s. of (6a) is a damping term due to an acoustic energy loss through the exhaust nozzle. The other damping mechanisms (particulate damping, viscosity, etc....) have been neglected for simplicity and are usually less important. The first term describes the specific mechanism causing the spontaneous instability by interactions between combustion and pressure fluctuations. It satisfies the Rayleigh criterion, the instability develops when the time average during a cycle of the product $\delta u_2 \delta p_2$ is positive. According to (4) and (5), the the longitudinal acoustic waves excited by the propergol combustion, are characterized by velocity fluctuations at both ends of the tube which are much smaller than the acoustic amplitude. Thus, the vibratory modes do not differ sensibly from the natural ones for the closed cavity ($\delta u_2 = 0$, $\delta u_0 = 0$). The fundamental mode corresponding to a half wavelength for which the pressure disturbances are the same at both ends, $\delta p_2 = \delta p_0 \equiv \delta p$, the leading order of (6a) yields:

$$d\mathcal{E}/dt = \delta p\,(\delta u_2 - \delta u_0). \tag{6b}$$

Denoting the characteristic acoustic time by $\tau_a \equiv L/c_2$, (L is the length of the tube), the order of magnitude of the acoustic energy per unit of cross-sectional area stored in the tube may be expressed in terms of the pressure fluctuation as $\mathcal{E} \approx \tau_a\,\delta p\,\delta u \approx \tau_a (\delta p)^2/\rho_2 c_2$. Then, the leading order of the linear growth rate of the acoustic oscillation, defined as $\tau_{ins}^{-1} \equiv \mathcal{E}^{-1} d\mathcal{E}/dt$, is, according to (4), (5) and (6b):

$$\tau_{ins}^{-1} = \tau_a^{-1}\,M_2\,\mathrm{Re}\{\mathcal{Z}(\omega) - \mathcal{Y}(\omega)\} \tag{6c}$$

where the angular frequency ω corresponds to the unperturbed acoustic mode. The nozzle acting as a damping process, one has $\mathrm{Re}(\mathcal{Y}) > 0$. The acoustic instability appears when $\mathrm{Re}(\mathcal{Z}) > \mathrm{Re}(\mathcal{Y})$, and the stability limits are defined by $\mathrm{Re}(\mathcal{Z}) = \mathrm{Re}(\mathcal{Y})$. When the characteristics of the nozzle are known, the problem is reduced to determine the acoustic admittance function $\mathcal{Z}$ which characterizes the burning response to pressure fluctuations.

There is an extensive literature on the evaluation of the admittance function of homogeneous solid propellants. Most of the existing analytical theories concern the same one dimensional model in which the condensated phase decomposes at the solid-gas interface by pyrolisis into premixed gases which are burned in the gas phase. Except for the recent work of Clavin and Lazimi[8], all of these analysis are based on the quasi steady state (Q.S) approximation of the gas phase indroduced by Zel'dovich[14]. The validity of the Q.S assumption lies on the small gas-to-solid density ratio which implies that the transit time in the gaseous flame is much shorter than the response time of the heat transfer in the condensated phase. Thus, when the combustion in the gas phase is stable, and when the acoustic frequency is sufficiently small, the combustion response is expected to be slaved by the condensated phase. In this case, orders of magnitude presented in (5) are correct.

The Q.S approximation is no more valid either at high frequencies, when the characteristic acoustic time becomes as small as the transit time across the gas flame (> 5000 Hz at a pressure of 10 atm.) or at high pressure burning ($\geq$ 50 atm.), when the gas-to-solid density ratio is not small enough to make the two relaxation times (in the solid and in the gas) sufficiently different. In such cases, the dominant coupling mechanism of vibratory combustion of solid propellant is different[8]. It is similar to the one developed in a one-dimensional gaseous flame propagating in a tube[6] and it involves the combustion temperature fluctuations. Then, orders of magnitude (5) have to be modified.

228

In the case of flames propagating in tubes, the flame region may be considered as a discontinuity of the flow velocity. The time derivative of the total energy per unit cross sectional area, $\mathcal{E}$, stored in the acoustic modes of the tube is related to the velocity jump across the flame yielding a relation similar to (6b),

$$d\mathcal{E} / dt = \delta p \, (\delta u_2 - \delta u_1), \tag{7a}$$

δp is the pressure fluctuation at the flame, δu_2 and δu_1 are the velocity fluctuations at the discontinuity on the downstream and upstream sides. Eq. (7a) is self explanatory if one notices that $(\delta u_2 - \delta u_1)$ times the cross sectional area of the tube, represents the time derivative of the volume variation of the reaction zone. Damping mechanisms have been omitted in (7). Then, the transfer function, defined here as

$$(\hat{u}_2 - \hat{u}_1) = Z(\omega) \, \hat{p}/\rho_1 c_1, \tag{7b}$$

may be computed by investigating the pressure response of the combustion rate.

Pressure variations have a direct effect on the burning rate through the pressure dependence of the prefactor in the Arrhenius laws controlling the pyrolisis and the reaction rates. This mechanism is retained in the quasi-steady state theory mentionned in the oreceding section. But, the burning response to pressure fluctuations is also influenced by the gas temperature oscillations produced by adiabatic compressions.When the temperature sensitivity is srong (high reduced activation energy) this indirect mechanism may be more important than the direct one. The corresponding transfer function $Z(\omega)$ has been calculated for gaseous flame[6].

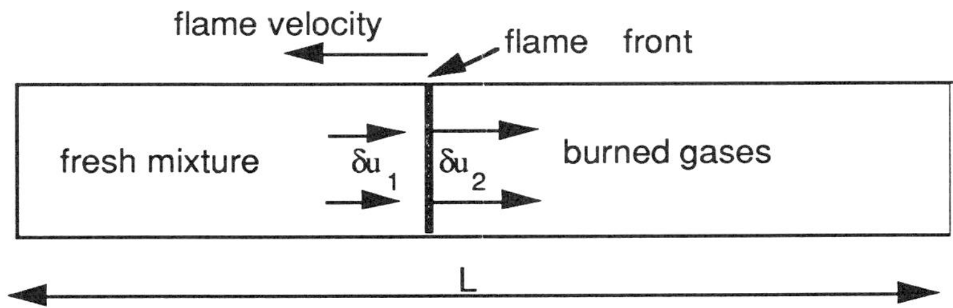

Figure 2. Planar flame propagating in a tube

The order of magnitude of the relative variations of gas temperature and pressure in acoustic waves is

$$\delta T/T \approx \delta p/p \approx M_2 \, \delta u/u_b. \tag{8a}$$

They induce a relative variation of the flame temperature T_f of a similar order of magnitude, $\delta T_f/T_f \approx M_2 \delta u/u_b$. As the combustion rate in the gas phase is controlled by an Arrhenius law, $B\exp(-E^{(g)}/2\mathcal{R}T_f)$, with a large activation energy $E^{(g)}/\mathcal{R}T_f \gg 1$, the acoustic induced fluctuation of the burning rate is of a relative order of magnitude larger than (5a) by a large factor $\beta \equiv E^{(g)}(T_f - T_u)/\mathcal{R}T_f^2 \gg 1$ where T_u is the temperature of the unburned mixture,

$$\delta u_b/u_b = O(\beta \, \delta p/p) \quad \Rightarrow \quad Z(\omega) = O(\beta M) \quad \text{and} \quad \tau_{ins}^{-1}/\tau_a^{-1} = O(\beta M). \tag{8b}$$

The resulting vibratory instability is stronger by a factor β (Zel'dovich number) than the one triggered through the pressure dependence of the prefactor B which yields, according to (5b,c):

$$\delta u_b/u_b = O(n_g\, \delta p/p) \quad\Rightarrow\quad Z(\omega) = O(n_g M), \quad \text{and} \quad \tau_{ins}^{-1}/\tau_a^{-1} = O(n_g M), \qquad (8c)$$

where M is the Mach number and n_g is the pressure exponent which is roughly one half of the reaction order. Typical vaues of n_g lie in the range (0.5, 1).

For a planar gaseous flame propagating in a tube, the order of magnitude estimate (8b) has been confirmed by a complete analysis[6]. The acoustic fluctuations of the gas velocity makes the position of the flame front oscillate. In the framework of the quasi isobaric approximation, this would not produce any direct modification of the flame structure. However, through the weak temperature modification associated with the compressibility of the gases, there is a pressure induced modification to the planar flame structure leading to a small relative fluctuation of the rate of heat release per unit cross section. Typical orders of magnitude are $\beta \approx 10$ and $M \approx 10^{-3}$. Thus, the characteristic order of magnitude of the instantaneous growth rate of such an acoustic instability is predicted to be small compared to the acoustic frequency $1/\tau_a \approx c_1/L$:

$$1/\tau_{ins} \approx \beta\, M/\tau_a \qquad (9a)$$

with $\beta M \approx 10^{-2}$. Nevertheless, through cumulative effects when the flame propagates on a distance of the order of the tube length L, the instability could be strong[6]. The residence time of the flame in the tube, defined as $\tau_{res} \equiv L/u_L$, being very long compared to the acoustic time, $\tau_{res}/\tau_a \approx 1/M$, one has: $\tau_{res}/\tau_{ins} \approx \beta$, and a perturbation of the flame position with an initial amplitude A_i can reach a final amplitude A_f which is increased by a very large factor $\exp\Gamma$,

$$A_f/A_i \approx \exp\Gamma \qquad \text{with} \qquad \Gamma \approx \tau_{res}/\tau_{ins} = O(\beta) \qquad (9b)$$

The complete analysis[6], including the damping mechanisms, determines the stability limits which show that there are large portions in the last part of the tube where a vibratory instability develops with the frequency of the fundamental acoustic mode. At sufficiently large Mach numbers, $M \geq 5.10^{-3}$, the flame responds in a quasi-steady approximation ($\tau_{L1} < \tau_a$) and the numerical value of the proportionality coefficient between Γ and β is found to be independent of the details of the flame structure and close to 10^{-1} which is too small to lead to noticeable effects. A strong acoustic vibratory instability is predicted to appear only in narrow bands of small Mach numbers ($M \approx 10^{-3}$), and only when a resonance phenomenon with the intrinsic thermal-diffusive instability of the flame structure at large Lewis number (Sivashinsky[15], Joulin and Clavin[16]) is involved. Such cases are not generic of premixed gas flames. The question to know whether or not the above described mechanism is involved in the strong acoustic instability observed in experiments of flame propagating in tubes (see the paper by Searby in this book), is still open. The planar vibratory instability is less important than the three-dimensional vibratory mechanism of cellular fronts (see the paper of Pelcé and Rochewerger in this volume) which is associated with the variation of flame area by acceleration instabilities. This last mechanism (or a similar one for curved fronts) may be responsible for the primary sound generation. But the strong secondary instability of a parametric nature, as described by Searby in this volume, appears on planar fronts. Such planar flames which are otherwhise unstable under the Darrieus-Landau hydrodynamic instabillity, are formed when the intensity of sound reaches an intermediate value. In most of the cases, the secondary acoustic instability develops at a slighly higher sound intensity. The planar mechanism of vibratory instability described above may be usefull to bridge the gap between these two thresholds. No definitive answer have been yet provided.

In homogeneous solid propellant burning, the planar flame fronts are most likely. But the presence of the solid-gas interface makes the selection of the dominant effect (pressure or temperature coupling) depending on the frequency range[8]. This phenomenon may be qualitatively understood in a non reactive example (with different scaling laws). Due to heat conduction toward the condensated phase, the gas temperature variations associated with pressure oscillations are damped in a gas layer at the solid-gas interface. With combustion, the thickness of this thermal layer varies as the inverse of the square (square root in non reactive case) of the frequency. Thus, at low frequencies, this thermal layer becomes larger than the thickness of the gas flame located at the interface. At such low frequencies, the gas flame structure is not influenced by the temperature fluctuations in the bulk of the burned gases. The flame temperature is slaved by the condensated phase and the burning response being triggered through the pressure exponents, the order of magnitude (8c) are valid as described by the quasy-steady state theory[14]. In the high frequency range, the opposite situation holds: variations of the reaction rate are triggered by the gas temperature oscillations of acoustic waves and become dominant as in planar flames propagating in tubes[6]. The order of magnitude (8b) is here valid. The full problem has been solved analytically very recently[8], showing that the solid propellant burning may be potentially unstable at high frequencies and not only in the low frequency range as predicted by the quasi-steady state approximation.

ACCELERATION COUPLING: CASE OF SPRAYS OR PARTICLE LADEN GASES

The acoustic fluctuations of the gas velocity makes the position of the flame front oscillate. If the effects of pressure and temperature fluctuations of acoustic waves are neglected, this will not produce any direct modification to the structure of planar fronts of premixed gas flames propagating in tubes. The case of cellular fronts is different[3], the accelerations which are are involved in the acoustic fluctuations, produce a modulation of the amplitude of the cells through what it is called the "Markstein-Taylor" mechanism. Thus the effective combustion rate by unit cross sectional area is modified even when the local modification of the inner structure of the flame is neglected.

The case of flames propagating in sprays or particle-laden gases is also very different from premixed gases. The accelerations may produce a direct modification to the inner structure of planar flame through inertia and drag of particles[7]. A velocity shift between liquid (or solid) particles and the surrounding gas is produced which induces fluctuations in the mass fraction of fuel and oxidizer feeding the reaction zone. The corresponding fluctuations of the reaction rate lead to a vibratory instability which is described, contrary to planar premixed gas flames, in the framework of the quasi isobaric approximation[7] (pressure and temperature fluctuations due to compressible effects in acoustic waves are neglected). A detailed analysis of this phenomena has been carried out in the framework of a simplified model of a monodisperse spray of lean liquid fuel in air, burning in the homogeneous regime[7]. This regime is valid when the initial mass fraction of liquid fuel is not too large ($<10^{-1}$) and when the droplets are sufficiently small and volatile to vaporize before they reach the combustion regime which is the typical case at ordinary conditions for droplets with radius smaller than 10 μm.

Orders of magnitude may be obtained by a simple analysis. In the low Reynolds number limit, the velocity shift $\delta u_l - \delta u$ between droplets and gases which is produced by the Stokes force in an acoustic wave with an angular frequency ω is

$$(\delta u_l - \delta u)/\delta u = i\omega\tau_{vis}/(1 + i\omega\tau_{vis})\qquad(10a)$$

where τ_{vis} is the characterisistic relaxation time associated with inertia and drag of particles

$$\tau_{vis} = (2/9)Pr(\rho_l/\rho_g)r^2D^{-1}\qquad(10b)$$

where Pr is the Prandtl number in gases, D is the thermal diffusion coefficient in gases and ρ_l/ρ_g is the liquid to gas density ratio. The validity domain of (10a) is $\omega\tau_{vis}<<1$. The velocity shift induced fluctuation of the flame temperature is given in the low frequency range by a quasi-steady balance of mass and energy across the flame yielding:

$$\delta T_f/(T_b - T_u) = -(\delta u_l - \delta u)/u_L.\qquad(10c)$$

The order of magnitude estimate of the linear growth rate of the resulting vibratory instability
is obtained in the same way as in (8b), to give:

$$\tau_{ins}^{-1} / \tau_a^{-1} \approx \beta\omega\tau_{vis} \tag{10d}$$

with $\omega=2\pi\tau_a^{-1}$. The comparison with the case of premixed gases (8b) shows that the
quantity $\omega\tau_{vis}$ replaces the Mach number M. For particles with a diameter of three microns
in a gas at ordinary conditions and enclosed in a tube of a typical length of one meter, the
drag of particles in the acoustic field is associated with the viscous Stokes force, and the
order of magnitude of $\omega\tau_{vis}$ is 10^{-1} which is one hundred times larger than the Mach
number. Thus, as shown by a detailed analysis[7], there are ranges of acoustic frequencies and
particle sizes for which planar flames propagating in spray or particle-laden gases are
predicted to experience a very strong planar vibratory instability, much stronger than the one
associted with the compressibility effects and, may be, stronger also than the
mutidimensional instability with cellular structures. An experimental program of research has
just started in our laboratory to asses such a specific property of two-phases mixtures which,
to the best of our knowledge, was up to now unknown.

ACKNOWLEDGEMENTS

P. Pelcé, G. Searby and P. Haldenwang are acknowledged for helpful discussions.
Partial financial support has been provided by PRC "Moteurs fusées" CNES/CNRS/SEP
and by DRET under grant number 88-210.

REFERENCES

1. Rayleigh J.W.S, (1878), Nature, 18, 319.
2. Strehlow R.A. (1979), Fundamentals of combustion. Robert E. Krieger.
3. Pelcé P., Rochewerger D. (1991) Vibratory instability of cellular flames propagating
   in tubes. To appear in J.F.M. See also this volume.
4. Searby G. (1991), Acoustic Instability in Premixed Flames.To appear in Combust.
   Sci. and Tech. See also this volume. Searby G., Rochewerger D. (1991) A parametric
   acoustic instability in premixed flames. To appear in J.F.M.
5. Clavin P., Siggia E.(1991), Turbulent premixed flames and sound generation.
   To appear in Combust. Sci and Tech.
6. Pelcé P. and Clavin P., (1982), Influence of hydrodynamics and diffusion upon the
   stability limits of laminar premixed flames. J. Fluid Mech.,124, pp. 219-237.
7. Clavin P., Sun J., (1991), Theory of acoustic instabilities of planar flames
   propagating in sprays or particle-laden gases. To appear in Combust. Sci. and Tech.
8. Clavin P., Lazimi D., (1991), Theoretical analysis of oscillatory burning of
   homogeneous solid propellant..To appear in Combust. Sci. and Tech.
9. Strahle, W. C., (1985) A modern theory of combustion noise in "Recent advances in
   the Aerospace Sciences" Plenum Press p 103.
10. Peters, N.,(1988), Laminar flamelet concepts in turbulent combustion, Twenty-first
    symposium on Combustion, Pittsburg, p. 1231.
11. Kerstein, A.,(1988) Fractal dimension of turbulent premixed flames, Combust. Sci.
    and Tech. 60 p. 441.
12. Landau L., Lifchitz, E.(1989) Mécanique des Fluides 2ème édition. Mir.p.416.
13. Crocco L., Cheng, S.I. (1956) Theory of combustion instability in liquid propellant
    Butterworths scientific publications, Agardograph 8. Appendix B.
14. Zel'dovich, Ya.B., (1942), On the combustion theory of powder and explosive;
    Zhurnal Eksperimental'noi i theoreticheskoi Fiziki 12 N°11-12, p.489.
15. Sivashinsky G.I. (1977) Nonlinear analysis of hydrodynamic instability in laminar
    flames. Acta Astronautica 4, p. 1177.
16. Joulin G., Clavin P., (1979), Linear stability analysis of nonadiabatic flames.
    Combust. Flame 35, p. 139.

EXPERIMENTAL INVESTIGATION OF ACOUSTIC INSTABILITIES

IN LAMINAR PREMIXED FLAMES

Geoffrey Searby

Laboratoire de Recherche en Combustion - CNRS URA D1117
Université de Provence Faculté de St. Jérôme
13397 Marseille Cedex 13 FRANCE

INTRODUCTION

It has been known for more than a century that flames can spontaneously produce acoustic oscillations in tubes and other enclosures (Mallard and Le Chatelier[1]). These self–excited acoustic oscillations can build up to very high intensities and their control is still a major problem in the design of modern liquid propellant rocket motors. Rayleigh[2] first gave the general criterion for acoustic amplification by any local heat source. This criterion states that an acoustic wave will be amplified if the time integral of the product of the pressure and heat release fluctuations is positive over a pressure cycle :

$$\int_{cycle} \delta p \; \delta q > 0$$

For this criterion to be fulfilled, there must exist a coupling between the acoustic wave and the heat source which modulates the instantaneous heat release in phase with the pressure. This paper presents an experimental and theoretical description of some aspects of the interaction of premixed methane flames with an acoustic wave. The coupling mechanism that we propose to explain the instability proceeds via the acoustic velocity field and is parametric in nature. It is shown that the observed characteristics of the interaction are well described by the proposed mechanism. Similar observations on propane flames have been published recently by Searby[3] and by Searby and Rochwerger[4] where the reader may find additional details.

SELF–EXCITED ACOUSTIC INSTABILITY IN TUBES

The spontaneous generation of sound by premixed flames propagating freely in a tube has already been studied by a number of authors. Kaskan[5] has observed flat flames and attributes the coupling mechanism to variation of the flame surface within the oscillating acoustic boundary layer. Markstein[6,7,8] has used a phenomenological model of flame front dynamics to show that a coupling can arise from variations of heat release produced by oscillations in the amplitude of cellular structures on the flame front. Leyer[9] has produced experimental evidence in favour of Markstein's calculations. Dunlap[10] and more recently Clavin et al.[11] have shown theoretically that the adiabatic temperature fluctuation caused by a pressure wave will modulate the heat release of a flame, but that the amplification of the pressure wave is small. The object of this paper will be to produce evidence that can distinguish between these different mechanisms.

Spontaneous acoustic instabilities have been observed using the apparatus represented in figure 1 (see Searby[3]). A premixed flame front propagates in an initially

Growth and Form, Edited by M. Ben Amar *et al.*
Plenum Press, New York, 1991

quiescent mixture contained in a Pyrex tube, closed at one end. In these experiments we have used methane-air mixtures at different equivalence ratios. The flame propagates downwards from the open end to the closed end. The tube used here is 120 cm long and 10 cm internal diameter. A piezo-electric pressure transducer (Kistler type 7261) is flush mounted at the closed end of the tube. The pressure signal is recorded on a digital oscilloscope.

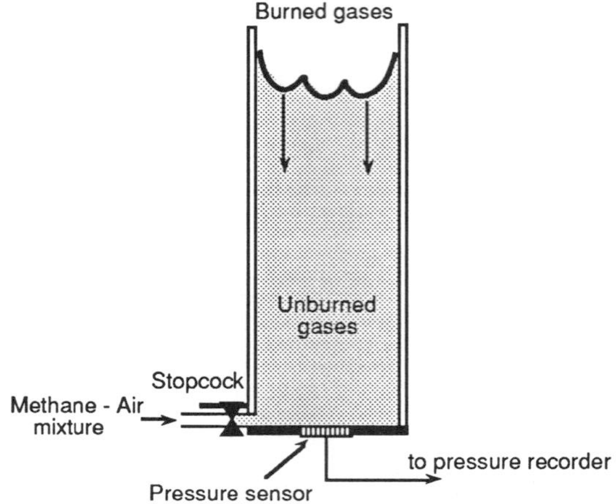

Fig. 1. Experimental apparatus for the observation of spontaneous acoustic instabilities.

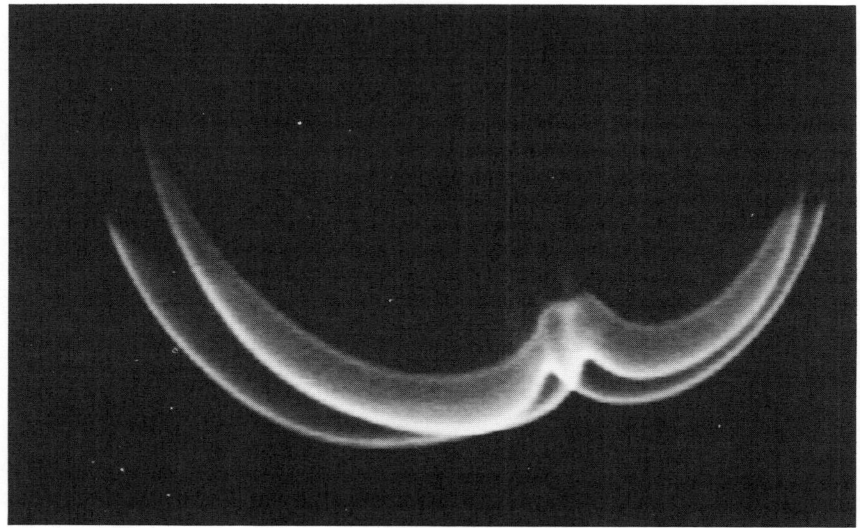

Fig. 2. Rich, cellular methane flame (ϕ = 1.4) in the regime of acoustic stability.

Different regimes can be observed depending both on the equivalence ratio and on the length of the burner tube. We observe :
- Firstly a regime of acoustic stability in which the flame is cellular and non vibrating.
- Secondly a regime of _primary_ acoustic instability in which the flame is quasi planar.
- Thirdly a regime of _secondary_ acoustic instability, in which the oscillations can reach very high amplitudes, and where the flame front develops pulsating cellular structures. This latter regime breaks down into an incoherent auto-turbulent regime at

sufficiently high acoustic amplitudes. These four regimes are described below. The mechanisms leading to the secondary instability are then investigated in more detail in a steady state experiment.

1) The regime of acoustic stability is observed for sufficiently slow methane flames on both the lean and rich sides of stoichiometry. The threshold flame speed is a function of both the resonant frequency of the tube and the magnitude of the acoustic losses, which vary with the tube diameter. For the 120 cm x 10 cm tube the lean threshold is at ≈8cm/sec (equivalence ratio ≈0.58) and the rich threshold is at ≈20.0 cm/sec (equivalence ratio ≈1.38). Below the threshold both the lean and rich flames propagate with large unsteady cellular structures arising from the Darrieus-Landau hydrodynamic instability (Darrieus[12], Landau[13], Clavin[14]). Figure 2 is a photograph, taken by a short exposure video camera, of the luminous emission from a slow rich methane flame. The combustion region propagates at about twice the laminar flame velocity, as would be expected from the ratio of the overall surface area of the flame compared to the cross section of the tube.

Fig. 3a. Edge view of the self-stabilised flat methane flame produced by the primary acoustic stability (ϕ=1.26).

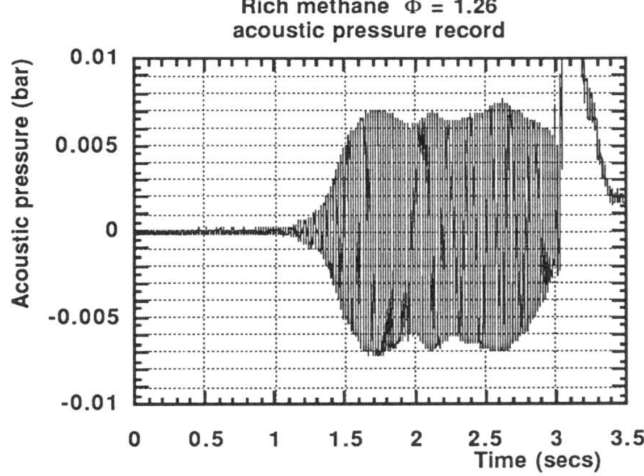

Fig. 3b. Acoustic intensity in the tube during the primary acoustic instability. The time origin is flame ignition. The apparent pressure pulse at 3.1 secs is caused by the thermal shock of the flame on the pressure transducer.

2) For slightly faster flames we observe a regime of _primary_ acoustic instability. In this regime "low intensity" acoustic oscillations are produced and the flame front becomes almost planar. The onset of the primary acoustical instability occurs close to the mid point of the tube which resonates in the 1/4 wavelength mode, the frequency is somewhat higher than the cold resonant frequency because of the increased sound speed in the hot burned gases and the phase jump of the acoustic wave at the flame front (see for

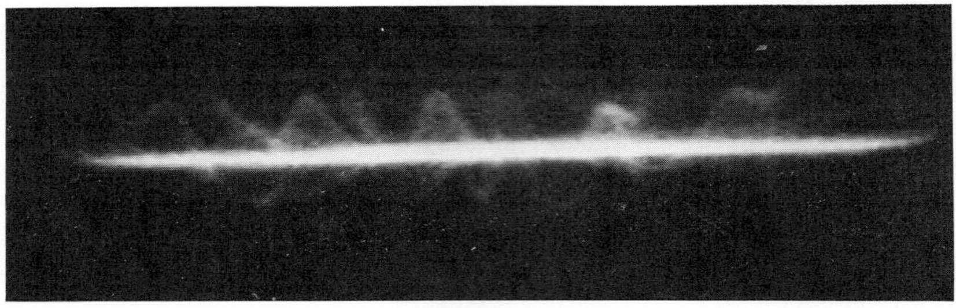

Fig. 4a. Side view of an early stage in the development of cellular structures associated with the secondary acoustic instability (methane $\Phi = 1.22$).

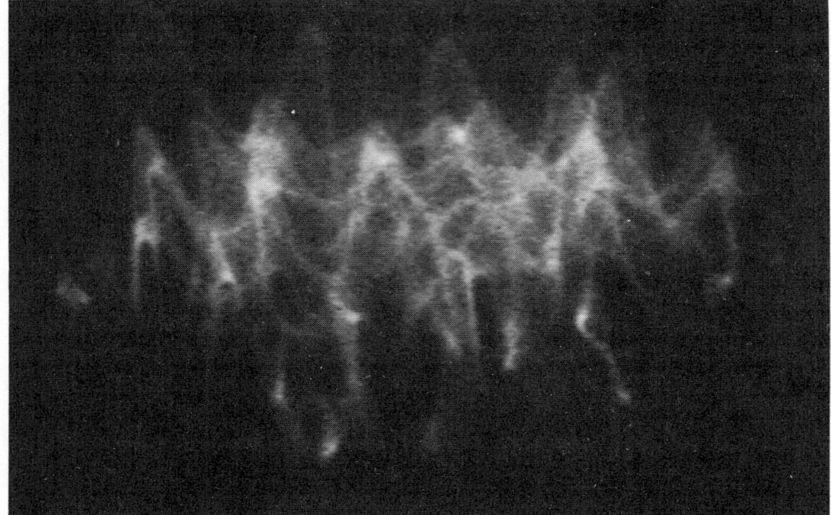

Fig. 4b. Same flame as in fig. 4a, 8×10^{-2} secs later

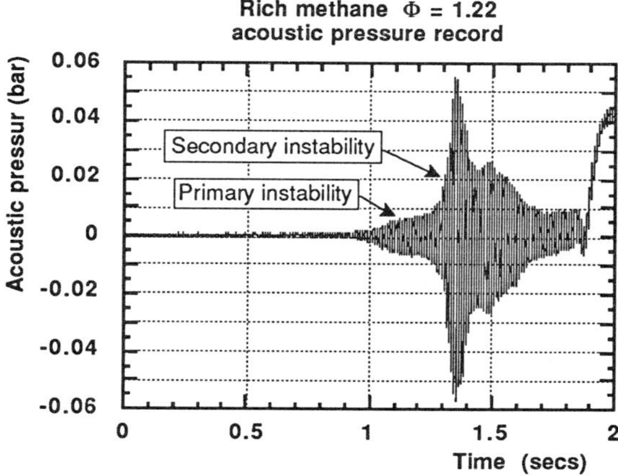

Fig. 4c. Acoustic intensity in the tube showing the secondary instability. Fig. 4a was taken at time 1.3 secs.

instance Clavin et al.[11]). Here the frequency is 130 Hz. The transition from a curved flame to a quasi planar flame occurs over a few acoustical cycles during the initial period of growth of the sound wave, after which the flame remains quasi planar until close to the end of the tube. In the 120 cm tube this primary instability is observed for the ranges :

$$\text{lean methane} \quad 8\text{cm/sec} < u_L < 14.5\text{cm/sec} \quad (0.58 < \Phi < 0.66)$$
$$\text{rich methane} \quad 20\text{cm/sec} < u_L < 33\text{cm/sec} \quad (1.38 > \Phi > 1.25)$$

These flat flames are the "vibrating flat flames" described by Kaskan[5]. Figure 3a shows a photograph of such an acoustically self–stabilised planar flame. Figure 3b shows the corresponding acoustic pressure in the tube. The rapid initial growth followed by a saturation of the sound level are characteristic of this primary instability.

The initial growth rate observed here for the acoustic pressure is 7.1 sec^{-1}. This is much greater than the growth rate expected for the 1–D mechanism of Clavin et al.[11], which is calculated to be of the order of 1 sec^{-1} for this flame, neglecting all acoustic losses (which are also of order 1 sec^{-1}). The mechanism for the initial stages of this primary instability is described by Pelcé and Rochwerger[15,16].

3) For even faster flames the _primary_ acoustic instability is followed by a _secondary_ instability . This secondary instability is accompanied by a sudden rise in the acoustic pressure level and by the rapid development of high amplitude cellular structures on the front. These cellular structures develop on the initially planar front produced by the primary instability. The structures increase the average flame surface and the overall burning rate of the combustion region. The corresponding increase in overall flame velocity can exceed a factor of 10. This regime is referred to as "blowdown" by Kaskan[5] and as "cellular" by Markstein[7] and Leyer[9]. Figures 4a and 4b show the development of the cellular structures in the early stages. Figure 4c shows the corresponding acoustic pressure in the tube. Note that figs 3b and 4c do not have the same scales. The primary and secondary stages of the instability are clearly visible in fig 4c. The initial growth rate of the secondary instability is ≈ 30 sec^{-1}

During the initial period of growth, the amplitude of these structures pulsates and passes through a minimum once per acoustic cycle. Closer observation reveals that the amplitude of the structures does not oscillate at the acoustic frequency but at _one half_ the acoustic frequency. On successive acoustic cycles, the fingers on the flame front are replaced by hollows and vice-versa. This period doubling of the response of the front to the acoustic excitation was first noticed by Markstein[6] who recognised it as being the characteristic signature of a parametric forcing in which a system responds to a periodic modulation of an internal parameter (see Landau and Lifchitz[17], for example). Markstein[6,7,8] has developed an analysis based on a phenomenological model of the flame front. Leyer[9] has studied a self–excited flame in a tube and confirms the main results of Markstein's work. This interpretation of the secondary acoustic instability has been studied in more detail by Searby and Rochwerger[4] using laminar flame theory developed recently. They do not treat the complete acoustic problem but simply the response of the flame front to the acoustic field. The cellular instability is created by the oscillating acceleration of the acoustic field acting on the density interface at the flame front in a manner similar to that of the Faraday[18] instability.We will reproduce here the mains lines of their development.

THE LAMINAR FLAME MODEL

The laminar flame model is that introduced by Clavin and Williams[20]. This model is valid in the limit of a high normalised activation energy, and in the limit of large scale flame wrinkling, $d/\lambda \rightarrow 0$, where d is the thickness of the diffusion zone associated with the flame front and λ is the length scale of wrinkling. These restrictions will be satisfied by the experimental conditions used below.

Using this model, Searby and Rochwerger[4] show that a spatial Fourier component, K, of a fluctuation of the flame front position $\xi(K,t)$ can be described by the following equation of evolution, see figure 5 :

$$A \, \ddot{\xi}_{tt}(K,t) \; + \; B \, \dot{\xi}_t(K,t) + C \, \xi(K,t) \; = \; 0 \tag{1}$$

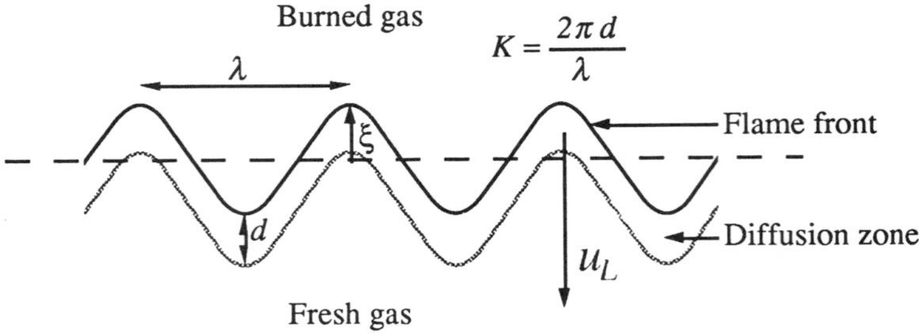

Fig. 5.   Notation used to describe a wrinkled flame front

In the following, unless specified otherwise, the units of length and time are normalised by the flame thickness, d, and the flame transit time, τ, respectively, so that K is a small quantity. The values of d and τ are defined by : $d = D_{th} / u_L$, $\tau = d / u_L$, where D_{th} is the thermal diffusivity of the fresh mixture and u_L is the laminar flame velocity. The dots denote time derivatives. The coefficients A, B and C are given by :

$$A = (2 - \gamma) + \gamma K (Ma - J/\gamma)$$

$$B = 2K + \frac{2}{1-\gamma} K^2 (Ma - J)$$

$$C = \frac{\gamma}{Fr} K - \frac{\gamma}{1-\gamma} K^2 \left\{ 1 + \frac{1-\gamma}{Fr}\left(Ma - \frac{J}{\gamma}\right) \right\}$$
$$+ \frac{\gamma}{1-\gamma} K^3 \left\{ h_b + (2 + \gamma)\frac{Ma}{\gamma} - \frac{2J}{\gamma} + (2Pr - 1)H \right\} \tag{2}$$

where $\gamma = (\rho_u - \rho_b)/\rho_u$ is the normalised gas expansion coefficient, ρ is the gas density, the subscripts u and b refer to the unburned and burned gases respectively. $\theta = (T - T_u)/(T_b - T_u)$ is the normalised temperature, $h(\theta)$ is the ratio of thermal diffusivity times density at temperature θ to its value in the unburned gases, h_b is the value of $h(\theta)$ in the burned gases, Pr is the Prandtl number (assumed to be independent of the temperature), the inverse Froude number, Fr^{-1}, is the dimensionless acceleration gd/u_L^2 where g is the acceleration of gravity,

$$H = \int_0^1 \left(h_b - h(\theta)\right) d\theta, \qquad J = \frac{\gamma}{1-\gamma} \int_0^1 \frac{h(\theta)}{1 + \theta\gamma/(1-\gamma)} d\theta \tag{3}$$

and Ma is the Markstein number, of order unity, which is a measure of the sensitivity of local flame velocity to curvature and stretch (see for instance Clavin[14]). This equation has the form of a damped harmonic oscillator. The damping, B, is always positive. The stability of the planar solution for the front depends on the sign of the coefficient C. This problem has been discussed by Clavin and Pelcé[21]. The sign of C depends on the values of the Froude number, the wave number of the wrinkling and on the diffusive properties of the mixture. However, except for very slow downward propagating flames, C is always negative for some range of wave numbers, implying that the planar front is unstable to wrinkling. The physical origin of the instability is hydrodynamic, as first described by Darrieus[12] and by Landau[13].

Consider an infinite quasi-planar laminar flame propagating downwards with respect to gravity and subjected to a planar acoustic field whose wave vector is normal to the average flame front and whose wavelength is very large compared to the thickness, d, of the diffusion zone. In the problem considered here, the flame front is periodically displaced by the velocity field of a planar sound wave. The front is invariant by translation but the Froude number in equ. (2) must be replaced by a term containing the

238

total dimensionless acceleration experienced by the front :

$$\frac{g\,d}{u_L^2} - \Omega_a\,U_a\cos(\Omega_a\,t)\tag{4}$$

where $\Omega_a = \omega_a\tau$ is the dimensionless acoustic frequency and $U_a = u_a\,/\,u_L$ is the dimensionless acoustic velocity at the front. Separating the coefficient C into a constant part C_0 and a time dependent part C_1, equ. (1) becomes :

$$A\,\ddot{\xi}_{tt}(K,t) + B\,\dot{\xi}_t(K,t) + \left[C_0 - C_1\cos(\Omega_a t)\right]\xi(K,t) = 0\tag{5}$$

with :

$$C_1 = \gamma K\,\Omega_a\,U_a\left\{1 - K\left(Ma - \frac{1}{\gamma}\right)\right\}$$

equation (5) is now in the form of a parametrically driven damped harmonic oscillator, with coefficients which depend on the wave length of wrinkling of the front. A phenomenological equation of the same form as (5) has been proposed previously by Markstein[19]. The contribution of Searby and Rochwerger[4] was to give theoretical expressions for the coefficients A, B and C which were obtained from a rigorous analysis of the flame structure. It is important to stress that Ma is the only parameter whose theoretical expression is sensitive to the details of the internal structure of the flame. All the other coefficients can be obtained from knowledge of the flame temperature and the thermodynamic properties of the gases independently of the detailed transport properties and chemical kinetics of the reactive mixture. In contrast with kinetic properties, parameters such as the gas expansion coefficient, γ, are easily measured. Thus the theoretical expressions (2) provide a convenient framework for an experimental comparison with equation (5) in which Ma can be treated as the only free parameter.

THE PARAMETRIC OSCILLATOR

Equation (5) is easily reduced to the well known Mathieu equation by means of the standard substitutions (McLachlan[22]) :

$$z = \frac{\Omega_a\,t}{2}\qquad\qquad \kappa = \frac{B}{\Omega_a\,A}\tag{6}$$

$$a = \frac{4\,A\,C_0 - B^2}{\Omega_a^2\,A^2}\qquad q = \frac{2\,C_1}{\Omega_a^2\,A}\qquad\qquad \xi = Y(z)\,e^{-\kappa z}e^{iKy}$$

to obtain:

$$Y'' + \left[a - 2\,q\cos(2z)\right]Y = 0\tag{7}$$

where the primes now denotes differentiation with respect to z. The Mathieu equation is similar to that of the harmonic oscillator, but with a restoring force that is a periodic function of time. The solutions, $Y(z)$ of (7) may be determined numerically by the continued fraction method (see for instance Abramowitz and Stegun[23]). For certain ranges of the parameters a and q, the solutions $Y(z)$ are found to increase exponentially with time. If the growth rate of $Y(z)$ is greater than κ, then the solutions for $\xi(t)$ will also be unstable. It must be remembered that a is a function of the wave number, K, so that the flame has no unique resonant frequency but a continuum of frequencies associated with a continuum of possible wavelengths of structures on the front.

The standard variables a and q do not form a convenient set for representing the dynamic properties of a flame front. We will choose instead to present the results in terms of the physical dimensionless variables Ω_a, U_a, K, Ma and Fr. Figure 6 shows a stability diagram for ξ, presented in the U_a, K plane for typical values of the reduced frequency, the Markstein number and the Froude number. The vertical coordinate should be thought of as the reduced acceleration $(\Omega_a u_a) / (\Omega_a u_l)$, which is the physical parameter driving the instability. The numerical values used in figure 6 for the various parameters are representative of the methane flames presented in the experimental study.

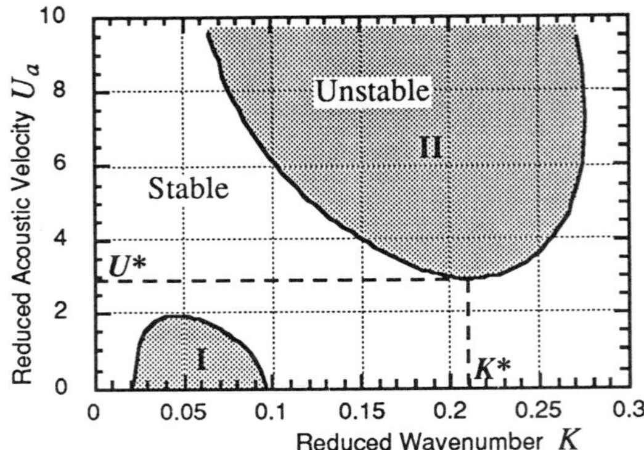

Fig. 6.   Typical stability diagram for a planar flame in an acoustic field.
This diagram is calculated for a rich methane flame with a flame
speed of 12.3 cm/sec ($\Phi = 1.49$) and a real frequency of 110 Hz,
corresponding to a reduced frequency $\Omega = 1.0$

As already noticed by Markstein[7], there are two distinct unstable regions, labelled I and II. The lower region extends down to zero amplitude of acoustic excitation where it corresponds to the well known Darrieus – Landau planar flame instability. It is stabilised at large wave numbers by diffusive effects and at very small wave numbers by the effect of gravity. In this domain the cellular structures on the front oscillate in amplitude at the acoustic frequency. We identify these oscillations of flame area in this lower unstable region as the mechanism leading to the 'primary' acoustic instability of Searby[3]. A theoretical analysis of this region is given by Pelcé and Rochwerger[15,16]. A remarkable feature is that the Darrieus–Landau instability can be re-stabilised by the oscillating acoustic acceleration and in general there exists a range of reduced acoustic velocities in which the planar flame is stable. In this example the re-stabilisation occurs when the local acoustic velocity is about 2 times the laminar flame velocity and ends when the acoustic velocity reaches about 3 times the laminar flame velocity.

In the upper unstable domain the structures on the flame oscillate at one half the acoustic frequency, it is the domain of parametric instability. The fact that a finite value of excitation is needed to excite this instability is directly related to the presence of a damping term in the original equation of evolution, equation (5). The oscillations of flame area in this upper unstable region can be identified as the mechanism leading to the 'secondary' acoustic instability of Searby[3].

EXPERIMENTS

We have made an experimental study of the threshold of excitation of the upper region, II, for methane flames in an imposed acoustic field. We have used the apparatus shown in figure 7. This is a steady state experiment in which a freely propagating flame is maintained stationary in the laboratory frame. The flame interacts with a standing acoustic field generated by a loudspeaker. Previous experiments have used either transitory or anchored flames.

Premixed methane–air mixtures were fed into the bottom of the burner and traversed a porous plate which killed turbulence and imposed a quasi top–hat velocity profile. The flame was maintained stationary by manually adjusting the gas flow rate to exactly equal the mass consumption rate of the flame. The tube was excited acoustically at one of its resonant frequencies by a loudspeaker mounted just below the porous plate. The burner was excited in either the 1/4 wavelength or the 3/4 wavelength longitudinal mode. In these modes there is a velocity node at the closed end of the burner and a pressure

240

node in the vicinity of the open end. It was verified that the velocity node was indeed situated close to to porous plate. A piezo–electric pressure sensor, placed adjacent to the porous plate, was used to measure the acoustic pressure level in the experiments with flames. Knowing the acoustic pressure at the bottom of the burner and the frequency of the standing wave, the phase and acoustic velocity at the flame front are easily calculated.

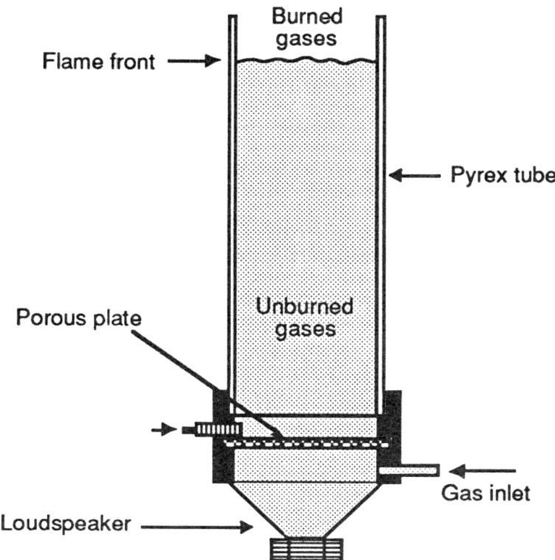

Fig. 7. Apparatus used for the steady state experiment

The flame front was maintained approximately one diameter from the burner exit. The flames were thus situated close to a pressure node, so that energy feedback into the acoustic wave was small and spontaneous acoustic oscillations were not observed. The frequency of excitation was re-adjusted to exact resonance of the tube and the acoustic level was slowly increased until the threshold, U_a^*, was reached and structures appeared at a finite wave number K^* (see figure 6).

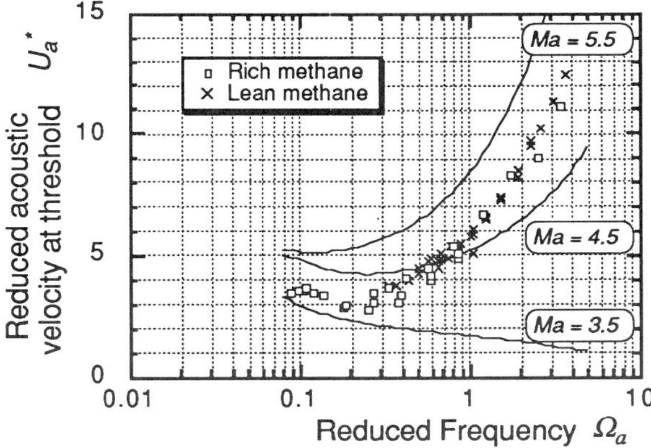

Fig. 8. Experimentally measured values of threshold of excitation for the parametric acoustic instability in methane flames, plotted in reduced coordinates. The full lines represent the theoretical thresholds calculated for 3 different values of the Markstein number.

This experiment was repeated for different equivalence ratios in both rich and lean methane and for different acoustic frequencies (tube lengths). The structures appeared abruptly, and for a single experiment, the level of threshold could be determined to within 1%, however the thermal gradients just ahead of the flame front produced slight deviations from the top–hat velocity profile and the associated velocity gradients caused experiment to experiment variations of up to 5%.

The experiments were carried out with tubes of various lengths . The frequency of excitation was varied from 43 Hz (200 cm tube, fundamental mode) to 265 Hz (100 cm tube, 1st harmonic). Typical acoustic velocities at the flame were of the order of 1m/s and the corresponding acoustic accelerations were of the order of 500 m/s^2.

The experimental results for the threshold of parametric instability are plotted in figure 8. All the results have been plotted in the reduced variables Ω_a and U_a. The reduced frequency was varied both by changing the frequency of excitation and by changing the flame velocity. It can be seen that in this representation the results of all measurements collapse onto a single curve, within experimental errors. The solid lines in figure 8 represent calculated values of the threshold, for three different values of Ma, using the analysis presented above. These curves are the locus of the minimum of the domain of parametric instability of which one point is shown in figure 6. The calculated position of this minimum is essentially a function of the reduced frequency Ω_a and of the Markstein number, Ma. It is also a function of the other mixture dependent properties such as Fr, γ, h_b, J, H . However although the flame speed varies considerably, (from ≈ 8 to ≈ 13 cm/sec in lean methane and from ≈ 10 to ≈ 30 cm/sec in rich methane) the quantities γ, h_b, J, H, are only slow functions of the burnt gas temperature and in fact do not vary very much over the range of mixtures used here. For example γ varies from 0.822 to 0.836 in the lean mixtures. For simplicity of presentation we have evaluated these quantities for a representative equivalence ratio $\Phi = 0.66$ (flame speed = 15 cm/sec) and for the purpose of plotting the thresholds in figure 8, we have assumed them to be constant at all flame speeds, thus permitting the theoretical results to be presented as a function of only a small number of parameters. The numerical values we have used are given in table 1.

Table 1

Numerical values used in calculations

Dimensioned constants

g	$= 981$ cm/sec^2	Acceleration of gravity
D_{th}	$= 0.22$ cm^2/sec	Thermal diffusivity in unburned gas
Pr	$= 0.679$	Prandtl number
γ	$= 0.836$	Normalised gas expansion factor
h_b	$= 3.41$	Normalised burned gas diffusivity
H	$= 1.06$	
J	$= 3.62$	

Dimensioned variables

d	$= D_{th}/u_L$	Flame thickness
τ	$= Dt_h/u_L{}^2$	Flame transit time

Dimensionless variables

Ω_a	$= \omega_a \tau$	Frequency of excitation
U_a	$= u_a/u_L$	Amplitude of excitation
K	$= 2\pi d/\lambda$	Wave number of structures on front
Fr	$= u_L{}^2/gd$	Froude number
Ma		Markstein number

The Froude number, Fr, varies very rapidly with the flame speed, so Ω_a and Fr are indeed independent parameters. However at high flame speeds, $1/Fr$ is a very small number and can be neglected. At low flame speeds $1/Fr$ is not a small number and the extent of the lower domain of instability does depend strongly on the value of Fr. Nevertheless, for the frequencies used here, the position of the minimum of the upper domain varies only slightly (less than 5%) when $1/Fr$ is varied from infinity to zero.

Since the experimental errors exceed 5%, we have set $1/Fr = 0$ for the purposes of calculating the thresholds in figure 8, allowing the theoretical results to be plotted as a function of only two independent variables, Ω_a, and Ma. It can be seen that, in this representation, the form of the theoretical thresholds is quite sensitive to the value used for the Markstein number (the only free parameter in the analysis). However the experimental results lie close to a single curve, indicating that the Markstein number of the rich and lean mixtures are similar. According to the theoretical analysis, the experimental results correspond to a Markstein number $Ma = 4.5 \pm 0.5$.

The experimental and theoretical results lie on curves with slightly different slopes. One reason for this is that we have assumed constant thermodynamic mixture properties in the calculation whereas the experiments were performed for a range of mixtures with a tendency to faster (hotter) flames at low reduced frequency and slower (cooler) flames at high reduced frequency. Taking variable mixture properties into account improves agreement a little but not completely. On the other hand it entails the use of a separate graph for each real acoustic frequency. For simplicity of presentation this is not shown here.

Searby and Quinard[24] have experimentally measured the value of the Markstein number for a number of fuels using a different technique. For methane–air mixtures they find that $Ma = 3.0 \pm 0.5$ in lean methane flames and $Ma = 4.0 \pm 0.5$ in rich flames. Their findings in rich methane are compatible with the values deduced here, however their findings for lean methane are outside our error estimates. For the moment this difference is not understood.

CONCLUSIONS

We have given an experimental presentation of the development of acoustic instabilities generated by premixed methane flames in tubes. We have paid particular attention to the secondary instability which was first identified by Markstein as a parametric instability and which has been analysed in detail by Searby and Rochwerger. The physical mechanism driving this instability is the periodic acceleration of the interface (flame) separating two regions of different density. An acoustic field of moderate intensity can first stabilise the natural Darrieus – Landau instability of premixed flames and then, at a higher intensity, produce a parametric cellular instability with a well defined threshold and associated with a well defined critical wave number.

Our experimental observations on lean, well controlled methane – air flames stabilised dynamically in an imposed acoustic field confirm these points. We have measured, experimentally, the thresholds of instability in rich and lean mixtures. Comparison between the experimental and theoretical results indicate that the Markstein number of methane is similar in both rich and lean mixtures with an average value of $Ma = 4.5 \pm 0.5$. The calculated results are in relatively good quantitative agreement with the experimental observations, indicating that the laminar flame theory is providing a quantitative description of the dynamics of flame fronts including acceleration effects. It remains to be shown that this parametric mechanism can explain the very high growth rate observed in the self – excited instabilities.

ACKNOWLEDGEMENTS

The author is grateful to P. Clavin and to P. Pelcé for their suggestions and enlightening discussions. This work was carried out in partial fulfilment of a contract D.R.E.T. N°. 88–210.

REFERENCES

1. Mallard, E.E and Le Chatelier, H., (1883), Recherches expérimentelles et théoriques sur la combustion des mélanges gazeux explosif., *Annls. Mines, Paris,* Partie Scientifique et Technique, Series **8**, N° 4, p. 274.

2. Rayleigh, J.W.S. (1878), *Nature*, **18**, 319.

3. Searby, G. , 1991, Acoustic instability in premixed flames, submitted to *Comb Sci. and Tech.* for publication (1990).

4. Searby, G. and Rochwerger, D., 1991, A parametric acoustic instability in premixed flames, to appear in *J. Fluid. Mech.*

5. Kaskan, W.E., (1953), An investigation of vibrating flames., *Fourth Symp. on Combustion*, Williams and Wilkins, Baltimore, pp. 575–591.

6. Markstein, G.H., (1953), Instability phenomena in combustion waves., *Fourth Symp. on Combustion*, Williams and Wilkins, Baltimore, pp. 44–59.

7. Markstein, G.H., (1964), *Nonsteady flame propagation*, Pergammon press.

8. Markstein, G.H., (1970), Flames as amplifiers of fluid mechanical disturbances. *Proc. Sixth National Congress of Appl. Mech.*, Cambridge Mass., pp. 11–33.

9. Leyer, J.C., (1969), Interaction between combustion and gas motion in the case of flames propagating in tubes., *Astronautica Acta*, **14**, pp. 445–451.

10. Dunlap, R.A., (1950), Resonance of flames in a parallel-walled combustion chamber. *Aeronautical Research Center. University of Michigan, Project MX833, Report UMM–43.*

11. Clavin, P., Pelcé, P. and He, L., (1990), One-dimensional vibratory instability of planar flames propagating in tubes. *J. Fluid Mech.*, **216**, pp.299–322.

12. Darrieus G., (1938), "Propagation d'un front de flamme", Unpublished work presented at La Technique Moderne, (1938), and at Le Congrès de Mécanique Appliquée, (1945).

13. Landau, L., (1945), On the theory of slow combustion., *Acta Physicochimica*, U.R.S.S. **19**, pp 77-85.

14. Clavin, P., (1985), Dynamic behavior of premixed flame fronts in laminar and turbulent flows. *Prog. Energy Combust. Sci.*, **11**, pp 1-59.

15. Pelcé P. and Rochwerger D., (1990), Vibratory instability of cellular flames propagating in tubes. Submitted for publication.

16. Pelcé, P. and Rochwerger, D., (1991), Sound Generated by cellular flames. This volume.

17. Landau, L. and Lifchitz, E., (1966) *Mécanique.* Translated from Russian, MIR Moscow.

18. Faraday, M., (1831), On the forms and states assumed by fluids in contact with vibrating elastic surfaces., *Phil. Trans.R. Soc. London*, **121**, pp. 319–340.

19. Markstein, G.H., (1951), *J. Aero. Sci.*, **18**, p. 428.

20. Clavin, P .and Williams, F. A., (1982), Effects of molecular diffusion and of thermal expansion on the structure and dynamics of premixed flames in turbulent flows of large scale and low intensity. *J. Fluid Mech.*, **116**, pp.251–282.

21. Pelcé, P. and Clavin, P. (1982), Influence of hydrodynamics and diffusion upon the stability limits of laminar premixed flames., *J. Fluid Mech.*, **124**, pp. 219–237.

22. McLachlan, N. W. (1951), *Theory and application of Mathieu functions.* Clarendon press, Oxford.

23. Abramowitz, M. and Stegun, I. (1972) *Handbook of Mathematical functions*, 9th printing, Dover, New York.

24. Searby G. and Quinard J., (1990), Direct and indirect measurements of Markstein numbers of premixed flames. *Combust. and Flame.* **82**, pp. 298-311.

SOUND GENERATED BY

CELLULAR FLAMES

Pierre Pelcé and Daniel Rochwerger

Laboratoire de Recherche en Combustion
Université de Provence, St Jerome
13397 Marseille Cedex 13, France

INTRODUCTION

It has been observed that flames propagating in tubes can spontaneously produce acoustic oscillations (Mallard and le Chatelier[1]). Rayleigh[2] gave a general criterion for acoustic amplification by any local heat source : When heat is released locally and periodically in a gaseous medium, an acoustic oscillation is amplified if the oscillating components of pressure and released heat flux are in phase (see Strehlow[3]).

Three basic mechanisms leading to the production of heat can be mentioned:
-The direct effect of pressure and temperature of incoming acoustic waves on the flame burning velocity (Dunlap[4]).
-The penetration of the flame edge into the acoustic boundary layer (Kaskan[5]).
-The variations of the area of an cellular flame caused by the acceleration of the acoustic velocity field (Markstein[6],Rauschenbakh[7]).

The first mechanism has been analysed in detail by Clavin[8] et al.They show that this coupling is weak in general and cannot overcome the acoustic damping in usual tubes.

The second mechanism is difficult to evaluate because it needs a difficult calculation of heat transfer around the flame and involves only a small part of the flame.

The third mechanism would be the most important[6] because it involves the whole surface of the flame. Rauschenbakh[7] performed a preliminary analysis of this mechanism and determined a transfer function giving birth to an instability. In this paper we will improve this analysis in order to make possible the comparison between theory and experiment.

In a first part we present the flame model and the basic assumptions that are used to solve the problem.

In a second part we determine the corresponding transfer function and in the last section determine the growth rate of an acoustic disturbance.

THE MODEL

When a cellular flame propagates at small Mach number in a tube, whose diameter is much smaller than its length, two different regions can be distinguished (Fig 1) :
-An acoustic region, outside the flame, where the flow is dominated by the one-dimensional acoustic waves. The size of this region is of the order of $\lambda = c / \omega$ where λ is the acoustic wavelength, c the sound velocity and ω the acoustic frequency.
-An incompressible region, called 'inner region' in the following, located around the flame, of thickness Λ, the wavelength of the cellular flame, where the flow can be assumed to be incompressible.

Growth and Form, Edited by M. Ben Amar *et al.*
Plenum Press, New York, 1991

The ratio of the sizes of these two regions , $\Lambda / \lambda \approx M \omega \Lambda / u_L$, is effectively a small parameter when the Mach number of the flame is small.

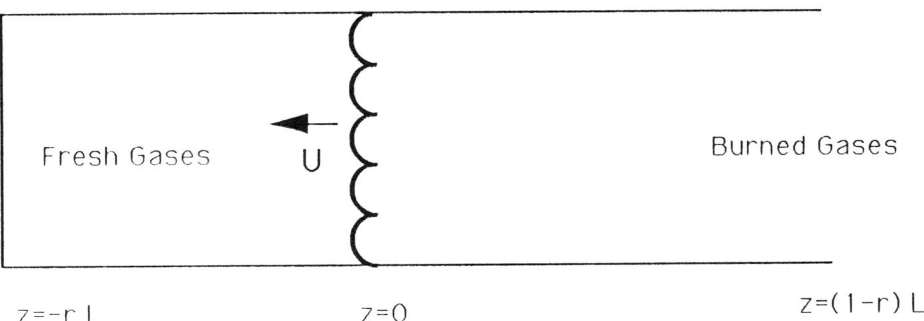

Fig.1. General configuration

Consider first the incompressible region with a frame where the flame is at rest, i.e.moving with respect to the laboratory frame with the velocity $U = dz_0 / dt$. Here, z_0 (t) which defines the new origin of the frame, is the average in space of the location of the flame. In this frame and in a region of size of order Λ around the flame, the velocity field w_i of the flow satisfies the incompressible Euler equation

$$\vec{\nabla}.\vec{w}_i = 0 \qquad (1a)$$

for the conservation of mass, and

$$\rho_i \left(\frac{\partial \vec{w}_i}{\partial t} + (\vec{w}_i.\vec{\nabla}) \vec{w}_i \right) = - \vec{\nabla} \pi_i \qquad (1b)$$

for the dynamics. Here $\pi_i = p_i - \rho_i (g + dU/dt) z$ is the effective pressure, p_i the pressure, ρ_i the density and g is the magnitude of the acceleration of gravity assumed positive when the flame propagates downwards.The index i designes the fresh mixture when $i = 1$ and the burned mixture when $i = 2$. At the interface, the following four boundary conditions must be satisfied:

$$\rho_1 (\vec{w}_1 - \vec{v}).\vec{n} = \rho_2 (\vec{w}_2 - \vec{v}).\vec{n} \qquad (2a)$$

for mass conservation,

$$\vec{w}_1 . \vec{\tau} = \vec{w}_2 . \vec{\tau} \qquad (2b)$$

for the equality of tangential velocities

$$\pi_1 + \rho_1 (g + \frac{dU}{dt}) z + \rho_1((\vec{w}_1 - \vec{v}).\vec{n})^2 =$$
$$\pi_2 + \rho_2 (g + \frac{dU}{dt}) z + \rho_2((\vec{w}_2 - \vec{v}).\vec{n})^2 \qquad (2c)$$

for conservation of the momentum component normal to the interface, and

$$(\vec{w}_1 - \vec{v}).\vec{n} = u_L \qquad (2d)$$

Here n and t are respectively the normal and tangent directions to the interface, v the local speed of the interface and u_L the relative velocity of the flame. For simplicity we neglect first the curvature effects of the flame, i.e. assume that the flame is an infinitely thin discontinuity surface propagating with constant velocity. This simplifying assumption allows us to discuss more clearly the reduction of the inner problem to a linear response problem. Then, curvature terms are reintroduced for the complete calculation of the transfer function.

Now consider the acoustic region where as a first step, the mean position of the flame is kept at rest in the laboratory frame by adjusting the mean flow velocity of the incoming fresh gases to the mean velocity of the flame. We first neglect all effects leading to dissipation of acoustic energy in the tube, i.e. viscous friction, conduction of heat to the tube walls and acoustic losses at the open extremity of the tube. Then the longitudinal acoustic field can be written simply as:

$$\delta p_{1,2} = \left\{ A_{1,2} \exp(i\frac{\omega}{c_{1,2}}z) + B_{1,2} \exp(-i\frac{\omega}{c_{1,2}}z) \right\} \exp i\omega t \qquad (3a)$$

$$\delta u_{1,2} = -\frac{1}{\rho_{1,2}c_{1,2}} \left\{ A_{1,2} \exp(i\frac{\omega}{c_{1,2}}z) - B_{1,2} \exp(-i\frac{\omega}{c_{1,2}}z) \right\} \exp i\omega t \qquad (3b)$$

where subscripts 1 and 2 denote unburned and burned mixtures respectively. At this spatial scale, the inner incompressible region appears as a discontinuity surface separating fresh and burned mixtures Since the flame propagates at small Mach number the fluctuations of acoustic pressures can be considered as equal on both sides of the discontinuity :

$$\delta p_1(0) = \delta p_2(0) \qquad (4)$$

To determine the jump conditions of velocity field across the discontinuity surface, one applies the mass conservation across the inner region i.e. integrates the incompressibility condition over the two volumes delimited by the contours shown on Fig.2.

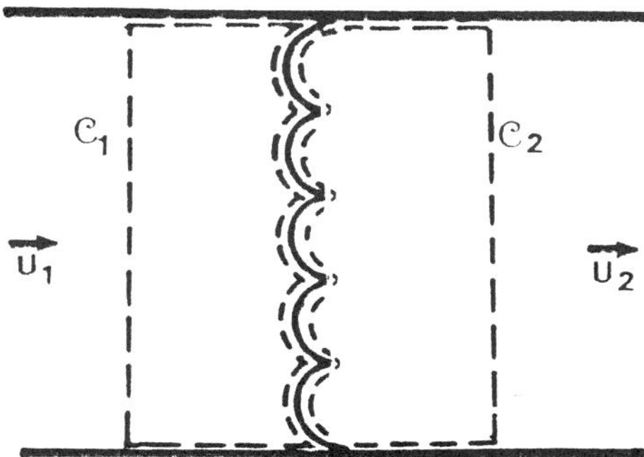

Fig. 2   Contours of integration in the incompressible zone

After integration of the incompressibility condition (1a) over the volume corresponding to the fresh mixture, one obtains the relative flame velocity as:

$$u_1 - U = u_L S \qquad (5)$$

where S is the surface of the flame relative to the area of the cross-section of the tube. After integration on the same equation on the second volume and with the help of the local mass conservation relation (3a) , one obtains the global mass conservation relation as:

$$\rho_1 (u_1 - U) = \rho_2 (u_2 - U) \tag{6}$$

The combination of these relations allows us to obtain the jump of the acoustic field across the flame as:

$$\delta u_2(0) - \delta u_1(0) = (\frac{\rho_1}{\rho_2} - 1) u_L \delta S \tag{7}$$

The feedback between a cellular flame and acoustics is now clear: From relation (6) the flame velocity is determined from the acoustic velocities. Then , relation (2c) determines how the flame shape and thus the surface is affected by the acceleration. Last, relation (7) determines how the acoustic velocities are modified by the variation of the flame surface.

We consider the classical configuration where the flame propagates from the open to the closed end of a tube sothat the appropriate boundary conditions at the tube extremities are:
-closed end in the fresh mixture,

$$z = - rL : \quad \delta u_1 = 0 \tag{8a}$$

-open end in the burnt gases

$$z = (1- r)L : \quad \delta p_2 = 0 \tag{8b}$$

where L is the total length of the tube and r the relative position of the flame.

THE TRANSFER FUNCTION

The complete problem is difficult to solve because the problem posed in the inner region is in general non-linear and time dependant. However the problem is simplified if one considers weakly cellular flames. Assume first that the flame is flat. Then, relation (2c) indicates that the flame shape is not affected by the acceleration ($z = 0$). The flame surface is constant and no acoustic energy is generated by the flame during its oscillation in the given acoustic field. Consequently the jump of acoustic velocities accross the discontinuity vanishes and the flame oscillates with the velocity $\delta U = \delta u_1 = \delta u_2$. Assume now that the flame is weakly cellular, $z = \xi_c(x)$ Then, relation (2c) shows that a hydrodynamic pressure jump across the flame $\delta \pi = (\rho_1 - \rho_2) (d (\delta u_1)/ dt) \xi_c(x)$ is generated which induces a local flow around the cells. This secondary flow modulates the flame shape and thus generates a jump of acoustic velocities across the discontinuity from relation (7). Thus, it is convenient to look for a solution in the inner region as:

$$z = \xi_c(x)+ \xi (x) \tag{9}$$

With the assumption that the steady flame is weakly cellular, the inner problem appears as a linear response problem, i.e. the perturbation of the flame shape is solution of a linear equation with a nonhomogeneous term determined by the acoustic forcing.
Once this linear problem is solved the jump of the acoustic velocities across the discontinuity is determined by relation (7) i.e.

$$\delta u_2(0) - \delta u_1(0) = (\frac{\rho_1}{\rho_2} - 1) u_L \frac{1}{\Lambda} \int_0^{\Lambda} \frac{d \xi_c}{d x} \frac{d \xi}{d x} dx \tag{10}$$

In order to solve the problem further it is necessary to overcome two difficulties.
First, we need to explicit the steady   solution for the cellular flame. With the assumption that the normal burning flame velocity is constant along the whole surface, only approximate solutions have been determined (Zeldovich[9] et al.).

Secondly, the linear inner problem admits general solutions (i.e. solutions of the homogeneous problem) whose amplitude grows exponentially with time with the Darrieus-Landau growth rate. These solutions overtake the particular oscillating solution after a finite time and the linear response problem posed above looses its meaning.

A more realistic situation can be considered if one takes into account the effects of flame thickness, i.e. essentially the dependance of the burning flame velocity on flame curvature and flow stretch. It is well known that in this case a flat flame can be stable for sufficiently low velocity. At threshold of instability , i.e. when the flame propagates with a critical velocity u_{L_c} determined by the diffusive characteristics of the reactive mixture, the flame becomes cellular with the marginal wavenumber k_c. The shape of the flame is simply

$$\xi_c(x) = a_0 \cos k_c x \tag{11}$$

where the flame amplitude a_0 is arbitrary but small

We first determine the perturbation of the flame amplitude $\xi = Q \cos (k_c x) \exp (i\omega t)$ in response to the fluctuating acoustic field $\delta u_1 \exp (i\omega t)$. After some calcultation we obtain:

$$Q = \frac{-i\Omega\, C (K_c)}{- \Omega^2 A (K_c) + i\Omega\, B (K_c) + D (K_c)}\, a_0 \frac{\delta u_1}{u_L} \tag{12}$$

where $\Omega = \omega d / u_L$ and $K = kd$ are respectively the dimensionless frequency and wave number adimensionalysed by the flame velocity ul and the flame thickness d. Here,[10] :

$$A (K) = (2 - \gamma) + \gamma (Ma - \frac{1}{\gamma} Log (\frac{1}{1-\gamma})) K \tag{13a}$$

$$B (K) = 2K + \frac{2}{1-\gamma} (Ma - Log (\frac{1}{1-\gamma})) K^2 \tag{13b}$$

$$C (K) = \gamma K (1 - K (Ma - \frac{1}{\gamma} Log (\frac{1}{1-\gamma})) \tag{13c}$$

and

$$D (K) = \frac{\gamma}{1-\gamma} K (\frac{gd}{u_L^2} (1 - \gamma) - K (1 + \frac{gd}{u_L^2} (1 - \gamma)(Ma - \frac{1}{\gamma} Log (\frac{1}{1-\gamma}))$$
$$+ K^2 (1 + \frac{2+\gamma}{\gamma} Ma - \frac{2}{\gamma} Log (\frac{1}{1-\gamma}))) \tag{13d}$$

where $\gamma = (\rho_1 - \rho_2)/\rho_1$ and Ma is the dimensionless Markstein length .

When the denominator of the r.h.s. of eqn. (12) vanishes, Ω and K are related by the dispersion relation for the disturbances of the planar flame The instability threshold of the planar flame is determined by the relations $\Omega (K_c) = 0$, $d\Omega / dK (K_c) = 0$.

Then, the transfer function $Tr = (\delta u_2 (0) - \delta u_1 (0)) / \delta u_1 (0)$ can be easily derived from eqns. (10) and (12) as :

$$Tr = \frac{1}{2} \frac{\gamma}{1-\gamma} (K_c A_0)^2 \frac{- i\Omega\, C (K_c)}{- A (K_c) \Omega^2 + B (K_c) i\Omega} \tag{16}$$

It is proportional to the small parameter $(K_c A_0)^2$ where $A_0 = a_0 / d$ is the dimensionless amplitude of the steady cellular flame. Real and imaginary parts of $Tr / (K_c A_0)^2$ are drawn on Fig.3 of rich ethylene-oxygen mixtures which correspond to Markstein numbers Ma = 3.

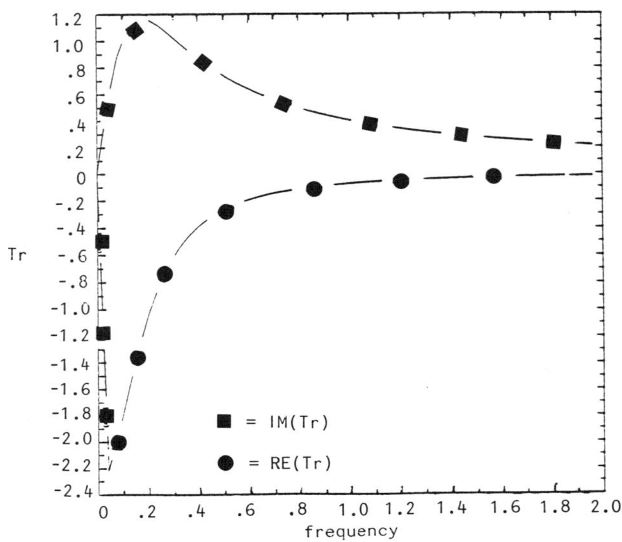

Fig.3. Real and imaginary parts of the transfer function,(Ma=3)

GROWTH RATE OF THE ACOUSTIC DISTURBANCE

By applying the boundary conditions (8) at the tube extremities and at the flame(16) for the acoustic field,one obtains the eigenmodes equation:

$$\frac{\rho_2 c_2}{\rho_1 c_1} \tan(rX) \tan\left((1-r)\frac{c_1}{c_2}X\right)(1+Tr) = 1 \tag{17}$$

where $X = \omega L/c_1$ is the dimensionless frequency .From eqn.(16), the transfer function is proportional to $(K_c A_0)^2$, which in our analysis is assumed to be a small parameter. It follows that the solutions of eqn. (17) can be expanded around the solutions for the free eigenmodes of the tube, X_0, which are given by

$$\frac{\rho_2 c_2}{\rho_1 c_1} \tan(rX_0) \tan\left((1-r)\frac{c_1}{c_2}X_0\right) = 1 \tag{18}$$

The X_0 are real numbers which characterize the acoustic frequencies of the tube when the flame is considered as a passive interface separating two different gaseous media.Writing $X = X_0 + \delta X$, one obtains at the first order in the power expansion of $(K_c A_0)^2$,

$$\mathrm{Im}(\delta X) = \frac{-\mathrm{Im}(Tr)\tan(rX_0)}{r(1+\tan^2(rX_0)) + \frac{\rho_2 c_2}{\rho_1 c_1}(1-r)\frac{c_1}{c_2}\left(1+\tan^2\left((1-r)\frac{c_1}{c_2}X_0\right)\right)\tan^2(rX_0)} \tag{19}$$

As the denominator of the r.h.s. of relation (19) is always positive, instability occurs if Im (Tr) tan (rX_0) is positive. If the fundamental tone is considered, tan (rX_0) is positive for all positions of the flame in the tube. It follows that this mode is always unstable since the imaginary part of the transfer function is positive. As in shown on Fig.6 the growth rate (19) calculated for $K_cA_0 = 1$ is maximum in the lower half of the tube.

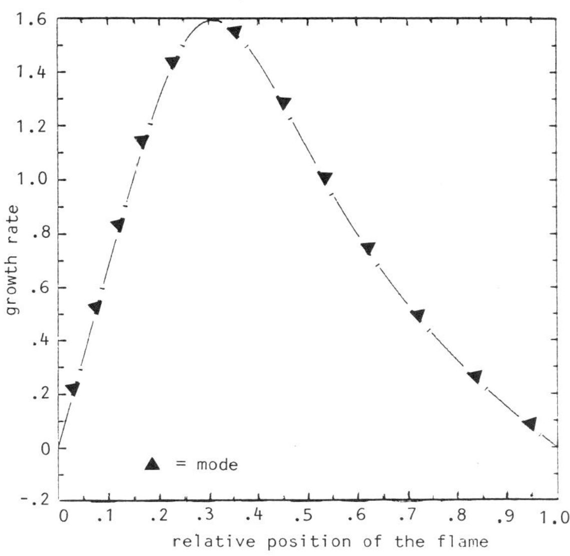

Fig. 4. Growth rate of the instability, (Ma=3)

The instability can develop only when the linear growth rate (19) is sufficiently large to overcome damping effects. There are two kinds of damping effects: heat transfer and viscous friction at the tube wall, acoustic radiation losses by the free extremity of the tube. In order to obtain an order of magnitude of the first mechanism, one evaluates the damping rate of the acoustic energy due to viscous friction (In gases the Prandtl number is of order unity, so the damping rate due to heat conduction to the wall is of the same magnitude as viscous damping). If δu is the amplitude of the acoustic velocity in the tube, the energy dissipated in the boundary layer per unit time and unit length of the tube is of order of ρv (δu)^2 R / h, with h $\approx$ (v / ω)$^{1/2}$. The acoustic energy stored in the acoustic mode per unit length of the tube is ρ (δu)^2 R^2. It follows that the corresponding damping rate is $1 / \tau_{dv} \approx$ (v ω)$^{1/2}$ / R. Another damping mechanism is due to the release of acoustic energy from the open end of the tube. The order of magnitude of the corresponding damping rate is (Rayleigh), $1 / \tau_{dr} \approx$ (R / λ)^2 ω . These rates increase with the harmonic number so that , the least damped mode is the fundamental tone, which is expected to be the first mode to become unstable, as has been observed in experiments. Consider that the damping due to the radiative losses dominates, which is the case in experiments[11]. Then , the vibratory instability occurs for the fundamental tone $\omega \approx c_1$ / L if the growth rate (19) dominates $1 / \tau_{dr}$, i.e. in order of magnitude:

$$(K_cA_0)^2 \tan (rX_0) > (\frac{R}{L})^2 \qquad\qquad (20)$$

CONCLUSION

We have determined the acoustic transfer function for a weakly cellular flame , i.e. when the propagation conditions are close to the threshold of the Darrieus-Landau instability. It is found that this transfer function is proportional to $(k_c a_0)^2$ where k_c and a_0 are respectively the wave number and the amplitude of the wrinkling of the flame at the instability threshold . It is multiplied by a frequency factor whose denominator is the dispersion relation for the disturbances of the planar flame front as is usual for the linear response problems. The consequences are that , as it is observed in experiments[11]:

-The primary sound is generated when the amplitude of spontaneous cellular structure of the flame is sufficiently large.

-The fundamental tone is the most unstable mode.

-Vibratory instability for the fundamental tone occurs in the lower half of the tube.

Some quantitative disagreement still remains on the stability limits of the vibratory instability.

Thus the mechanism of variation of the flame area caused by the acceleration the acoustic velocity field appears as a good candidate for the explanation of the generation of primary sound by flames propagating in tubes.

Much work is still needed to validate this scenario quantitatively: One has to control experimentally the wavelength of the cellular flame; One needs to determine transfer functions for cellular flames with a relative amplitude of wrinkling of order unity.

REFERENCES

1. Mallard ,E.E and Le Chatelier,H (1883) Recherches expérimentales et théoriques sur la combustion des mélanges gazeux explosifs, Annales Mines Paris, Partie scientifique et technique,Série 8 n°4,p. 271.
2. Rayleigh J.W.S, (1878), Nature, 18, 319.
3. Strewlow R.A. (1979) Fundamentals of combustion. Robert E. Krieger.
4. Dunlap R.A.,(1950), Resonance of flames in a parallel-walled combustion chamber. Aeronautical Research Center.University of Michigan , Project MX833, Repoort UMM-43.
5. Kaskan W.E.,(1953), An investigation of vibrating flames. Fourth Symp. on Combustion, Williams and Wilkins ,Baltimore , pp. 575-591.
6. Markstein G.H.,(1970), Flames as amplifiers of fluid mechanical disturbances.Proc Sixth National Congress of Appl. Mech., Cambridge Mass., pp. 11-33.
7. Rauschenbakh B.V.,(1961), Vibrational Combustion, Fizmatgiz , Moscow.
8. Clavin P.,Pelcé P. and He L.,(1990), One -dimensional vibratory instability of planar flames propagating in tubes. J. Fluid Mech., 216, pp. 299-322.
9. Zeldovich Ya.B., Istratov A.G., Kidin N.I. and Librovich V.B.,(1980), Flame Propagation in Tubes : Hydrodynamics and Stability, Combustion Science and Technology ,24, pp. 1-13.
10. Pelcé P. and Clavin P. ,(1982), Influence of hydrodynamics and diffusion upon the stability limits of laminar premixed flames. J. Fluid Mech.,124,pp. 219-237.
11 Searby G. (1990),Acoustic Instability in Premixed Flames, Preprint.

SIMULATION OF FLAME FRONTS BY SOURCES OF FLUID VOLUME

P.L. García-Ybarra, J.C. Antoranz and J. L. Castillo

Departamento de Física Fundamental, U.N.E.D.
Apartado Correos 60.141, 28080-Madrid, Spain

INTRODUCTION

Gaseous combustion waves are self propagating exothermic chemical reactions. In the low Mach number limit, these waves are the usual premixed flames where a gas mixture, in an initially unstable molecular configuration at a given temperature, reaches its thermodynamical equilibrium at the flame temperature. Heat release by combustion reaction diffuses towards the fresh gas whose volume increases by thermal expansion. Each heated fluid element expands, acting like a piston that pushes the surrounding fluid. A propagating planar flame front is rendered unstable by such a phenomenon because when it becomes slightly corrugated, crests and valleys push themselves away each other. The linear stability analysis of the planar propagation was performed, independent and almost simultaneously, by Darrieus[1] and Landau[2] who calculated the potential flow induced in the fresh gas by an infinitesimally distorted flame front. By relating the gas thermal expansion effects to appropriate hydrodynamic jump conditions through the flame, they found that the burned gas is in rotational motion and that each Fourier component of the front distortion grows exponentially with a positive growing rate σ_{DL} that increases linearly with the corrugation wave number k. Explicitly this result is

$$\sigma_{DL} = u\,k\,\frac{\sqrt{1+\rho^{-1}-\rho}\,-1}{1+\rho} \tag{1}$$

where u is the flame burning velocity, with respect to the fresh gas, and $\rho \equiv \rho_2/\rho_1$, $(0 < \rho \le 1)$, with ρ_1 and ρ_2 being the unburnt and burnt gas densities, respectively. It is interesting to note the two limit forms of Eq. (1) for nearly isothermal flames ($\rho \approx 1$) and for high temperature flames ($\rho \approx 0$), respectively,

$$\sigma_{DL}(\rho \approx 1) \to u\,k\,(1-\rho)/2 \tag{2}$$

$$\sigma_{DL}(\rho \approx 0) \to u\,k/\rho^{1/2} \tag{3}$$

More detailed calculations show that a premixed flame may be stabilized by the action of both gravity and diffusive effects inside the flame thickness.[3-5] However, Darrieus-Landau original work contains most of the features of real flames and, in the following, we will restrict our discussion to this level of description by neglecting the internal flame structure and assimilating the flame front to a hydrodynamic jump advancing with a constant normal speed.

The flow will be assumed inviscid and we will not consider the rotational part of the burnt gas velocity, focusing attention in the fresh gas potential flow only. In fact, a generic

Growth and Form, Edited by M. Ben Amar *et al.*
Plenum Press, New York, 1991

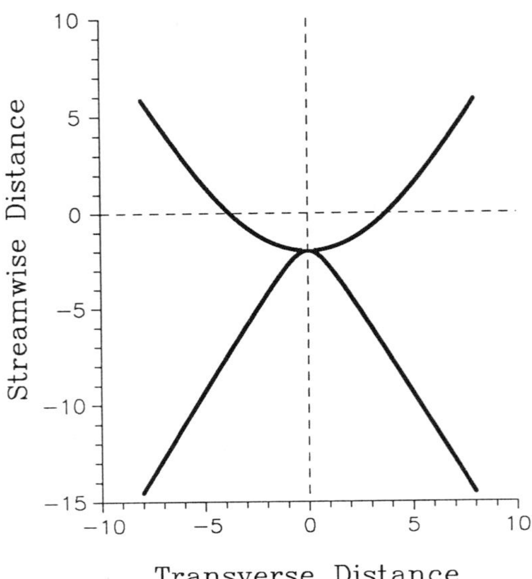

Fig. 1. The two stationary shapes that can adopt a line advancing with a constant normal velocity $u = 0.5\ v_\infty$ in the flow field generated by a point source of strength $2\pi L v_\infty$ located at the origin and a uniform vertical flow of magnitude v_∞ directed upwards. (L is the length unit).

property of potential flows is responsible for the two more usual non-planar flame shapes, namely, the conical Bunsen flame and the curved flame propagating in a duct. Both have been experimentally observed under the same flow conditions.[6] They correspond just to a particular case of the solution to the general problem of finding the locus of a surface propagating with a constant normal velocity in a slightly diverging flow. For instance, in the flow generated by a point source of fluid volume immersed in an uniform velocity field, the two possible solutions to this problem are depicted in Figure 1 (in 2-D). The lower branch is easily recognized as the typical shape of the Bunsen burner flame whereas the upper one corresponds to the shape of a flame propagating in a tube and also to each unitary cell in which a planar flame breaks up when it becomes unstable. Although, flows induced by actual flames are undoubtedly more complex, this baby example reveals the use of volume source distributions as a promising tool for describing non-planar flame shapes. Sivashinsky and Clavin[7] studied the non-linear hydrodynamic instability of a flame front in the nearly isothermal limit ($\rho \approx 1$). They developed a potential (non-rotational) model for the flame front description that leads to the same kind of non-linear differential equation as the exact theory but with slightly different coefficients. Therefore, the qualitative behaviour of actual flames is retained by the potential model at least for low thermal expansions. Here, we will consider only potential flows and study the behaviour of surfaces formed by sources of fluid volume moving with a constant normal velocity.

In the next sections, first, the linear stability of a line source of fluid volume with constant strength density is compared to Darrieus-Landau result for a flame, Eq. (1). Then, a discrete version of the problem is presented by means of a cellular automata. Finally, the conclusions are summarized and future research lines indicated.

LINEAR STABILITY OF A MOVING LINE SOURCE

We wonder up to which extent the dynamical effects associated with a non planar flame front can be mimicked by a moving line source. These sources should generate fluid volume at a constant rate to simulate locally the volume increase due to thermal expansion in the pseudo equivalent onedimensional flame. First, let us calculate the velocity field induced by a distribution of sources located along a segment of length 2s parallel to the x-axis (y=y´). By integrating along the segment the well known formulae of the velocity field associated to a point source,[8] we get

$$v_x = \frac{m}{2\pi} \int_{-s}^{s} \frac{x - x\,'}{(\mathbf{r} - \mathbf{r}\,')^2}\, dx\,' = -\frac{m}{4\pi} \ln \frac{(x - s)^2 + (y - y\,')^2}{(x + s)^2 + (y - y\,')^2} \tag{4}$$

$$v_y = \frac{m}{2\pi} \int_{-s}^{s} \frac{y - y\,'}{(\mathbf{r} - \mathbf{r}\,')^2}\, dx\,' = -\frac{m}{2\pi} \left(\tan^{-1} \frac{x - s}{y - y\,'} - \tan^{-1} \frac{x + s}{y - y\,'} \right) \tag{5}$$

for the horizontal and vertical components of the velocity field, respectively. Here, $\mathbf{r} = (x,y)$ is the position vector and m is the linear source strength density (volume generation of fluid per unit length and time). When the length of the segment is extended to infinity, the limit of the above expressions is, simply,

$$\lim_{s \to \infty} v_x = 0 \tag{6}$$

$$\lim_{s \to \infty} v_y = \begin{cases} \dfrac{m}{2} & (y > y\,') \\[2em] -\dfrac{m}{2} & (y < y\,') \end{cases} \tag{7}$$

i.e., the velocity field is vertical, of magnitude m/2, and discontinuous through the line source, y=y´. To simulate a flame front moving downwards with velocity u and density ratio ρ, we just add an uniform velocity field of magnitude u+m/2, directed vertically upwards and choose m = u (1-ρ) / ρ, to preserve mass conservation. Up to this (trivial) point, the mimic is perfect and we can proceed further to study the linear stability of the constructed line source.

Let us assume that the line source location, taken initially on the origin along the x-axis (y´=0), is infinitesimal and sinusoidally disturbed becoming

$$y\,' = \varepsilon \sin kx\,', \qquad \text{with} \qquad \varepsilon \ll 1 \tag{8}$$

The new induced velocity in the vertical direction is given by the integral expression of Eq. (5) plus the added uniform velocity. Retaining only the first correction in ε and performing the integral when s→∞, we get

$$v_y = u + \frac{m}{2} + \frac{m}{2\pi} \int_{-\infty}^{\infty} \frac{y - \varepsilon \sin k\,x\,'}{(\mathbf{r} - \mathbf{r}\,')^2}\, dx\,' = u + \frac{m}{2} \left[\varepsilon\, k\, e^{ky} \sin kx + O(\varepsilon^2) \right] \tag{9}$$

for $y < y\,'$ (fresh gas). Furthermore, the normal propagation velocity of the line source remains unaltered, then the line must move with the local value of the velocity disturbance, *i.e.*,

$$\frac{\partial y\,'}{\partial t} = \frac{m}{2}\, k\, \varepsilon \sin kx\,' \tag{10}$$

using here Eq. (8), a differential equation for ε is obtained that shows an exponential explosion, $\varepsilon \approx \exp(\sigma t)$, with the growing rate

$$\sigma = u\, k\, \frac{1 - \rho}{2\,\rho} \tag{11}$$

When this result is compared with the Darrieus-Landau growing rate for the flame instability, Eq. (1), it appears that both expressions coincide in the weak expansion limit, whereas for large expansions the line source exhibits a stronger instability,

$$\sigma(\rho \approx 1) = u\,k\,\frac{1-\rho}{2} = \sigma_{DL}(\rho \approx 1) \tag{12}$$

$$\sigma(\rho \approx 0) = \frac{u\,k}{2\,\rho} > \sigma_{DL}(\rho \approx 0) \tag{13}$$

In fact, the Darrieus-Landau dispersion relation consists of two different contributions, a term coming from the potential part of the flow field which coincides exactly with the line source dispersion relation, Eq. (11), and a second term that accounts for the rotational part of the flow. This means that the mechanism of flame instability is just the same as the moving line source instability we study here. The velocity field associated with the moving line source may verify the same jump conditions through the sources as the velocity field across a flame front. However, the continuity condition for the pressure disturbance is violated in general (except for the trivial case $\rho=1$). Only by consideration of a rotational velocity contribution this condition can be fulfilled. Eq. (12) is in agreement with Sivashinsky and Clavin's calculation[7] who also noticed the stabilizing character of the rotational flow induced in the burnt gas of actual flames. Here, the inequality indicated in Eq. (13) shows clearly this point.

Flames may be, then, only approximately described by moving line sources. The lower the difference between burnt and unburnt gas densities, the more accurate the description would be. But, in general, only a *similar* qualitative behaviour can be expected.

The extension of this idea to the analytic description of well developed non-planar flame shapes seems to be not so straightforward. Nevertheless, in the next section a simple discrete model is depicted that lies on the same idea and allows the simulation of arbitrary shape fronts.

DISCRETE SIMULATION BY A CELLULAR AUTOMATON

The use of cellular automata to simulate fluid motion has received a strong stimulus since the promising results obtained recently.[9] The automaton we present here is based on the model introduced by Saïd and Borghi[10] where each basic cell represent a macroscopic fluid element instead of a microscopic unit. The automaton elements are assumed to occupy the sites of a square mesh where no empty sites are allowed and no more than a particle per site may exist. In that way incompressibility is guaranteed by construction. Only two kinds of fluid elements, burnt or unburnt, are considered (represented by different colours, white and black, let us say). The complex interaction between actual reactive fluid elements is here oversimplified and split into two distinguished and consecutive steps: burning and expansion.

The burning step simulates the advance of the chemical reaction by an interaction rule among neighbours. Thus, each burnt particle belonging to the burnt-unburnt border, transforms randomly a number of unburnt neighbours into burnt elements per step. To insure isotropy in spite of the square mesh, different probabilities are assigned to each hierarchy of neighbours.

During the expansion step, every burnt cell generated in the previous burning step induces, around itself, a fixed number of new burnt cells, according to the prescribed expansion. The new cells take the sites of the surrounding neighbours by pushing them away along some chosen expansion directions. For each cell, the expansion directions are taken randomly although preserving the isotropy. The displaced cells move on those directions by taking, in their turn, the places of their neighbours and so on. This procedure follows up to the automaton boundaries, in such a way that a long range interaction (infinite range, in fact) among particles is introduced, simulating the actual pressure effects. This expansion step provides the purely hydrodynamical destabilizing mechanism theoretically described in the previous section. Although the diffusive nature of a flame front is not taken into account by this automaton, as we mention in the introduction, the noise induced by the randomness in the automaton rules produces a thickening of the front and thus a stabilizing effect on the small wavelengths very similar to the one associated to the global effect of heat and mass diffusion in lean mixtures flames of heavy fuels.

Fig. 2. Evolution of a plane front. Black and white points correspond to unburnt and burnt elements, respectively. Time runs from up to down and then from left to right. Mesh size is 300x100. Probabilities for burning are one for first neighbours and one fourth for second neighbours. Expansion is one-to-five with one hundred and eighty allowed expansion directions.

Fig. 2 depicts a typical run of this automaton with a planar front as initial condition. As it is shown, the planar front becomes unstable and breaks up in small bulges which finally lead to two slightly distorted large humps. It is worthwhile to notice that this general picture agrees qualitatively with the more recent numerical integration of the Sivashinsky's flame front evolution equation.[11]

The resulting cellular states are now being characterized by means of front length measurement, velocity propagation, deformation amplitude, etc. Also, other initial as well as boundary conditions are currently being investigated.

CONCLUSIONS

Flame front dynamics has been shown to have many common features with the dynamics of volume source fronts. In particular, the hydrodynamic jumps through a flame front are recovered by a volume source front except for the condition of vanishing pressure jump. This fact has its origin in the potential character of the flow associated with a moving volume source front in constrast to the corresponding flow field generated by unsteady flame fronts, which is rotational in the burned gas side. The linear stability analysis of a volume source front leads to a dispersion relation similar to the dispersion relation of flame fronts. In fact, both relations collapse in the weak thermal expansion limit, where the rotational part of the flow plays a negligible role.

We have developed a cellular automaton which is, in fact, a discrete version of the volume source front model. It permits to simulate, with the help of a small personal computer, the unsteady evolution of 2-D fronts of any shape. For the moment we have shown that this automaton exhibits the main features shown by numerical integration of the up-to-day accepted equation describing flame behaviour (Sivashinsky's equation).

The described automaton can be implemented to simulate flame propagation in turbulent flows by including the presence of randomly located vortices which induce disordered although spatio-temporally correlated motions of the basic cells. On the other hand, the extension to 3-D simulations is conceptually straightforward. It will require the use of more computational power which can be really achieved with today available mean-size computers. Further results along these lines will be published elsewhere.

ACKNOWLEDGEMENTS

The authors acknowledge financial support by the European Community under contract JOUE 42-C (EDB) and by Dirección General de Investigación Científica y Técnica (DGICYT, Spanish Ministry of Education and Science) under project PB88-0159.

REFERENCES

1. G. Darrieus, Propagation d'un front de flamme. Essai de théorie des vitesses anormales de déflagration par développement spontané de la turbulence. Presented at *La Technique Moderne* (1938) and at *Le Congrès de Mécanique Appliquée* (1945).
2. L. Landau, On the Theory of Slow Combustion, *Acta Physicochimica* U.S.S.R, Vol. XIX, N°1:77 (1944).
3. P. Pelcé and P. Clavin, Influence of Hydrodynamics and Diffusion upon the Stability Limits of Laminar Premixed Flame, *J. Fluid Mech.* 124:219 (1982).
4. P. Clavin and P. García-Ybarra, The Influence of the Temperature Dependence of Diffusivities on the Dynamics of Flame Fronts, *J. Méc. Théor. Appli.* 2:245 (1983).
5. P.L.García-Ybarra, Flame Front Stability with General Intermolecular Interaction Potential, *Prog. Astro. Aero.* 95:115 (1984).
6. R. M. Fristrom, Definition of Burning Velocity and a Geometric Interpretation of the Effects of Flame Curvature, *Phys. Fluids* 8:273 (1965).
7. G.I. Sivashinsky and P. Clavin, On the Nonlinear Theory of Hydrodynamic Instability in Flames, *J. Physique* 48:193 (1987).
8. G.K. Batchelor, "An Introduction to Fluid Dynamics," Cambridge University Press, Cambridge (1983).

9. B. Hasslacher, Discrete Fluids, *Los Alamos Science* (Los Alamos National Laboratory) Special Issue 15:175 (1987).
10. R. Saïd and R. Borghi, A Simulation with a Cellular Automaton for Turbulent Combustion Modelling, *in*: "22th. Symposium (International) on Combustion," The Combustion Institute, Pittsburgh (1990). In the press.
11. S. Gutman and G.I. Sivashinsky, The Cellular Nature of Hydrodynamic Flame Instability, *Physica D* 43:129 (1990).

EXCITABLE MEDIA

VORTEX INTERACTIONS IN THE COMPLEX GINZBURG-LANDAU EQUATION

Christian Elphick

Physics Department
Universidad Tecnica F. Santa Maria
Valparaiso Casilla 110-V, Chile

Ehud Meron

Department of Chemical Physics
Weizmann Institute of Science
Rehovot 76100, Israel

1. INTRODUCTION

Extended non-equilibrium systems exhibit a variety of phenomena that have similar counterparts in equilibrium systems. Examples include super and subcritical instabilities, the counterparts of which are second and first order phase transitions, defect dynamics in convective patterns which have much in common with the analogous dynamics in crystals [1], and defect mediated turbulence which is reminiscent of two-dimensional melting [2]. The analogy between equilibrium and non-equilibrium phenomena, however, do not go too far. One distinctive feature whose consequences have been studied recently in a number of contexts [3,4] is the non-variational nature of the equations which govern non-equilibrium systems. The absence of a free-energy or a Liapunov functional makes the study of these equations more difficult and calls for different analytical tools. In this paper we study vortex dynamics in non-equilbrium two-dimensional systems; we show how vortex interactions can be derived in the absence of a free-energy functional, and discuss new qualitative features that arise in this case.

We will confine ourselves to a particular model system, namely, the complex Ginzburg-Landau equation (hereafter CGL) in two space dimensions

$$\partial_t Z = \mu Z - (1 + i\alpha)|Z|^2 Z + (1 + i\beta)\vec{\nabla}^2 Z, \tag{1}$$

where $Z(x, y, t)$ is a complex field and μ, α and β are real constants. The CGL equation describes spatio-temporal modulations of non-equilibrium homogeneous states that oscillate periodically in time [5]. As an example, one may think of an oscillating chemical reaction taking place in a thin solution layer. Concentrations of chemical species, $\vec{U} = (U_1, U_2, ...)$, are then given by

$$\vec{U}(\vec{x}, t) = \vec{U}_0 + Z(\vec{x}, t)\vec{A}\exp(i\omega_0 t) + c.c., \tag{2}$$

Growth and Form, Edited by M. Ben Amar *et al.*
Plenum Press, New York, 1991

where $\vec{U}_0$ and $\vec{A}$ are constant vectors and ω_0 is the frequency of the uniform oscillations.

When $\alpha = \beta = 0$ the CGL equation is derivabale from a free energy:

$$\partial_t Z = -\frac{\delta \mathcal{F}}{\delta Z^*}, \tag{3a}$$

where

$$\mathcal{F} = \int d\vec{x} \left[|\vec{\nabla} Z|^2 + \frac{1}{2}(|Z|^2 - \mu)^2 \right]. \tag{3b}$$

This form of free-energy functional has been studied in the context of thin superconductor films [6] and leads to Coulombic type interactions; vortices attract or repell each other with forces that are inversely proportional to the inter-vortex distances.

In section 2 we will consider this special case and describe two methods of deriving vortex interactions. The first exploits the fact that a free-energy functional exists, and therefore is applicable to variational systems only. The second method is more general and involves a direct reduction of the CGL equation to equations of motion for vortex positions. This method will be used in section 3 to derive vortex interactions in the non-variational case. The new qualitative features that arise in this case are discussed in section 4.

2. THE VARIATIONAL CASE

The simplest non-trivial solutions of equations (3) describe uniform states and are given by $Z_\phi = \sqrt{\mu} \exp(i\phi)$, where ϕ is an arbitrary constant phase. These solutions are also the most stable ones as they correspond to the absolute minimum of the free-energy, $\mathcal{F} = 0$. In addition, equations (3) have vortex solutions that correspond to maps from closed loops in the (x, y) plane into the circle of uniform states $\{Z_\phi | 0 \le \phi < 2\pi\}$ [7]. Each solution is characterized by its winding number (or topological charge), n, defined as the number of times one winds around the circle of uniform states upon completing one turn in the (x, y) plane. The vortex solutions assume the form

$$Z_n(\vec{x}) = f(r) \exp(in\theta), \tag{4a}$$

where r and θ are polar coordinates and the amplitude f is given asymptotically by

$$\begin{aligned} f(r) &\sim r^{|n|} & r &\to 0, \\ f(r) &\sim \sqrt{\mu} + O(1/r^2) & r &\to \infty. \end{aligned} \tag{4b}$$

For winding numbers, $n \neq 0$, the phase, $\varphi \equiv \arg(Z_n)$, is multivalued at the vortex core ($r = 0$). This singularity is reflected by the relation

$$n = \frac{1}{2\pi} \oint \vec{\nabla}\varphi \cdot d\vec{x}, \tag{5}$$

where the integral is taken over a closed contour that contains the core. The modulus $|Z_n|$ is approximately constant ($\sqrt{\mu}$) everywhere in the plane except at the immediate vicinity of the vortex core ($r \sim O(\mu^{-1/2})$ where it goes to zero to guarantee that Z_n remains single valued.

In order to evaluate the interaction between a pair of vortices we consider the free energy of interaction

$$\mathcal{F}_{\text{int}} = \mathcal{F}(Z_{n_1,n_2}) - \mathcal{F}(Z_{n_1}) - \mathcal{F}(Z_{n_2}), \tag{6}$$

where Z_{n_1,n_2} is a two-vortex solution. We will assume that the two vortices are well separated so that $|\vec{x}_2 - \vec{x}_1| \gg \mu^{-1/2}$, where $\vec{x}_1$ and $\vec{x}_2$ are the positions of the vortex cores. We now approximate a two-vortex solution by a product of two single-vortex solutions

$$Z_{n_1,n_2}(\vec{x}, t) \approx \mu^{-1/2} Z_{n_1}[\vec{x} - \vec{x}_1(t)] Z_{n_2}[\vec{x} - \vec{x}_2(t)], \tag{7}$$

and use this form in (6) to obtain the leading order contribution

$$\mathcal{F}_{\text{int}} = -\frac{2}{\mu} \int_0^\Lambda dr\, r \int_0^{2\pi} d\theta\; Z_{n_1}^* Z_{n_2}^* \vec{\nabla} Z_{n_1} \cdot \vec{\nabla} Z_{n_2} = 4\pi\mu n_1 n_2 \ln(\Lambda/\rho) + \text{Constant}, \tag{8}$$

where Λ is the size of the system and $\rho = \frac{1}{2}|\vec{x}_2 - \vec{x}_1|$. The interaction force between the two vortices is obtained by deriving $\mathcal{F}_{\text{int}}$ with respect to ρ. This leads to a Coulombic $1/\rho$ type force.

The dynamics of a vortex pair can be obtained from the energy evolution equation

$$\frac{d\mathcal{F}}{dt} = \frac{d\mathcal{F}_{\text{int}}}{dt} = -2 \int |\partial_t Z_{n_1,n_2}|^2. \tag{9}$$

The range of integration in (9) deserves special consideration. Integration over the entire system leads to logarithmic divergences. This is due to the fact that according to (7) the phase field is rigidly translated with the vortex cores. Far away from the cores, however, we do not expect the phase field to be significantly affected by core translations. We rectify this by limiting the integration range to the immediate vicinities of the vortex cores [8]. Using the asymptotic forms (4b) we find

$$2\pi\mu\dot{\vec{x}}_i = -\vec{\nabla}_{\vec{x}_i}\mathcal{F}_{\text{int}}, \qquad i = 1, 2,$$

or, using (8),

$$\dot{\vec{\rho}} = \frac{n_1 n_2}{\rho} \hat{\rho}, \tag{10}$$

where $\hat{\rho}$ is a unit vector along the inter-vortex axis, $\vec{\rho} = \rho\hat{\rho}$ and we have assumed that $|n_1| = |n_2| = 1$. Thus vortices with opposite winding numbers attract each other while those with equal winding numbers repell.

We will derive now equation (10) using a different method which exploits the continuous symmetries of the system [9]. The starting point is the same two-vortex solution (7) which we write as

$$Z_{n_1,n_2}(\vec{x}, t) = \mu^{-1/2}\left[Z_{n_1}[\vec{x} - \vec{x}_1(t)] Z_{n_2}[\vec{x} - \vec{x}_2(t)] + C(\vec{x}, t)\right], \tag{11}$$

where C is a small correction term. To be more specific we introduce a small parameter $\epsilon \sim \rho^{-1}$ and choose the following orders of magnitude: $\dot{\vec{x}}_i \sim \epsilon$, $i = 1, 2$, $C \sim \epsilon^2$ and

$\partial_t C \sim \epsilon^4$. This choice is consistent with (10). When we insert (11) into the CGL equation we obtain to leading order

$$\mathcal{L} C = I, \tag{12a}$$

where

$$\mathcal{L} = -\mu - \vec{\nabla}^2 + \mu^{-1}\left(2|Z_{n_1}|^2 |Z_{n_2}|^2 + Z_{n_1}^2 Z_{n_2}^2 \mathcal{G}\right), \tag{12b}$$

and

$$I = Z_{n_2}\dot{\vec{x}}_1 \cdot \vec{\nabla} Z_{n_1} + Z_{n_1}\dot{\vec{x}}_2 \cdot \vec{\nabla} Z_{n_2} + 2\vec{\nabla} Z_{n_1} \cdot \vec{\nabla} Z_{n_2}. \tag{12c}$$

Here, $\mathcal{G}$ is the complex conjugation operator, $\mathcal{G} U = U^*$.

Equation (3) is invariant under space translations. Thus, if $Z(\vec{x})$ is a solution so is $Z(\vec{x} + \delta\vec{x})$. Using this property for small $\delta\vec{x}$ we obtain from (3)

$$\left(-\mu - \vec{\nabla}^2 + 2|Z|^2 + Z^2\mathcal{G}\right)\vec{\nabla} Z = 0. \tag{13}$$

Inserting (11) into (13) we find that $\mathcal{L}\vec{\nabla}\left(\mu^{-1/2} Z_{n_1} Z_{n_2}\right) = O(\epsilon^2)$. This result implies that the operator $\mathcal{L}$ is nearly singular and that certain solvability conditions should be satisfied in order for C to remain a *small* correction term. To see how these conditions arise we define first a scalar product in the space of complex-valued functions that decay to zero at infinity:

$$(U, V) = Re \int dx\, dy\, U^* V. \tag{14}$$

We then find that $\mathcal{L}$ is self adjoint, $\mathcal{L} = \mathcal{L}^\dagger$. Taking the scalar product of I with $\mu^{-1/2} Z_{n_1} Z_{n_2}$ and using (12a) we obtain

$$\left(I, \mu^{-1/2}\vec{\nabla}(Z_{n_1} Z_{n_2})\right) \sim O(\epsilon^4). \tag{15}$$

Now, the left hand side of (15) contains terms of $O(\epsilon)$. A consequence of (15) is that these terms must cancel each other. A violation of this condition would imply that our assumption about the smallness of C cannot be satisfied. Equating the $O(\epsilon)$ terms in (15) we find

$$Re \int_0^{2\pi} d\theta_1 \int_0^{r_0} dr_1 r_1 \left[Z_{n_2}^* \dot{\vec{x}}_1 \cdot \vec{\nabla} Z_{n_1}^* + 2\vec{\nabla} Z_{n_1}^* \cdot \vec{\nabla} Z_{n_2}^*\right] Z_{n_2} \vec{\nabla} Z_{n_1} = 0, \tag{16}$$

where the same reasoning about the integration range in (9) has been used here as well. In (16), r_0 is the radius of the vortex core, $\theta_1 = \theta(\vec{x} - \vec{x}_1)$ and $r_1 = |\vec{x} - \vec{x}_1|$. A similar integral holds for the second vortex core. Evaluations of these integrals lead to the same vortex equations (10) that have been derived using the free energy functional $\mathcal{F}$.

We presented in this section two methods of deriving vortex equations. The first exploits the existence of a free energy functional while the second utilizes the translational invariance of the system. The methods lead to the same result and in this respect are equivalent. However, only the second method is suitable for studying vortex dynamics in non-variational systems.

266

3. THE NON-VARIATIONAL CASE

We consider now the CGL equation for general α and β values for which a free energy functional does not exist. A significant consequence of α and β being unequal is that the stationary single-vortex solution (4) is replaced by a rotating spiral-vortex solution. For given values of α and β, the frequency of rotation and the asymptotic spiral wavenumber are uniquely selected. The direction of rotation is determined by the sign of the winding number. The explicit form of the spiral-vortex solution for $\alpha \neq \beta$ is [5,10]

$$Z_n(\vec{x}, t) = f(r) \exp[in\theta + i\omega t + is(r)], \tag{17a}$$

where $f(r)$ and $s(r)$ are given asymptotically by

$$\begin{aligned} f(r) &\sim r^{|n|}, & s(r) &\sim r^{|n|+1} & r &\to 0 \\ f(r) &\sim f_\infty + c/r, & s(r) &\sim pr + q\ln r & r &\to \infty. \end{aligned} \tag{17b}$$

The quantities ω, f_∞, c, and q are given by

$$\omega = -\alpha\mu + (\alpha - \beta)p^2 \qquad f_\infty = \sqrt{(\mu - p^2)}, \tag{18a}$$

$$c = p(1 + \beta^2)/[2A_0(\alpha - \beta)] \qquad q = -(1 + \alpha\beta)/[2(\alpha - \beta)], \tag{18b}$$

whereas p, the selected asymptotic wavenumber, should be evaluated either approximately [10] or numerically [11]. The limit $\beta \to \alpha$ deserves special attention and will be considered below.

To study vortex dynamics we have to resort to the second method that has been described in section 2, namely, the method that exploits the continuous symmetries of the system. The starting point is again an approximate two-vortex solution in the form of a product of two single-vortex solutions. In section 2 we used the symmetry of space translations to introduce translational degrees of freedom, $\vec{x}_1$ and $\vec{x}_2$ and to determine their dynamics. The CGL equation has another continuous symmetry that has not been exploited in section 2, namely, the $U(1)$ invariance $Z \to Z \exp(i\Phi)$. We use this symmetry to introduce additional degrees of freedom in the form of a phase field, $\Phi(\vec{x}, t)$. We thus write a two-vortex solution as

$$Z_{n_1, n_2}(\vec{x}, t) = f_\infty^{-1} \left[Z_{n_1}[\vec{x} - \vec{x}_1(t)] Z_{n_2}[\vec{x} - \vec{x}_2(t)] + C(\vec{x}, t) \right] e^{i\Psi(\vec{x}, t)}, \tag{19}$$

where $\Psi = \Phi - \omega t - p|\vec{x} - \vec{R}|$ with $\vec{R} = \frac{1}{2}(\vec{x}_1 + \vec{x}_2)$. We proceed now along the lines of section 2 to derive equations of motion for the vortex positions. In doing so we assume the same orders of magnitude for ρ^{-1}, $\dot{\vec{x}}_i$, C and $\partial_t C$ as in section 2. In addition we assume $p \sim O(\epsilon)$ $\vec{\nabla}\Phi \sim O(\epsilon)$ and $\partial_t \Phi \sim O(\epsilon^2)$. The resulting vortex equations take the form [12,13]

$$\begin{aligned} \frac{1}{2}\dot{\vec{x}}_i &= \beta\vec{\nabla}\Phi_{|\vec{x}_i} + n_i\vec{\nabla}\Phi_{|\vec{x}_i} \times \hat{\ell} + (n_in_j + \beta q)\frac{\vec{x}_i - \vec{x}_j}{|\vec{x}_i - \vec{x}_j|^2} \\ &\quad -(n_j\beta - n_iq)\frac{(\vec{x}_i - \vec{x}_j) \times \hat{\ell}}{|\vec{x}_i - \vec{x}_j|^2}, \qquad i,j = 1,2 \quad i \neq j, \end{aligned} \tag{20}$$

where $\hat{\ell}$ is a unit vector perpendicular to the (x, y) plane.

Before discussing the consequences of equations (20) for arbitrary α and β values let us consider the limit $\beta \to \alpha$ and, in particular, the case $\alpha = \beta = 0$ which has been studied in section 2. In this limit the CGL equation assumes the variational form

$$\partial_t z = -(1 + i\alpha)\frac{\delta \mathcal{F}[z]}{\delta z^*}, \tag{21}$$

where $z = Z\exp(i\alpha\mu t)$ and $\mathcal{F}$ is given by (3b). Equation (21) has vortex solutions in the form of (4). Comparing (4) with (17) we deduce that the limit $\beta \to \alpha$ amounts to setting q and p equal to zero. To obtain the vortex equations for the case $\alpha = \beta = 0$ we have to put in (20) $\beta = q = 0$. This leads to

$$\frac{1}{2}\dot{\vec{x}}_i = n_i \vec{\nabla}\Phi|_{\vec{x}_i} \times \hat{\ell} + n_i n_j \frac{\vec{x}_i - \vec{x}_j}{|\vec{x}_i - \vec{x}_j|^2} \qquad i,j = 1,2 \quad i \neq j. \tag{22}$$

Equations (22) generalize the vortex equations that have been derived in section 2 to situations where vortices appear on top of underlying periodic patterns. In the absence of such patterns one may take $\Phi = \text{Constant}$ and (22) reduce to (10). Very often, however, vortices appear as defects in periodic patterns. Dislocations in convective roll patterns provide a well studied example [1]. The phase gradient term on the right hand side (rhs) of (22) is responsible for forces that are exerted by these patterns. For a pattern of the form $\Phi = \vec{k}\cdot\vec{x}$ the phase gradient term describes a climbing force that acts in a direction perpendicular to the wave vector $\vec{k}$. This force is analogous to the Peach-Köhler force that acts on dislocations in crystals.

4. DYNAMICS OF VORTEX PAIRS

We can examine now the effect of α and β being unequal on vortex dynamics. The first point to note in this respect is that the interaction force between two vortices does not act only along the axis that connects the two vortex cores but acquires in addition a component perpendicular to that axis (last term on the rhs of (20)). We shall see below that this component is responsible for rotational and drift motions of vortex pairs. Another point that deserves attention is the appearance of a phase gradient term (first term on the rhs of (20)) that describes gliding forces, that is, forces that act along the wave vectors of underlying periodic patterns.

It is important to stress that the primary role of the phase field Φ is to correct the phase (argument) of the zero'th order two-vortex solution, $f_\infty^{-1} Z_{n_1} Z_{n_2}$, for vortex interaction effects. This approximate form already includes the spiral phase fields of the individual vortices but do not include, for example, the sink of counter propagating traveling waves at the interface between the vortices which are sufficiently apart from each other [14]. Thus, even in the absence of a background field Φ may not be zero. The possible forms that Φ may assume should appear as solutions of a *phase equation*, a partial differential equation that Φ satisfies. An approximate derivation of this equation is given in [12]. The phase equation follows from a solvability condition that pertains to the $U(1)$ invariance of the CGL equation.

In studying the dynamics of vortex pairs it is convenient to express equations (20) in terms of the relative coordinates $\vec{\rho} = \frac{1}{2}(\vec{x}_2 - \vec{x}_1)$ and the "center of mass" coordinates $\vec{R} = \frac{1}{2}(\vec{x}_2 + \vec{x}_1)$. For equal winding numbers we find

$$\dot{\vec{R}} = \beta\vec{\Delta}_+ + \vec{\Delta}_+ \times \hat{\ell}$$

$$\dot{\vec{\rho}} = \beta\vec{\Delta}_- + \vec{\Delta}_- \times \hat{\ell} + \frac{\beta q + 1}{\rho}\hat{\rho} + \frac{\beta - q}{\rho}\hat{\phi}, \qquad n_2 = n_1 = 1, \qquad (23a)$$

whereas for opposite winding numbers we obtain

$$\dot{\vec{R}} = \beta\vec{\Delta}_+ + \vec{\Delta}_- \times \hat{\ell} - \frac{\beta + q}{\rho}\hat{\phi}$$

$$\dot{\vec{\rho}} = \beta\vec{\Delta}_- + \vec{\Delta}_+ \times \hat{\ell} + \frac{\beta q - 1}{\rho}\hat{\rho} \qquad n_2 = -n_1 = 1. \qquad (23b)$$

Here $\hat{\phi} = \hat{\ell} \times \hat{\rho}$ and $\vec{\Delta}_\pm = \vec{\nabla}\Phi_{|\vec{R}+\vec{\rho}} \pm \vec{\nabla}\Phi_{|\vec{R}-\vec{\rho}}$.

Let us ignore first the phase field corrections. Inserting $\vec{\Delta}_\pm = 0$ in (23) and using the relation $\dot{\vec{\rho}} = \dot{\rho}\hat{\rho} + \rho\dot{\phi}\hat{\phi}$ we find for vortices with equal winding numbers: $\dot{\vec{R}} = 0$, $\dot{\phi} = (\beta - q)/\rho^2$ and $\dot{\rho} = (\beta q + 1)/\rho$. For vortices with opposite winding numbers we find: $\dot{\vec{R}} = -(\beta + q)/\rho\hat{\phi}$, $\dot{\phi} = 0$ and $\dot{\rho} = (\beta q - 1)/\rho$. Thus, in both cases vortices can attract or repell, depending on the signs of $\beta q \pm 1$. In the case of equal winding numbers, however, the vortices *rotate* about a fixed point in space, while vortices with opposite winding numbers *drift* in a direction perpendicular to the inter-vortex axis (or $\hat{\rho}$). Such motions have recently been observed in numerical simulations on the CGL equation [15].

We will not consider here explicit forms of the phase field Φ and, instead, refer the reader to reference [12]. We would like only to point out a possible consequence of $\vec{\Delta}_\pm$ being nonzero. It is evident from (23) that in this case stationary solutions of the equation for $\dot{\rho}$ cannot be ruled out. A stable solution $\rho = \rho_s$ of $\dot{\rho} = 0$ would then describe a *bound* vortex pair, while an unstable solution of this form would suggest the existence of a critical distance $\rho = \rho_s$ at which exchange of interaction occurs, that is, attractive interaction for $\rho < \rho_s$ becomes repulsive for $\rho > \rho_s$. Bound vortex pairs and exchange of interactions have recently been observed in numerical simulations [14-16].

In summary, the introduction of complex coefficients into the Ginzburg-Landau equation has significant effects on vortex dynamics. Vortices with equal winding numbers rotate about fixed points in space, vortices with opposite winding numbers drift in directions perpendicular to the inter-vortex axes, vortices can form bound pairs and vortices can exchange interactions at critical distances.

ACKNOWLEDGMENTS

EM is a Recipient of a Sir Charles Clores Fellowship.

REFERENCES

1. G. Tesauro and M. C. Cross, "Climbing of dislocations in nonequilibrium patterns", Phys. Rev. A 34:1363 (1986) and references therein.
2. P. Coullet, L. Gil and J. Lega, "Defect-mediated turbulence", Phys. Rev. Lett. 62:1619 (1989).
3. S. Fauve and O. Thual, "Solitary waves generated by subcritical instabilities in dissipative systems", Phys. Rev. Lett. 64:282 (1990).

4. P. Coullet, J. Lega, B. Houchmanzadeh and J. Lajzerowicz, "Breaking chirality in nonequilibrium systems", Phys. Rev. Lett. 65:1352 (1990).

5. Y. Kuramoto, *Chemical Oscillations, Waves and Turbulence*, Springer, Berlin (1984).

6. P. G. de Gennes, *Superconductivity of Metals and Alloys*, Benjamin, New York (1966).

7. G. Toulouse and M. Kléman, "Principles of a classification of defects in ordered media", J. Phys. Lett. 37:L-149 (1976).

8. Corrections due to the response of the far phase field to vortex motion have been studied by E. Bodenschatz, W. Pesch and L. Kramer, "Structure and dynamics of dislocations in anisotropic pattern-forming systems", Physica D 32:135 (1988), and by L. M. Pismen and J. D. Rodriguez, "Mobility of singularities in dissipative Ginzburg-Landau equation", preprint (1990).

9. This method has proved successful in a number of contexts. See for example C. Elphick, E. Meron and E. A. Spiegel, "Spatiotemporal complexity in Traveling Patterns", Phys. Rev. Lett. 61:496 (1988), and references therein.

10. P. S. Hagan, "Spiral waves in reaction-diffusion equations", SIAM J. Appl. Math. 42:762 (1982).

11. E. Bodenschatz, A. Weber and L. Kramer, "Structure and dynamics of spiral waves and of defects in traveling waves", University of Bayreuth preprint (1989).

12. C. Elphick and E. Meron, "Dynamics of phase singularities in two-dimensional oscillating systems", preprint (1990).

13. The same vortex equations have been derived independently by S. Rica and E. Tirapegui, "Interaction of defects in two-dimensional systems", Phys. Rev. Lett. 64:878 (1990); "Analytical description of a state dominated by spiral defects in two-dimensional extended systems", preprint (1990).

14. E. Bodenschatz, A. Weber and L. Kramer, "Structure and Dynamics of Spiral Waves and of Defects in Traveling Waves", preprint (1989).

15. H. Sakaguchi, "Numerical study of vortex motion in the two dimensional Ginzburg-Landau equation", Prog. Theor. Phys. 82:7 (1989).

16. T. Bohr, A. W. Pedersen, and M. H. Jensen, "Transition to Turbulence in a Ginzburg-Landau Model", Nordita preprint (1989).

VELOCITY SELECTION IN TWO-DIMENSIONAL EXCITABLE MEDIA: FROM SPIRAL WAVES TO RETRACTING FINGERS

Alain Karma

Physics Department
Northeastern University
Boston MA 02115, USA

ABSTRACT

Spiral waves in excitable media are typically observed to rotate at a unique angular frequency ω around an effective hole region of radius r_0. We investigate the selection problem for r_0 and ω in the large r_0 limit of the free-boundary problem of wave propagation in excitable media for the case where the recovery variable is non-diffusive. Analytical forms are derived for the dependence of r_0 and ω on the usual small parameter ε (which measures the abruptness of excitation) and a parameter Δ which measures the excitability of the medium. Where spiral wave solutions cease to exist we find a new class of solutions to the free-boundary problem analogous in shape to the Saffman-Taylor fingers of viscous hydrodynamics. These solutions correspond physically to wave fronts with a retracting tip structure.

I. INTRODUCTION

Theoretical progresses in understanding wave patterns in excitable media[1-6] have come mainly from the study of two-variable reaction-diffusion models of the form

$$\varepsilon \frac{\partial u}{\partial t} = \varepsilon^2 \nabla^2 u + f(u,v) \tag{1a}$$

$$\frac{\partial v}{\partial t} = \gamma \varepsilon \nabla^2 v + g(u,v) \tag{1b}$$

where the function $f(u,v)$ and $g(u,v)$ describe the nonlinear excitable kinetics of specific systems and ε is a small parameter which measures the abruptness of excitation. Models of the form (1) have the important advantage of capturing, at least qualitatively, most of the important features of wave dynamics seen in experiment while remaining mathematically and numerically tractable. The parameter γ is fixed for a given system. In the BZ reaction it corresponds to the ratio D_v/D_u of diffusion coefficients of the two components and is about unity while in models of neuromuscular tissues it is usually taken equal to zero. The two functions f and g need to have particular forms in order for the model to posses excitable

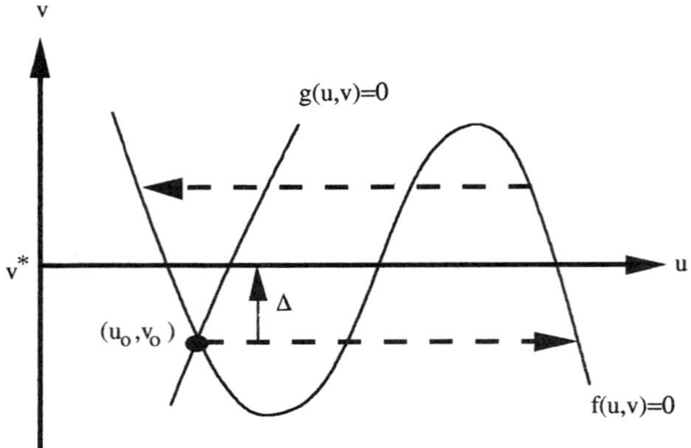

Fig. 1. Typical phase plane for an excitable media.

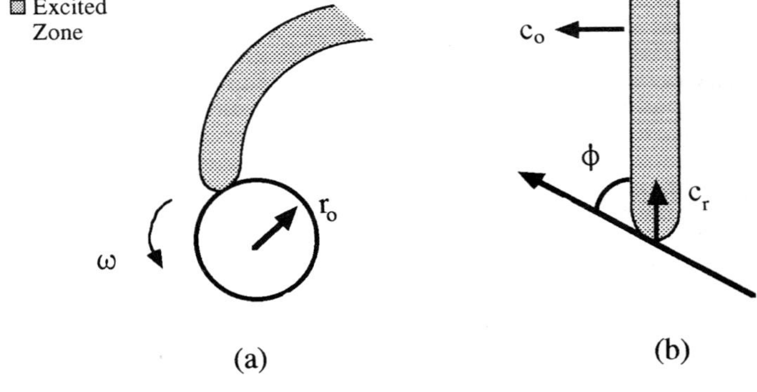

Fig. 2

kinetics. Most generally, the u-nullcline f(u,v)=0 is and N shaped function and the v-nullcline g(u,v)=0 is monotone intersecting the u-nullcline at only one point (u_0,v_0) as illustrated in Fig. 1.

The equilibrium point (u_0,v_0) is linearly stable but excitable in the sense that a finite amplitude perturbation can cause the system to execute a large excursion in phase space before returning to equilibrium. It is convenient to measure the excitability of the system by the parameter Δ, defined here in terms of Fig. 1 as $\Delta=v^*-v_0$ where v^* corresponds to the value on the v-axis at which the propagation velocity $c(v)$ of the planar pulse structure vanishes (i.e. $c(v^*)=0$). When Δ is small v_0 is close to v^* and a larger amplitude perturbation of the rest state is needed to cause an excitation. The velocity of the isolated planar pulse $c_0=c(v_0)$ is most generally a monotonously increasing function of Δ.

It is important to note that, for small Δ, Δ scales as the velocity c_0 ($c_0 \cong \alpha \Delta$ where α is a constant of order unity which is fixed by the form of the function $f(u,v^*)$). In this study we will be mainly interested in a region of parameter space (i.e. of the (ε,Δ) plane) where both ε and Δ are small. Although we have found more convenient here to parametrize the system of PDE's (1) by ε and Δ, it follows trivially from the proportionality of c_0 and Δ that our parametrization is equivalent in the parameter range of interest to one in terms of ε and c_0. It is sometimes helpful to associate Δ to a velocity when interpreting the results derived here.

An essential problem of excitable media lays in predicting the unique values of ω and r_0 characterizing steadily rotating spirals (Fig. 2a) for given values of the control parameters Δ and ε. As will be discussed below, we have found that in the range of parameters where spiral waves cease to exist the tip of a broken wave-front does not start to rotate around an effective hole region, but rather starts retracting at a constant velocity c_r giving birth to a finger-shaped wave which shrinks in a direction perpendicular to the propagation direction (Fig. 2b). In this case the velocity selection problem lays in predicting the value of c_r or equivalently the angle ϕ (see Fig. 2b) between an axis parallel to the wave-front and an axis parallel the tangential velocity of the tip structure. It will turn out that spiral and retracting wave solutions are connected in the sense that the wave-tip tangential velocity c_t remains continuous as the boundary line in the (ε,Δ) plane which separates the two type of solutions is crossed. As this boundary line is approached by decreasing Δ or increasing ε, r_0 diverges and the product $c_t=\omega r_0$ approaches the planar pulse velocity c_0. As this boundary line is approached by increasing Δ or decreasing ε, ϕ approaches $\pi/2$, $c_t=c_0\,\csc(\phi)$ approaches c_0 and $c_r=c_0\,\text{ctg}(\phi)$ vanishes.

In their approach to spiral wave dynamics, Keener and Tyson[5,6] derived a well known curvature relation between ω, r_0, and c by considering the rigid rotation of the front portion of the excitation wave with a boundary condition imposed on a fictitious inner radius r_0. By combining this relation with the plane wave dispersion relation $c = \sigma(2\pi/\omega)$ they obtained a unique relation between ω and r_0. This relation is particularly useful to determine ω using a value of r_0 extracted from experiment but is insufficient to determine ω and r_0 independently. With regard to the selection problem, it should be noted that situations can arise where ω takes on a discrete set of values[7,8]. In such situations, one is still left with the selection problem of determining r_0 independently of ω even if, as was first explained by Winfree[7], several values of ω are possible for a given value of r_0 due to oscillatory relaxation of the excitable kinetics.

Fife[9] showed how the problem of spiral wave dynamics could be expressed in terms of a free-boundary problem (FBP) and conjectured that the uniquely observed values of ω and r_0 should be obtainable by solving this FBP. This FBP is analogous to the one governing the solidification of a pure substance[10] - v playing the role of a temperature - but differs in that different boundary conditions are obeyed by the slow variable v at the boundary dividing the excited zone and refractory zone, and that the governing equations for v in the two phases are more complicated than the simple diffusion equation. Fife considered the diffusive case $\gamma=1$ and proposed that the added effect of diffusion would tend to smooth out the discontinuities in v which arise in the FBP formulation when taking the limit $\varepsilon \longrightarrow 0$ keeping Δ of $O(1)$. In particular, diffusion permits v to remain close to v^* along the boundary and therefore that the requirement $1/\kappa \sim \varepsilon/|v^*-v_{tip}| >> \varepsilon$ be satisfied even with Δ of $O(1)$. This requirement is necessary for the FBP formulation to be valid (i.e. for the boundary to remain curved on a lengthscale which is much larger than its thickness). Furthermore, Fife proposed that the variations of v along the boundary should scale as

$$v^* - v \quad \sim \quad \varepsilon^{1/3},$$

and that, accordingly, spiral wave solutions to the FBP should have wavelengths and frequencies scaling respectively as $\varepsilon^{2/3}$ and $\varepsilon^{-1/3}$. Kessler and Levine[11] examined some subtleties of the Fife scaling associated to the plane wave dispersion relation and expressed the FBP in terms of integro-differential equations using the Green's function technique developed previously for solidifcation problems[12]. Solutions of this diffusive FBP are currently awaited.

Recently, Pelce and Sun[13] put forward a wave-front interaction model. The starting equations of this model are identical to those of the FBP but for the non-diffusive case $\gamma=0$. Because the slow field v is non-diffusive this FBP is numerically and analytically more tractable than its diffusive counterpart. By solving numerically this FBP, Pelce and Sun showed for the first time that spiral wave solutions only existed for a unique value of ω and r_0 and calculated the variations of these two quantities with ε and Δ. They also found that r_0 diverges as the excitability of the medium approaches a minimum value Δ_{min} at fixed ε (equivalently when ε approaches a maximum value ε_{max} at fixed Δ) and Δ_{min} to be a monotonously increasing function of ε for values of ε ranging approximatively between 0.1 and 7. However, the values of ε and Δ studied by these authors lay outside the domain in the (ε,Δ) plane where solutions of the FBP represent a self-consistent quantitative description of the underlying system of PDE's (1). These solutions have the advantage of capturing some essential aspect of the velocity selection problem but suffer from the limitation that they can not be linked quantitatively to solutions of the system (1).

The domain in the (ε,Δ) plane where solutions of the non-diffusive ($\gamma=0$) FBP represent a self-consistent quantitative description of the system (1) is more restricted than the corresponding domain for the diffusive case ($\gamma=1$). In both cases, we must obviously have $\varepsilon<<1$ for a sharp boundary dividing the excited and refractory phases to exist. However, in the absence of diffusion, the value of v at the spiral tip (defined to be the point where the normal velocity of the boundary vanishes) must equal the equilibrium value v_0 and consequently the curvature of the boundary at tip $\kappa_{tip} \sim c_0/\varepsilon \sim \Delta/\varepsilon$ is comparable to ε^{-1} for Δ of $O(1)$. Since the FBP formulation is only valid if $1/\kappa_{tip} >> \varepsilon$, we therefore have the requirement that $\Delta<<1$ for the non-diffusive FBP to be valid. There exists the additional requirement[13] that, for the front and back interfaces to remain well separated, the width of the excited region for the planar pulse must be much larger than ε. This in turn implies $c_0\Delta \sim \Delta^2 >>\varepsilon$ or $\Delta>>\varepsilon^{1/2}$. For the non-diffusive case we therefore have a scenario where the FBP can only be valid in a domain of the (ε,Δ) plane where $\varepsilon<<1$ and $\varepsilon^{1/2}<<\Delta<<1$, while for the diffusive case there is in principle the possibility that this domain extends up to values of Δ of order unity (some additional subtleties of the diffusive case are discussed by Kessler and Levine in Ref. 11).

For the non-diffusive case, the condition $\varepsilon^{1/2}<<\Delta<<1$ is very restrictive and the question arises as to whether spiral wave solutions to the FBP can be found in this domain. One indication that solutions may exist comes from the aforementioned numerical finding of Pelce and Sun[13] that r_o diverges as Δ approaches a minimum value $\Delta_{min}(\varepsilon)$. In particular, their results indicated that $\Delta_{min}(\varepsilon)$ vanishes as $\varepsilon \longrightarrow 0$ and therefore that the above condition of validity may be satisfied for solutions in the neighborhood of the line $\Delta_{min}(\varepsilon)$ in the (ε,Δ) plane.

Recently, Karma investigated the small ε limit of the line $\Delta_{min}(\varepsilon)$ in the non-diffusive FBP and demonstrated[14] that $\Delta_{min}(\varepsilon)$ vanishes as $\varepsilon^{1/3}$. An important consequence of this result is that solutions on and in the neighborhood of the line $\Delta_{min}(\varepsilon)$ (i.e. solutions with Δ of $O(\varepsilon^{1/3})$) do indeed satisfy the condition of validity of the FBP $\varepsilon^{1/2}<<\Delta<<1$. The validity of the FBP formulation was investigated by comparing the value of $\Delta_{min}(\varepsilon)$ predicted from the solution of the FBP with that obtained from the full numerical simulation of a specific (Fitz-Hugh Nagumo) reaction-diffusion model of excitable kinetics for a value of $\varepsilon=10^{-3}$. The value of Δ_{min} was extracted numerically by extrapolating the value of Δ at which the radius of tip trajectories r_o diverged. A good quantitative agreement was found between the two values, within the limitations of the accuracy of the numerical simulation and higher order corrections in powers of ε to the prediction of the FBP. Keener and Tyson[15] had found previously that the predictions of their curvature relation (ω as a function of r_o) compared relatively well with the results of the numerical simulation of Pertsov *et al.*[16]. The comparison of the values of Δ_{min} has the new ingredient that it establishes a direct link between a solution of the complete FBP, within its domain of validity, and a solution of the system (1).

In this proceeding, we report some new results which pertain to solutions of the non-diffusive FBP which lay on both sides of the line $\Delta_{min}(\varepsilon)$ in the (ε,Δ) plane. Only the case where r_o is infinite was previously considered in determining the small ε scaling form of the line $\Delta_{min}(\varepsilon)$[14]. In section II we describe the results of numerical simulation of the Fitz-Hugh-Nagumo model of excitable kinetics. These numerical results illustrate how wave propagation evolves from spiral rotation to retracting finger propagation when the line $\Delta_{min}(\varepsilon)$ is crossed in parameter space. Finger and spiral wave solutions of the FBP are considered respectively in section III and IV, and concluding remarks are given given in section V. The details of the calculations and results presented here will be reported elsewhere[17].

II. NUMERICAL SIMULATION

The boundary line $\Delta_{min}(\varepsilon)$ in the (ε,Δ) plane corresponds to the line where spiral waves rotate on an infinite radius. If this line is approached by decreasing Δ at fixed ε or increasing ε at fixed Δ, the radius r_o of tip trajectories diverges and $\Delta_{min}(\varepsilon)$ thus corresponds to the lower limit in the (ε,Δ) plane where spiral wave propagation is possible. As Δ is decreased below this line at fixed ε or ε increased at fixed Δ, the broken-end of the wave front (what used to be the spiral tip) retracts at a constant velocity c_r. This velocity increases as one moves further away from the infinite radius limit line. It is convenient here to define a single control parameter

$$B \equiv (g^*/\alpha^2) \, \varepsilon \, / \, \Delta^3$$

where g^* is a constant of order unity (defined in the next section) characterizing the kinetics of the slow variable v and α (also of order unity) is the proportionality constant between the velocity of the planar pulse structure and Δ (recall $c_o = \alpha \Delta$ for small Δ). It will come out of the analysis of the FBP that in some domain of the (ε, Δ) plane wave propagation is governed entirely by B. In other words changes in ε and Δ which lead to the same change in B will have the same effect on r_o, ω or ϕ. In terms of B, the infinite radius limit line of spiral wave propagation corresponds to a constant value $B = B_c$ (the value $B_c = 0.535$ comes out of the analysis of the FBP[14]). Spiral waves solutions exist for $B < B_c$ and retracting finger solution for $B > B_c$.

We performed numerical simulations of a simple Fitz-Hugh-Nagumo form of excitable kinetics. This kinetics is given by

$$f(u,v) = 3u - v - u^3$$
$$g(u,v) = u - \delta.$$

For this choice of kinetics we have

$$v^* = 0,$$
$$(u_o, v_o) = (\delta, 3\delta - \delta^3)$$
$$\Delta = \delta^3 - 3\delta$$
$$g^* = 2\sqrt{3}$$
$$\alpha = 1/\sqrt{2}.$$

It is important to note that the particular choice of excitable kinetics is not important here (i.e. the particular choice of the N-shaped and monotonous functions f and g). The only important parameters in the theory (FBP) are ε, Δ and B. An other choice of kinetics would only lead to different numerical values of g^* and α, and a different definition of Δ.

The system of PDE's (1) with $\gamma = 0$ and the above choice of f and g was integrated numerically using an ADI (alternating direction implicit) finite difference algorithm. Zero-flux boundary conditions were used at the edges. We performed the test for $\varepsilon = 10^{-3}$ which required using large 512x512 square lattices. With $dx = 0.333 \, \varepsilon$ and B near B_c, the length of the simulation box $L = 512 \, dx$ is about 8 times the width of the excited region of the planar pulse.

Fig. 3 shows the results of 4 independent runs on 512x512 lattices performed with increasing values of B (decreasing values of Δ): (a) B=0.35 (b) B=0.45 (c) B=0.48 and (d) B=0.70. The infinite radius spiral wave occurs for $B_c \equiv g^* \varepsilon/(\alpha^2 \Delta_{min}^3) = 0.45$ and is shown in fig. 3b. Note that this value of B_c differs somewhat from the theoretical value $B_c = 0.535$. This small difference is actually expected to occur since higher order corrections to the theoretical value of B_c should be proportional to $\varepsilon^{1/3}$ and therefore comparable to the difference between the two values. Also, with the values of dx and dt used in our simulations we can not expect an accuracy greater than a few percent. Testing the convergence of the FBP formulation beyond the present study consists a formidable task requiring the use of even smaller ε and larger lattices. The thick solid lines in each

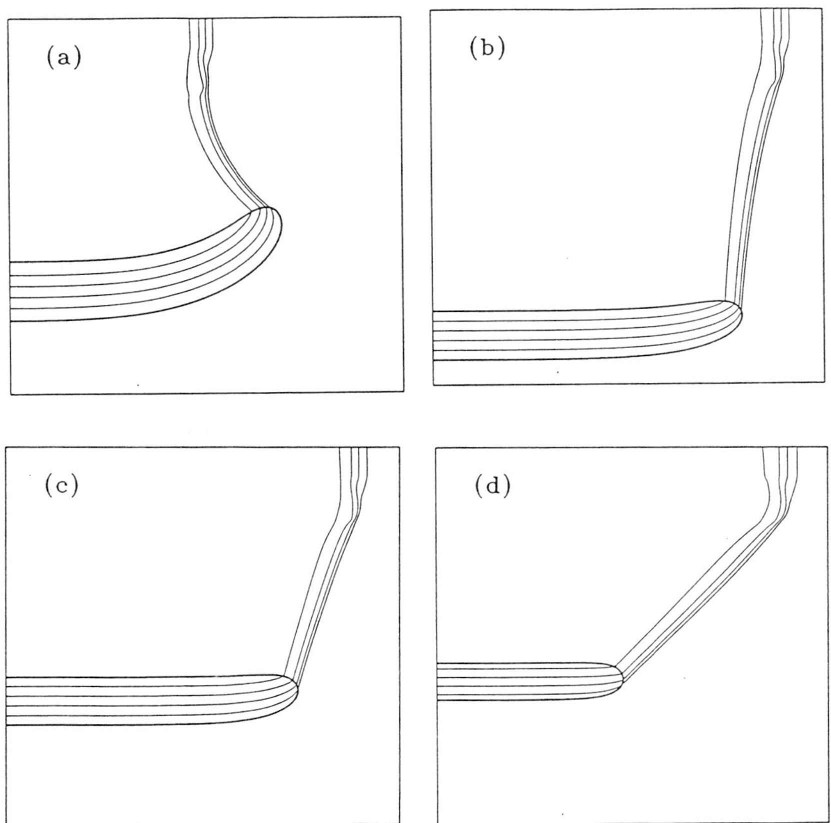

Fig. 3

figure represent the boundary (u=0 isocontour of the fast variable) and the solid lines crossing the boundary represent 4 isocontours of the slow variable v=0.15, 0.05, −0.05, and -0.15 with the boundary value of v increasing in value away from the spiral tip. Since the variable v is non-diffusive the trajectory of the spiral tip is very close to that of the v=−0.15 isocontour. For the same reason, the irregularity in the v-contours at the top of each figure represents the scares left by the transient during which the system adjusts to its steady-state from the initial condition.

In all 4 runs the waves move in the downward direction and a planar pulse structure with a broken-end placed where the irregularity of the v-contours appear was used as initial condition. Fig. 3a corresponds to a large r_o spiral wave solution the curvature of tip trajectories being seen clearly in the v-isocontours. In Fig. 3b the wave is essentially rotating on the infinite radius and translates at constant velocity c_o. Figs. 3c,d corresponds to retracting waves. The planar pulse continues to propagate but the borken end of the pulse retracts at constant velocity. The tip tangential velocity c_t increases continuously from a value less than c_o to one greater than c_o as B increases from a value below B_c to a value above B_c. As B increases above B_c the angle ϕ between the horizontal axis and the line of tip trajectory (v-isocontours) decreases continuously from its value $\phi=\pi/2$ at $B=B_c$. The value of ϕ corresponding to Fig. 3d is nearly 45º.

It is apparent from Fig. 3a that a complete cycle of spiral wave rotation can not be completed given the size of the simulation box. In this particular run, the wave will be absorbed at the right boundary after completing only about one quarter of a complete cycle. It follows that the dynamics of the front of the wave is not affected by the refractory tail and that the measured value of r_o may differ from the value that would be measured if several cycles of rotation were simulated. However, this problem is not so limiting since these simulations were performed in the first place to determine the value B_c at which r_o is infinite. Clearly for large enough radius, the effect of the refractory tail is negligible. Therefore, the value of B_c determined by extrapolating the value of B at which r_o diverges in our zero-cycle simulation needs to be the same as the value that would be obtained in many-cycle simulations. In previous work[14], we had determined B_c from r_o. Here, we have determined B_c by extrapolating the value of B for fingers at which ϕ equals $\pi/2$ and obtained a value in quantitative agreement with the previously measured value. For retracting fingers the issue of the effect of the refractory tail is not present since the wave always propagate in a region at equilibrium.

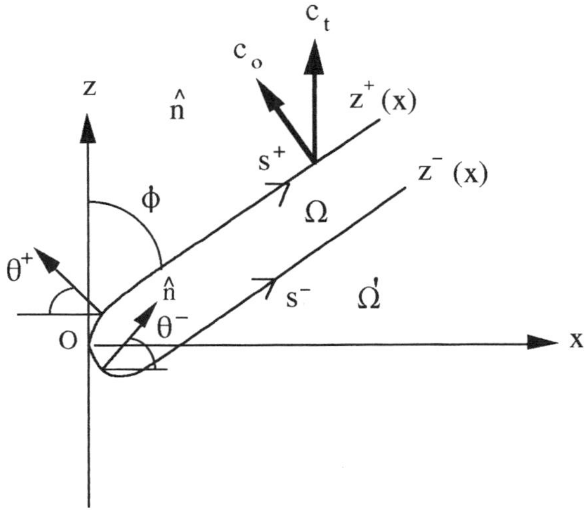

Fig. 4

III. RETRACTING FINGER SOLUTIONS (B>B$_c$)

The problem of velocity selection for retracting waves consists in determining the value of c_t or equivalently of ϕ which is selected for given values of B (i.e. Δ and ε). To understand the origin of the equations entering the FBP[6,9,11,13] let us first consider the shape of a planar pulse propagating in the unperturbed medium at constant velocity $c_o \equiv c(v_o)$. The pulse consists of a narrow front region of width ζ of $O(\varepsilon)$ which results form the abrupt rise of excitation, an excited region of width W, a narrow back region of width ζ, and a refractory region during which the system slowly returns to equilibrium. During the duration of excitation, the system evolves along the right-most branch of the u-nullcline spending a total time T_e of $O(\Delta)$ on this branch. The width $W = c_o T_e$ is therefore of $O(c_o\Delta)$ >> ζ in the limit ε<<1. The front and back regions can therefore be regarded as sharp boundaries separating the enclosed domain of excitation Ω from the outside region Ω'. The problem of wave propagation becomes a free-boundary problem where the normal velocity of the boundary c_n depends on v and the local interfacial curvature κ via the relation[5] $c_n = c(v) - \varepsilon \, \kappa$. The field v evolves according to eqn. 1b with u replaced by $h_e(v)$ inside Ω and $h_r(v)$ outside; $u=h_e(v)$ and $u=h_r(v)$ correspond respectively to the right-most and left-most branch of the u-nullcline.

The parametrization used for the FBP of finger solutions is shown in Fig. 4. In terms of this parametrization the equations of the FBP can be written:

$$c_o - \varepsilon \, \kappa^+ = c_t \sin(\theta^+) \tag{2a}$$

$$- c(v^-) - \varepsilon \, \kappa^- = c_t \sin(\theta^-) \tag{2b}$$

$$c_t \, \partial_z v + g(h_e(v),v) = 0 \tag{2c}$$

where v is obeys (2c) in Ω, (2a) and (2b) express the normal velocity relation of the shape-preserving front $(z^+(x))$ and back $(z^-(x))$ boundaries respectively, and v^- is the value of v evaluated on the back boundary. Here, θ^+ (θ^-) measures the angle between the x-axis and the outward (inward) pointing normal on the front (back) boundary, and $\kappa^\pm = \pm \, d\theta^\pm/ds^\pm$ is the curvature measured along the arclength coordinate $s^\pm$ of each boundary. The recovery variable ahead of the front is spatially uniform and takes on its equilibrium value $v=v_o$. Furthermore, in the present situation the field v in the refractory region does not affect the motion of boundaries. Consequently, only spatial variations of v within the enclosed domain Ω in Fig. 4 need to be considered (eqn. 2c). It is also convenient to define the relative tangential velocity

$$U = c_t \, / \, c_o = \csc(\phi) \tag{3}$$

We first rewrite (2) in a form where only the ratio $B = g^* \varepsilon/(\alpha^2 \Delta^3)$ and U appear explicitly. The solution of (2c) inside Ω is simply given implicitly by:

$$z = - c_t \int dv/g(h_e(v),v) + Cte \tag{4}$$

Since Δ scales as $\varepsilon^{1/3}$<<1 the function $g((h_e(v),v)$ in the denominator of the integrand can be replaced by a constant value: $g((h_e(v),v)=g^*+O(\varepsilon^{1/3})$ where $g^*\equiv g(h_e(v^*),v^*)$. Performing the integral for z over the appropriate limits we obtain at once $v = v_o - (g^*/c_t)(z-z^+(x))$ in Ω, and $v = v^- = v_o-(g^*/c_t)(z^-(x)-z^+(x))$ on the back boundary.

Substituting the expression for v^- in the velocity relation $c(v^-)=\alpha(v^*-v^-)$ and using the definitions $\Delta=v^*-v_0$ and $c_0=\alpha\Delta$, the equation for the back boundary can be rewritten $-\alpha\Delta\{1 + (U\alpha\Delta^2/g^*)^{-1}(z^-(x)-z^+(x))\} - \varepsilon\,\kappa^- = \alpha\Delta U\sin(\theta^-)$. Finally, performing the scale change $(X, Z^{\pm}) = (\alpha\Delta/\varepsilon\; x,\; \alpha\Delta/\varepsilon\; z^{\pm})$ and $K^{\pm}=(\alpha\Delta/\varepsilon)^{-1}\kappa^{\pm}$, the system (2) can be rewritten in the form

$$1 - K^+ = U\sin(\theta^+) \tag{5a}$$

$$-1 - (B/U)\{Z^-(x)-Z^+(x)\} - K^- = U\sin(\theta^-). \tag{5b}$$

The equation for the front boundary (5a) can easily be integrated to obtain analytical expressions for $Z^+(\theta^+)$ and $X(\theta^+)$ which implicitly define the front $Z^+(X)$. This integrability of the equation for the front boundary was noted previously in Ref. 13. In the special case where $U=1$, $Z^+(X) \longrightarrow \ln(X)$ as $X \longrightarrow \infty$. For $U>1$, $Z^+(X) \longrightarrow \mathrm{ctg}(\phi)\,X$ as $X \longrightarrow \infty$. Finally, the requirement that solutions of (5b) obey the boundary conditions $Z^-(0)=0$ and $Z^-(X) \longrightarrow \mathrm{ctg}(\phi)\,X$ as $X \longrightarrow \infty$ fixes a unique relation between U and B. We determined numerically the relation between B and U for values of U up to 1.5. For U near 1 we find

$$B = B_c + K(U-1)\quad,\quad B_c \cong 0.535 \text{ and } K \cong 0.64. \tag{6}$$

Further away from $U=1$ we would not expect B to be necessarily linear in U-1 since this term only represent the first term in a power series expansion in U-1 of the complete function B(U). However, numerically we found the form (6) to remain a very good approximation to the function B(U) for values of U up to U=1.5. Finally, eqn. (6) implies that the retracting velocity c_r vanishes according to a square-root behavior $c_r \approx c_0[2(B-B_c)/K]^{1/2}$.

III. SPIRAL WAVES IN THE LIMIT OF VERY LARGE RADIUS $(B<B_c)$

To investigate the scaling properties of spiral waves forming for $B<B_c$ we follow a procedure similar to the one outlined in the previous section. Details of the calculation will be reported elsewhere[17] and we only present the results here. The FBP of the spiral wave rotation in the limit of very large radius can be expressed in terms of three parameters. The scaled tangential velocity U which takes the form

$$U = \omega r_0/c_0, \tag{7}$$

the scaled radius of tip trajectories

$$\rho = r_0 c_0/\varepsilon, \tag{8}$$

and the control parameter B. Two relations are needed to determine ρ and U uniquely and thus r_0 and ω in terms of B. The first is similar to the curvature relation of Keener and Tyson[6] and is obtained by solving the equation for the front boundary with the boundary conditions (BC) that the front be tangent to the circle of tip trajectories at the spiral tip. This BC is exact for the non-diffusive case $\gamma=0$. Keener and Tyson used instead the BC that the front be perpendicular to the circle of tip trajectories at the spiral tip. This boundary condition leads to a quantitatively different curvature relation. We find that in the limit $\rho\gg1$ the curvature relation (expressed in terms of U and ρ) takes on the form

$$U = 1 - b\,\rho^{-\nu}, \tag{9}$$

280

with $v \cong 0.65$ and $b \cong 2.33$ for the exact tip boundary condition, and, within the numerical accuracy we used, the same value of v and $b \cong 1.17$ for the boundary condition adopted by Keener and Tyson. Interestingly, for $\rho=1$ (where (9) is no longer valid) we find that the value $U \cong 0.22$ determined using the exact BC is not too far from the value $U \cong 0.28$ determined using the BC of Keener and Tyson. In a parameter range where ρ is of order unity, the curvature relation derived by these authors therefore represents a relatively good quantitative approximation to the true curvature relation obtained using the exact BC. In general however the exact BC should be used and correspondingly the value $b \cong 2.33$ in (9) for very large radius solutions. Here we obtained (9) by numerically solving the eqn. governing the front boundary of the spiral wave. Pelce and Sun[13] also considered the large radius limit of spiral waves solutions and showed that, to lowest order in $1/r_0$, the solution of the tip region could only be matched to that of the tail region if $U=1$. There is the interesting possibility that a generalization of this matching procedure to higher orders in $1/r_0$ could be used to extract the form (9) analytically.

The second relation turns out to be exactly (6) with U defined by (7) instead of (3). The reason for this is that the FBP in the tip region (front+back+tip) depends on B, U and ρ^{-1} which is small in the tip region. We must therefore have:

$$B = B_c + K\,(U-1) + O(\rho^{-1}) \tag{10}$$

Furthermore, since form (9) $U-1$ is of order $\rho^{-v} >> \rho^{-1}$, the terms of order ρ^{-1} on the right-hand side of (10) are negligible and relation (10) becomes identical to (6) to lowest order in ρ^{-1}. An other interpretation of (10) is that the tangential velocity of the wave tip (U) must vary continuously as B is varied past B_c.

Combining, eqns. (9)-(10) we obtain at once that ρ diverges as

$$\rho = [(B_c - B)/Kb]^{-1/v} \tag{11}$$

and ω vanishes as $1/\rho$ (since $U=1$ to lowest order in $1/\rho$). Finally in deriving (11) we have neglected the effect of the refractory tail left behind the wave and assumed ω to be sufficiently small for the slow variable v to have essentially returned to its fixed point value v_0 on a the time scale of order $1/\omega$. Since this return takes place on a time scale of order unity, we have the requirement that (11) only be strictly valid for $\omega <<1$. This corresponds to values of $r_0 >> c_0 \sim \epsilon^{1/3}$ or $\rho >> \epsilon^{-1/3}$.

V.   CONCLUSIONS

In conclusion, we have found the first self-consistent solutions to the free-boundary problem of wave propagation in a domain of parameter space where the free-boundary formulation of the non-diffusive case $\gamma=0$ is valid. The solution to this problem leads to uniquely determined values of ω and r_0 as previously found in the wavefront interaction model[13]. The lower limit line $\Delta_{min}(\epsilon)$ (i.e. $B=B_c$), below which spiral wave propagation ceases to exist, scales as $\epsilon^{1/3}$. For $B<B_c$ the effective hole radius r_0 decreases with increasing B (with decreasing ϵ or increasing Δ) according to the power law (11). For $B>B_c$ there exist a new class of solution corresponding to finger-shaped waves which retract at constant velocity. The retracting velocity vanishes as the square root of $B-B_c$ as B_c is approached. The predictions of the FBP formulation are found to be in relatively good quantitative agreement with the results of numerical simulation of the full reaction diffusion equations, at least for the determination of B_c.

Two important problems which remain to be elucidated pertain to the phenomenon of meander and the effect of diffusion of the slow field v on spiral wave propagation. The meandering quasiperiodic motion of the spiral tip has been observed experimentally in the BZ reaction[18-20] and reaction diffusion models of excitable kinetics[3,8,21-23], and was recently shown to be related to an instability (Hopf bifurcation) of rigidly rotating spiral wave solutions of the reaction diffusion models[8,23]. In the Fitz-Hugh-Nagumo model this instability was found[8] to occur by increasing the excitability Δ beyond a maximum value $\Delta_{max}(\varepsilon)$ but the scaling with ε of this stability boundary remains so far unknown. The elucidation of this scaling is crucial to determine the domain in (ε,Δ) plane where stable rigid rotors can be found. Finally, a better understanding of the effect of diffusion of the slow field v is currently needed in order to better characterize the patterns observed in the BZ reaction. In particular, one may expect the domain of existence of spiral wave solutions to be very sensitive to the value of γ if solutions obeying the Fife scaling are to exist for values of Δ and γ of order unity. The results for the non-diffusive case presented here may provide a starting point for understanding this more complex case.

Acknowledgments

I wish to thank Pierre Pelce for many fruitful discussions. This research is supported by Northeastern University through a grant from the Research and Scholarship Development Fund supplemented by funds from NATO.

REFERENCES

[1] R. J. Field and M. Burger, *Oscillations and Traveling Waves in Chemical Systems* (Wiley, New York, 1985).

[2] A.T. Winfree, *When Time Breaks Down* (Princeton University Press, New Jersey, 1987).

[3] V.S. Zykov, *Simulation of Wave Processes in Excitable Media* (Manchester University Press, New York, 1987).

[4] A.S. Mikhailov and V.I. Krinsky, Physica **9D**, 346 (1983).

[5] J. P. Keener and J. J. Tyson, Physica **21D,** 307 (1986).

[6] J.J. Tyson and J. P. Keener, Physica **32D**, 327 (1988).

[7] A. T. Winfree, to appear in Physics Letters A.

[8] A. Karma, Phys. Rev. Lett. **65**, 2824 (1990).

[9] P. C. Fife, J. Stat. Phys. **39**, 687 (1985).

[10] J. S. Langer, Rev. Mod. Phys. **52**, 1 (1980).

[11] D. A. Kessler and H. L. Levine, Physica **39D,** 1 (1989).

[12] G. Nash and M. E. Glicksman, Acta Metall. **22**, 1283 (1974).

[13] P. Pelce and J. Sun, (to appear in Physica **D**).

[14] A. Karma (to be published).

[15] J. J. Tyson and J. P. Keener, Physica **29D**, 215 (1987).

[16] A. M. Pertsov, E. A. Ermakova, and A.V. Panfilov, Physica **14D**, 117 (1984).

[17] A. Karma (to be published).

[18] A.T. Winfree, Science **175**, 634 (1972).

[19]S.C. Muller, T. Plesser, and B. Hess, Physica **24D**, 87 (1987).

[20]G.S. Skinner and H.L. Swinney (to appear in Physica D).

[21]W. Jahnke, W.E. Skaggs, and A.T. Winfree, J. Phys. Chem. **93**, 740 (1989).

[22]E. Lugosi, Physica **40D**, 331 (1989).

[23]D. Barkley, M. Kness, and L. Tukerman, Phys, Rev, **A42,** 2489 (1990).

STEADILY ROTATING SPIRALS

IN EXCITABLE MEDIA

Pierre Pelcé and Jiong Sun

Laboratoire de Recherche en Combustion
Université de Provence, St Jerome
13397 Marseille Cedex 13, France

INTRODUCTION

The problem of the formation of spiral waves in two-dimensional excitable media is one of the classical problems of nonequilibrium pattern-forming system which remains unsolved. It is clear now , from numerical simulations that reaction diffusion sytems are a good physical basis for the explanation of the formation and of the characteristics of the spiral waves. However, the mathematical understanding of these solutions is far from being complete.

The simplest model from which one can start consists of a set of two coupled reaction diffusion equations with two very different time scales, one for the trigger variable c_1 which varies on the fast scale, the other for the recovery variable c_2 , on the slow scale. Numerical simulations of this model exhibit steadily rotating fields , with a well determined frequency ω, the region where the gradient of c_1 is sharp being located on a spiral shape. In order to understand these solutions and determine the frequency of rotation of the spiral, one simplifies the model and assimilates the regions where c_1 varies rapidly to a closed moving contour which delimits an excited region. Outside this contour , in the refractory region , the slow recovery variable decays until the occurence of another excitation. Moreover, aroud the center of rotation of the spiral lies an unexcited region with a well determined radius r_0.

In a first part we describe this simplified model and discuss how it can be determined from the set of two coupled reaction-diffusion equations. In a second part, we study the solutions of the model corresponding to steadily rotating spirals. We determine uniquely the angular velocity ω and the hole of radius r_0 as a function of the control parameters of the model, i.e. ε the ratio between fast reaction time and refractory time and δ the excitability.

When the hole radius r_0 is large, front and back are decoupled, i.e. the shape of the front can be found independantly of the back. This is due to the fact that when the tip of the spiral moves on a circle of large radius , the recovery variable c_2 has time to come back to its steady value. In that case, an asymptotic semi-analytical solution for the spiral shape can be found.

THE MODEL

The basis of W.F.M. is the classical piece-wise linear model for the following reaction-diffusion system:

$$\varepsilon \frac{\partial c_1}{\partial t} = \varepsilon^2 \Delta c_1 + f(c_1, c_2) \tag{1}$$

Growth and Form, Edited by M. Ben Amar *et al.*
Plenum Press, New York, 1991

$$\frac{\partial c_2}{\partial t} = c_1 - \delta \tag{2}$$

where $f(c_1, c_2)$ is the piece-wise linear function drawn on Fig.1:

$$f(c_1, c_2) = \left\{ \begin{array}{ll} -c_1 - c_2 + 1 \quad, & 0 < c_1 \\ -c_1 - c_2 - 1 \quad, & c_1 < 0 \end{array} \right\} \tag{3}$$

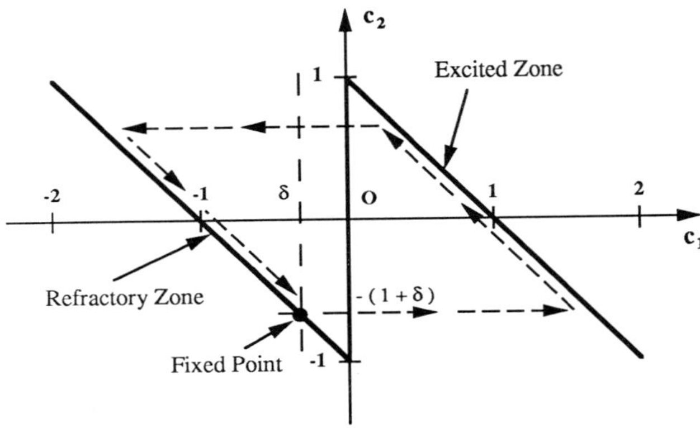

Fig.1 Piecewise linear model.

and δ a real and negative number larger than -1 which characterizes the excitability of the system. Here, $\varepsilon = \tau / T$ is the ratio between fast reaction time τ and the refractory time T. Time is scaled with T and lengths by the diffusive length $(D\tau)^{1/2} / \varepsilon$, where D is the diffusive coefficient of the trigger variable c_1. It is assumed here that only the trigger variable can diffuse in the system.

It is well known that when ε is small , the reaction-diffusion system admits one-dimensional solutions corresponding to the propagation of a sequence of stable pulses. Each pulse of the sequence is characterized by two waves (the front and the back) propagating with the same velocity separated by an excited region. In each wave , the concentration of the recovery variable c_2 can be assumed constant . The propagation velocity $c(c_2)$ of the wave can be found as a function of c_2 after integration of eqn.1 in a frame moving with constant velocity $c(c_2)$, in the new space variable $\xi = (x - c(c_2)t) / \varepsilon$. In the case of the piece-wise linear model (3), this velocity is found as

$$c(c_2) = -\frac{2c_2}{\sqrt{1 - c_2^2}} + O(\varepsilon) \tag{4}$$

As introduced in Zykov[1] book p.5, an important quantity caracterizing the front is the maximum relative rate of increase of c_1 , $E_{max} / A = |(\partial c_1 / \partial t)|_{max}/ A$,where A is the amplitude of the wave. In the case of the simple piece-wise linear model defined above, this quantity is found as $|c_2|/\varepsilon$. Its inverse corresponds to the transit time of the wave. In the case of the single pulse $c_2 = -(1 + \delta)$, sothat $E_{max} / A = |1 + \delta|/\varepsilon$ .When the pulse is isolated, the duration of the pulse (transit time of the excited region) is found as

$D = - \text{Log} | \delta |$. Experimentalists are more familiar with the quantities E_{max} / A and D. The model is defined with the two independant parameters ε and δ. In the following we will use alternatively these two kinds of parameters.

When the pulse propagates in an inhomogeneous field c_2, the equiconcentration waves of the trigger variable c_1 become distorted and the normal velocity (4) of the waves are modified by transverse concentration fluxes. It is well known that when the curvature radius of the wave is large compared to its thickness, the normal velocity of the wave is modified proportionally to the curvature as

$$\vec{v}.\vec{n} = c (c_2) - \varepsilon \kappa + O(\varepsilon) \tag{5}$$

where $c (c_2)$ is determined by relation (4). When the curvature increases, the propagation velocity $v.n$ decreases up to a critical value v_{cr} reached for a critical value of the curvature κ_{cr} for which stationary spread of the wave becomes impossible (see Zykov[1] book p.117, 143). At the leading order in ε, $\kappa_{cr} = 1/\varepsilon (c (c_2) + O(\sqrt{\varepsilon}))$ and $v_{cr} = O(\sqrt{\varepsilon})$.

We will now explain the model studied in the paper. We call this model " Wave Front Interaction Model " (W.F.M.) since it describes the motion of two fronts , the front and the back , which move with a normal velocity given by relation (5). We assume a zero value for the critical velocity , so that $\kappa_{cr} = c (c_2) / \varepsilon$ and use relation (5) whatever the value of the curvature. These two waves interact with a relaxational field c_2 which satisfies the equation

$$\frac{\partial c_{2\pm}}{\partial t} = - c_{2\pm} \pm 1 - \delta \tag{6}$$

where the subscript + (resp. -) means that the relaxational field c_2 is calculated in the excited region (resp. refractory region). This last equation is simply deduced from relation (2) with the additional relation $c_1 = \pm 1 - c_2$ which holds respectively in excited and refractory regions. The common end point of the two curves must have a zero normal velocity. This model appears as a free-boundary problem in the spirit of the one proposed by Fife[2] and Tyson and Keener[3] , but simpler and thus more tractable since the diffusion coefficient of the recovery variable is assumed here to be zero. It contains two control parameters , the ratio between fast reaction time and refractory time ε , a real positive number and the excitability δ, a real negative number larger than -1. Two basic questions can be posed about this model:

-What are the solutions of the model, as for instance the steadily rotating spiral waves?

-What is the validity of the model, i.e. for which conditions the solutions of W.F.M. are solutions of the initial reaction-diffusion sytem?

It is clear from the beginning that a first condition is that the two waves will be well separated, which imposes that $0 < \varepsilon \ll 1$. Secondly, that the waves exist everywhere around the excited region and that their curvature radius is much larger than their thickness. For nonzero ε, the former condition is violated in a neighbourhood of the tip because the normal velocity of the wave cannot vanish at the tip. But , as ε goes to zero, the size of this neighbourhood goes to zero too, since the critical velocity v_{cr} is of order $\sqrt{\varepsilon}$. The latter condition can be written simply $c \ll 1$. From relation (4), this condition can be satisfied if δ goes to -1. Thus , from the conditions that

-The waves are well separated,

-The planar pulse structure is stable,

-The curvature radius of the waves is large compared to their thickness,

the validity range of W.F.M is determined as : $0 < \varepsilon \ll 1; \sqrt{\varepsilon} \ll c \ll 1$.

STEADILY ROTATING SPIRALS

We look for solutions of W.F.M. corresponding to clockwise spirals rotating at constant angular velocity ω , around a hole of radius r_0. Consider a polar coordinate system (r, θ) rotating with constant angular velocity ω. Then , the shapes of the front and the back , respectively $\theta_F (r)$ and $\theta_B (r)$ satisfy eqn.5 , i.e.

$$\frac{\omega r}{\sqrt{1 + \Psi^2_{F,B}}} = \pm c \, (c_{2F,B}) - \varepsilon \left(\frac{\dfrac{d\Psi_{F,B}}{dr}}{\sqrt{1 + \Psi^2_{F,B}}^3} + \frac{\Psi_{F,B}}{r\sqrt{1 + \Psi^2_{F,B}}} \right) \qquad (7)$$

Here , $\Psi_{F,B} = r \, d\,\theta_{F,B}(r)/dr$, the sign + (resp. -) corresponds to the subscript F (resp. B). and c (c_2) is determined by relation (4). The two curves meet tangentially to the circle of radius r_0 sothat $\Psi_F(r_0) = -\Psi_B(r_0) = -\infty$, or from eqn. (7) their curvature reaches their critical value $\kappa_{cr} = c\,(c_{2F,B})/\varepsilon$. At large distance from the tip, front and back behave as phase-shifted Archimedean spirals, i.e. $\theta_B(r) = k\,r = \theta_F(r) + $ constant , where k is a positive number . In each part of the domain delimited by the curves, the concentration of the recovery variable c_2 satisfies

$$\omega \frac{dc_{2\pm}}{d\theta} = -c_{2\pm} \pm 1 - \delta \qquad (8)$$

where the subscripts + and - are respectively associated to the excited and refractory regions. On the front , $c_{2+}(\theta_F(r)) = c_{2-}(\theta_F(r)) = c_{2F}$ and on the back, $c_{2+}(\theta_B(r)) = c_{2-}(\theta_B(r)) = c_{2B}$.

From Eqn.(8), one can deduce c_{2F} and c_{2B} as :

$$c_{2F} = 1 - \delta + \frac{2\left(\exp\left(\dfrac{\theta_B - \theta_F - 2\pi}{\omega}\right) - 1 \right)}{\left(1 - \exp\left(-\dfrac{2\pi}{\omega}\right)\right)} \qquad (9)$$

$$c_{2B} = 1 - \delta + \frac{2\left(\exp\left(-\dfrac{2\pi}{\omega}\right) - \exp\left(\dfrac{\theta_F - \theta_B}{\omega}\right) \right)}{\left(1 - \exp\left(-\dfrac{2\pi}{\omega}\right)\right)} \qquad (10)$$

<u>Numerical Results</u>

For the details of the resolution of the system composed by eqns. (7), (9),(10) see the full paper by Pelcé and Sun[4]. For convenience we introduce the radial coordinate scaled with the tip radius $R = r\,\kappa_{tip} = r\,c/\varepsilon$.

<u>Angular velocity ω and hole radius r_0</u>

Hole radius r_0

The smaller is the ratio between fast reaction time and refractory time, the smaller is the radius of the circle around which the spiral tip rotates at constant angular velocity Fig.2. This curve diverges at the limiting value $\varepsilon_{max} = 6.8$. This means that solutions for steadily rotating spirals exist only for values of ε less than 6.8. For this particular value of δ, large hole radii are found for values of ε which are not small so that results obtained from W.F.M. may differ from the one obtained from the complete reaction-diffusion system. On the other hand solutions corresponding to small radii which are obtained for small values of ε may be in good agreement with the one obtained from complete simulations of the reaction-diffusion system.

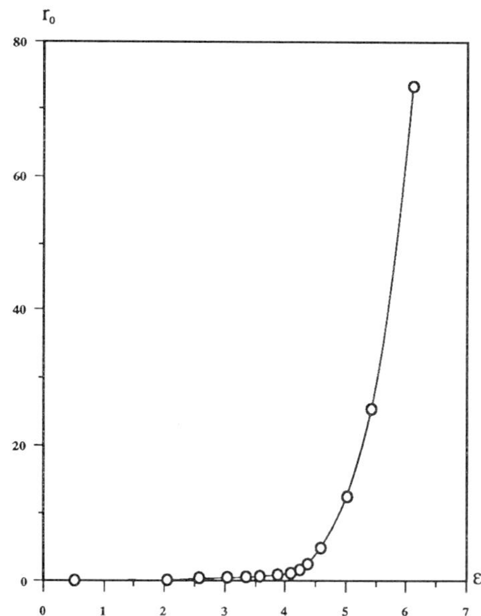

Fig.2 Hole radius r_0 as a function
of ε ($\delta = -0.1$).

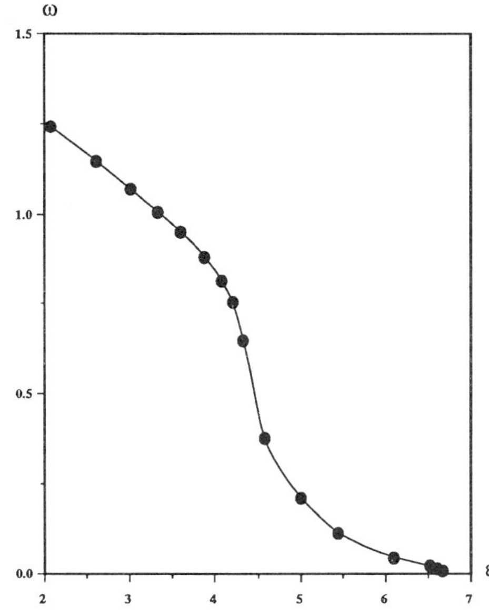

Fig.3 Angular velocity ω as a function
of ε ($\delta = -0.1$).

Angular velocity ω

It is a decreasing function of ε Fig.3.

<u>The spiral shape</u>

Spiral shapes are drawn on Fig.4 for $R_0 = 1.$ and $R_0 = 10.$ for the same value of the excitability $\delta = -.1$.

<u>The concentrations c_{2F} and c_{2B}</u>

On Fig.5 we draw the concentrations c_{2F} and $- c_{2B}$ as a function of the radial distance R. As expected the two curves have the same limit at large distance from the tip determined by relation (10). The two concentrations are equal to the fixed point value $c_2 (-1- \delta)$ at $R = R_0$ and increase as the radial distance increases.

<u>The tangential velocity</u>

It is found to increase as R_0 increases and saturate to the velocity of the planar pulse Fig.6. As R_0 is an increasing function of ε, smaller ratio between fast reaction time and refractory time leads to smaller tangential velocity.

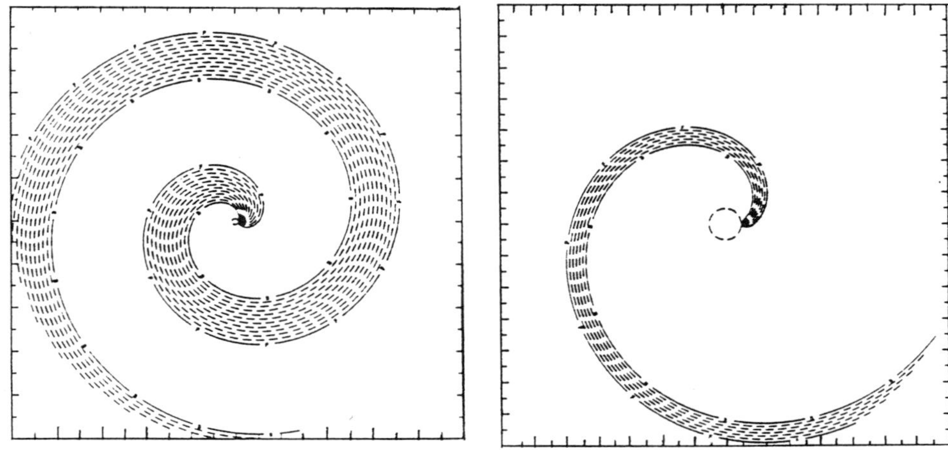

Fig.4 Shapes of spirals ($\delta = - 0.1$): left $R_0 = 1.$; right $R_0 = 10.$

<u>Steadily rotating spirals in the diagram ($E_{max}/A , D$)</u>

It is more convenient to draw the steady state curves for rotating spirals in the diagram (maximum relative rate of increase of the trigger variable c_1, pulse duration D) (Fig.7). This kind of diagram is well discussed in the Zykov book[1] and very useful as far as tip meandering problem is posed in the cardiologist context. For this, we take different values of ε and δ , determine the corresponding values of E_{max}/A and D and compute the hole radius R_0 for these values in a similar way as what was done in section II.a) . Then, curves of equiradius are drawn in the diagram ($E_{max}/A, D$) . As was found for the approximate solutions of Zykov[1], for a fixed hole radius, the maximum relative rate of increase of the trigger variable c_1 decreases when the pulse duration D increases. On this diagram, we draw the "validity line" of W.F.M. which limits the region where $0 < \varepsilon < .5$.

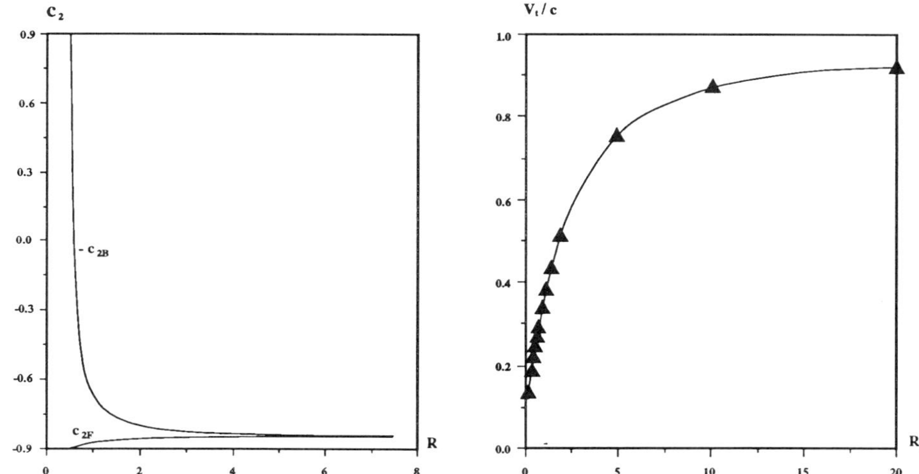

Fig.5 The concentrations c_{2F} and $-c_{2B}$ function of the radial distance R (δ = - 0.1 , R_0 = .5)

Fig.6 Tangential velocity of the spiral tip as a function of R_0 (δ = - 0.1).

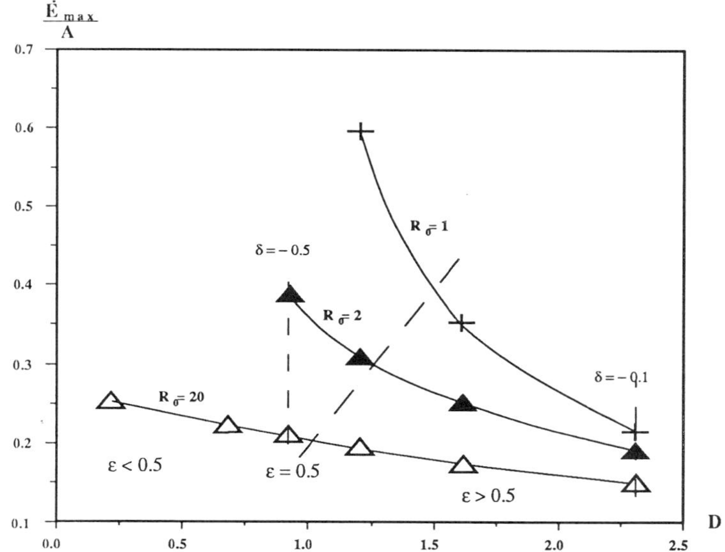

Fig.7 Steady state rotating spirals in the (E_{max} / A, D) diagram. The dashed lines limits the region where $0 < \varepsilon < .5$.

<u>The large R0 limit</u>

In the large R_0 limit, the shape of the spiral can be separated into two regions: A core region , of size of order unity , moving at uniform tangential velocity $v_t = c \, \Omega R_0$; A far-tip region where curvature effects are negligible .

<u>The tip region</u>

Define the the spatial variable $z = R - R_0$, which varies of order unity in the tip region. Rescale the variables $\theta_{F,B}$ as $u_{F,B} = R_0 (\theta_{F,B} - \theta_{tip})$. Then, at leading order in $1 / R_0$, equations for front and back are respectively :

$$\frac{\Omega R_0}{\sqrt{1 + (\frac{du_F}{dz})^2}} = 1 - \frac{\frac{d^2 u_F}{dz^2}}{((1 + (\frac{du_F}{dz})^2))^{3/2}} \tag{11}$$

and

$$\frac{\Omega R_0}{\sqrt{1 + (\frac{du_B}{dz})^2}} = - \frac{c (c_{2B})}{c} - \frac{\frac{d^2 u_B}{dz^2}}{((1 + (\frac{du_B}{dz})^2))^{3/2}} \tag{12}$$

Here , one anticipated that ω is small , of order $1/ R_0$, so that , from relations (9) and (10), $c_{2F} = - (1 + \delta)$ and $c_{2B} = 1 - \delta - 2 \exp [(u_F - u_B) / \omega R_0]$.
Contrary to the general case, where the equations for the front and the back are coupled, eqns. (11) and (12) are decoupled in the large R_0 limit. In this case, the period of rotation of the tip of the spiral is very large so that the recovery variable has time to come back to its steady value before another excitation of the medium occurs. After the change of variable $du_F / dz = \tan\Phi$, one can integrate eqn. (11) as

$$z = \int_{-\frac{\pi}{2}}^{\Phi} \frac{\cos \Phi'}{1 - \Omega r_0 \cos \Phi'} d\Phi' \tag{13}$$

There is no evident analytical solution for the back so that eqn. (12) must be integrated numerically.

<u>The far tip region</u>

Define the spatial variable $Z = (R - R_0) / R_0$ which varies of order unity in this region. Then, at leading order in $1 / R_0$, equations for the front and the back become:

$$\sqrt{1 + \Psi_F^2} = \Omega R_0 (1 + Z) \tag{14}$$

$$- \frac{c (c_{2B})}{c} \sqrt{1 + \Psi_B^2} = \Omega R_0 (1 + Z) \tag{15}$$

where here too, $c_{2B} = 1 - \delta - 2 \exp [(\theta_F - \theta_B) / \omega]$. A simple solution of eqns. (14) and

(15) can be obtained if the difference between θ_F and θ_B is a constant which satisfies the relation $\delta = - \exp (- \Delta\theta / \omega)$. In this case, $c_{2B} = 1 + \delta$ sothat $c (c_{2B}) = -c$, i.e the transverse structure of the spiral is the one of a single pulse. Thus, it is sufficient to integrate eqn. (14) to obtain the shape of the counterclockwise spiral as

$$\theta_F = \int_{const}^{Z} \sqrt{ \frac{ (\Omega r_0)^2 (1 + Z')^2 - 1 }{ (1 + Z')^2 } } \, dZ' \tag{16}$$

This is the equation of the involute of a circle, shape already discussed earlier by Winfree[5] in the context of spirals rotating rigidly in excitable media.

Matching of the solutions

As solution (16) must be valid up to Z close to zero in order to be matched to the inner solution one has $\Omega R_0 \geq 1$. But, from relation (13), the shape of the spiral goes out from the inner region with the fixed value $\Psi_F = - \text{arctg} (((\Omega r_0)^2 - 1)^{1/2})$ and can match the clockwise spiral determined by relation (16) only if Ψ_F has a positive value, i.e. if $\Omega r_0 = 1$. It is equivalent to say that, in the large R_0 limit, the tangential velocity of the spiral tip is equal to the velocity of the planar pulse, c.
Then, one rewrites eqn.(12) as :

$$\frac{d^2 u_B}{dz^2} = -(1 + (\frac{du_B}{dz})^2)(1 + \frac{c(c_{2B})}{c} \sqrt{ 1 + (\frac{du_B}{dz})^2 }) \tag{17}$$

where $c_{2B} = 1 - \delta - 2 \exp [(u_F - u_B) / \omega R_0]$, $u_F (z)$ determined by relation (13), and with the boundary conditions $du_B / dz (0) = +\infty$, $du_B / dz (+\infty) = 0$. One finds a solution only if $\omega R_0 \approx 2.5$. It follows that for large R_0 and $\delta = -.1$, the limiting value of ε is 6.8. This limiting value ε_{max} decreases when δ decreases as is shown in Fig.8. Thus, as smaller values of ε is obtained for smaller values of δ, better quantitative agreement between W.F.M. and the initial reaction-diffusion system is expected for smaller value of the excitability.

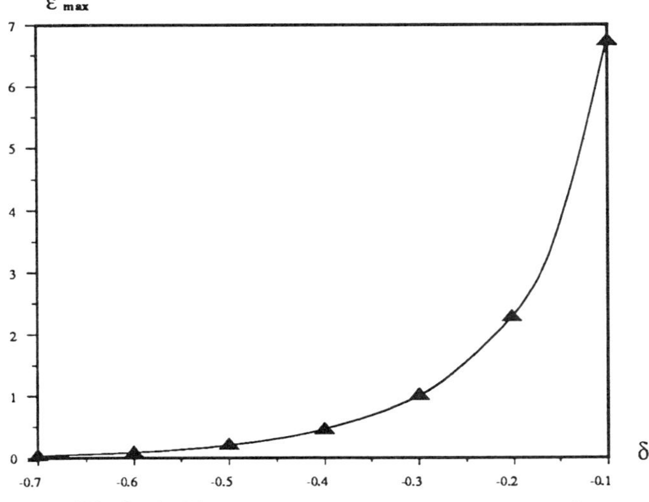

Fig.8 Limiting value of ε, ε_{max} as a function of δ

CONCLUSION

In this paper, we develop a simplifying model for excitable media , in which two curves , the front and the back , interact with a relaxational field c_2. This model can be derived from the initial reaction-diffusion model when ε , the ratio between fast reaction time and refractory time , is small and when the curvature radius of the two waves is large compared to their thickness. In this paper we determine the steady solutions of the model which corresponds to steadily rotating spiral waves. It is shown that for given control parameters δ and ε the angular velocity of the spiral and the radius of the circle around which the spiral tip rotates can be determined uniquely. Thus the tangential velocity of the spiral tip can be determined , which can be an important step in the understanding of meandering. The shape and the size of the spiral core can be determined. Furthermore, a semi-analytic determination of the solution can be performed in the large hole radius limit. This last solution is valid only if the excitability δ is smaller than -.5 , case for which one can have simultaneously a large hole radius and a small value of ε.

Acknowledgments: This research is supported by a collaborative research grant NATO.

REFERENCES

1.   V.S. Zykov , Modelling of wave processes in excitable media (Manchester University Press , Manchester , 1988).
2.   P. C. Fife, Propagator-controller systems and chemical patterns, in: Non-Equilibrium Dynamics in Chemical Systems, C. Vidal and A. Pacault, eds. (Springer, Berlin, 1984), pp.76-88.
3.   J.J. Tyson and J.P. Keener, Physica D (Amsterdam) 32, 327 (1988).
4.   P.Pelcé and J.Sun, To appear in Physica D , March (1991).
5.   A.T. Winfree, Science (1972) , 175 , 634.

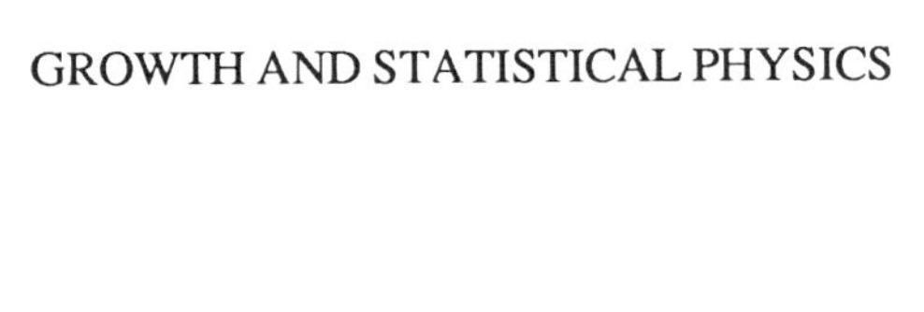

GROWTH AND STATISTICAL PHYSICS

DIFFUSION CONTROLLED GROWTH PHENOMENA:

FROM SMOOTH INTERFACES TO FRACTAL STRUCTURES

A. Arneodo[a], F. Argoul[a], Y. Couder[b] and M. Rabaud[b]

[a]Centre de Recherche Paul Pascal, Avenue Schweitzer
33600 Pessac, France
[b]Laboratoire de Physique Statistique, 24 rue Lhomond
75231 Paris Cedex 05, France

INTRODUCTION

Diffusion-controlled growth phenomena have recently attracted considerable attention[1-11]. Notable examples[1-12] of pattern formation in diffusive systems include dendritic crystal growth, electrochemical deposition, viscous fingering and diffusion-limited aggregation (DLA). Roughly two different types of pattern morphologies are observed corresponding to isotropic and anisotropic growths[12]. Anisotropic growths are characterized by the existence of preferential directions of growth. The term anisotropy must be used with care however, since the anisotropy of growth of the pattern is not necessarily due to some intrinsic microscopic anisotropy in the physical properties of the system[1-12] but can also result from some external fields or perturbations[9-20].

Experimental and theoretical efforts have been mainly focused in two directions. On the one hand, the shape and selection mechanism of nonlinear *stable curved fronts* were investigated and analytical solutions were found, e.g., the parabolic needle crystal[1-7,21] and the Saffman-Taylor (ST) fingers in linear[3-6,22] and sector-shaped channels[23]. On the other hand, the very unstable patterns have been mainly considered from the point of view of their *fractal structure*[8-13,24-31]. These patterns are actually fractal only in a limited range between a smallest and a largest length scale. In this regard, they differ from mathematically defined fractals which cover an infinite range of length scales[32,33]. In the systems that we will discuss, the smallest scale ℓ_{min} can be linked with the instability process or with the computational technique, the largest scale ℓ_{max} being imposed by either the geometrical boundary conditions or, in free growth, the overall size reached by the pattern. We can therefore tune the extent of the fractal range and go from compact objects when the two length scales are close to each other, to structures with a large fractal range when the two scales are far from each other. The aim of the present work is to study the structure of complex fractal patterns in the experiments listed above and to compare them with the stable solutions. Limiting ourselves to patterns grown in a Laplacian (viscous fingers and DLA clusters) or nearly Laplacian (crystalline dendrites) fields, we investigate their statistical properties and show that, as the fractal range opens up, some properties of the compact forms are retained by the fractal structure[34-36]. Since in Laplacian forming systems the motion of one of the boundaries (the interface) depends on the other boundary conditions fixed by the cell shape, the choice of the geometry is crucial[12]. We will concentrate here on geometries in which stable smooth solutions are known, i.e., the linear and the sector-shaped geometries.

Growth and Form, Edited by M. Ben Amar *et al.*
Plenum Press, New York, 1991

Isotropic Laplacian fractal patterns are commonly regarded as being the result of consecutive tip-splitting and screening instabilities[3,8-11,24-31,37,38]. The role of the effective anisotropy in stabilizing the tip of the dendrites against tip-splitting has been clearly identified[8-11,29,39]. Experiments[12-17,40-42] as well as simulations[29,39,43-45] of the Witten and Sander DLA model[27] have revealed a transition from apparently random DLA like fractal patterns to more regular dendritic fractal patterns when the effective anisotropy is enhanced. In the present work, we carry out a finite-size scaling analysis of this anisotropy-induced morphological transition[36]. An analytical expression for the fractal dimension D_F of anisotropic DLA clusters as a function of the effective anisotropy is derived. This relation strongly suggests that, to some extent, the fractal properties of unstable anisotropic patterns grown in Laplacian fields are already contained in the smooth ST solutions[22].

ISOTROPIC GROWTHS

We will limit ourselves here to the two simplest cases of isotropic Laplacian pattern forming systems: the Saffman-Taylor instability[22,46] and the formation of clusters in the numerical DLA model[27,28]. These two systems are very different not only because of their deterministic or stochastic nature but also from the existence of surface tension in viscous fingering which has no obvious counterpart in DLA. Nevertheless the similarity of their ruling laws is well known[3,47].

The instability giving rise to ST viscous fingering[22] occurs at the interface between two fluids moving between narrowly spaced solid plates. The interface is unstable when the less viscous fluid forces the most viscous fluid to recede. The flow of the fluid is dominated by the viscous dissipation on the plates and the mean velocity in the cell plane is proportional to the pressure gradient $\vec{V} = -(b^2/12\mu)\vec{\nabla}p$, where b is the cell thickness and μ the viscosity of the most viscous fluid. Because of the incompressibility of the fluids, the pressure field obeys a Laplace law $\Delta p = 0$. Surface tension has a stabilizing influence[3]; this is taken into account by adding a boundary condition for the pressure jump at the interface: $\partial p = T\kappa$, where T is the surface tension and κ the local curvature of the meniscus in the plane of the cell. The linear analysis[46] of the stability of a plane interface moving at velocity $\vec{V}$ gives the wavelength of maximum instability: $l_c = \pi b(T/\mu V)^{1/2}$. The capillary length l_c is the characteristic small length scale of viscous fingering.

In the numerical model of randomly walking particles introduced by Witten and Sander[27], the probability P of visit of a site obeys a Laplacian law $\Delta P = 0$ and the normal velocity of growth of a region of the interface is $V_n \propto (\vec{\nabla}P)_n$. As was first noted by Paterson[47], the equations of growth of DLA are similar to those of ST fingering in the limit of zero surface tension. In DLA it would thus appear that all length scales are unstable; but the computation of DLA clustering is usually done on a lattice and the lattice-mesh size l_u introduces a smallest length scale l_{min}.

A. Linear cells

In the configuration initially chosen by Saffman and Taylor[22], the fluids move in a very long linear channel of width W. These boundaries somewhat simplify the problem by imposing translational invariance. The control parameter is usually defined[3,12] as $1/B = 12\pi^2(W/l_c)^2$. Experiments[12] show that the two characteristic length scales can be identified as $l_{min} = l_c$ and $l_{max} = W$. The control parameter represents their ratio according to the scaling relation $l_{max}/l_{min} = (12\pi^2 B)^{-1/2}$. If the experiment is done in a given cell so that l_{max} is constant, l_{min} can be chosen at will by changing the velocity. For $l_{max}/l_{min} < 8$, a single finger is observed, scaled on the cell width (Fig. 1a). In the upper part of its stability range the relative width λ of the finger tends asymptotically towards 0.5. Saffman and Taylor[22] had found a one-parameter (λ) family of solutions for the interface shape at $T = 0$. These were fingers parametrized by their relative width λ

which could take any value from 0 to 1:

$$y = \frac{W(1-\lambda)}{2\pi} \ln\left[\frac{1}{2}\left(1 + \cos\frac{2\pi y}{\lambda W}\right)\right]. \tag{1}$$

The selection of the observed asymptotic finger width $\lambda = 0.5$ was understood rather recently after numerical investigations[48,49] and analytical works[50-52]. It results from the action of surface tension acting as a singular perturbation. For larger values of $l_{max}/l_{min}(> 8)$ we get very unstable fingers (Fig. 1b) that have a fractal appearance similar to those of DLA clusters grown in a strip (Fig. 1c). In contrast to the stable fingers, only a little is known about the nonlinear geometrical complexity of the unstable fingers. The main results of our study of isotropic fractal patterns have been announced in Ref. 34.

Our experimental set-up was designed to reach very large values of $1/B$ which implies large applied pressures. We performed the viscous fingering experiments in a linear cell made of float glass 19 mm thick. The cell thickness b was either 0.25 mm or 0.125 mm and the width in the region of measurement was $W = 10$ cm. We used a silicon oil Rhodorsyl 47 V 500 with $T = 21 \ 10^{-3}$ N/m and $\mu = 0.48$ kg/m.s. The air was injected at constant pressure.

The DLA clusters were computed in strip geometries of width $W = 32, 64, 128$ and 256 lattice units. We used an on-square lattice algorithm and reflecting lateral walls[34]. M was chosen large enough so that the characteristic size of the aggregate (along Ox) be much larger than W. As in the experiments, a numerical trick was used to initiate the growth from the cell axis: the very first random walking particle sticks onto a needle (a few lattice units long) centered on the cell axis; this reduces considerably the period of the selection regime[53] among DLA trees grown from the linear substrate, which is out of the scope of the present analysis.

In both experiments and simulations we wanted, after a large number N of independent runs, to measure the mean occupancy of each site of the channel. The results are reported with the following conventions: we choose Ox along the cell axis and Oy across the cell. In viscous fingering we limited ourselves to the measurement of the mean occupancy across the cell. We chose a section of the cell where the pattern had finished its evolution and built a histogram of the transverse occupancy by air in all the runs. A division by N gave the mean occupancy $r_T(y)$ in this section. In the stable case, $r_T(y)$ is a square profile with probability of occupation 1 in the central half of the channel and 0 elsewhere. When averaging over $N = 75$ realizations of the type shown in Fig. 1b, one gets the transverse profile shown in Fig. 2; it is smoother than a step profile, has a maximum value $r_{max} < 1$ at the center ($y = 0$) and decreases to zero at the walls ($y = \pm W/2$). This tranverse profile becomes smoother and smoother when increasing $1/B$ and the width of the regions which are never visited (along the walls) shrinks. The limiting profile of the histogram is obtained for DLA, where it is suprisingly well fitted by $r_T(y) = r_{max}\cos^2(\pi y/W)$ as illustrated in Fig. 2 (see also Fig. 10b). Moreover the computational technique lends itself to the measurement of $r(x,y)$ at every point of the strip. The histogram of occupancy along the axis of the linear strip is shown in Fig. 3. Except in the initial region and in the tip region, r_L is constant. The stability of r_{max} shows that in the region where the growth has ceased, the cell translational invariance imposes itself to the occupancy profile. Scaled on W, the mean length of the cluster x_{tip}/W is thus proportional to its mass M. The fall-off of r_L in the tip region of width Δx_{tip}, corresponds to the active part of each pattern and to the dispersion of the tip position. Our results are consistent with a dependence $\Delta x_{tip}/W \sim (x_{tip}/W)^{1/2}$; this is explained naturally if the growth process, on the scale W, can be considered as the successive addition of n independent bunches of a fixed number of particles having different configurations and thus different lengths so that the dispersion of the tip positions be proportional to $\sqrt{n}$.

The fundamental result about these histograms of mean occupancy is that *the width at midheight of the transverse profiles is half the channel width $\lambda = 0.5$*. In other words, going from the stable finger to unstable patterns, the occupancy rate becomes smeared

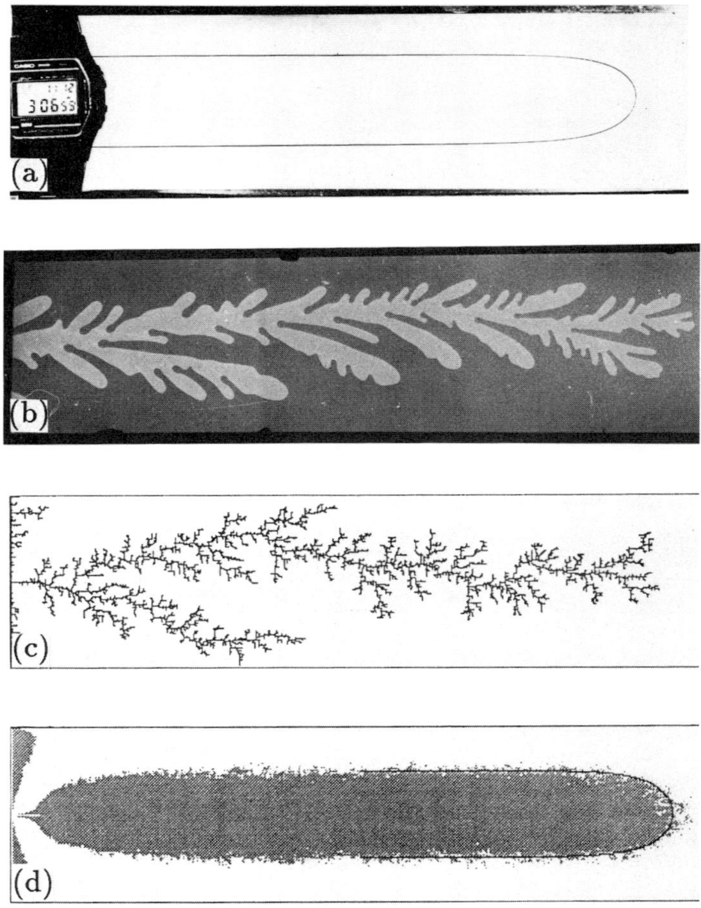

Fig. 1 (a) A stable ST finger $\lambda = 0.5$. (b) An unstable ST finger for $l_{max}/l_{min} = 50$. (c) A DLA cluster with 6000 particles grown in a strip of width $W = 128$. (d) Region of the strip in (c) with mean occupancy above average ($r \geq r_{max}/2$); 510 aggregates of the type shown in (c) were grown to obtain this repartition.

out but its width at midheight is preserved. The results are in fact even more specific: Fig. 1d shows all the points of the strip where r is larger than $r_{max}/2$ in a series of 510 DLA simulations. The limit of this region could also have been found by determining the points on each side of the cell axis which satisfy:

$$y_m^{\pm}(x) \; = \; \frac{1}{r_{max}} \int_0^{\pm W/2} r(x,y)dy \; . \tag{2}$$

All the histograms that we obtained for isotropic patterns were such that the same mean profile was found by the two procedures ($r \geq r_{max}/2$ or $r \geq r(y_m)$). As shown in Fig. 1d *the region of large occupancy is very well fitted by the Saffman-Taylor analytical solution (1) for $\lambda = 0.5$.* (We have checked that other sections at different levels of occupancy have different widths but are not fitted by other ST solutions). This is a rather suprising result as it means that *the selection of a solution survives its instability.* As it could have been a mere coincidence, we checked it by investigating sector-shaped geometries.

B. *Sector-shaped cells*

A variant of the Hele-Shaw geometry was introduced for viscous fingering where the cell has a sector shape[23] of angle θ_0. The finger can move either from the apex to the periphery (divergent finger, $\theta_0 > 0$), or from the periphery to the apex (convergent finger, $\theta_0 < 0$). For a ST front moving at constant velocity, l_c remains constant. The other length scale, the local width $W(x) = x\theta_0$ of the cell is a function of the distance x of the front to the apex. In this geometry the ratio l_{max}/l_{min} and thus the control parameter $1/B$ changes spontaneously with the distance of the active front to the cell apex. Previous investigations showed that near the apex, there is a region where the finger is stable and tends to occupy a finite fraction $\lambda(\theta_0)$ of the cell angular width[23]. For all convergent fingers and for divergent fingers up to $\theta_0 < 20°$, an asymptotic minimum angular relative width λ_m was found to depend linearly on θ_0. In cells with $\theta_0 > 20°$, the asymptotic width of divergent fingers was not observed because they destabilized at lower values of $1/B$ into DLA-like patterns[12,23] as shown in Fig. 4a.

We performed viscous fingering experiments in four sector-shaped cells ($\theta_0 = 23°, 45°, 60°$ and $90°$). Measurements were done under the same experimental conditions as for the linear cells. For divergent cells air was injected at a controlled pressure, while for convergent fingers oil was pumped out of the apex and an experimental expedient was used: at rest we created a dip in the meniscus at the periphery of the cell so that the finger would originate from this point on the cell axis. Fig. 4a shows an unstable ST pattern obtained in a cell with $\theta_0 = 90°$. A DLA cluster is shown for comparison in Fig. 4b. In our DLA simulations conducted in sector geometry[34,54,55], the cell boundaries were approximated by a staircase structure and the particles were locally reflected normally to these boundaries. Fig. 5 shows the tranverse histograms of mean occupancy measured on an arc of circle in a series of viscous fingering experiments and DLA simulations. These histograms have a similar shape for DLA and viscous fingers. For convergent patterns (Figs 5a and 5c) the peak is rather narrow and there is a large region on each side where $r = 0$. For divergent patterns (Figs 5b and 5d) the histogram has a large plateau at r_{max} with a somewhat abrupt fall-off to zero. For all angles $-90° < \theta_0 < 90°$, the width at midheight is best fitted by a linear relation[34] (Fig. 6):

$$\lambda_m \; = \; 0.5 + (3.4 \pm 0.2)\,10^{-3}\theta_0 \; . \tag{3}$$

It is remarkable that for all convergent channels and for divergent ones with $\theta_0 < 20°$, these mean widths concide with the asymptotic widths of stable fingers[23]. For $\theta_0 > 20°$ the same identity is not observed because the stable fingers destabilize before reaching their asymptotic width as previously mentionned. If one keeps extrapolating relation (3) for larger θ_0 values, the mean width λ_m is expected to reach 1 (the region of large occupancy fills the entire cell), for a critical value of the sector-cell angle $\theta_0 \approx 4\pi/5$. Our preliminary results suggest that this critical angle marks a transition in the shape of the mean occupancy profile which then presents several lobes[54]. Note that the presence

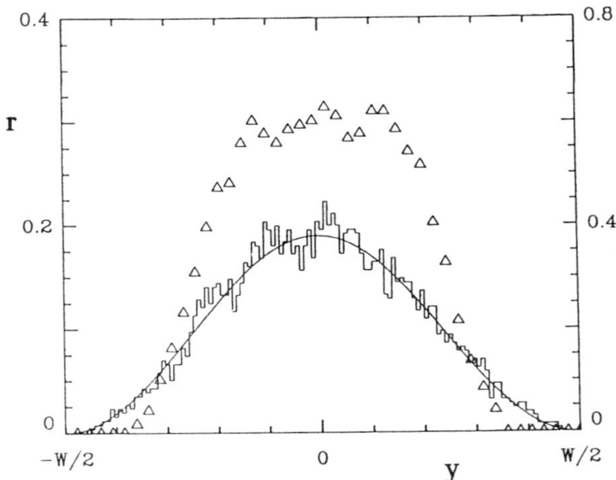

Fig. 2 Histograms showing the mean occupancy across a section of the linear cell for $N = 75$ unstable ST fingers (Δ) at $l_{max}/l_{min} = 50$ (scale of $r(x,y)$ on the right) and $N = 510$ DLA clusters of mass $M = 6000$ grown in a strip of width $W = 128$ (scale on the left). The solid line corresponds to a fit of the DLA transverse histogram by $r(x,y) = r_{max}\cos^2(\pi y/W)$.

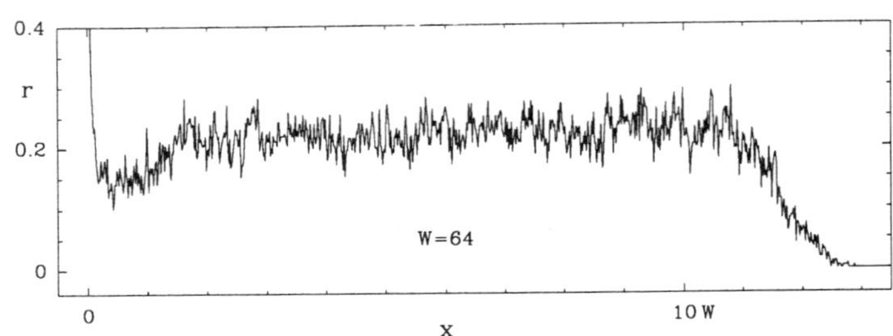

Fig. 3 Histogram of the mean occupancy along the axis of the linear strip. $N = 255$ DLA clusters of mass $M = 6000$ were grown in a strip of width $W = 64$.

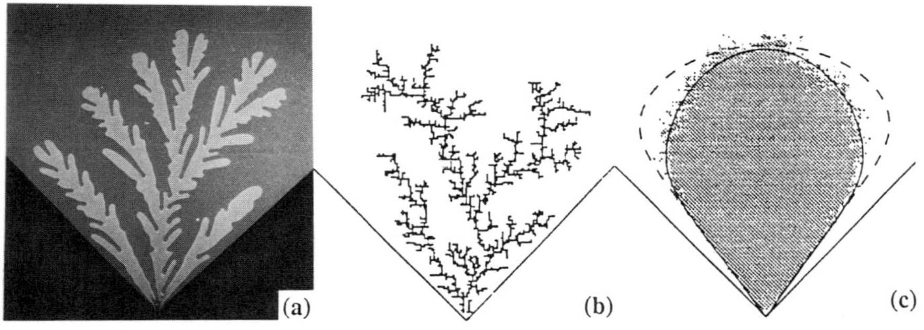

Fig. 4 (a) Unstable viscous fingers and (b) DLA cluster grown in a cell of angle $\theta_0 = 90°$. (c) In gray, region of large occupancy ($r \geq r_{max}/2$) for $N = 420$ DLA clusters of mass $M = 2000$. The continuous line is the analytical solution given by Eq.(4) for $\lambda = 0.82$. The dashed line is the conformal transform of the ST solution $\lambda = 0.82$ (from Ref. 34).

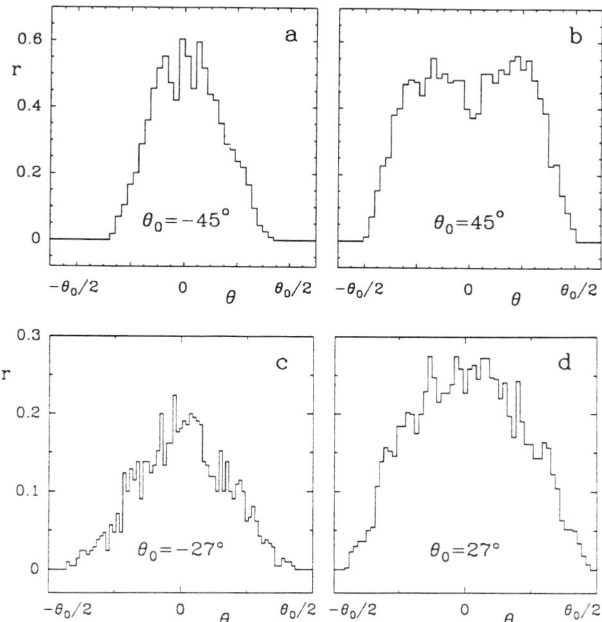

Fig. 5 Histograms showing the mean occupancy across a section of an angular cell of
angle θ_0. (a) $N = 75$ unstable viscous patterns grown in a convergent cell ($\theta_0 =
-45°$). (b) $N = 55$ unstable viscous patterns grown in a divergent cell ($\theta_0 = 45°$).
(c) $N = 255$ DLA clusters grown in a convergent sector geometry ($\theta_0 = -27°$).
(d) $N = 255$ DLA clusters grown in a divergent sector geometry ($\theta_0 = 27°$).

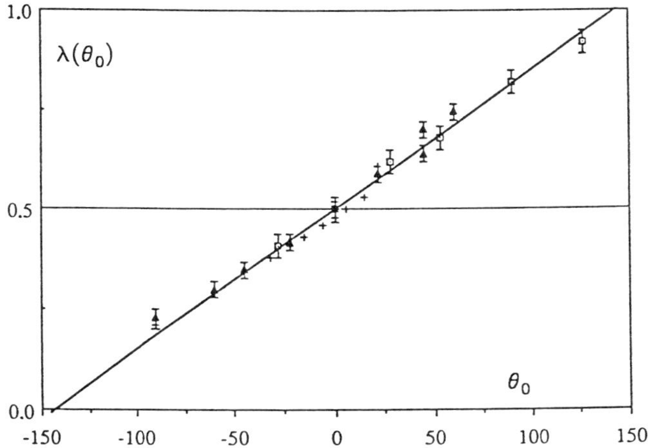

Fig. 6 Width of the region of large occupancy in sector-shaped cells as a function of
the cell angle θ_0. (▲) Unstable ST fingers; (□) DLA clusters, (+) recall of
the minimum width observed for stable fingers in these cells. The solid line
corresponds to the best linear fit (relation (3)) (from Ref. 34).

of a five-fold symmetry in diffusion-limited aggregation has already been suggested in previous works[31,56,57].

Divergent cells with $\theta_0 = 90°$ are of particular interest because a family of analytical solution is known[23], giving the finger shape in the absence of surface tension. (Similar results have been recently obtained by M. Ben Amar[58] for cells of arbitrary angle). These fingers form a self-similar counterpart to the ST solutions. Their equations are also parametized by the relative angular width λ:

$$x = \frac{1}{\sqrt{2}} \cos \alpha \left[(\tan \alpha)^a + (\tan \alpha)^{1-a} \right],$$

$$y = \frac{1}{\sqrt{2}} \cos \alpha \left[(\tan \alpha)^{1-a} - (\tan \alpha)^a \right] \tan \left[\frac{\lambda \pi}{4} \right], \tag{4}$$

where $\lambda \epsilon [0,1]$, $a = (1 - \lambda)/2$, $\alpha \epsilon [0, \pi/2]$, and Ox is along the finger axis. Fig. 4 shows two realizations of unstable patterns and the region of large occupancy $(r > r_{max/2})$. The shape of the latter is very well fitted by the analytical form (4) and differs markedly from the profile that would have been obtained from the conformal transformation of the ST fingers. Moreover its angular width $\lambda_m = 0.82 \pm 0.02$ agrees very well with the recent theoretical prediction[59] $\lambda = 0.85$ for the asymptotic relative width of stable fingers in the limit of zero surface tension.

For DLA clusters we have also computed radial histograms along the bissector of the cell[54] . For this geometry, r_{max} is a decreasing function of the distance x to the apex. This is due to the increase of the ratio l_{max}/l_{min} with the distance of the front region from the cell apex. When defining the mean finger shape from either Eq.(2) or the cut at midheight $(r \geq r_{max}/2)$, we get the same mean finger of constant relative width $\lambda(\theta_0)$ but at different stages of growth.

Altogether these results confirm once more the similarity between the fractal structure of viscous fingers and DLA patterns[3,24−29,37,38,47]. This suggests that the detailed noise mechanism does not play an important role[31]. It is not such a surprise that the cell geometry should determine the large-scale shape of the profile of mean occupancy. It is a very striking result, however, that the selected solution should be the same as the stable one[34]. In sector-shaped cells, when the structure diverges from the apex, it builds up a fractal structure in a larger range of scales between l_c and $x\theta_0$. During this build up it retains the same sensitivity to both the large and the small scales. In other terms, *the selection action of the microscopic length scale acts through the entire range, up to the largest scale of the pattern.* In that respect our results bring the clue that the role of the viscous-finger capillary length scale is played by the lattice-mesh size in DLA clustering[34].

ANISOTROPIC GROWTHS

Since the selection mechanism of stable patterns was shown to be dramatically affected by the presence of anisotropy[1−7,12], an interesting issue was to extend our statistical analysis to anisotropic fractal patterns. We will limit ourselves here to anomalous viscous fingers[14−16] and anisotropic noise-reduced DLA clusters[29,39,43−45,60] grown in a strip geometry. We will start our study by investigating dendritic crystal growth[1−12]; in nearly Laplacian conditions crystalline dendrites develop side branches that give them a fractal structure. The main results of this work have been announced in Ref. 35.

A. Crystallines dendrites

The rough growth of monocrystals in an undercooled solution gives rise to the characteristic crystalline dendrites. These grow along the main axis of the crystal and have a parabolic tip. Their sides destabilize into lateral branches which compete and, far from the tip, form a complex pattern. The anisotropy of the growth is due to the

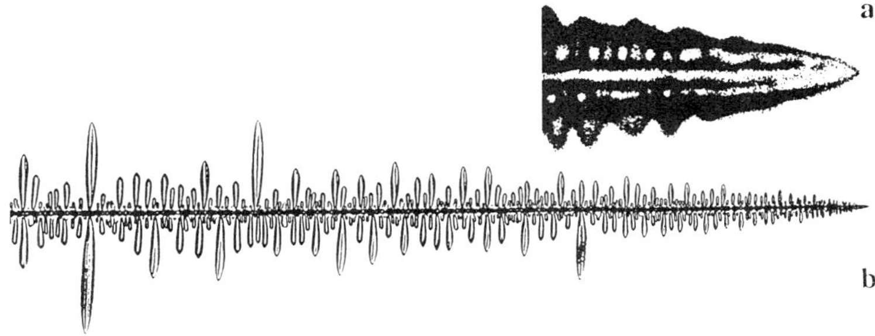

Fig. 7 (a) Tip region (of lenght 20ρ) of a dendrite showing its stable parabolic tip and the initial destabilization. (b) Photograph of a region of length 500ρ of the same dendrite (here $\rho = 2.5\,\mu m$) (from Ref. 35).

anisotropy of the surface tension of the crystalline structure[1-7]. Experimental[61,62] and theoretical[63-67] efforts have resulted in the understanding of the selection mechanism of the parabola[21]. The characteristic length scale l_{MS} of the instability of a planar crystallization front growing at velocity V is given by the linear analysis of Mullins and Sekerka[1,68] $l_{MS} \propto (d_0 D/V)^{1/2}$, where D is a diffusion constant and d_0 a capillary length. The radius of curvature of stable parabolic dendrites is proportional to l_{MS} so that $\rho^2 V = $ const; the coefficient of proportionality is a decreasing function of the surface tension anisotropy ϵ (for a cubic crystal $\rho = 0.5\epsilon^{-7/8} l_{MS}$ in the limit of very weak anisotropy[67]). The selection mechanism of the tip of a dendrite is thus governed by the anisotropy of the surface tension[63-67].

Our main goal in this section is to study the general structure of an experimentally obtained complex dendrite, from the point of view of its fractal structure and of its relation to the analytical parabola[35]. Fractal side branching is generally observed when the impurity diffusion field has a length scale $l_D = D/V$ large as compared to the observed region of the dendrite. The screening-off between different branches is then the same as in a Laplacian field. However, l_D remains small relatively to the distance to the lateral walls of the cell so that the medium can be considered as infinite.

We grew ammonium bromide crystals in conditions similar to those described in Ref. 62, but from a solution with a different concentration. As the thermal diffusivity is three orders of magnitude larger than the mass diffusivity, the system is limited by mass diffusion for which $D = 2.\,10^{-9} m^2/s$ and we work at Péclet number Pe $\sim 10^{-3}$. Extensive measurements were done on two dendrites (1 and 2) grown respectively at velocity $V_1 = 0.6\,\mu m/s$ and $V_2 = 2.6\,\mu m/s$. The impurity diffusion lengths were $l_{D1} = 3.3\,mm$ and $l_{D2} = 0.78\,mm$. The cell thickness $e = 30\,\mu m$ was small as compared to l_D, so that the diffusion field could be considered as two-dimensional. Near the tip, however, the situation remained three dimensional because the tip radius $\rho < e$. One single, well-oriented germ (two [100] axes in the cell plane) initiated the growth and we waited long enough to get a sufficiently developed dendrite far away (more than a centimeter) from the boundaries and from the the other main arms grown from the germ. A well-oriented initial seed in a flat cell usually gives rise to four arms. When the diffusion length is large, these arms interact with each other and grow in a petal shape[39]. It is only the front part of each arm which can be considered as parabolic. In our experiments we analyzed 2 mm of a dendrite when it had grown 1 cm away from the seed. By means of a 35 mm camera and a videocamera, we first recorded the detailed shape of the dendrite tip (Fig. 7a). In this region we determined by image processing the parabola which was the best fit to the stable part of the profile extending over a length $\Delta x \sim 4\rho$. We found $\rho_1 = 3.8 \pm 0.1\,\mu m$ and $\rho_2 = 1.7 \pm 0.1\,\mu m$ respectively. The corresponding values $\rho_1^2 V_1 = 7.4 \pm 0.2\,\mu m^3/s$ and $\rho_2^2 V_2 = 7.5 \pm 0.2\,\mu m^3/s$ are in good agreement with each other. We simultaneously recorded the unstable dendrite in a large region behind the

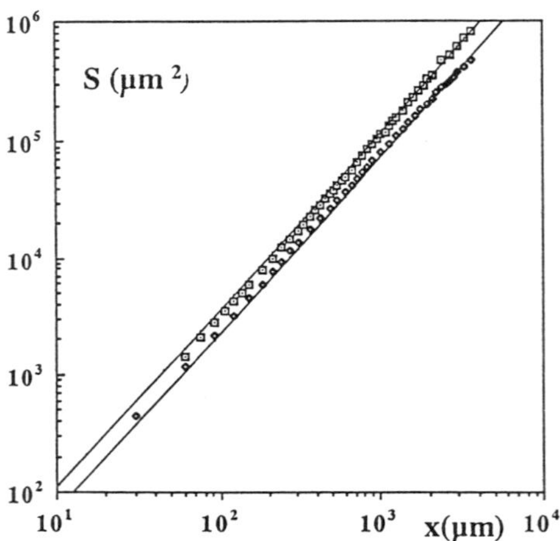

Fig. 8 Logarithmic plot of the surface $S(x)$ of dendritic patterns as a function of the distance x to the tip: ($\square$) $S_1(x)$; ($\diamond$)$S_2(x)$. The solid lines correspond to the surfaces of compact parabolic dendrites with the same radius of curvature $\rho_1 = 3.7\,\mu$m and $\rho_2 = 1.7\,\mu$m as the fractal dendrites 1 and 2, respectively (from Ref. 35).

tip (Fig. 7b). The hierarchy of sizes of the side branches makes the dendrites fractal objects eventhough due to anisotropy, they are compact along their axis. This fractal character is directly related to the competitive interaction in a long-range field. We image processed a part of the dendrite far from the tip, analyzed it by the box-counting method, and found a fractal dimension $D_F = 1.58\pm0.03$ in agreement with the dimension found numerically for four-fold anisotropic DLA clusters in previous studies[29,39,43–45].

Along the line of our approach of isotropic fractal patterns, we performed statistical measurements on the whole structure of these fractal dendrites[35]. For each value of the distance x to the tip, we measured the dendrite area $S(x)$ from the tip to x. Fig. 8 shows a logarithmic plot of $S(x)$ as a function of x. Over three orders of magnitude of length scales, $S(x) \propto x^{3/2}$, as would have been expected if the dendrites had had a smooth parabolic shape. Furthermore, in the range $10\rho < x < 1000\rho$, the coefficient of proportionality gives $\rho_1 = 3.7 \pm 0.1\,\mu$m and $\rho_2 = 1.7 \pm 0.1\,\mu$m, respectively, the same values as for the stable tip. The mean parabolic shape occupied by the unstable dendrite is thus the parabola of the stable tip. A parabolic dependence of the unstable dendrite could be expected from Ivantsov's simple qualitative argument[21]: the tip of the dendrite moves at constant velocity so that its position is proportional to the time t; the growth of the lateral sides results from diffusive process and their position moves as $t^{1/2}$. However, there is more to our result. We find once again that *the selected solution for the mean of the fractal object and for the compact one are the same, here a parabola with the same ρ.*

B. *Unstable anisotropic viscous fingers and noise-reduced DLA clusters*

We wish now to extend our study to unstable patterns in viscous fingering and DLA when a preferential direction of growth exists. Viscous fingering usually creates isotropic patterns. Several global[13] and local[14–20,42] means have been used to create preferential directions of growth. These are well observed in the circular configuration but their selective effect can be measured quantitatively in linear cells[15,16] in which they result into anomalously narrow ST fingers with a parabolic tip scaled on $l_c \propto V^{-1/2}$. As for dendrites, the relation $\rho^2 V = $ const is satisfied and the finger exhibits dendritic side branches. In our experiments[35] we used linear cells, giving one of their plates a periodic structure in two perpendicular directions by stretching over it a thin nylon tulle cloth of

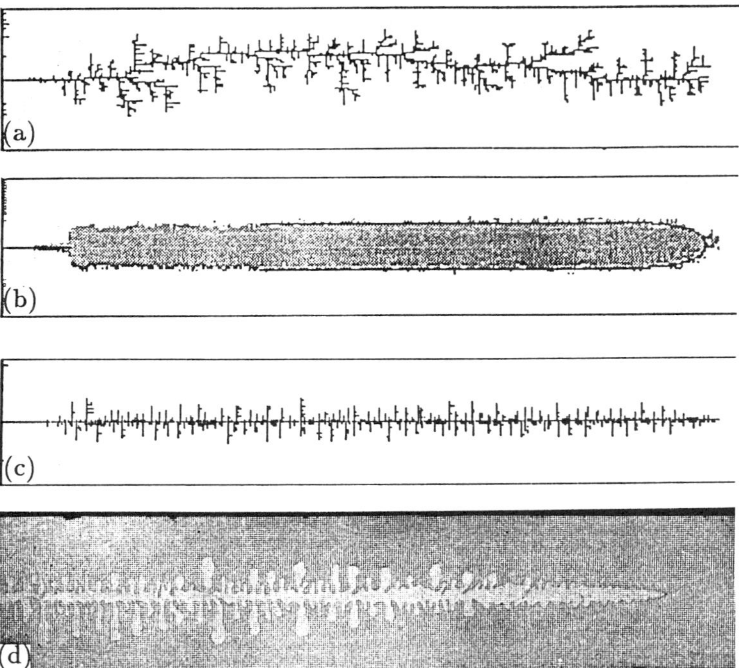

Fig. 9 Anisotropic patterns in linear cells. (a) A noise-reduced DLA cluster with $m = 3$ grown in a strip of width $W = 64$; the lattice is parallel to the strip. (b) The points of the strip where the occupancy rate is larger than $r(y_m)$ are represented in gray. This repartition was obtained from the analysis of 250 aggregates of the type shown in (a). The solid line is the shape of the analytical ST finger (Eq.(1)) of width $\lambda = 0.38$. (c) A noise-reduced DLA cluster width $m = 15$. (d) An unstable viscous finger in a cell of width $W = 10.5\,\mathrm{cm}$ and thickness $b = 0.5\,\mathrm{mm}$, when an axis of easy growth is along the channel.

thickness 0.2 mm. A typical unstable finger is shown in Fig. 9d when the direction of the weaving is along the cell axis. The observed patterns are similar to those of anisotropic DLA. We will here limit ourselves to the discussion of the numerical results for DLA clusters; they have exact counterparts in unstable ST growth.

In DLA growth, a transition from an apparently randomly ramified geometry to a more regular dendritic morphology can be observed when either introducing anisotropy at a microscopic level in the sticking rule[29,69,70] or enhancing the anisotropy of the underlying lattice by reducing the noise of the random walk[29,39,43–45]. Here we will concentrate on noise-reduced DLA clusters grown on a square lattice in a strip of width W between two reflecting walls, the lattice being parallel to the cell axis. Noise reduction is introduced into the DLA model in a natural way[60]: each lattice site adjacent to the cluster has a counter and a random walker only sticks onto it when a number m of arrivals has been reached. For $m = 1$ we recover the ordinary DLA model[27]. When m is increased, the axis of the square lattice become preferential directions of growth and the DLA clusters are progressively transformed into dendritic fractal patterns as illustrated in Figs 9a ($m = 3$) and 9c ($m = 15$) for a strip of width $W = 64$. We have reproduced our statistical analysis[35,36] for noise-reduced DLA clusters with different m values, grown in strips of width $W = 32, 64, 128$ and 256 lattice units. The occupancy rate $r(x, y)$ has different profiles for isotropic and anisotropic structures. In both cases, however, the longitudinal profiles are constant in the inactive region of interest. On the contrary, the transverse profiles for noise-reduced DLA clusters (Fig 10c and 10d) are narrower than for DLA clusters (Figs 10a and 10b), and there is a large region near the

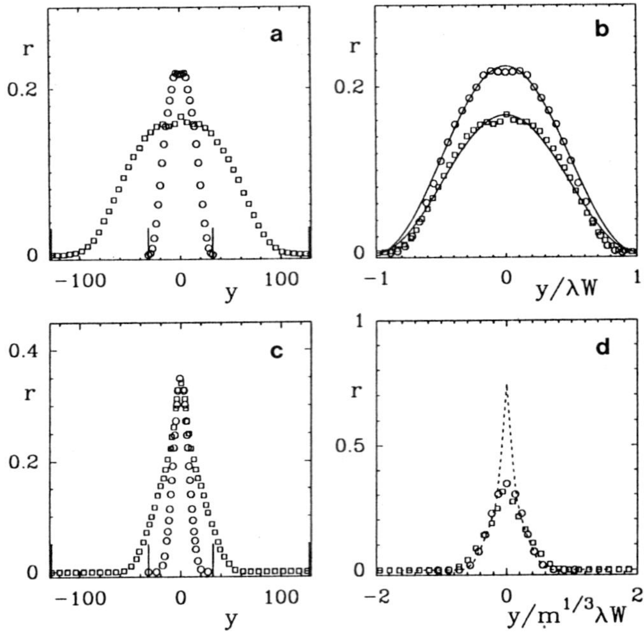

Fig. 10 Histograms of the mean tranverse occupancy across a strip of width $W = 64$ $(\circ)$ and 256 $(\square)$ for 250 noise-reduced DLA clusters. *Isotropic growth* $(m = 1)$: (a) r *vs* y; (b) r *vs* $y/\lambda W$ and its fit by $r(x,y) = r_{max} \cos^2(\pi y/W)$. *Anisotropic growth* $(m = 4)$: (c) r *vs* y; (d) r *vs* $y/m^{1/3}\lambda W$, where the dashed line corresponds to the histogram for $m = 15$, $W = 64$ (from Ref. 36).

walls where $r = 0$; in fact the larger m, the narrower the tranverse occupancy profile.

The mean width of the occupancy probability distribution is found by determining the points $y_m^\pm$ satisfying condition (2). For isotropic patterns we have checked that the values $y_m^\pm$ so obtained coincide with those deduced from a cut at midheight $(r_{max}/2)$. For anisotropic patterns they differ. In the case of dendrites, for instance, $r_{max} = 1$ near the axis; two transverse histograms at x_1 and x_2 only differ by their fall-off at large y (small r) because only the large side branches keep growing. Relation (2) accounts for this evolution while a cut at midheight does not. Condition (2) is thus the most general procedure to obtain the mean profile. Unlike the isotropic case, for a given m, the relative width of the region of large occupancy depends on the width of the cell. For instance for $m = 3$, we found $\lambda = 0.38 \pm 0.02$ for $W = 64$, $\lambda = 0.27 \pm 0.02$ for $W = 128$ and $\lambda = 0.21 \pm 0.02$ for $W = 256$. Fig. 9b shows all the points of the strip of width $W = 64$ where r is larger than $r(y_m)$. The limit of this region is well fitted by the ST analytical solution (1) of corresponding λ. As illustrated in Fig. 11a, when increasing W (decreasing $1/W \propto B^{1/2}$), λ actually decreases from $1/2$ (isotropic case) to 0 (anisotropic case) at a rate which is m-dependent. In Fig. 11b we show that the dependence of λ on m and W can be represented by the finite-size scaling form:

$$\lambda = (y_m^+ - y_m^-)/W = \Lambda\left(1/m^{3/2}W\right), \tag{5}$$

where the universal cross-over scaling function $\Lambda(u)$ behaves as

$$\Lambda(u) \propto \begin{cases} \frac{1}{2} & \text{if } u \gg u_c \text{ (isotropic growth)} \\ u^{1/2} & \text{if } u \ll u_c \text{ (anisotropic growth)} \end{cases} \tag{6}$$

where u_c is a finite critical value. This cross-over behavior indicates that *the anisotropy*

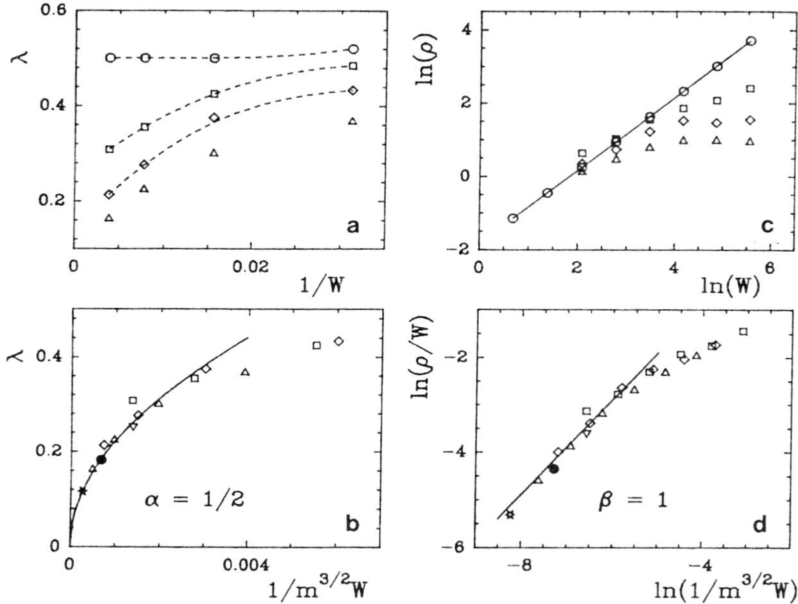

Fig. 11 Characteristics of the mean occupancy profile of 250 noise-reduced DLA clusters grown in a strip of width W. The symbols correspond to different values of $m = 1$ (o), 2 ($\square$), 3 ($\diamond$), 4 ($\triangle$), 5 ($\triangledown$), 8 ($\bullet$) and 15 ($\ast$). (a) λ vs $1/W$. (b) λ vs $1/m^{3/2}W$; the solid line corresponds to the scaling behavior $\Lambda(u) \propto u^\alpha$ with $\alpha = 1/2$ for $u = 1/m^{3/2}W \ll u_c$ (Eqs (5) and (6)). (c) $\ln\rho$ vs $\ln W$. (d) $\ln(\rho/W)$ vs $\ln(1/m^{3/2}W)$; the solid line corresponds to the scaling behavior $\mathcal{R}(u) \propto u^\beta$ with $\beta = 1$ for $u \ll u_c$ (Eqs. (8) and (9)) (from Ref. 36).

in the growth process can be controlled by either tuning the noise-reducing parameter m or changing the width W of the strip through the (effective anisotropy) scaling variable $m^{3/2}W$. Large effective anisotropy can thus be reached by either increasing m for a given cell width W or increasing W for a given value of m. This result is not so surprising since in Laplacian forming systems the motion of one of the boundaries (the interface) depends on the other boundary conditions fixed by the shape of the cell and its characteristic size[12].

Another characteristic feature of ST fingers[22] is the radius of curvature at the tip which can be deduced from the analytical expression (1):

$$\rho = \lambda^2 W/\pi(1 - \lambda).\tag{7}$$

In Fig. 11c we present the radius of curvature at the tip of the mean occupancy profile of 250 noise reduced DLA clusters versus W, for different values of m. For $m = 1$, the observation that $\lambda = 1/2$ implies that the radius of curvature scales linearly on W ($\rho = W/2\pi$), i.e., on the largest length scale of the system. For a given $m > 1$, ρ increases initialy with W up to some critical width (which depends on m) where it reaches a constant value. For instance, for $m = 3$ we found (in lattice units) for the three strips investigated $\rho_{64} = 4.6\pm0.5$, $\rho_{128} = 4.3\pm0.6$, $\rho_{256} = 4.7\pm0.8$, respectively. *For anisotropic patterns, the width of large occupation is thus selected by its radius of curvature at the tip*[35]. This is precisely the selection mechanism of stable anomalous viscous fingers[15,16] and parabolic needle crystals[63-67]. Experiments performed with different values of m

show that ρ satisfies the following finite-size scaling behavior:

$$\rho = W\,\mathcal{R}(1/m^{3/2}W)\,,\tag{8}$$

where the scaling function $\mathcal{R}(u)$ has the properties:

$$\mathcal{R}(u) \propto \begin{cases} \frac{1}{2\pi} & \text{if } u \gg u_c \text{ (isotropic growth)} \\ u & \text{if } u \ll u_c \text{ (anisotropic growth)} \end{cases}\tag{9}$$

For large effective anisotropy $(m^{3/2}W \gg 1/u_c)$, this implies that $\rho \propto m^{-3/2}$, independently of W. At large values of m, the cut-off value induced by the grid mesh size is reached and the aggregates, like dendrites, are compact at the tip. This morphological evolution manifests itself in the transverse occupation profile by a systematic increase of r_{max} towards 1 (Fig. 10d) due to the anisotropy-induced stabilization of the tip of the aggregate against tip-splitting. Although there is a priori no surface tension in the problem, the noise-reducing parameter m for DLA clusters grown on a square lattice[36] $(\rho \propto m^{-3/2})$ can thus be seen as playing a role analogous to surface tension anisotropy in crystal growth[63-67] $(\rho \sim \epsilon^{-7/8}$, where ϵ characterizes the presence of a four-fold anisotropy).

CROSS-OVER FROM DLA TO DENDRITIC FRACTALS

The next step in our demonstration is to show that the geometrical properties of anisotropic Laplacian fractal patterns are in turn amenable to a finite-size scaling description[36]. From the translational invariance of the occupancy probability distribution along the growth axis, one deduces readily that the mass has a one-dimensional component in the Ox direction which behaves like:

$$M_L(x) \propto x^{D_L} \quad \text{with } D_L = 1\,,\tag{10}$$

where D_L is the longitudinal partial dimension.

Obviously, the fractality of the patterns has to come from the direction perpendicular to the growth axis[39,71]. The computation of the area of the transverse occupancy profile (Fig. 10) as a function of the width of the strip gives the transverse partial dimension: $M_T(W) \propto W^{D_T}$. From the definition (Eq.(2)) of the region of large occupancy, the area of the transverse occupancy profile is equal to the area of a step function profile of width λW and height r_{max}:

$$M_T(W) = r_{max}\lambda W \propto W^{D_T}\,.\tag{11}$$

Let us point out that the data presented in Fig. 12a provide evidence for a general relationship between r_{max} and ρ:

$$r_{max} \propto \rho^{-1/3} \text{ for } \rho \gg 1\,.\tag{12}$$

A. DLA clusters

Large mass DLA clusters grown on a square lattice are known to display preferential directions of growth[8-11,72-74]. Let us mention that the values of W considered in this study are relatively small so that the DLA case $(m = 1)$ actually mimics isotropic growth. From the observation that $\lambda = 1/2$ or equivalently $\rho = W/2\pi$ (Eq.(7)), one deduces from Eq.(12) that:

$$r_{max} \propto W^{-1/3}\,.\tag{13}$$

Namely the transverse occupancy profile has a constant relative width $\lambda = 1/2$ but its

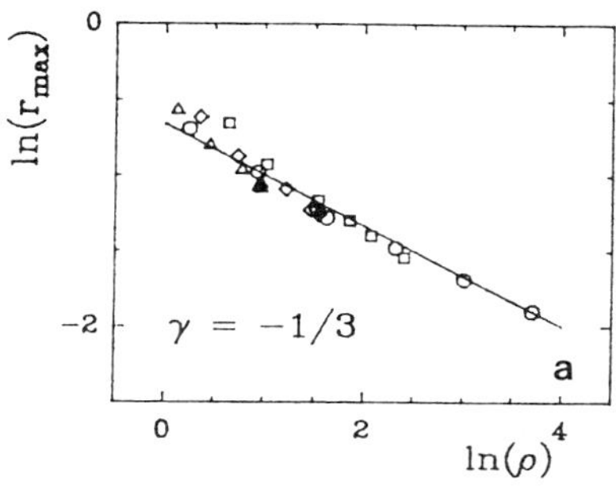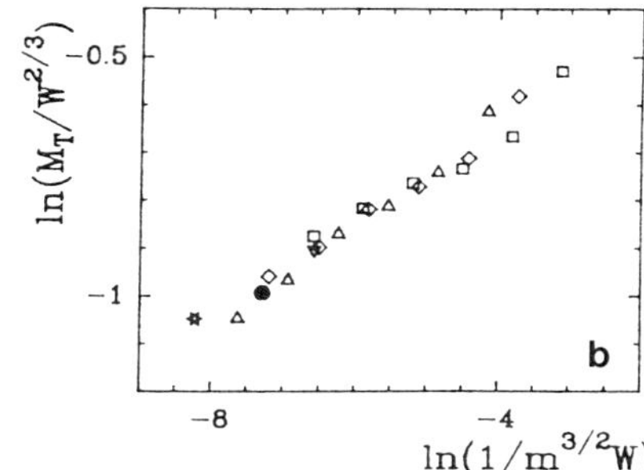

Fig. 12 Characteristics of the transverse occupancy profile of 250 noise-reduced DLA clusters grown in a strip of width W. The symbols are defined in Fig. 11. (a) $\ln r_{max}$ vs $\ln \rho$; the solid line provides evidence for the power-law behavior (12). (b) $\ln(M_T/W^{2/3})$ vs $\ln(1/m^{3/2}W)$ (from Ref. 36).

maximum r_{max} decreases with W according to Eq.(13) (Figs. 10a and 10b). Then form Eq.(11) one obtains:

$$M_T(W) \propto W^{D_T} \quad \text{with} \quad D_T = 2/3 \cdot \tag{14}$$

Combining Eqs.(10) and (14) one gets a value of the fractal dimension[36]:

$$D_F = 5/3 \quad (DLA\ clusters) , \tag{15}$$

which matches perfectly the mean-field prediction $D_F = (d^2 + 1)/(d + 1)$ for diffusion-limited aggregation in $d = 2$ dimensions[75,76].

B. *Noise-reduced DLA clusters*

Unlike DLA clusters, the transverse occupancy profile of noise-reduced DLA clusters has a relative width that depends on W while its height r_{max} is constant (Fig. 10c). As seen in Fig. 10d, for a given m, the transverse profiles obtained for different W actually collapse on a same curve when these profiles are represented at the scale of their own mean width λW. The way r_{max} increases towards 1, with increasing m, is given by Eq.(12) (Fig. 12a).

Along the line of our finite-size scaling analysis, one can reasonably propose the scaling ansatz[36]

$$M_T(m, W) = W^{2/3} \mathcal{M}(1/m^{3/2}W) , \tag{16}$$

where the scaling function $\mathcal{M}(u)$ behaves as

$$\mathcal{M}(u) \propto \begin{cases} \text{const} & \text{if} \quad u \gg u_c \quad \text{(isotropic growth)} \\ u^{\delta} & \text{if} \quad u \ll u_c \quad \text{(anisotropic growth)} \end{cases} \tag{17}$$

Eqs (16) and (17) account for a cross-over from $D_T = 2/3$ (isotropic growth, $D_F = 5/3$) to $D_T = 2/3 - \delta$ (dendritic growth, $D_F = 5/3 - \delta$) when increasing the effective anisotropy $m^{3/2}W$. The data collapse predicted by Eq. (16) is numerically tested in Fig. 12b. Although the existence of the scaling function $\mathcal{M}(u)$ is clearly revealed, the numerical results gathered in this figure do not provide an asymptotic estimate of the exponent δ (a value $\delta \sim 0.1$ can be extracted from the data; as discussed below this estimate is misleading). In order to investigate this power-law regime in the limit of large effective anisotropy, one thus needs to extend our statistical analysis to larger values of m and W; this is far beyond our computational potentiality.

Fortunately, there is an analytical alternative[36] which relies on the observation that $r_{max} \propto f(\rho)$ (Fig. 12a) no longer depends on W as soon as ρ does not depend on W. From the scaling relation (8) and the results in Fig. 11d, this happens for $Y = W/2\pi\rho \gg Y_c$ where $Y_c \simeq 3$. Let us remark that $Y = 1$ corresponds to isotropic growth, while the limit $Y \propto m^{3/2}W \rightarrow +\infty$ corresponds to the dendritic limit at large effective anisotropy. The fractal properties of noise-reduced DLA clusters are thus contained in the Y-dependence of λ via $M_T(Y) \propto \lambda(Y)Y \propto Y^{D_T}$ (Eq.(11)). Since $D_L = 1$ (Eq.(10)), one gets:

$$\lambda(Y) \propto Y^{D_F(Y)-2} \quad \text{with} \quad Y = W/2\pi\rho \, . \tag{18}$$

But $\lambda(Y)$ is known from the expression (7) of the radius of curvature at the tip of the analytical ST solutions:

$$\lambda(Y) = -1/4Y + (2/Y + 1/4Y^2)^{1/2}/2 \, . \tag{19}$$

This leads to the following differential equation for $D_F(Y)$ for $Y > Y_c$:

$$\frac{dD_F(Y)}{dY} + \frac{D_F(Y)}{Y \ln Y} = \frac{1}{Y \ln Y}\left[1 + \frac{4Y}{1 + 8Y - \sqrt{1 + 8Y}}\right] \, . \tag{20}$$

A straightforward integration yields the following analytical expression[36] for $D_F(Y)$

$$D_F(Y) = 1 + \frac{1}{\ln Y}\left[(D_F(Y_c) - 1)\ln Y_c + \ln\frac{\sqrt{1 + 8Y} - 1}{\sqrt{1 + 8Y_c} - 1}\right] \, , \tag{21}$$

where the constant of integration $D_F(Y_c)$ is a free parameter. In Fig. 13, we compare Eq.(21) with direct box-counting measurements of the fractal dimension of noise-reduced DLA clusters grown in a strip of width $W = 512$, with values of m ranging from 1 to 15. The box-counting procedure was applied in the range of scales $l_u \ll l \ll W$. With an adequate choice of $D_F(Y_c)$, the analytical expression (21) provides a very good fit of the experimental box-counting dimensions when measured in the inactive region of the patterns. Let us point out that this analytical prediction also accounts for the fractal dimension of the patterns in the active region of growth at the tip for values of $Y > Y_c$. For isotropic growth $(Y < Y_c)$ the fractal dimension in the active region $(D_F \simeq 1.57 \pm 0.02)$ is significantly smaller than in the inactive region as already noticed in previous works[77].

Considering the $Y \rightarrow +\infty$ limit in Eq.(21) leads to the following asymptotic value for the fractal dimension of dendritic fractals[36]:

$$\lim_{Y \rightarrow +\infty} D_F(Y) = 3/2 + O(1/\ln Y) \quad \textit{(dendritic fractals)} \, . \tag{22}$$

The very slow logarithmic convergence to the asymptotic value $D_F = 3/2$ makes this limit quite inaccessible in numerical experiments; this may explain the difficulties encountered in previous studies to approach this asymptotic prediction[8-11] (e.g. the dendrite in Fig. 7 has a fractal dimension $D_F = 1.58 \pm 0.03$). Finally let us mention that from the relation $D_F = 3/2 = 5/3 - \delta$, one predicts the value $\delta = 1/6$ for the power-law exponent of the scaling function $\mathcal{M}(u)$ in Eq.(17).

In summary, we have presented a scaling picture for the morphological transition between two geometrical phases for unstable patterns grown in Laplacian fields: isotropic DLA clusters $(D_F = 5/3)$ and anisotropic dendritic fractals $(D_F = 3/2)$. This finite-size scaling analysis based on simulations of the noise-reduced version of the DLA model is likely to generalize to various models of Laplacian growth as well as to experimental situations. Results in Refs. 78 and 79 indicate that for a given value of the anisotropy a cross-over is observed from the stable $\lambda = 0.5$ ST finger to the stable $\lambda = 0$ parabolic needle finger when W/l_c is increased. This cross-over is likely to be amenable to a similar scaling description. But different ranges of anisotropy and W/l_c values are actually explored in the case of stable and unstable patterns. A quantitative comparative study is currently in progress. This anisotropy-induced morphological transition is actually controlled by the effective anisotropy of the system. *For any non zero microscopic*

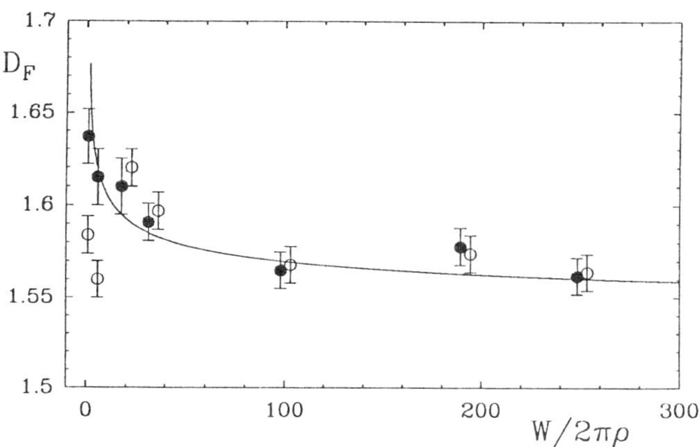

Fig. 13 Box-counting measurements of the fractal dimension D_F of noise-reduced DLA clusters grown in a strip of width $W = 512$, with values of m ranging from $m = 1$ to 15: inactive region ($\bullet$), active region ($\circ$). The solid line corresponds to the analytical prediction (21) with $D_F(Y_c = 3) = 1.64$ (from Ref. 36).

anisotropy, the fractal geometry of unstable patterns can cross-over into a dendritic fractal structure by simply enlarging the width of the cell. This observation confirms the crucial role played by the geometry of the system in which Laplacian growth phenomena take place[12].

CONCLUSION

In conclusion we have shown that unstable isotropic and anisotropic Saffman-Taylor fingers and DLA clusters, eventhough fractal, have a mean shape which reflects the shape of the smooth analytical solution. Moreover, the selection of a particular finger width, generally ascribed (for stable fingers) to the effect of surface tension, survives the instability of the structure. In the isotopic case, the mean solution is selected by the largest length scale of the system (the channel width). In the anisotropic case, it is selected by the radius of curvature at the tip of the mean profile. This radius ρ is proportional to the smallest length scale (l_{MS} for dendrites, l_c for viscous fingers and l_u for DLA clusters). The proportionality coefficient is a decreasing function of the anisotropy (ϵ for dendrites, m for noise-reduced DLA). When it is small, ρ is of the order of l_{MS}, l_c or l_u, and a stable region exists at the tip; the fractal structure only results from the growth away from the tip. When it is large, the tip itself is unstable; the selected radius of curvature is not observed on a given realization, but it still exists as a statistical property. The morphological transition from DLA to dendritic fractals is actually governed by the effective anisotropy of the system which can be tuned independently by changing the characteristic size of the system. We have applied a finite-size scaling analysis to account for the cross-over from isotropic DLA clusters ($D_F = 5/3$) to dendritic fractals ($D_F = 3/2$). An analytical expression for the fractal dimension D_F of anisotropic DLA clusters has been derived which suggests that, to some extent, the fractal properties of unstable anisotropic Laplacian aggregates are already contained in the stable anomalous Saffman-Taylor fingers.

ACKNOWLEDGMENTS

The material reported here has relied on collaborative works with G. Grasseau, V. Hakim and J. Maurer. We benefitted from useful discussions with M. Ben Amar. This work was supported by the Centre National des Etudes Spatiales under contract (N°90/CNES/215) and the Direction des Recherches Etudes et Techniques (DRET) under contract (N°89/196). Two of us (A.A and F.A.) would like to thank the hospitality of the Center for Nonlinear Dynamics at the University of Texas (Austin) where part of this work was done.

REFERENCES

1. J. S. Langer, Rev. Mod. Phys. 52:1 (1980).
2. S. C. Huang and M.E. Glicksman, Acta Metall. 29:701 (1981).
3. D. Bensimon, L. P. Kadanoff, S. Liang, B. I. Shraiman and Chao Tang, Rev. Mod. Phys. 58:977 (1986).
4. J. S. Langer, in "Chance and Matter", J. Souletie, J. Vanimenus and R. Stora, eds, North-Holland, Amsterdam (1987).
5. D. A. Kessler, J. Koplik and H. Levine, Adv. Phys. 37:255 (1988).
6. P. Pelcé, "Dynamics of Curved Fronts", Academic Press, Orlando (1988).
7. J. S. Langer, Science 243:1150 (1989).
8. W. Guttinger and D. Dangelmayr, eds, "The Physics of Structure Formation", Springer-Verlag, Berlin (1987).
9. H. E. Stanley and N. Ostrowsky, eds, "Random Fluctuations and Pattern Growth", Kluwer Academic Publisher, Dordrecht (1988).
10. J. Feder, "Fractals", Pergamon, New-York (1988).
11. T. Vicsek, "Fractal Growth Phenomena", World Scientific, Singapore (1989).
12. Y. Couder, to appear in the Proceedings of the Summer School "Chaos, Order and Patterns" held in Como (1990).
13. E. Ben Jacob, R. Godbey, N. D. Goldenfeld, J. Koplik, H. Levine, T. Muller and L. M. Sander, Phys. Rev. Lett. 55:1315 (1985).
14. Y. Couder, O. Cardoso, D. Dupuy, P. Tavernier and W. Thom, Europhys. Lett. 2:437 (1986).
15. Y. Couder, N. Gerard and M. Rabaud, Phys. Rev. A 34:5175 (1986); A 37:935 (1988).
16. G. Zocchi, B. E. Shaw, A. Libchaber and L. P. Kadanoff, Phys. Rev. A 36: 1894 (1987).
17. M. Matsushita and H. Yamada, J. Cryst. Growth 99:161 (1990).
18. D. C. Hong and J. S. Langer, Phys. Rev. Lett. 56:2032 (1986); Phys. Rev. A 36:2325 (1987).
19. R. Combescot and T. Dombre, Phys. Rev. A 39:3525 (1989).
20. H. Thomé, R. Combescot and Y. Couder, Phys. Rev. A 41:5739 (1990).
21. G. P. Ivantsov, Dokl. Acad. Nauk SSSR 58:567 (1947).
22. P. G. Saffman and G. I. Taylor, Proc. R. Soc. London, Ser. A 245:312 (1958).
23. H. Thomé, M. Rabaud, V. Hakim and Y. Couder, Phys Fluids A 1: 224 (1989).
24. S. N. Rauseo, P. D. Barnes and J. V. Maher, Phys. Rev. A 35:1245 (1987).
25. A. R. Kopf-Sill and G. M. Homsy, Phys. Fluids 31:242 (1988).
26. Y. Couder, in Ref. 9, p. 75.
27. T. Witten and L. M. Sander, Phys. Rev. Lett. 47:1400 (1981); Phys. Rev. B 27:5686 (1983).
28. P. Meakin, in "Phase Transition and Critical Phenomena", C. Domb and J.L. Lebowitz, eds, Academic Press, Vol.12, Orlando (1988).
29. J. Nittmann and H. E. Stanley, Nature (London) 321:663 (1986).
30. F. Argoul, A. Arneodo, G. Grasseau and H.L. Swinney, Phys. Rev. Lett. 61:2558 (1988).
31. F. Argoul, A. Arneodo, J. Elezgaray, G. Grasseau and R. Murenzi, Phys. Lett. A 135:327 (1989); Phys. Rev. A 41:5537 (1990).
32. B. B. Mandelbrot, "The Fractal Geometry of Nature", Freeman, San Francisco (1982).
33. A. Arneodo, F. Argoul, J. Elezgaray and G. Grasseau, in "Nonlinear Dynamics", G. Turchetti, ed., World Scientific, Singapore (1989) p. 130.
34. A. Arneodo, Y. Couder, G. Grasseau, V. Hakim and M. Rabaud, Phys. Rev. Lett. 63:984 (1989); in "Nonlinear Evolution of Spatio-Temporal Structures in Dissipative Continuous Systems", F.H. Busse and L. Kramer, eds, Plenum (1990) p. 481.
35. Y. Couder, F. Argoul, A. Arneodo, J. Maurer and M. Rabaud, Phys. Rev A 42:3499 (1990).

36. A. Arneodo, F. Argoul, Y. Couder and M. Rabaud, Anisotropic Laplacian Growths: from diffusion-limited aggregates to dendritic fractals, preprint (December 1990) to be published.
37. L. M. Sander, P. Ramanlal and E. Ben Jacob, Phys. Rev. A 32:3160 (1985).
38. E. Meiburg and G. M. Homsy, Phys. Fluids 31:429 (1988).
39. J. P. Eckmann, P. Meakin, I. Procaccia and R. Zeitak, Phys. Rev. A 39:3185 (1989); Phys. Rev. Lett. 65:52 (1990).
40. Y. Sawada, A. Dougherty and J. P. Gollub, Phys. Rev. Lett. 56:1260 (1986).
41. D. Grier, E. Ben-Jacob, R. Clarke and L. M. Sander, Phys. Rev. Lett. 56:1264 (1986).
42. V. Horvath, T. Vicsek and J. Kertész, Phys. Rev. A 35:2353 (1987).
43. J. Kertész and T. Vicsek, J. Phys. A 19:L 257 (1986).
44. P. Meakin, Phys. Rev. A 36:332 (1987).
45. P. Meakin, J. Kertész and T. Vicsek, J. Phys. A 21:1271 (1988).
46. R. L. Chuoke, P. Van Meurs and C. Van der Pol, Tr. A.I.M.E. 216:188 (1959).
47. L. Paterson, Phys. Rev. Lett. 52:1621 (1984).
48. J. W. Mc Lean and P. G. Saffman, J. Fluid. Mech. 102:445 (1981).
49. J. M. Vanden-Broeck, Phys. of Fluids 26:2033 (1983).
50. R. Combescot, T. Dombre, V. Hakim, Y. Pomeau and A. Pumir, Phys. Rev. Lett. 56:2036 (1986); Phys. Rev. A 37:1270 (1988).
51. B. Shraiman, Phys. Rev. Lett. 56:2028 (1986).
52. D. C. Hong and J. Langer, Phys. Rev. Lett. 56:2032 (1986).
53. C. Evertz, Phys. Rev. A 41:1830 (1990).
54. G. Grasseau, Ph. D. Thesis, Bordeaux (1989).
55. H. Kondo, M. Matsushita and S. Ohnishi, "Science on Form": Proceedings of the first International Symposium for Science on Form, S. Ishizaka, ed., K.T.K. Scientific Publishers, Tokyo (1986).
56. I. Procaccia and R. Zeitak, Phys. Rev. Lett. 60:2511 (1988).
57. G. M. Dimino and J. H. Kaufman, Phys. Rev. Lett. 62:2277 (1989).
58. M. Ben Amar, preprint (1990).
59. E. A. Brener, D. A. Kessler, H. Levine and W. J. Rappel, Europhys. Lett. 13:161 (1990).
60. C. Tang, Phys. Rev. A 31:1977 (1985).
61. A. Dougherty and J. P. Gollub, Phys. Rev. A 38:3043 (1988).
62. J. Maurer, P. Bouissou, B. Perrin and P. Tabeling, Europhys. Lett. 6:67 (1989).
63. D. A. Kessler, J. Koplik and H. Levine, Phys. Rev. A 33:3352 (1986).
64. M. Ben Amar and Y. Pomeau, Europhys. Lett. 2:307 (1986).
65. A. Barbieri, D. C. Hong and J. S. Langer, Phys. Rev. A 35:1802 (1987).
66. B. Caroli, C. Caroli, C. Misbah and B. Roulet, J. Phys. A 48:547 (1987).
67. M. Ben Amar, Phys. Rev. A 41:2080 (1990).
68. W. W. Mullins and R. F. Sekerka, J. Appl. Phys. 35:444 (1964).
69. R. C. Ball, R. M. Brady, G. Rossi and B. R. Thompson, Phys. Rev. Lett. 55:1406 (1985).
70. M. Matsushita and H. Kondo, J. Phys. Soc. Jpn 55:2483 (1986).
71. L. Pietronero, A. Erzan and C. Evertz, Phys. Rev. Lett. 61:861 (1988); Physica A 151:207 (1988).
72. R. C. Ball and R. M. Brady, J. Phys. A 18:L 809 (1985).
73. P. Meakin, Phys. Rev. A 33:3371 (1986).
74. P. Meakin, R. C. Ball, P. Ramanlal and L.M. Sander, Phys. Rev. A 35:5233 (1987).
75. M. Tokuyama and K. Kawasaki, Phys. Lett. A 100:337 (1984).
76. K. Honda, H. Toyoki and M. Matsushita, J. Phys. Soc. Jpn 55:707 (1986).
77. J. Feder, E. L. Hinrichsen, K. J. Maloy and T. Jossang, Physica D 38:104 (1989).
78. D. Kessler, J. Koplik and H. Levine, Phys. Rev. A 34:4980 (1986).
79. M. Ben Amar, private communication.

COMPETING STRUCTURES IN DLA AND VISCOUS FINGERING

V. Hakim

Laboratoire de Physique Statistique
24, rue Lhomond 75231 Paris, France

ABSTRACT

We summarize some recent work aiming at the characterization of coherent structures in unstable viscous fingering and diffusion-limited aggregation. A simplified needle model has been found which gives a good description of the probability field for n-fold symmetric anisotropic growth for small n. It is pointed out that it also explains why the aggregates have n equal branches for n<6 and why this can no longer be the case for n>6. Isotropic growth has been investigated in confined and radial geometries. In confined geometries, the mean shape of aggregates and unstable viscous fingers is well described by the known classical smooth shapes. This is unfortunately not explained by a simple mean-field theory and the basic difficulties are summarized. In order to describe the growing structures in the radial geometry, the overlap between two growth sites is defined. This gives a quantitative measure of the notion that two growth sites belong to the same branch. For isotropic growth in the radial geometry, the average overlap is found to decrease with the number N of particles. This indicates that the number of branches increases with N. Moreover the overlap function and its fluctuations obey a scaling law suggesting that a mean branch shape exists in the limit $N \rightarrow \infty$ but that it fluctuates from aggregate to aggregate.

1. INTRODUCTION

Simple dynamical rules on a "microscopic" scale can create spatially complex structures on wider scales. This is easily seen on pictures of fluid motion [1] but a detailed understanding is more difficult to gain. Viscous fingering in two-fluid hydrodynamics [2] and Witten and Sander's diffusion-limited-aggregation [3] are simple prototypic examples of the phenomenon and, as such, have attracted the attention of many people during the last few years. In both cases, complex interfaces are created two examples of which are shown on fig. 1. One approach to understanding these patterns is to assume that they are spherically symmetric in an averaged sense. This is for example what has been done in ref. [4] in a mean field approach. A spherically symmetric solution to mean-field equations has been obtained. However, it has then been realized [5] that this radially symmetric solution is unstable for the same reason that the primary interface is (i.e. outward pointing bumps grow faster). This suggests that structures may be created at all scales and that perhaps, it would be better to try to understand the complex patterns in terms of competing structures than as a spherically symmetric density decreasing with radial distance. The first step in this program is to identify the competing structures. In the present paper some recent work toward that aim will be summarized. One important feature of the two phenomena is that the anisotropy of the dynamical process on the "microscopic" scale plays a very important role even on the large scale structures. We will show that three cases of increasing difficulty should be distinguished:

- the anisotropy is less than six fold symmetric
- the anisotropy is more than six fold symmetric
- there is no anisotropy.

They will will be discussed in turn.

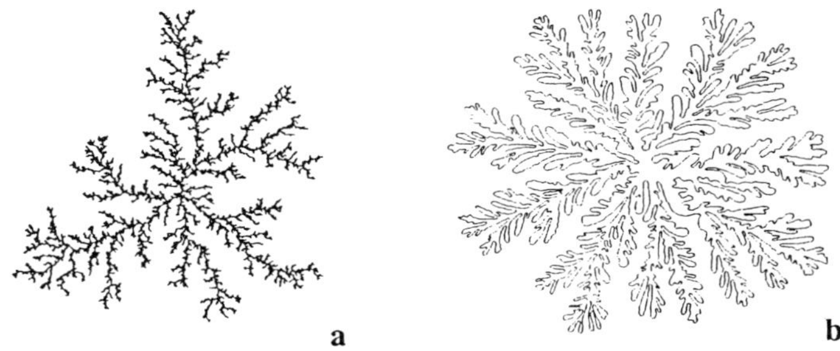

Figure 1 a) an aggregate of 10^4 particles grown using the algorithms described in ref.(17).b) an unstable viscous finger in radial geometry (courtesy of Y. Couder)

2. DIFFUSION CONTROLLED DYNAMICS : DLA AND VISCOUS FINGERING

For the two processes, we are considering the motion of the interface is controlled by a Laplacian field outside the interface. For DLA, the aggregate grows by successive aggregation of random walkers which are released far from the aggregate. Therefore the growth is controlled by the probability field for the random walkers which satisfies the stationary diffusion equation

$$\nabla^2 P = 0 \tag{1}$$

outside the aggregate with P = 0 on the aggregate since particles are absorbed by the aggregate. Viscous fingering appears when air is injected into oil in an effective 2D geometry (two closely-spaced glass plates). The Laplacian field which controls the motion of the interface is the pressure P. It obeys equation (1) due to the incompressibility of the oil. The effective boundary condition on the interface is well approximated by a 2D Laplace's law (see ref. (2c) for a more careful discussion)

$$P = -\frac{T}{R} \tag{2}$$

where T is surface tension. The motion of the interface is determined by the conservation of oil and is thus proportional to the pressure gradient (in this viscosity dominated 2D geometry the pressure is proportional to the velocity potential).

$$v_n \, \alpha - \mathbf{n}.\nabla P \tag{3}$$

where v_n is the normal velocity of the interface and $\mathbf{n}$ the local normal to the interface. The growth of the aggregate in D.L.A. is given by a stochastic version of (3). The effect of surface tension is to create a characteristic small scale in viscous fingering below which the interface is flat. For DLA, this characteristic small scale is the radius of the sticking particle. This deep analogy between the two processes was first pointed out by Paterson [6] and explained nicely in [7].

3. ANISOTROPIC GROWTH

Large scale structures are not immediately apparent for a DLA cluster grown with an isotropic sticking rule (see fig. 1). On the contrary for low anisotropy sticking rule large scale structures are clearly visible. For example, an anisotropic cluster grown using the "antenna" method [8] is shown on fig. 2a. The particles diffuse as in the isotropic case but when they contact the cluster they are rotated toward fixed directions depending on the contact sector. In the case shown where a three-fold symmetric anisotropy was introduced, three main branches are apparent. It is therefore tempting to identify them with large scale competing structures and to approximate them simply by outgoing needles (fig.2b). The field satisfies then Laplace's equation everywhere outside the needles and is equal to zero on the needles. Several authors [8,9] have shown that this gives a rather accurate zeroth-order description of the probability field. If this picture is taken seriously, then two related simple questions can be asked [10] :

i) It is seen on fig. 2 that the three branches have approximately the same length. How is this compatible with Mullins Sekerka's instability which predicts than a longer branch will grow faster than a shorter one ?

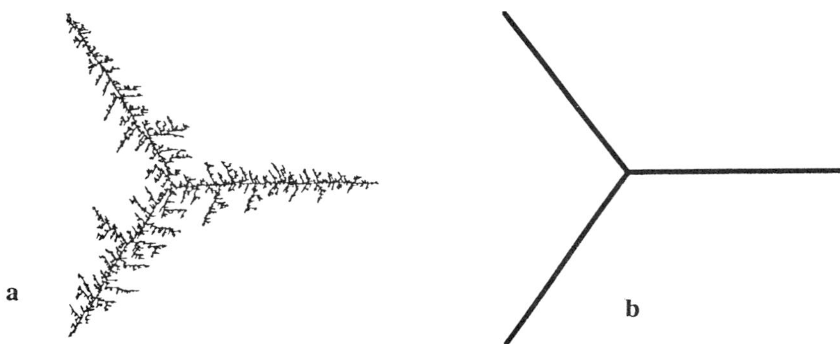

Figure 2a) An aggregate of 10^4 particles grown with the antenna method (three-fold symmetric anisotropy). b) A simple model of the aggregate large-scale structure.

ii) A three fold symmetric anisotropy gives three main branches of equal length. Will an n-fold symmetric anisotropy give n-branches of equal length ?

The answer to i) is easy when one has realized that the symmetric visual aspect of fig. 2 is really a statement about length ratio. If l_1 and l_2 are the lengths of two branches fig. 2 shows that $\frac{l_1}{l_2} \sim 1$. This is of course not in contradiction with Mullins Sekerka's instability which shows that if l_1 is larger than l_2, l_1 grows more rapidly than l_2. However, the ratio $\frac{l_1}{l_2}$ can decrease or increase. The answer to ii) can then only be obtained by a detailed computation of the respective growth rates. In order to model it, we transform the static needle description into a dynamic one and consider a model where n needles of different lengths grow along fixed directions regularly angularly spaced by $2\pi/n$. The growth velocity of a needle is proportional to the field at its tip (see fig. 3). Then, a linear stability analysis show that the fixed point with n equal branches is stable only if $n \leq 5$. For n=6, long branches alternating with short branches are possible, the asymptotic ratio between the two lengths is arbitrary and depends on the initial condition. For n>6 the symmetric fixed point is unstable and some needles disappear in this simplified model. In this calculation we found it convenient to compute exactly the field via a conformal transformation which maps the exterior of the unit circle on the complex plane cut along the needles. While the model is extremely simplified it explains quite well that more than 6 symmetric branches are never observed in DLA simulation. A detailed investigation of the seven-fold anisotropy case may also be worth pursuing theoretically although experiments seem difficult to perform.

4. ISOTROPIC GROWTH

When the growth rules are isotropic, large scale structures are more difficult to determine and two approaches have been followed:

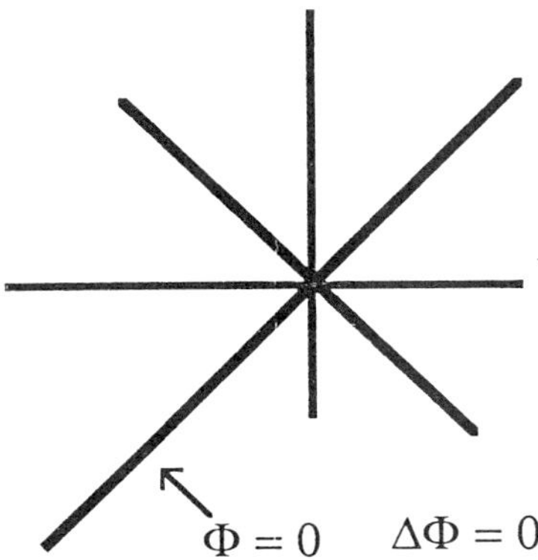

Figure 3. A simple dynamic model for the growth of the main branches of an anisotropic aggregate. The potential is equal to zero on the needles and obey Laplace equation. The velocity of a needle tip is proportionnal to the field E there (more precisely to the limit E(r)√r when r→rtip)

- The first is to look for them in confined geometries. It is then possible to compare the average shape of the aggregates to classical smooth structures which are known in these geometries.

- In the radial case, it is difficult to find directly the shape of an average large scale structure. A quantitative measure of distance between two growth sites has been introduced in order to give a quantitative meaning to the qualitative notion that "two sites belong to the same structure".Then a number of questions concerning large clusters can be asked and answered. Is the number of structures well defined asymptotically or does it increase with the number of particles ? Is there a unique mean structure or does it fluctuate from sample to sample ?

a) Confined geometries

Viscous fingering experiments can be performed not only in radial geometry but also in cells of various shapes. For example Saffman and Taylor used a long linear cell in their classical experiment[2]. When the speed of injection of air is not too high, they discovered that a smooth finger of air moves through the cell at constant velocity. The experiment can also be done in sector-shaped cells and then smooth self-dilating fingers are observed [11]. It was suggested [12] that in these geometries the mean shape of fast unstable viscous fingers and DLA aggregates can reflect the smooth classical shapes. This has now been observed [13] as shown in figure 4 where many aggregates grown in a strip with reflecting walls have been superimposed and the mean occupancy of each point of the strip computed. Points where the mean occupancy is higher than half the maximal mean occupancy fill a domain whose boundary is accurately fitted by the smooth classical Saffman-Taylor finger. The same result has also been obtained in sector shaped cells and has now been extended to more complicated situations [14].

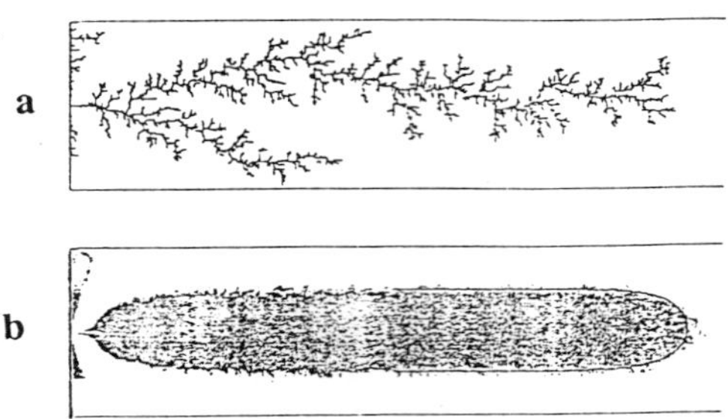

Figure 4 a) A DLA cluster of 6000 particles grown in a strip of width 128 b) 510 such aggregates have been superimposed. The points where the mean occupancy is higher than half the maximal mean occupancy are represented in gray. The continuous line is the selected analytical Saffman-Taylor shape.

It seems natural to try to understand this finding through mean field equations where a binary occupancy is replaced by a continuous density. Unfortunately the study of Witten and Sander's mean field equation [4] in a strip geometry led to the conclusion that they do not explain what has been observed [15]. Firstly in a 1D geometry (a planar front moving in the x-direction) they led to a singularity in finite time. This means that the leading part of the front starting from x=1 reaches x=+∞ in a finite time. The density decreases like x^{-2} from x=1 to the leading part of the front. Unfortunately, a strip geometry does not seem to cure this pathology. In a Fourier decomposition of the density along the transverse y direction the smallest k-component is the fastest and for sufficiently large x is coupled to a transversally uniform diffusion field as in the 1D case. Therefore, the conclusion is that Witten and Sander's mean field equation do not produce a steady state or a front spreading like $\sqrt{t}$ in a strip geometry. It seems that neglecting completely the correlation between the density and the diffusing field is too crude an approximation. The collection of aggregates then moves according to a mean value of the diffusion field and leading aggregates move more rapidly than those in the rear . This stretches the mean front more and more. A better mean field approximation with more fields may still exists but it remains to be found.

b) Radial geometry

Although the large scale structure in strips or sector-shaped cells are not yet well understood, computer experiments and simulation demonstrate their existence as described above. It is then a possibility that in a radial geometry the radial symmetry is spontaneously broken and that the pattern consists of a certain number of structures ("branches") competing with each other [12]. The question then arises of characterizing these structures. Is their number well defined as the number of particles tends to infinity? What is the mean shape and does it fluctuate from sample to sample ?

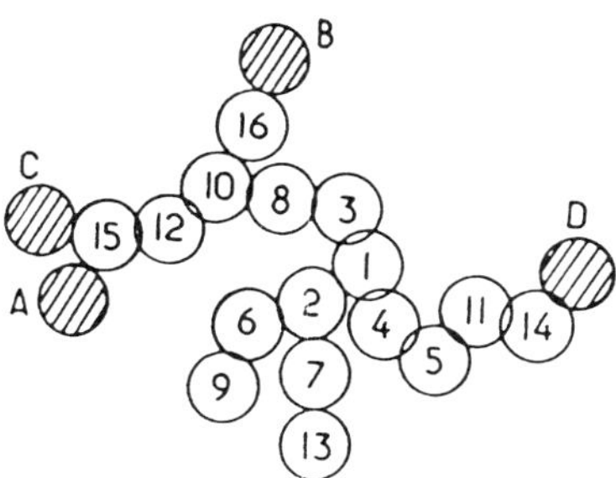

Figure 5 . An aggregate with its particles (open circles) numbered according to their order of arrival. The overlaps of the two pairs (A,B) and (C,D) of test particles (hatched circles) are respectively $q_{AB}=10/15$ and $q_{CD}=1/15$

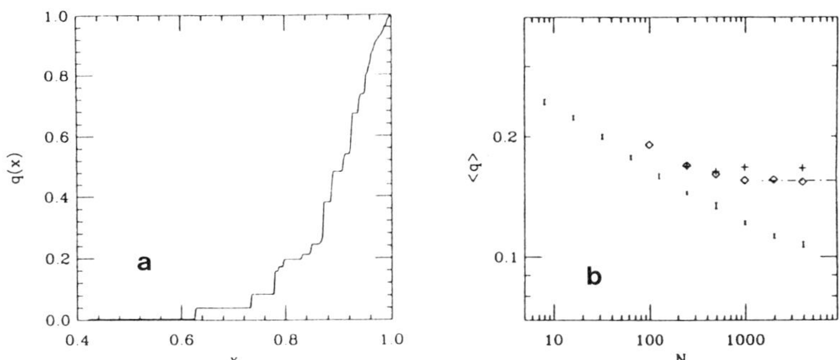

Figure 6 . a) The overlap function q(x) of one aggregate of 500 particles computed using 10^4 test pairs b) Average overlap versus mass N for aggregate with biaxial anisotropy (+), three-fold symmetric anisotropy (squares) and isotropic sticking rule (error bars). The dash-dotted line is the analytic prediction of the three-needle model .

Taking some inspiration from the physics of spin glasses [16] a natural notion of distances and of "overlap" between two growth sites can be introduced [17]. The quantity so defined gives a quantitative meaning to the qualitative notion that two sites belong to the same branch and can be used to answer the questions asked above.

Let us consider an aggregate where all particles are numbered by their rank of arrival. Now, let two test particles diffuse independently until they come into contact with the aggregate with particle A and B. If h is the rank of the closest common ancestor of A and B (see fig. 5) we define the overlap q_{AB} of the pair A, B as $q_{AB} = \frac{h}{N}$ where N is the total number of particles in the aggregate. When the results of throwing many test pairs are collected, the histogram of values of q, or the probability distribution of q, $P_a(q)$ is obtained for the particular aggregate that is being tested. It is convenient to study the integrated distribution $x_a(q) = \int_0^q P_a(q) \, dq'$ or the inverse function $q_a(x)$ which gives the value of q such that $q_{AB} \leq q$ for a fraction x of the test pairs. The overlap function $q_a(x)$ for one aggregate is shown in figure 6 and one sees that more than sixty percent of the overlap are almost zero.

 One can then compute the mean value of q_a for an aggregate and its average $<q>$ over many samples. If the number of branches is finite when $N \rightarrow \infty$ then one expects that $<q>$ tends to a constant in this limit. This is indeed what is observed for the growth with three-fold symmetric anisotropy (see fig.6b) and the constant can be accurately computed using the simplified needle model described in section 3. On the contrary, in the isotropic case the mean value of q decreases slowly with N (fig. 6b), like $\frac{A}{\log N}$. The data for the whole averaged function $P(q) = <P_a(q)>$ and for the fluctuation around the average can be interpreted with a simple model. We assume that the aggregate consists of a certain number of branches. If two test particles fall on different branches then their overlap is almost zero. If they fall on the same branches then they have a certain probability distribution $P^*(q)$ of having an overlap q ($P^*(q)$ is the average of contributions of different branches in the aggregate).Thus

$$P(q) = \delta(q) (1 - \varepsilon) + \varepsilon P^*(q)$$

where ε is the probability for the two test particles to fall on the same branch and depends on the number of particles in the aggregate.This model makes two simple predictions:

i) the average value of $<q>$ and $<q^2>$ should scale in the same way with N since:

$$<q> = \varepsilon(N) \int_0^1 q \, P^*(q) \, , \quad <q^2> = \varepsilon(N) \int_0^1 q^2 \, P^*(q) \, .$$

A test of this relation is shown in figure 7a.

ii) Since x(q) is given by

$$x(q) \; = \; 1 - \varepsilon + \varepsilon \int_0^q P*(q) \; = \; 1 - \varepsilon \int_q^1 P*(q)$$

then the function $\dfrac{1 - x(q)}{<q^2>} = F(q)$ should be a function of q independent of N. This is well verified as shown in figure 7b where it should also be noticed that its fluctuations do not tend to zero when the number of samples is increased.

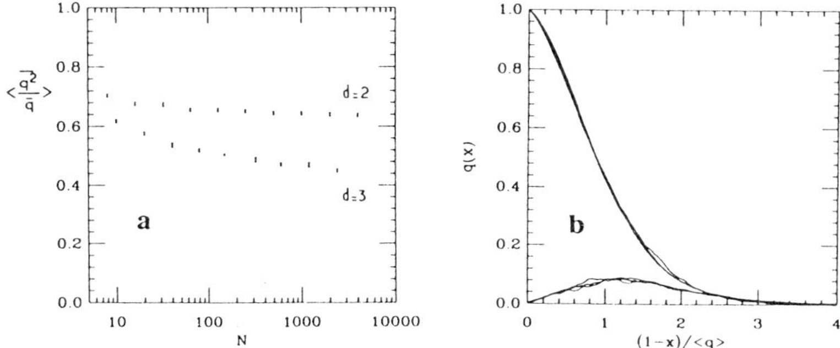

Figure 7. a) Mass dependence of $<q^2/q>$ b) The overlap function and its fluctuation (lower curve) versus the scaling variable for 2D-aggregates of mass N=250 (400 samples), 1000 (200 samples) and 4000 (100 samples).

The function F(q) can be analytically computed in two cases. It can be computed for the needle model of section 3 and the result is an accurate description of the numerically computed F(q) for anisotropic growth (fig.8a). On the Cayley tree DLA can be solved exactly [18]. Using the techniques developed in ref. (18) P(q) can be computed and one has the simple result

$$F(q) \; = \; \frac{1}{q} - 1$$

This can be compared with the numerical curves in dimension 2 and 3 (fig.8b).

The conclusion of this quantitative study seems to be that the number of branches or competing large scale structures increases slowly with the number of particles for isotropic DLA growth. Moreover the mean structure fluctuate from aggregate to aggregate. It is a challenge to understand this more precisely.

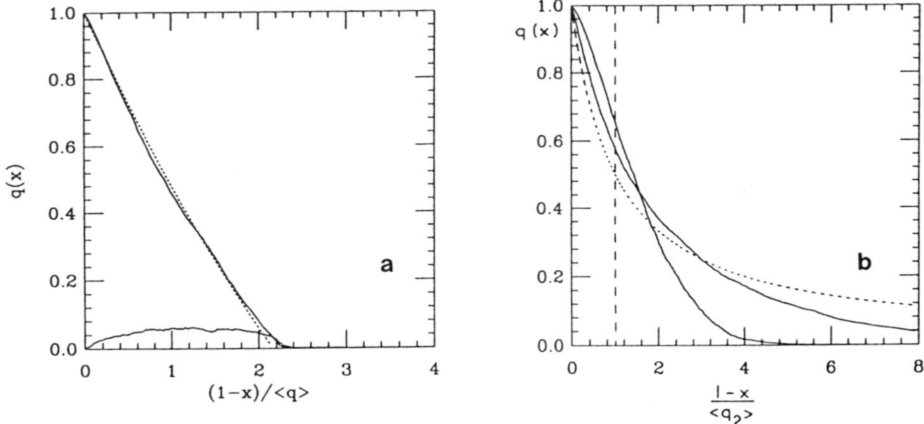

Figure 8 . a) Average overlap function for 50 anisotropic aggregates of 1000 particles with three-fold symmetric antennas. The dotted line is the analytic prediction of the needle model. b) Overlap function for dimensions 1 (vertical dashed line), 2, 3 (both full lines) and analytical result for the Cayley tree in the limit of infinite branching ratio.

5) CONCLUSION

It is clear that much remain to be done before a good understanding of these ramified aggregates is obtained. For low order anisotropic growth, the simplified needle model gives a good description but it remains to turn it into a full theory (for a recent attempt see [19]). Anisotropic growth with $n \geq 7$ merits some more investigation. It is quite surprising that the average shapes in confined geometries mimic accurately the selected classical shapes. A sound mean field theory of this phenomenon would be very interesting to find and may well give a different perspective on the selection of the classical shapes. In radial geometry, the number of branches increases and an average shape seems to exist. It is possible to understand this in relation with the sector shaped geometries ? Let us hope, that nice solutions and perhaps new surprises will be discovered in the near future.

Acknowledgments: It is a pleasure to thank M. Benamar, P. Pelcé and P. Tabeling for inviting me to this interesting and stimulating conference. I am also very grateful to A. Arnéodo, Y. Couder, B. Derrida, M. Rabaud and J. Vannimenus with whom this work has been done.

REFERENCES

(1) M. Van Dyke, *"An album of fluid motion"* (Parabolic, Stanford CA, 1982).

(2) P. Pelcé, *"Dynamics of curved fronts"* (Academic, Orlando, Fl, 1988) b) L. Kadanoff (these Proceedings) c) S. Tanveer (these Proceedings).

(3) T. Witten and L.M. Sander, Phys. Rev. Lett. <u>47</u>, 1400 (1981) see also L. Pietronero (these Proceedings).

(4) R. Ball, M. Nauenberg and T. Witten, Phys. Rev. <u>A29</u>, 2017 (1984).

(5) M. Nauenberg and L.M. Sander, Physica <u>123 A</u>, 360 (1984).

(6) L. Paterson, Phys. Rev. Lett. <u>52</u>, 1621 (1984).

(7) L. P. Kadanoff, J.Stat. Phys. <u>39</u>, 267 (1985).

(8) J.P. Eckmann, P. Meakin, I. Procaccia and R. Zeitak, Phys. Rev. <u>A39</u> , 3185 (1989).

(9) H.G.E. Hentschel, F. Family, Faraday Disc. Chem. Soc. <u>83</u>, 139 (1987). K. Kassner, F. Family, Phys. Rev. <u>A39</u> , 4797 (1989).

(10) B. Derrida and V. Hakim (1990, unpublished).

(11) H. Thomé, M. Rabaud, V. Hakim and Y. Couder, Phys. Fluids <u>A1</u> , 224 (1989).

(12) V. Hakim, Phys. Rep. <u>184</u> , 259 (1989).

(13) A. Arnéodo, Y. Couder, G. Grasseau, V. Hakim and M. Rabaud, Phys. Rev. Lett. <u>63</u> , 984 (1989).

(14) Y. Couder, F. Argoul, A. Arnéodo, J. Maurer and M. Rabaud, Phys. Rev. <u>A42,</u> 3499 (1990). See also A. Arneodo (these proceedings).

(15) V. Hakim (1989, unpublished).

(16) M. Mézard, G. Parisi and M. Virasoro "*Spin glass theory and beyond*" World Scientific (Singapore 1987).

(17) B. Derrida, V. Hakim and J. Vannimenus, Phys. Rev. <u>A43</u> (1991, in press).

(18) J. Vannimenus, B. Nickel, V. Hakim, Phys. Rev. <u>B30</u>, 391 (1984).

(19) J.P. Eckmann, P. Meakin, I. Procaccia and R. Zeitak, Phys. Rev. Lett. <u>65</u>, 52 (1990).

EXPERIMENTAL EVIDENCE FOR SPATIO-TEMPORAL CHAOS

IN DIFFUSION-LIMITED GROWTH PHENOMENA

F. Argoul[1], A. Arneodo[1], J. Elezgaray[2] and H.L. Swinney[3]

[1] CRPP, Avenue Schweitzer, 33600 Pessac, FRANCE
[2]MSI, Cornell University, Ithaca, NEW-YORK 14583
[3]Department of Physics, University of Texas, Austin, TEXAS, 78712

INTRODUCTION

Interfacial growth processes are generally fascinating nonequilibrium phenomena; physical, chemical, biological and geophysical systems exhibit similar rich morphologies[1-4]. The main characteristic feature of these systems is the existence of a diffusion field that drives them far from equilibrium. In this context, the diffusion limited aggregation (DLA) model introduced by Witten and Sander[5] in 1981 has played a major role since it has stimulated a variety of renewed experiments and numerical simulations[1-4]. But, despite the apparent simplicity of the DLA model, there is still no rigourous theory for diffusion-limited growth processes. Many important theoretical questions remain unanswered; in particular, it is still an open question whether the geometrical complexity of DLA clusters is a product of the randomness in the growth process[6,7] or the result of a proliferation of deterministic tip-splitting instablities[8-11]. Indeed, most previous works[5,12-21] have mainly focused on the geometrical properties of growing aggregates. The fractal geometry of DLA clusters has been analyzed using powerful mathematical techniques such as the computation of the generalized fractal dimensions and $f(\alpha)$ spectrum[22,23] and very recently, the application of the wavelet transform[24,25]. In a previous work, we have succeeded in elucidating the conjectured self-similarity[23-25] of DLA clusters. These aggregates are homogeneous fractals; that is, on a range of physical scales, the mass locally behaves as a power-law of the length scale with a unique scaling exponent $\alpha \sim 1.60$, which is independent of the point chosen on the cluster. DLA clusters are thus statistically self-similar: the generalized fractal dimensions are all equal $D_q = 1.60 \pm 0.02$. In an experimental study[23,25] we have shown that electrodeposition clusters, grown in the limit of weak current and small concentrations of electroactive species are also statistically self-similar. Within the experimental uncertainty their fractal dimensions $D_q = 1.63 \pm 0.03$ are in very good agreement with D_q for DLA clusters. Moreover, it has been recently realized that this geometrical self-similarity of diffusion--controlled aggregates is intimately related to the non-homogeneous distribution of the velocity field along the cluster boundary. The same mathematical tools have been applied to characterize the multifractal properties of the growth probability distribution (or sticking probability) of DLA clusters[11,26-32]; its $f(\alpha)$-spectrum is non-trivial, and the dimensions D_q decrease with increasing q. Experimental evidence for the multifractal nature of the growth probability distribution has been also reported in Refs 33, 34.

All these numerical and experimental analyses have only produced invariants of the diffusion-limited growth process, i.e., asymptotic statistical quantities which do not change in time. Therefore, they fail to provide any insight into the dynamical properties of the growth process. Moreover, since these fractal aggregates retain in their geometrical structure the full memory of their past evolution, space and time are intimately

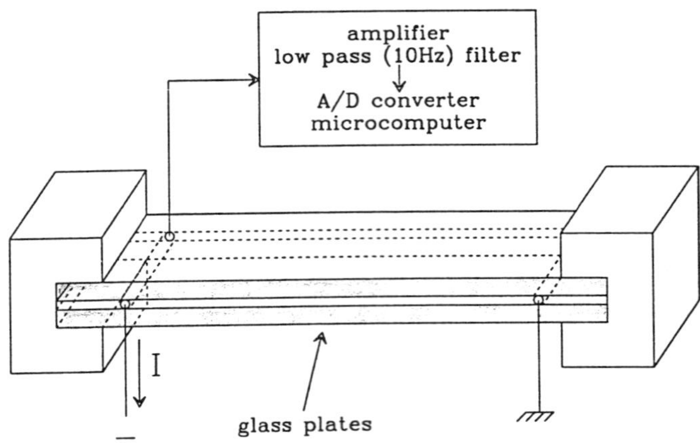

Fig. 1 Experimental set-up: the two glass plates, the two zinc electrodes and their clamping system are shown schematically. The left electrode (cathode) is maintained at a lower potential than the electrolyte; thus one expects a reduction reaction to occur, inducing the growth of a metallic zinc cluster.

connected[32]. Clearly a fundamental step has to be taken in the dynamical analysis of the spatio-temporal structuration of diffusion-controlled growing interfaces. New quantities that will characterize the dynamical deformation of the DLA cluster boundary during the growth have to be defined and computed.

In the first part of this paper, we emphasize electrochemical systems[35] as the paradigm for theoretical studies of the statics and dynamics of fractal growing patterns. Measurements of the impedance of the cell in a two-electrode experiment during the growth process yield direct information on the dynamics[36]. Our measured time-series reveal the richness of the dynamics[9,10,23,25,37-48] associated with fractal growth phenomena. In a second part, we report some preliminary results of numerical computations of the capacitance of a DLA cluster as it grows. We propose a first attempt of interpretation of the complexity of this signal in terms of the electrostatic spatial screening which governs the selection between branches of a growing DLA aggregate. The sharpness of this internal screening competition, together with the screening induced by the boundaries of the system, provide a clue to the highly nonlinear character of the selection mechanism.

THE ELECTRODEPOSITION EXPERIMENT

The experimental set-up (Fig. 1) consists of two parallel zinc electrodes (0.25 mm diam.) separated by a distance of 120 mm and confined between two rectangular glass plates (width 30 mm). The spacing between the electrodes is filled by capillarity with an aqueous solution of $ZnSO_4$ (0.05 M < $[ZnSO_4]$ < 0.30M). The system is illuminated with white light from below and photographed from above with a 35 mm camera (magnification from 1x to 10x). During the growth, the current is maintained constant and we record the voltage signal with a high resolution (16 bits) and high speed A/D board coupled with an external offset and amplifier (low-pass filter, $f_c \sim$ 10Hz). This signal provides a measurement of the conductivity of the medium (global impedance of the electrochemical cell including electrodes and electrolyte). Since the conductivity is expected[46] to fluctuate as the structures grow on the surface of the cathode, the temporal evolution of the voltage across the cell should provide interesting information about the screening effects and selection processes that govern electrochemical deposition.

Electrochemical deposition[35-48] is well suited for the studies of pattern formations[1-4] since, by varying the concentration of metal ions, the conductivity of the electrolyte and

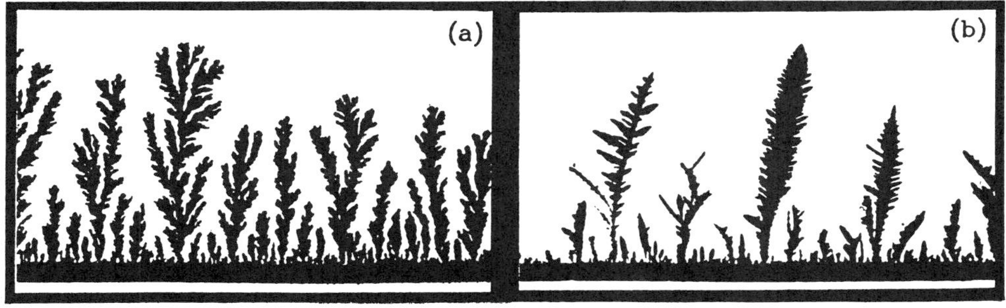

Fig. 2 Two characteristic morphologies observed in electrodeposition of zinc (a) DLA like metallic clusters: $[ZnSO_4]=0.05$ M, I=0.24 mA, j=1 mA/cm^2. (b) Highly anisotropic metallic clusters (dendritic): $[ZnSO_4]=0.05$ M, I= 1.5 mA, j=6.4 mA/cm^2.

the applied constant current (or voltage), one can explore a wide variety of morphologies including dense radial[42], dendritic[40,41] and DLA like fractal[23,25,37−39,41] patterns. The competition[36] between the reaction kinetics at the interface and transport processes such as chemical diffusion and migration is likely to be at the origin of the drastric changes in the morphology. At small ionic concentration and low current, where the diffusion length becomes much larger than the characteristic size of the pattern, one observes DLA-like fractal patterns[9,23,39] (Fig. 2a). Increasing both the concentration of electroactive species and the intensity of the current flowing through the cell, one progressively witnesses the invading of the structure by dendritic (directional) branches[47]. As the control parameter is varied the youngest generation branches (the smallest ones) get preferential direction of growth and then progressively the older generations are "contaminated", until the dendritic crystal shape shown in Fig. 2b is finally reached. Therefore, the local electric field which varies linearly with the current density propagates at larger scales the local anisotropy of the zinc metal (hcp hexagonal-close-packed).

Such large scale dendritic structures have not been observed in copper electro-deposition[8,9]. This suggests that the interfacial deposition kinetics must also play a significant role in the selection of the shape of the global structure. Only preliminary explanations have been proposed to understand the morphological transition from the apparently randomly ramified DLA structures to well-ordered (anisotropic) dendritic patterns in electrodeposition experiments[9]. From an electrochemical point of view, the recording of the potential of the cathode is a classical way[32,36,46] to get quantitative information on the local kinetics. This is an additional motivation for the time-series analysis we describe in the next paragraph. The main results of our experimental study have been announced in Ref. 49.

We will focus first on the early stages of the growth in a dendritic regime shown in Fig. 3. After some induction period observed immediately after turning on the applied current, many zinc trees emerge from the cathode. The initial number of trees is found to depend essentially upon the zinc sulfate concentration, the current intensity I and the cell aspect ratio. The significant feature is that very quickly these trees start competing. Successively, most of them stop growing as they enter in the shade of the surviving trees. At the beginning of the growth, the trees look like spikes. When the selection process is under way, the surviving trees start to ramify and their width is constantly increasing. Because the diffusion length is larger than the width of the cell, the zinc trees are not only influenced by their immediate neighbourhood, but also by distant trees. Thus, depending upon the experimental conditions, cooperative phenomena such as the simultaneous extinction of several trees, can be observed. This screening-induced selection process is clearly illustrated in Fig. 3, where only one tree survives this "struggle for life", 7 min after the beginning of the growth. Note that the internal structure of

Fig. 3 Dendritic zinc metal trees emerging from a zinc cathode in a 0.1 M $ZnSO_4$ solution with an applied current of 1.5 mA, j=6.4 mA/cm^2. The dendritic (anisotropic) character of the structure, as well as the sharpness of the selection process between different trees emerging from the cathode is striking in this succession of pictures.

these trees is dendritic. Since the diffusion length is a decreasing function of the external electric field, we focused on the low current regimes to keep the diffusion length large enough with respect to the cell width.

In the induction regime, the measured cell impedance initially decreases sharply and then increases slowly. This behavior is likely to correspond to (i) the formation of the electrochemical double layer[36,50] at the cathode and anode interfaces (polarization of the electrodes) and (ii) the initial transient regime from a three to a two-dimensional growth process (a rather flat interface starts moving very slowly from the cathode). Apparently, when the voltage exceeds some critical value, oscillations emerge from this quiescent regime, as shown in Fig. 4a. They are observed simultaneously to the selection process previously discussed (see Fig. 3). They actually appear when the instability of the cathode develops into a forest of zinc trees. A strong correlation exists between the coherence time of the oscillations and the recursive character of the selection process. Since the interface of the cathode is moving, the amplitude of the oscillating signal displays a slow drift while a shift is detected in its fundamental frequency. These observations indicate that the internal control parameters evolve during the growth. Indeed, this oscillatory regime is a transient phenomenon which turns off as soon as the selection process between the zinc trees is over.

When adjusting the control parameters, in such a way that the electrodeposits display a rather ordered dendritic structure (Fig. 3), the recorded oscillations appear to be coherent over many cycles (Fig. 4). Once the drift inherent to growth processes with moving interface is removed by Fourier filtering the low frequencies ($f \sim 0$), these oscillations turn out to be periodic. We have observed both "quasi" harmonic oscillations (Fig. 4b) which resemble the oscillations of small amplitude that emerge from a Hopf bifurcation[51,52], and relaxation oscillations (Fig. 4c) which are often observed in homogeneous non-equilibrium chemical systems[53,54]. Both these oscillations have approximately the same characteristic frequency $f \sim 25$ mHz, but the relaxation oscillations provide evidence for enhanced nonlinear interactions between the zinc trees that

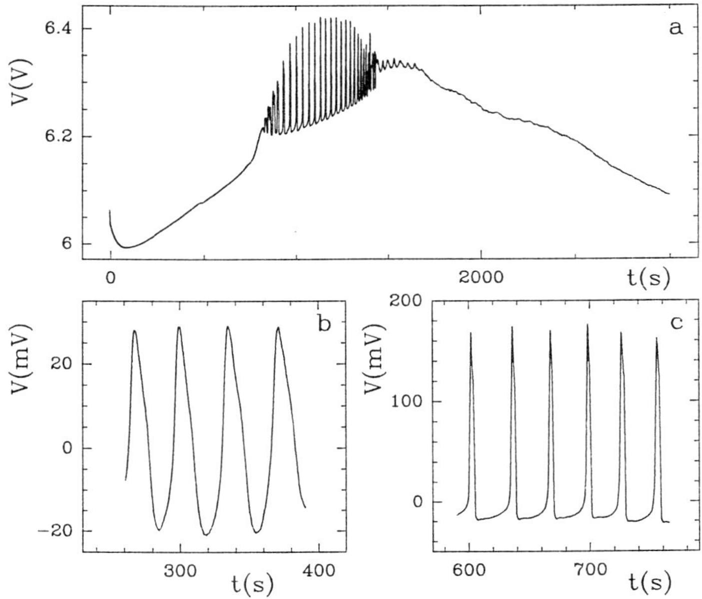

Fig. 4 Time-series recorded from cell voltage measurements at fixed current intensity. (a) The full time evolution for the parameter values [ZnSO$_4$]=0.25 M, I=1.5 mA (j=6.4 mA/cm^2). (b) "Quasi" sinusoidal periodic oscillations: [ZnSO$_4$]=0.15 M, I=0.9 mA (j=3.82 mA/cm^2). (c) Relaxation oscillations: [ZnSO$_4$]=0.25 M, I=1.5 mA (j=6.4 mA/cm^2). In (b) and (c), the drift in the signal has been removed by Fourier filtering the low frequencies ($f \sim 0$) (from Ref. 49).

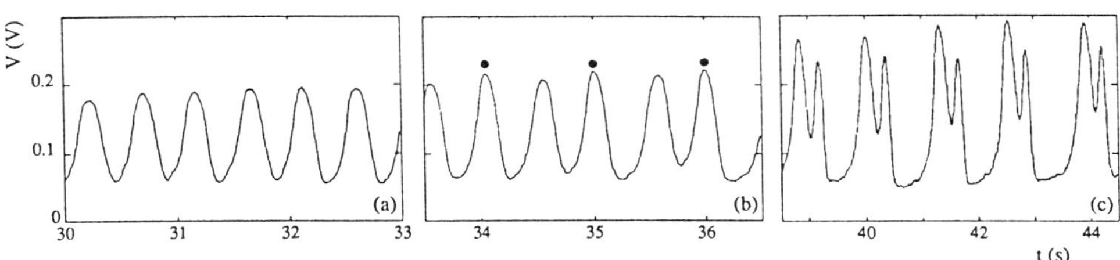

Fig. 5 Experimental evidence for a period-doubling bifurcation in the electrochemical deposition process for the parameter values corresponding to [ZnSO$_4$]=0.2 M, I=2.1 mA (j=8.9 mA/cm^2). (a), (b) and (c) represent different stages of the electrochemical process (from Ref. 49).

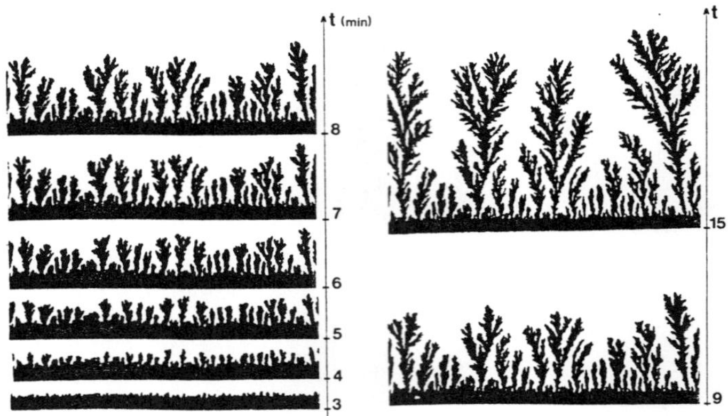

Fig. 6 Hybrid dendritic-DLA zinc-metal trees (about 15 mm long after 15 min of growth) photographed during the early stages of the chemical electrodeposition process. These zinc trees are grown from 0.1 M $ZnSO_4$ (aq) with an applied current of 0.5 mA (j=2.1 mA/cm^2) (from Ref. 49).

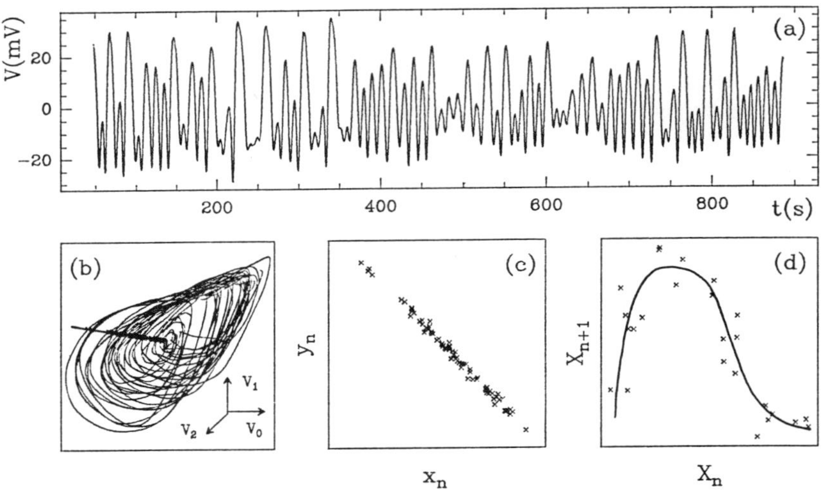

Fig. 7 A chaotic regime extracted from potential measurement during the screening induced selection process shown in Fig. 6. (a) The filtered signal V(t) vs time (the drift has been removed by Fourier filtering the low frequencies $\left(f \sim 0\right)$). (b) A three-dimensional phase portrait $\left(V_0{=}V(t),\ V_1{=}V(t{+}\tau),\ V_2{=}V(t{+}2\tau)\right)$ reconstructed from the time-series in (a) using the time-delay method with $\tau = 2.5s$. (c) A Poincaré section constructed by the intersections (x_n, y_n) of negatively directed trajectories with the plane passing through the line sketched in (b). (d) A one-dimensional map obtained by plotting as ordered pairs (X_n, X_{n+1}), where $X_n = x_n \cos \theta + y_n \sin \theta$ and $\theta = 85^o$; the hand-drawn curve suggests the existence of a unimodal 1-D map (from Ref. 49).

lead to an abrupt extinction of the growth of the screened trees (sudden death induced by screening). Note that the characteristic frequency of the recorded oscillations is at least two orders of magnitude smaller than the characteristic frequency of convective motions that are observed at the growing tips of the zinc trees. This observation argues against any interpretation of the macroscopic nonlinear selection process in terms of hydrodynamic instabilities[55].

When moving the system away from the dendritic morphology, toward more disordered highly ramified fractal patterns, period-doubling bifurcations[52,56-59] (Fig. 5) are observed as the precursor to chaotic dynamics. In Fig. 6, we show a hybrid dendrite-DLA morphology whose selection mechanism produces the chaotic time-series shown in Fig. 7. The nonperiodicity of the recorded time-series is analyzed using well-known techniques such as phase portraits, Poincaré sections and one-dimensional maps[52,56-59]. Fig. 7b shows a three-dimensional phase portrait reconstructed from the time-series of Fig. 7a using the time delay method[60,61]. It looks strikingly similar to the phase portraits of strange attractors displayed by low-dimensional dynamical systems[56-59] (e.g. Rossler's strange attractors). Rather than analyze the phase portrait directly, it is easier to look at the Poincaré section formed by the intersections of the trajectories with a plane approximately normal to the orbits. As shown in Fig. 7c, the points on the Poincaré plane lie to a good approximation along a smooth curve. The fact that the Poincaré section is not a scatter of points indicates that the irregular oscillations are not stochastic. Moreover, it demonstrates the low-dimensional nature of this chaotic state: the orbits lie approximately on a (multi-folded) two-dimensional sheet in the phase-space. Further insight into the dynamics can be achieved by constructing a one-dimensional map; a plot of X_{n+1} vs X_n is shown in Fig. 7d, where X is some coordinate in the Poincaré plane. Within the experimental resolution, the data appear to fall on a smooth curve, indicating that the dynamics is governed by a deterministic law[52,56-59]; for any X_n, the map gives the value X_{n+1} at the next intersection. The hand-drawn curve suggests the existence of a unimodal 1-D map, the hallmark of deterministic chaos[62]. The existence of an underlying unimodal 1-D map is indirectly confirmed by the identification of the first steps of the period-doubling cascade[62-65] illustrated in Fig. 5. The scatter of points around the curve is largely a consequence of the transitory character of this selection process. The experimental points plotted in Fig. 7d correspond to the initial part of the time series in Fig. 7a. Further recording displays an overall shift of the 1-D map with a slight decrease of its height, as a consequence of the internal control parameter evolution during the growth. This observation sets a fundamental physical limitation to any noise reducing procedure for decreasing the scatter of points around the hand-drawn curve in Fig. 7d.

The strong evidence for deterministic spatio-temporal chaos in the growth of hybrid DLA-dendritic metallic clusters led us to investigate situations much closer to the DLA limit, i.e., smaller currents and smaller metal cation concentrations. Doing so, we unfortunately faced additional technical difficulties: the amplitude of the oscillations decreased to the order of a few millivolts, comparable to the noise level. Moreover, these oscillations became increasingly complex, which made it difficult to discriminate the high frequency part of the signal from the noise. We thus decided to focus first on the low-frequency dynamics, imposing a 10Hz low-pass filtering on the voltage signal before digitizing it. This filtering probably removed the three-dimensional regime at small scales of the growth dynamics and the two-dimensional dynamics at intermediate scales. We will discuss these finite size effects in the second part of this paper where we will report a comparative study of DLA simulations.

In Fig. 8, we show the analysis of a "close-to" DLA limit regime (as far as our electronic device provides reliable measurements). We again stress the fact that the signal has been low-pass filtered at 10Hz. For this regime, no large scale oscillations have been observed during the selection mechanism and the signal of Fig. 8a has been extracted from the quiescent (filtered) part of the global voltage recording if we refer to the example of Fig. 4a. This experiment is crude (the filtering and shielding techniques need to be refined), so one cannot trust the high frequency range of the power spectrum (Fig. 8b). Nevertheless, it appears clearly in Fig. 8c that the reconstructed phase portrait is very reminiscent of low dimensional chaotic attractors such as those illus-

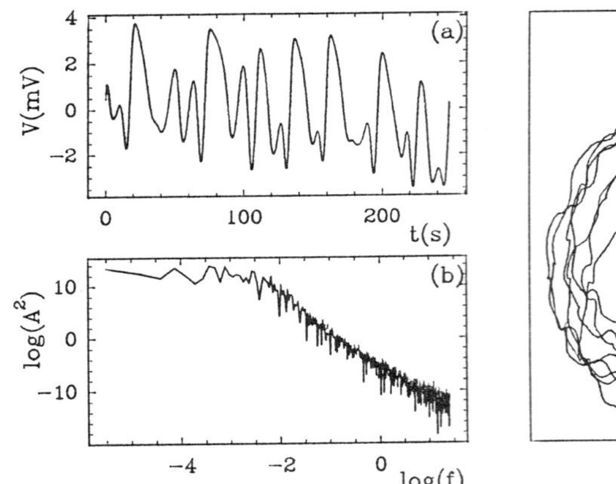
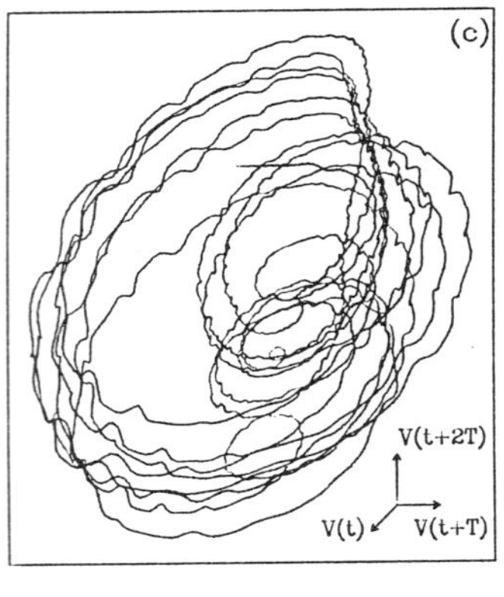

Fig. 8 Dynamics associated with the growth of DLA-like electrodeposits of zinc: $[ZnSO_4]$=0.05 M, I=0.5 mA, j=2.1 mA/cm^2. (a) Filtered time series recorded from cell voltage measurements. (b) The corresponding power spectrum. (c) The reconstructed phase portrait using the time-delay method presented in Fig. 7.

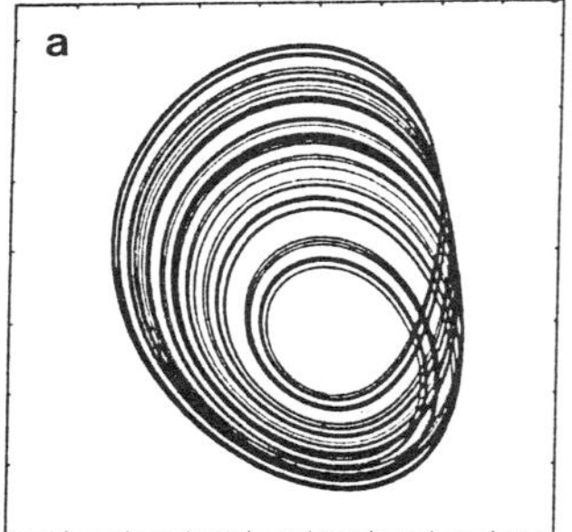

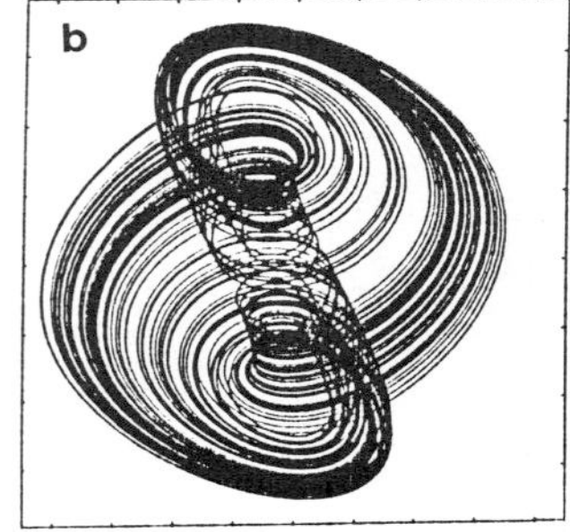

Fig. 9 Chaotic attractors exhibited by the third-order differential system: $\dddot{A} + \mu_2 \ddot{A} + \mu_1 \dot{A} + \mu_0 A = f(A, \dot{A}, \ddot{A})$, where f is a nonlinear function (for more details see Ref. 66).

trated in Fig. 9. As far as the large scale dynamics (small frequencies) is concerned, we thus have strong evidence that diffusion-controlled electrodeposition is spatio-temporal deterministic chaos rather than a stochastic process.

Of course these results are very preliminary; they need to be confirmed by the application of dynamical systems techniques (Poincaré and first return maps, Lyapunov exponents[52,56−59]) on longer time series. Moreover, by removing the 10 Hz low-pass filtering constraint, one can hope to answer the challenging question of the transition from low-dimensionality chaotic behavior to higher dimensionality spatio-temporal chaos or turbulence. This transition is of fundamental importance: from a physical point of view, one can hope to get further insight into the universal character of fractal growth processes; from a chemical point of view, one can expect to get quantitative information about the competition between the reaction and the transport processes, which may improve our understanding of electrochemical deposition.

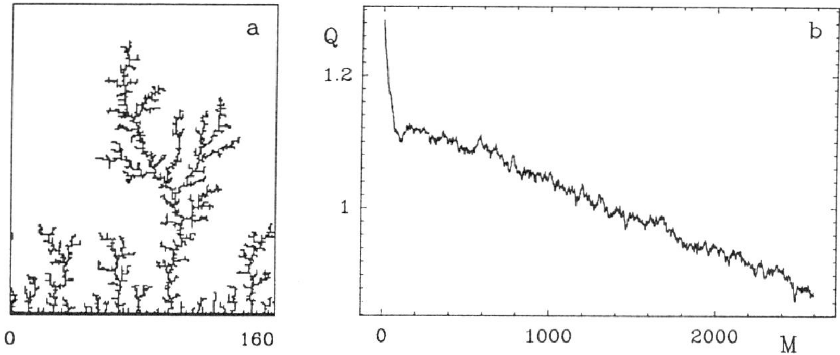

Fig. 10 (a) A 2495 particle DLA cluster computed using the Green's function technique as explained in the text. (b) A plot of the global charge along the boundary of the cluster ($\propto$ capacitance) as a function of its mass M. The perimeter W of the semi-infinite cylinder is 160 pixels.

CAPACITANCE COMPUTATION IN DLA SIMULATIONS

Our experimental results show that there is an intimate relationship between the spatial structure of the electrodeposition aggregates and their dynamical (time varying) properties[32,49]. We know that the DLA model[5] provides a good quantitative description of the asymptotic (time independent) geometry of these electrodeposition clusters[23,25]; it is then reasonable to ask whether the same is true for the dynamical quantities describing the growth process[50]. The numerical results we present here suggest strongly that this is indeed the case. However, one major assumption has to be made since it is not clear from the standard formulation of the DLA growth process what is the model parameter playing the role of time. Should the mass of the aggregate correspond to its age (namely, the time interval between the sticking of two particles is constant), or should one considers that the time between each arrivals is proportional to the distance covered by the corresponding random walker? Intuitively, it seems that the fixed Δt hypothesis corresponds to a fixed current electrodeposition experiment (the rate of arrival of particles on the cathode is constant). On the other hand, this hypothesis seems to be close to a stochastic formulation of the continuous deterministic equations describing the so-called quasi-stationary approximation[5]:

$$\left. \begin{array}{ccc} \Delta\Phi & = & 0 \\ \vec{v}.\vec{n} & = & \vec{\nabla}\Phi.\vec{n} \end{array} \right\} \tag{1}$$

where Φ is the electrostatic potential, $\vec{n}$ is the outward normal to the boundary and $\vec{v}$ is the local speed. In other words, the sticking probability (per time unit) is proportional to the local charge. In the following, we will then stick to the Δt=constant hypothesis, and we will assume that the DLA dynamical process can be compared to the fixed current electrodeposition experiment. We will also suppose that the aggregate is a perfect conductor, so that its potential is constant. At each time step, the sticking probability distribution is:

$$p_i = \rho_i / \sum_j \rho_j \ , \tag{2}$$

where ρ_i is the charge at the site i; thus, if Φ_i is the electrostatic potential at the site i, one can write

$$\Phi_i = \sum G_{ij}\rho_j \ , \tag{3}$$

where G_{ij} is the two-dimensional Green's function[12,27,29,67-69] between the two points i and j. Our working hypothesis $\Phi_i = constant$, $\forall i$ on the aggregate, implies that the knowledge of the ρ_i's amounts to the inversion of a $M \times M$ Green's matrix[70], where M is the mass of the DLA cluster.

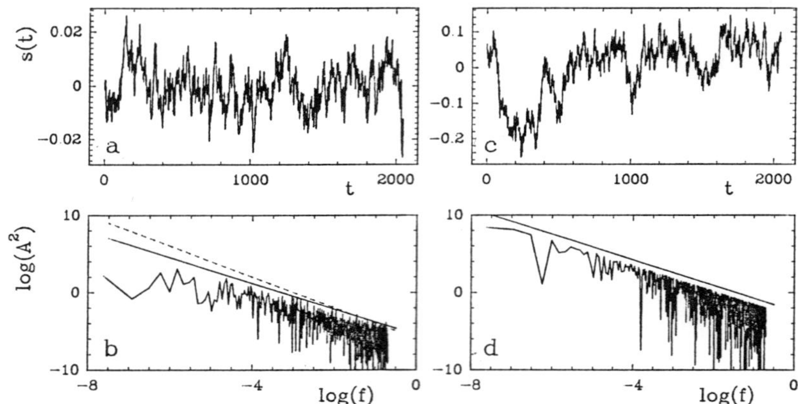

Fig. 11 (a) The fluctuations of the global charge on the boundary of a DLA cluster grown in a 160-pixel perimeter cylinder, as a function of time. (b) The power spectrum of the signal in (a); the continuous line corresponds to a -5/3 slope, while the dashed line corresponds to a -2 slope. (c) Fractional brownian motion with the same power spectrum exponent $\beta = 5/3$ as in (b). (d) The power spectrum of the signal in (c) with the -5/3 slope.

Fig. 10a illustrates a DLA cluster grown using this Green's function technique. The computation of a 2495 particle cluster was performed on a Cray X-MP/14SE with a 10^{-9} accuracy in the estimation of the ρ_j. Φ was computed on a semi-infinite discrete cylinder (periodic boundary conditions: $W = 160$ pixels). In Fig. 10b, we show the corresponding evolution of the global charge on the boundary of the cluster as a function of its mass. With the assumption that the mass of the cluster is proportional to time, as discussed before, Fig. 10b describes the time evolution of the charge Q of the cluster. From the quasistatic character of the DLA model, Q is linearly related to the capacitance C of the DLA cluster. Looking at the time series of Fig. 10b, one observes an average decrease of the charge of the cluster, which indicates that its global propensity to grow decreases progressively. During this rather short period of growth, the charge Q behaves linearly with M on the average, and we will focus in the following on the fluctuations of Q around this linear behavior. These fluctuations are shown in Fig. 11a. The corresponding power spectrum in Fig. 11b displays, above some critical frequency f_c ($\log f_c \sim -4$), a power law behavior $S \sim f^{-\beta}$ with an exponent β close to 5/3 and definitely smaller than 2. This value 5/3 is likely to be intimately connected with the fractal dimension of the DLA cluster itself, which has been theoretically predicted[71,72] (mean-field analysis) to be $D_f = 5/3$.

An interesting issue is now to characterize the nature of the fluctuations in Fig 11.a, with the ultimate goal to answer the fundamental question about their deterministic or stochastic character. Despite the fact that the short length of this numerical signal makes any rigourous study very hazardous, we have carried out the analysis of its fractal properties. More precisely, a fractal signal $s(t)$ can generally be considered as self-affine, if it displays the following scaling property: there exists an exponent H such that the function $r^{-H}s(rt)$ is statistically invariant[73]. This means that if the time is scaled by a factor r, the signal has to be rescaled by r^{-H} to recover the same dynamics. Is the charge signal in Fig. 11a a self-affine function? Let us compare this signal with a theoretical self-affine fractal: the one-dimensional fractional brownian motion[74,75]. By definition, if a variable $s(t)$ represents a fractional brownian motion, then the difference $s(t_2) - s(t_1)$ is Gaussian distributed and its mean square deviation $< (s(t_2) - s(t_1))^2 >$ is proportional to $|t_2 - t_1|^{2H}$, where H is called the self-affinity exponent. The power-law exponent of the power spectrum is simply $\beta = 2H + 1$. In Figs 11c and 11d we illustrate the time series

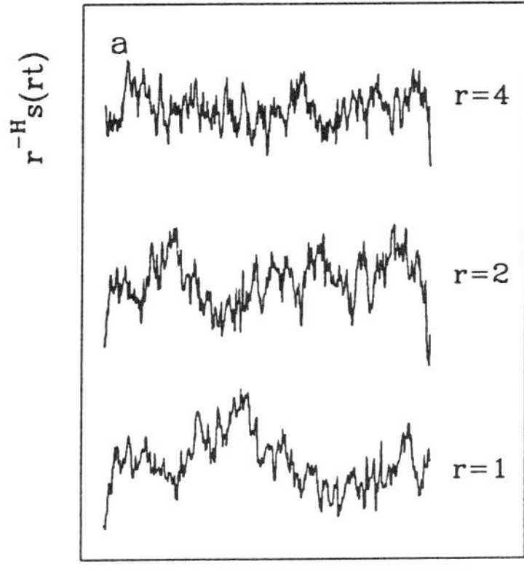
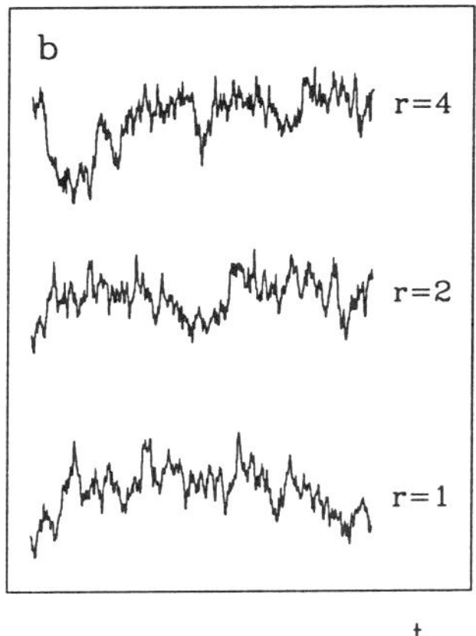

Fig. 12 Rescaled signals $r^{-H}s(rt)$ versus t, in order to test their self-affine properties. (a) A charge signal extracted from DLA simulations in a 160-pixel perimeter cylinder. (b) Fractional brownian motion with $H = 1/3$.

and power spectrum of fractional brownian motion with the same exponent $\beta = 5/3$ as the DLA signal and which corresponds to a self-affinity exponent $H = (\beta - 1)/2 = 1/3$. We clearly observe in the power spectra that, if the DLA charge signal is self-affine, it displays a clear power-law behavior only at high frequencies $f > f_c$ ($\log f_c \sim -4$). A visual test of the self affinity of the DLA signal is shown in Fig. 12, where its rescaled function $r^{-H}s(rt)$ is plotted in Fig. 12a and compared to the fractional brownian motion in Fig. 12b. Again, the DLA signal does not seem to fit this self-affine description over the whole range of scales investigated.

A more rigourous way to evidence the loss of self-affine properties is to plot the square root of the square mean deviation: $< X_\ell^2 >^{1/2} = < (s(t + \ell) - s(t))^2 >^{1/2}$ versus the increment ℓ. This technique has already been applied to turbulent signals[76,77]. For a self-affine signal, $< X_\ell^2 >^{1/2}$ is expected to behave like ℓ^H. In Fig. 13, we apply this "increment" method to our DLA charge signal and to the corresponding fractional brownian motion. For increments larger than a critical value $\ell_c \sim e^{3.5}$, the DLA signal (Figs 13a and 13b) deviates significantly from the power-law behavior of the corresponding fractional brownian motion with self-affinity exponent $H = 1/3$ (Figs 13c and 13d). Above this critical increment value, $< X_\ell^2 >^{1/2}$ no longer diverges as a power-law but tends to saturate, indicating some confinement of the dynamics. This critical increment value is found to depend on the width W of the cell (perimeter of the cylinder), where the DLA simulations have been performed. This saturation suggests the existence of strong correlations in the charge signal as a result of long-range screening interactions between the growing DLA trees via the periodicity of the boundary conditions. If one focuses on the low-frequency dynamics by filtering the frequencies larger than f_c, one gets the signal shown in Fig. 14a. The filtering operation we have used is illustrated in the power spectrum shown in Fig. 14b. Obviously, the total length is insufficient to carry out a reliable quantitative analysis and the reconstructed phase portrait in Fig. 14b needs more trajectories to be more convincing. Nevertheless, this phase portrait looks similar to the experimental phase portraits observed in the electrodeposition experiment (Fig. 8c) and in low-dimensional dynamical systems (Fig. 9)[66]. This analysis needs, however, to be extended to longer time series and to different channel widths in order to be conclusive. This work is currently in progress at the Centre de Recherche Paul Pascal and should lead to the characterization of this large-scale DLA dynamics in terms of spatio-temporal chaos[78].

As far as the intrinsic short-range dynamics is concerned, its self-affine character remains to be understood. The possible Gaussian nature of the increment distribution needs to be studied in order to push further the comparison with fractional brownian

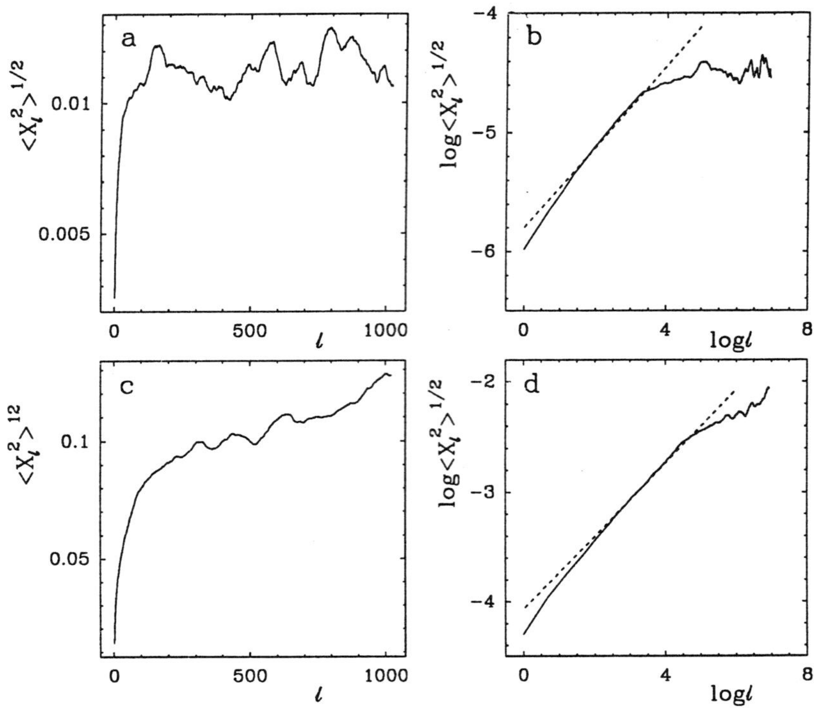

Fig. 13 The "increment" method applied to the DLA charge signal (a) and to the fractional brownian motion (c). $< X_\ell^2 >^{1/2} = < (s(t+\ell) - s(t))^2 >^{1/2}$. The dashed lines in the corresponding log-log plots in (b) and (d) respectively correspond to a self-affinity exponent $H = 1/3$.

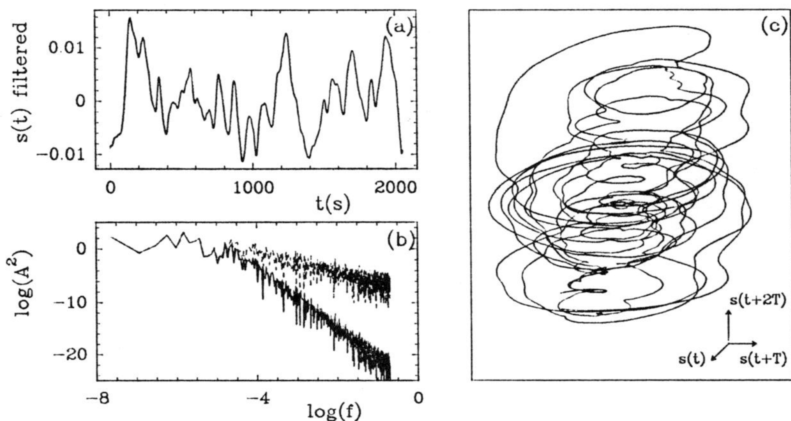

Fig. 14 (a) The DLA charge signal of Fig. 11a after a low-pass filtering ($\log f \sim -4$). (b) Illustration of the filtering operation in the power spectrum; the upper trace is unfiltered, the lower trace is filtered. (c) Reconstructed phase portrait (time delay method) from the filtered time series in (a).

motion. Again, this issue will require extensive simulations in large cells and for long periods of growth. Moreover, by extending the Green's function techniques to variants of the DLA model, e.g., the η-model[1-4,69], which accounts for a transition from dense to spiky morphology, or some anisotropic versions[1-4] of the DLA model where the sticking rule is modified in order to mimic the transition from DLA to dendritic fractals, one can hope to relate the self-affinity exponent of the small-scale dynamics to the fractal properties of the growing cluster (which actually displays self-similar properties at scales smaller than the width of the cell). We hope to elaborate on this point in a forthcoming communication.

ACKNOWLEDGMENTS

We are very grateful to T.C. Oppe for fruitful discussions. This work was supported by the Centre National des Etudes Spatiales under contrat (N° 90/CNES/215) and the Direction des Recherches Etudes et Techniques (DRET) under grant (N° 89/196). A. A and F. A were also supported by the US Department of Energy under contrat DE-FG05-88ER13821. J. E. is also partly supported by the U.S. Army Research Office through the M.S.I. of Cornell University.

REFERENCES

1. W. Guttinger and D. Dangelmayr, eds, "The Physics of Structure Formation", Springer-Verlag, Berlin (1987).
2. H. E. Stanley and N. Ostrowsky, eds, "Random Fluctuations and Pattern Growth", Kluwer Academic Publisher, Dordrecht (1988).
3. J. Feder, "Fractals", Pergamon, New York (1988).
4. T. Vicsek, "Fractal Growth Phenomena", World Scientific, Singapore (1989).
5. T. Witten and L. M. Sander, Phys. Rev. Lett. 47:1400 (1981); Phys. Rev. B 27:5686 (1983).
6. H. E. Stanley, in Ref. 1, p.210.
7. J. Nittman and H. E. Stanley, Nature 321:663 (1986); J. Phys. A 20:L1185 (1987).
8. L. M. Sander, in "Fractal in Physics", L. Pietronero and E. Tosati, eds, North-Holland, Amsterdam (1986).
9. L. M. Sander, in Ref. 1, p.257.
10. L. M. Sander, P. Ramanlal and E. Ben-Jacob, Phys. Rev. A 32:3160 (1985).
11. P. Ramanlal and L. M. Sander, J. Phys. A 21:L995 (1988).
12. L. A. Turkevich and H. Scher, Phys. Rev. Lett. 55:1026 (1985).
13. R. C. Ball, Physica A 140:62 (1986).
14. T. C. Halsey, P. Meakin and I. Procaccia, Phys. Rev. Lett. 56:854 (1986).
15. M. Matsushita, K. Konda, H. Toyoki, Y. Hayakawa and H. Kondo, J. Phys. Soc. Jpn. 55:2618 (1986).
16. P. Meakin, Phys. Rev. A 27:1495 (1983).
17. P. Meakin and Z. R. Wasserman, Chem. Phys. 91:391 (1984).
18. P. Meakin and L. M. Sander, Phys. Rev. Lett. 54:2053 (1985).
19. R. C. Ball and R. M. Brady, J. Phys. A 18:L809 (1985).
20. P. Meakin, Phys. Rev. A 33:3371 (1986).
21. P. Meakin, R. C. Ball, P. Ramanlal and L. M. Sander, Phys. Rev. A 35:5233 (1987).
22. P. Meakin and S. Havlin, Phys. Rev. A 36:4428 (1987).
23. F. Argoul, A. Arneodo, G. Grasseau and H. L. Swinney, Phys. Rev. Lett. 61:2558 (1988).
24. A. Arneodo, F. Argoul, J. Elezgaray and G. Grasseau, in "Nonlinear Dynamics", G. Turchetti, ed., World Scientific, Singapore (1989) p. 130.
25. F. Argoul, A. Arneodo, J. Elezgaray, G. Grasseau and R. Murenzi, Phys. Lett. A 135:327 (1989); Phys. Rev. A 41:5537 (1990).
26. P. Meakin, Phys. Rev. A 34:710 (1986); 35:2234 (1987).
27. C. Amitrano, A. Coniglio and F. di Liberto, Phys. Rev. Lett. 57:1016 (1986).
28. P. Meakin, A. Coniglio, H. E. Stanley and T. A. Witten, Phys. Rev. A 34:3325 (1986).
29. Y. Hayakawa, S. Sato and M. Matsushita, Phys. Rev. A 36:1963 (1987).

30. J. Lee and H. E. Stanley, Phys. Rev. Lett. 61:2945 (1988).
31. R. C. Ball and M. Blunt, Phys. Rev. A 39:3591 (1989).
32. F. Argoul, A. Arneodo, J. Elezgaray and G. Grasseau, in "Mesures of Complexity and Chaos", N. B. Abraham, A. M. Albano, A. Passamante and P. E. Rapp, eds, Plenum Press, New York (1989) p.433.
33. S. Ohta and H. Honjo, Phys. Rev. Lett. 60:611 (1988).
34. M. Blunt and P. King, Phys. Rev. A 37:3935 (1988).
35. A. R. Despic and K. I. Popov, in "Modern Aspects of Electrochemistry", B.E. Conway and J.O'M Bockris, eds, Plenum Press, New York (1972).
36. A. J. Bard and L. R. Faulkner, "Electrochemical Methods, Fundamentals and Applications", Wiley, New York (1980).
37. R. M. Brady and R. C. Ball, Nature (London) 309:225 (1984).
38. M. Matsushita, M. Sano, Y. Hayakawa, H. Honjo and Y. Sawada, Phys. Rev. Lett. 53:286 (1984).
39. M. Matsushita, Y. Hayakawa and Y. Sawada, Phys. Rev. A 32:3814 (1985).
40. Y. Sawada, A. Doughterty and J. P. Gollub, Phys. Rev. Lett 56:1260 (1986).
41. D. G. Grier, E. Ben-Jacob, R. Clarke and L. M. Sander, Phys. Rev. Lett. 56:1264 (1986).
42. D. G. Grier, D. A. Kessler and L. M. Sander, Phys. Rev. Lett. 59:2315 (1987).
43. G. L.M.K.S. Kahanda and M. Tomkiewicz, J. Electrochem. Soc. 136:1497 (1989).
44. D. P. Barkey, R. H. Muller and C. W. Tobias, J. Electrochem. Soc. 136:2199 (1989); 136:2207 (1989).
45. J. R. Melrose and D. B. Hibbert, Phys. Rev. A 40:1727 (1989).
46. R. M. Suter and P. Wong, Phys. Rev. B 39:4536 (1989).
47. D. G. Grier, K. Allen, R. S. Goldman, L. M. Sander and R. Clarke, Phys. Rev. Lett. 64:2152 (1990).
48. J. N. Chazalviel, "Some electrochemical aspects of the generation of fractal electrodeposits", preprint (1990).
49. F. Argoul and A. Arneodo, J. Phys. France 51:2477 (1990).
50. T. C. Halsey and M. Leibig, "Electrodeposition and diffusion-limited aggregation", preprint (1989).
51. J. E. Marsden and M. Mc Cracken, "Hopf Bifurcation and its Applications", Applied Math. Sci. 19, Springer-Verlag, New York (1976).
52. J. Guckenheimer and P. Holmes, "Nonlinear Oscillations, Dynamical Systems and Bifurcations of Vector Fields", Springer-Verlag, Berlin (1984).
53. C. Vidal and A. Pacault, eds, "Nonlinear Phenomena in Chemical Dynamics", Springer-Verlag, Berlin (1981).
54. C. Vidal and A. Pacault, eds, "Nonequilibrium Dynamics in Chemical Systems", Springer-Verlag, Berlin (1984).
55. L. D. Landau and E. M. Lifschitz, "Fluid Mechanics", Pergamon, New York (1975).
56. P. Cvitanovic, ed., "Universality in Chaos", Hilger, Bristol (1984).
57. H. Bai-Lin, ed., "Chaos", World Scientific, Singapore (1984).
58. H. G. Schuster," Deterministic Chaos", Physik-Verlag, Weinheim (1984).
59. P. Bergé, Y. Pomeau and C. Vidal, "Order within Chaos", Wiley, New York (1986).
60. N. H. Packard, J. P. Crutchfield, J. D. Farmer and R. S. Shaw, Phys. Rev. Lett. 45:712 (1980).
61. F. Takens, Lect. Notes Math. 898:366 (1981).
62. P. Collet and J. P. Eckmann, "Iterated Maps of an Interval as Dynamical Systems", Birkhauser, Boston (1980).
63. M. J. Feigenbaum, J. Stat. Phys. 19:25 (1978); 21:669 (1979).
64. P. Coullet and C. Tresser, J. Physique Colloq. France 39:C5 (1978).
65. C. Tresser and P. Coullet, C. R. Acad. Sci. 287:577 (1978).
66. A. Arneodo and O. Thual, in Ref. 1, p.313.
67. T. Morita, J. Math. Phys. 12:744 (1971).
68. F. Spitzer, "Principles of Random Walk", Springer-Verlag, Berlin (1976).
69. L. Niemeyer, L. Pietronero and H. Wiesman, Phys. Rev. Lett. 52:1033 (1984).

70. We have used the NSPCG package for solving large sparse linear systems by iterative methods which was implemented on the Cray by T.C. Oppe, W.D. Joubert and D.R. Kincaid at the Center for Numerical Studies of the University of Texas at Austin.

71. M. Tokuyama and K. Kawasaki, Phys. Lett. A 100:337 (1984).

72. K. Honda, H. Toyaki and M. Matsushita, J. Phys. Soc. Jpn 55:707 (1986).

73. B. B. Mandelbrot, Physica Scripta 32:257 (1985).

74. B. B. Mandelbrot, "The Fractal Geometry of Nature", Freeman and Co, New York (1982).

75. H. O. Peitgen and D. Saupe, eds, Springer-Verlag, New York (1988).

76. F. Anselmet, Y. Gagne, E. J. Hopfinger and R. A. Antonia, J. Fluid. Mech. 140:63 (1984).

77. B. Castaing, Y. Gagne and E. J. Hopfinger, Physica D 46:177 (1990).

78. F. Argoul, A. Arneodo and J. Elezgaray, in preparation.

WHY NATURE MAKES FRACTALS

L. Pietronero

Dipartimento di Fisica, Università di Roma "La Sapienza"

Piazzale Aldo Moro 2, 00185 Roma, Italy

ABSTRACT

The concept of fractal geometry allows to look at nature in a new perspective and to consider irregularities as intrinsic entities. The main problem in this field is to understand the microscopic origin of these structures. The first step in this direction is to formulate models of fractal growth based on physical processes. This has been achieved since a few years. In addition one should be able to formulate a theory of fractal growth analogous to the Renormalization Group for critical phenomena. The attempts to apply RG methods to fractal growth turned out to be rather problematic. Recently we have introduced a new theoretical framework based on a Fixed Scale Transformation that exploits also a different invariance property than RG. This new method allows to understand the origin of fractal properties in these models and to compute analytically the value of the fractal dimension. In this paper we present a critical discussion of the field and point out the main open problems.

1. INTRODUCTION

A system has fractal properties if, by magnifying a portion of it, one obtains a structure of the same complexity as the original one at larger scale [1]. This implies that such a structure is non analytic in every point and that fluctuations are strong at all scales. Fractal geometry allows to describe in quantitative mathematical terms those systems in which irregularity represents an intrinsic property [2,3]. This allows to pose these problems within a scientific framework but it does not provide a theory for the origin of these properties. It should be noted however that the broader framework of fractal geometry (with respect to analytic concepts) even at the phenomenological level, can give rise to substantial change in the interpretation of the same experimental data. An interesting example of this is provided by the properties of the galaxy distributions that turn out to be fractal to all observational scales while before the same data were considered in agreement with a homogeneous distribution[4].

In this paper however we intend to discuss the present understanding of the origin of fractal structures in physical phenomena[2,3]. This question will be addressed mainly from the point of view of the model of Diffusion Limited Aggregation [5] and the Dielectric Breakdown Models [6]. These models are considered as the prototypes fractal growth models based on physical processes and will be described in Sect.2 In Sect.3 we summarize their properties as obtained from extensive computer simulations.

In Sect.4 we discuss the properties of the growth probability from the point of view of multifractality.

In Sect.5 we discuss the analogies and differences of these models with respect to equilibrium critical phenomena. In particular we are going to show why the application of the standard Renormalization Group (RG) concepts to these problems is, in general, quite problematic.

In Sect.6 we describe a new theoretical framework for fractal growth fundamentally different than the RG. This is the Fixed Scale Transformation method that allows to understand the origin of fractal properties in these models and to compute the fractal dimension analytically. This method can also be applied to a variety of other problems like the fractal structure of percolating and Ising clusters and several others.

In Sect.7 we summarize the situation and outline the main open problems.

2. THE BASIC MODELS: DLA AND DBM

There are very many examples of structures that are more or less fractal and it is not conceivable to try to make a general theory that may explain all of these structures. The idea is instead to concentrate on some specific case and hope that the theoretical concepts eventually developed for this case can then be modified and applied also to other situations. This is for example what has happened for the critical properties of second order transitions. For this class of problems the key model has been played by the Ising model that has been crucial to develop the theoretical ideas that led to the Renormalization Group (RG). Afterwards these ideas could be extended to virtually any other model [7].

For fractals the first growth model based on a well defined physical process has been Diffusion Limited Aggregation (DLA) [5]. This was then generalized by the Dielectric Breakdown Model (DBM) [6] that also makes clear the mathematical nature of the phenomenon. This is based on an iterative stochastic process in which the growth probability is modulated by the solution of Laplace equation with appropriate boundary conditions. These models can explain the origin of fractal structures in a variety of process like dielectric breakdown, dendritic growth, viscous fingers and various others. They are considered by most of the workers in the field as the prototype fractal growth models analogous to the Ising model for phase transitions [2,3,8].

For this reason they have been studied very extensively and it seems that this activity has given rise to more than 1000 papers. Most of this activity has been based on computer simulations and a massive amount of data has been produced. One cannot say however that a clear picture has yet emerged because these data sometimes in conflict with each other and, in the lack of appropriate theoretical concepts, not always the right question has been asked to the computer.

<u>The Dielectric Breakdown Model</u> [6]

Consider, for simplicity, a square lattice in which the central point represents an electrode with potential $\phi = 0$, while the second electrod with $\phi = 1$ consists of a circle at infinity. The bonds on which breakdown has already occurred constitute the pattern at a given time that is considered equipotential. It is possible then to compute the local field around this structure by solving the Laplace equation

$$\nabla^2 \phi = 0 \qquad (1)$$

with the boundary conditions of constant potential on the grown structure and a different value of the potential at infinity. The growth probability p_j for each bond (j) connected to the structure is then related to the local field

$$p_j = \frac{|\nabla \phi|_j^n}{\sum_j |\nabla \phi|_j^n} \quad , \qquad (2)$$

where η is a parameter that modulates the randomness of the process. For $\eta =1$ the process coincides with DLA, in the sense that DLA corresponds to a Monte Carlo realization of the DBM process. This probability distribution is then used to select one bond that becomes part of the pattern. The process continues by iterating this procedure and the result is a spontaneous generation of a fractal structure.

Laplacian fields are common ingredients in diverse physical processes and realistic models for these processes often incorporate stochasticity. For these reasons these models are now believed to capture the essential features of pattern formation in semingly different phenomena [2,3,8].

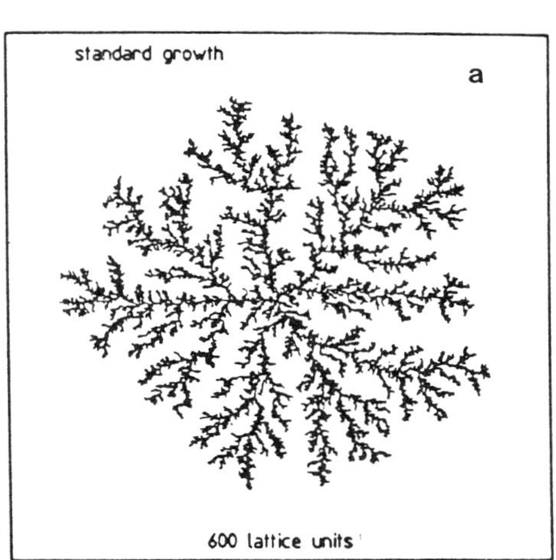

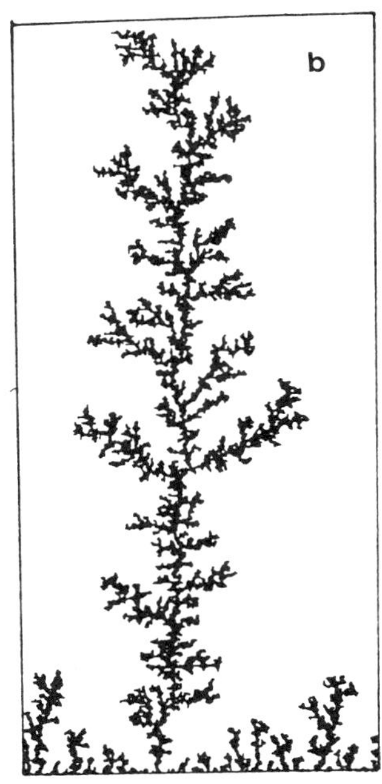

Fig.1. (a) Radial DLA growth on a square lattice. At relatively small scale the shape is approximative circular. (b) Growth between two parallel planes. Given the periodic boundary conditions used this growth occurs on the surface of a cylinder. The two geometrics give rise to roughly the same type of structure. However small but persistent differencies have been detected for various properties. The cylinder geometry has various conceptual advantages, for example it defines a unique growth direction.

<u>Radial and Cylinder Geometry</u>

The previous discussion refers to radial geometry that leads to structures of the type shown in Fig.1a. One can also define the growth process starting from a base line and proceeding towards a far away line with different potential. This is shown in Fig.1b. Since in practice the length of the base line is finite and one uses periodic boundary conditions, topologycally the growth occurs on the surface of a cylinder [9,10].

The two processes give rise to basically similar structures, however we are going to see that there are small but persistent differences among them and the process cannot be considered as strictly universal in this respect. The cylinder geometry offers conceptual advantages for a theoretical discussion because it defines a unique growth directions and it allows to vary independently properties that are instead intrinsically linked in the radial geometry.

<u>The Mathematical Nature of the Problem</u>

In Fig.2 we show schematically the mathematical nature of the growth process. It consists in an irreversible stochastic process with long range couplings both in space and time. One may notice that it is not possible to define a statistical weight to a given configuration but one has to know its entire history. For this and other reasons the concepts of the RG theory do not apply naturally to this problem.

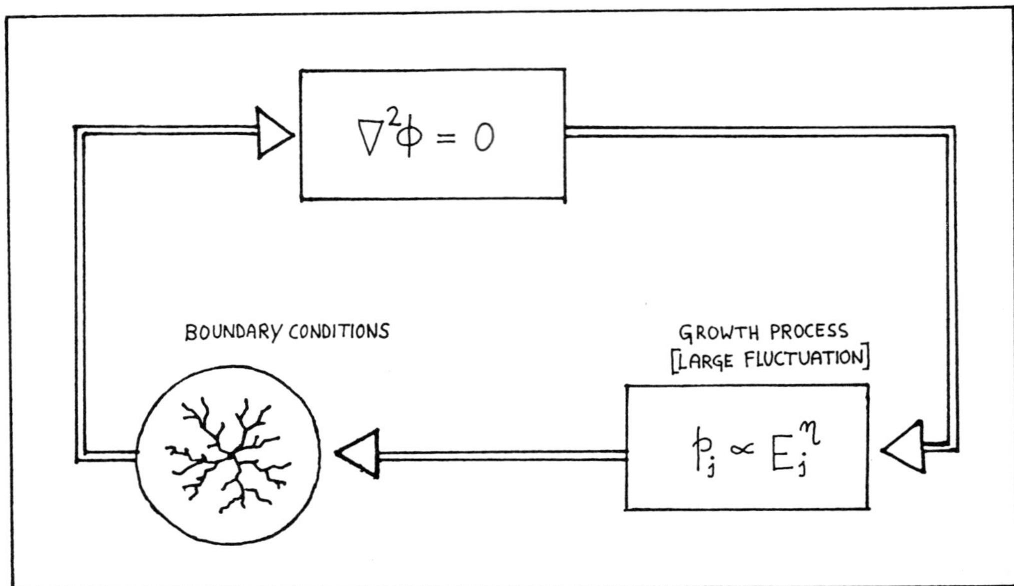

Fig.2. The mathematical nature of the DBM (or DLA) growth process corresponds to an inrreversible iterative dynamics with long range correlations. No statistical weight can be defined for a given configuration without taking into account its entire history.

3. GENERAL PROPERTIES

The most characteristic feature of these models is clearly that they give rise to fractal structures. We are going to see however that as soon as one tries to make this statement more refined and quantitative, a number of problems appear, some of which are still open.

In order to discuss the fractal dimension one should make clear that this is a property that refers to the <u>frozen</u> part of the structure, namely the one that is fixed asymptotically and that will not be modified by further growth. For Fig.1a (radial) this zone consists of a circular area smaller than the perimeter of the structure shown, while for the cylinder case (Fig.1b) this is the entire structure somewhat below the growing profile. This discussion is necessary to clarify one of the most debated and confused effects in this field:

<u>The Lattice Anisotropy Effect</u>

If the growth process is defined on a square lattice, at small scales one observes a reasonably circular shape while at larger scales the structure becomes cross shaped

(Fig.3) [11]. If one measures the fractal dimension D by the mass length ratio by computing the gyration ratio one observes that until the shape is about circular one has $D \simeq 1.70$ while

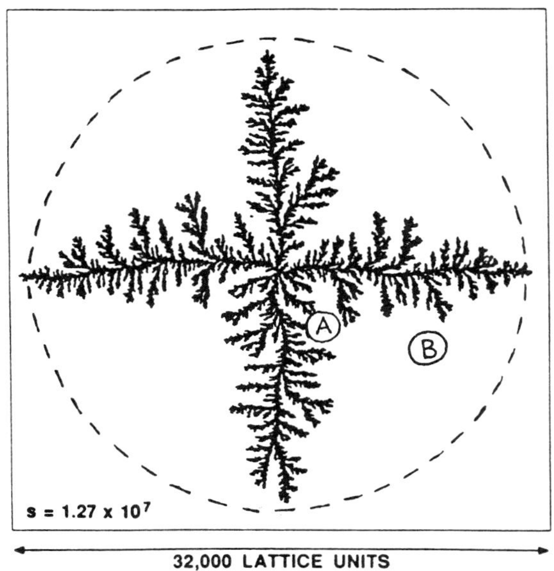

Fig.3. At very large scale DLA on a square lattice becomes cross-shaped. In this case the determination of the fractal dimension by the mass-length relation (gyration radius) becomes somewhat misleading because it mixes two different effects: One is the structure left behind by the growing interface (this is the real fractal growth problem). The other one is the instability of the velocity of the interface along various directions. The mixing of this two effects implies that two empty regions of different nature, like A and B, play the same role in the determination of the fractal dimension. [This figure is from Ref.(11) and we thank P.Meakin and S.Tolman].

for larger sizes and non circular shapes the value drops to about $D \approx 1.5$ for the largest sizes ($N \simeq 10^6$). This result has given rise to a lot of discussion about the effect of lattice anisotropy on the asymptotic value of the fractal dimension [11].

In our opinion however this discussion has been confused by the following effect. The mass length ratio measures the scaling of the filling of the portion of space occupied by the structure as a function of its size. In Fig.3 the radius of the structure corresponds to the maximum extension of the branches. This defines therefore the volume enclosed by the dashed circle. Within this circle however one can identify empty regions of totally different nature. For example the empty region A is <u>within</u> the structure and the growing interface has already passed through it, so if the structure would grow further most likely the region A will remains empty asymptotically. The region B instead is empty because the front of the structure has not yet arrived there. In the mass length (gyration radius) determination of the fractal dimension these two types of empty regions play the same role while they have different nature. Therefore the gyration radius mixes the real fractal properties of the structure (those within the growing interface) with the eventual overall shape of the entire system. This implies that for non circular shapes one should use other methods to define the fractal dimension.

<u>Two different Problems</u>

From this discussion and from various other data it is clear that the overall shape of the growing interface and the structure produced by the growth process are <u>two different problems</u>. From the recent analysis of Arneodo et al [12] one can actually conjecture that the average growing interface of DLA may be governed by the same equations as those of the interface of the deterministic Saffman-Taylor problem that produces instead compact structures.

The fractal side of the problem should instead be defined in the following way. Given a region of space in which the growing interface passes, what are the asymptotic properties of the structure that is <u>left behind</u>? It is only in this sense that we will discuss the problem in the following. One may notice that also for the effect of noise reduction and for the determination of the width of the growing interface it would be useful to make a clear distinction between the two problems.

<u>The Fractal Dimension</u>

For two dimensional DLA the various methods to define the fractal dimension give the following results (the data are from various authors reviewed in Ref.(11):

(a) Radial Geometry

- off lattice $D = 1.715$ ($N \simeq 10^6$)
- square lattice (circular shape; small sizes) $D = 1.71$
- density-density correlation $D = 1.66 \div 1.68$
- box-counting
$D(q=0)=1.61$; $D(q=1)=1.65$; $D(q=2)=1.65$
- box counting
  $Dq(-15 \le q \le 15) = 1.60$

(b) Cylinder Geometry (from Refs.(9,10)).

- Box counting (scaling regime) $D = 1.66$

- Box counting (intersection of steady state regime) $D \simeq 1.60$.

If DLA would produce a fractal structure with universal properties all these values should coincide. To some degree they do because the fluctuations in the observed values of D are of the order of 5%. However these data are from rather accurate simulations and these fluctuations may also carry a deep message that at the moment is not quite clear. In summary:

- The fractal dimension is roughly universal, however it shows small but persistent variations as function of the method of analysis and the of the geometry.

- For DBM the value of D as a function of the parameter η varies continuously from $D(\eta=0)=2$ (Eden limit) to $D (\eta \to \infty) = 1$.

<u>Anisotropy and Self-Affine Properties</u>

Just by looking at the structures (Fig.1) it is evident that the structure is connected in the growth direction but not connected in the direction perpendicular to it. This gives rise to different correlation propertied in these directions [11]. Detailed studies for the cylinder geometry show in addition that the nature of the clusters is self-affine [9,10]. Of course also in such a case it is possible to define a fractal dimension but its meaning is not as general as in the case of isotropic scaling. These effects are a warning that the structure has fractal properties but not of the simplest type. This may actually be a possible reason for the fluctuations observed in the value of D as function of geometry and method of analysis.

Dependence of D on the Space Dimension

Until now we have discussed only growth in a two dimensional embedding space (d=2). Simulations have actually been performed up to d=8. The results for DLA are [11]

d =	2	3	4	5	6	7	8
D =	1.71	2.48	3.4	4.3	≈ 5.3	≈ 6.2	≈ 7.2

and refer to a determination of D by the mass length relation. For large values of d the fractal dimension D(d) seems to be in agreement with the relation [13]

$$\lim_{d \to \infty} D(d) = d - 1 \quad . \tag{3}$$

An important conclusion we can draw from these data is the <u>absence of an upper critical dimension</u>. This is a fundamental difference with respect to usual critical phenomena and the statistics of self-avoiding polymers for which such a dimension exists and it is of basic importance for the development of RG methods [7].

4. MULTIFRACTALITY

Another concept that has played a role in this field is that of multifractality [14]. This is a generalization of the concept of fractal in which one considers the possibility that a continuum distribution of different singularities can be present instead of a single type as in the case of a simple fractal. This extension does not cover however other possible complications like self-affinity. Multifractals arise naturally in self-similar distribution even as given from a simple multiplicative process [15]. Clearly the growth probability in DLA and DBM is a distribution with some sort of self-similar properties so it was natural to expect multifractal features. One has to say however, that even if multifractals appear to describe some features of this probability, it is also clear that they do not provide a complete description nor it is evident whether it is really useful to look at the problem from this perspective. In this respect it should be made clear that, despite some claims that multifractals should be the key concept to build a theory of fractal growth, not much has been achieved in this direction.

Since the introduction of multifractality there has been a large amount of work aimed at the characterization of the growth probability in DLA and DBM in terms of a multifractal spectrum. After the initial attempts that pointed towards a relatively simple description [16], it has recently become clear that the situation is actually much more complex and the problem is, at the moment, quite controversial [17-20]. Most of these studies refer to radial geometry. Our opinion however is that the cylinder geometry is more appropriate to study DLA and DBM from a numerical point of view but also to formulate unambiguous theoretical concepts. We have therefore studied the properties of the growth probability for the cylinder geometry. The main results are [21]:
- The projection of the growth probability along the growth direction is <u>exponential</u> (Fig.4). Considering that the radial probability distribution obtained by Plischke and Racz [22] was <u>gaussian</u> this implies that the region of small probability is strongly non universal.

- There is strong evidence for a phase transition in the multifractal spectrum at q=1 (Fig.5). This is due to the fact that the growth probability consists of two separate regions. The growing zone with simple multifractal properties and rather universal character and the frozen region whose properties are strongly geometry dependent and give rise to anomalous multifractal properties. Notice that this division can be interpreted geometrically only at the largest scale, while for smaller scales it concerns the singularity exponents.

- The growth process is determined only by the regular growing zone. The anomalous part is irrelevant with respect to the growth process even though it is determined by it. Different geometries induce small differences in the fractal structure that however have a large effect on the regions with small probabilities.

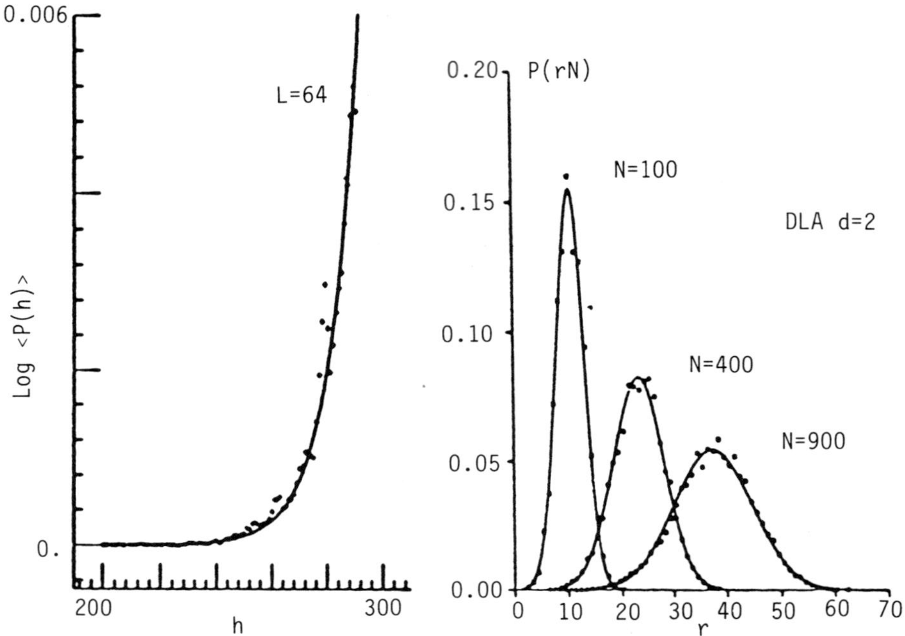

Fig.4. The behavior of the projection of the growth probability on the direction of growth for cylinder geometry (left, Ref.(21)) and for circular geometry (right (Ref.22)). The variable on the x-axis is the height from the bottom line in the former case, while it is the distance from the center of mass in the second one.

- It is possible to generalize the method of the Fixed Scale Transformation (see Sect.6) to compute and analytically the regular part of the multifractal spectrum. The separation between the growing and the frozen zone appears natural in this method and the non universal properties of the small probability regions can be related to the high order terms in the FST.

5. WHY RENORMALIZATION GROUP IS PROBLEMATIC

The fact that the models produce structures with self-similar properties raised the expectation that their understanding could be achieved following the ideas developed for

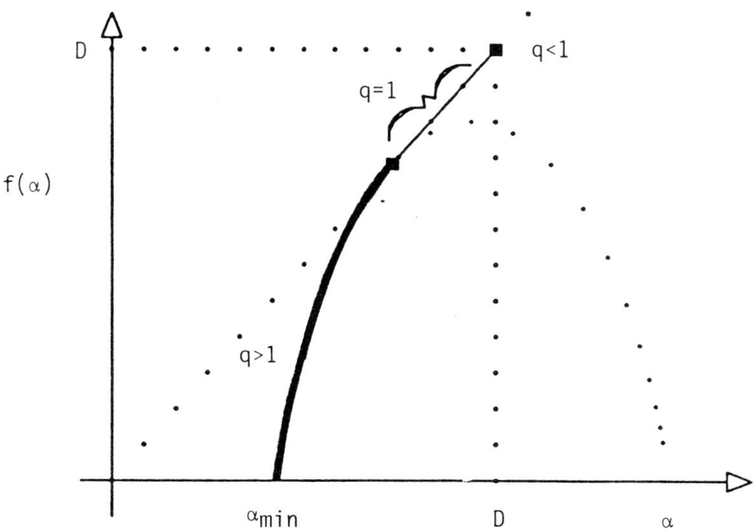

Fig.5. First order phase transition at q=1 in f(α) spectrum (Ref.(21)). The α<1 part corresponds to the values q>1, the straight line from α=1 to α=D corresponds to the point q=1 and the point α = f = D to the q<1 part of the τ(q) curve.

critical phenomena. There are however important differences between these two problems and the application of RG ideas to fractal growth turned out to be quite problematic even though some of the approaches are interesting [23]. The main characteristic of fractal growth that are different from usual critical phenomena are:

-   The process is irreversible and cannot be described in Hamiltonian terms. In addition it is not possible to assign a statistical weight, like the Boltzmann factor to a given configuration because the entire history of the of the process has also to be considered.

-   There is no upper critical dimension.

-   Universality exists only in weak form and in this respect the problem is not just to compute very accurately a couple of independent exponents. After all the fractal structures one observes in nature are quite various.

-   The process is intrinsically critical in the sense that the system evolves spontaneously towards a self-similar structure. It does not require the adjustment of a parameter (like the temperature in phase transition).

    The basic idea of RG methods is to integrate over some "internal" degrees of freedom in order to define new "coarse grained" variables and then to iterate this process until it reaches a fixed point. This procedure is not very appropriate for fractal growth because, as we have seen, the fractal features refer to the properties of a structure that is frozen, namely it does not grow any more if the process is continued. This implies that the asymptotic limit is reached when the growing interface is very far from the considered structure. It is clearly problematic to describe this situation by integrating internal degrees of freedom of any type. A more detailed discussion of this point can be found in Ref.(24). A practical consequence of this problem is that it is impossible with RG methods to recover the trivial Eden limit, namely that in DBM one has D=2 for η = 0. This is a fundamental problem because it shows that <u>RG methods are not able to distinguish between fractal and non fractal structures</u>. We are going to see in the next section that the Fixed Scale Transformation can properly deal with these problems.

6. NEW APPROACH: THE FIXED SCALE TRANSFORMATION

Here we briefly describe a new approach called Fixed Scale Transformation [24,25] that is based also on a different invariance property than the Renormalization Group. The starting point is to look at the fractal structure as the attractor of a dynamical system. One should define iterative equations whose fixed point corresponds to the fractal structure.

The first important problem is therefore to define an appropriate **subject** that should appear in these iterative equations. It is useful in this respect to consider the two dimensional case with boundary conditions consisting of two parallel lines. The discussion is easily generalizable to higher dimensions. This geometry has the conceptual advantage of defining a unique growth direction. The intersection of a given fractal structure of dimension D with a line perpendicular to the growth direction gives rise to a set of points of dimension $D'=(D-1)$. By analyzing this set of points with a procedure of box covering we assign a black dot to a box if this contains some point of the set and a white dot otherwise. The elementary process by which a black box is subdivided into two leads to two possible configurations indicated as type 1 (one black and one white sub-box) and type 2 (both sub-boxes black). The corresponding probabilities in this process of fine graining are indicated by C_1 and C_2 respectively. The average number of black sub-boxes that appear at the next level of fine graining from one black box is [24,25]

$$\langle n \rangle = \sum_i n_i C_i = C_1 + 2 C_2 \tag{4}$$

It is easy to show that the fractal dimension of the whole structure is related to the values of C_1 and C_2 by

$$D = 1 + \frac{\ln \langle n \rangle}{\ln 2} \tag{5}$$

It is clear therefore that the natural subject for the iterative fixed point problem should be the **distribution of elementary configurations** $\{C_i\}$ that appear from one scale to the next in the process of fine graining.

<u>Fixed Scale Transformation.</u> We have now to define the appropriate iterative equations for the distribution $\{C_i\}$. From the point of view of the analogy with critical phenomena one would attempt to define the iteration via a renormalization scheme. The present problem however has the peculiarity of being **intrinsically critical**; namely there is no order parameter that leads to scale invariance only for a particular value of it. The growth rules are scale invariant by construction. This leads to a higher degree of symmetry :
- The structure is invariant by **scale transformation** in the sense that the values of (C_1, C_2) that are obtained via a fine graining from scale l to scale l/2 are the same as those from scale l' to scale l'/2.
- In addition there is also an invariance with respect to the **dynamical evolution at the same scale**. Namely if one considers two different sections of the original structure and performs a fine graining analysis at the same scale for the two different sets one obtains the same distribution (C_1, C_2) that should also be identical to the previous case of scale change.

Note that this discussion holds for a homogeneous as well as a self-affine structure. We have now a choice for the formulation of the iterative process: One can use a renormalization transformation based on the first invariance property , or one can define the iteration from the dynamical evolution at the same scale and then use the fixed point value of the (C_1, C_2) distribution at all other scales via Eqs.(4) and (5).In view of the discussion in the previous section this second possibility is more convenient for this problem.

We define a **Fixed Scale Transformation** (FST) that corresponds to the dynamical evolution at the same scale. This gives rise to an iterative equation of type [24,25]

$$\begin{pmatrix} C_1^{(k+1)} \\ C_2^{(k+1)} \end{pmatrix} = \begin{pmatrix} M_{1,1} M_{2,1} \\ M_{1,2} M_{2,2} \end{pmatrix} \begin{pmatrix} C_1^{(k)} \\ C_2^{(k)} \end{pmatrix} \tag{6}$$

where the matrix element $M_{i,j}$ defines the conditional probability to have a configuration of type i followed by one of type j in the growth direction. Note that this refers to a **"frozen structure"** that has already grown to its asymptotic state and in which no more growth will occur in the future.

The practical calculation of $M_{i,j}$ can be done by starting with a configuration of type i and consider the growth processes that can lead to a subsequent configuration j. For example

$$M_{1,2} = \sum (\text{growth processes leading from 1 to 2}) = p_\alpha + (1 - p_\alpha)\, p_\beta + \cdots \tag{7}$$

where p_α represents in this case the probability that the first non trivial growth process leads to a configuration of type 2, while p_β corresponds to the second order term and so on. More details about the explicit construction of Eq.(7) can be found in Ref. (24).

Once the matrix elements are defined and considering that

$$\begin{cases} M_{11} + M_{12} = 1 \\ M_{12} = M_{22} = 1 \end{cases} \tag{8}$$

the fixed point condition gives

$$C_1 = \left(1 + \frac{M_{12}}{M_{21}}\right)^{-1} \tag{9}$$

and from this, via Eqs.(4,5) one obtains the fractal dimension D.

The **mathematical condition for the generation of fractal structures** can be now given by studying the value to which the series of Eq.(7) will converge. If this value is less than one this implies that there is a finite probability that the growth process will leave sites asymptotically empty. In view of the scale invariance of the Laplace equation and of the whole growth process, this conclusion holds at any scale and therefore holes of all scales will be generated.

In view of the screening properties of Laplace equation one can see that the convergence of this series is quite fast and indeed to a number smaller than one. In the limit $\eta \to 0$ the screening is eliminated, the convergence is then slower and the series converges to one, leading to a compact structure with dimension 2.

<u>Fluctuations of Boundary Conditions and Void Distribution.</u> In order to compute explicitly the matrix elements one has to specify the boundary conditions for the process of conditional growth. This is a new type of problem that usually does not appear in the Renormalization Group method in which one integrates over the internal degrees of freedom and boundary conditions play no role. In our problem the boundary conditions instead have an important effect. In fact the growth of a structure will be very different if it is isolated or surrounded by other growing structures. In principle therefore one has to take into account all these possibilities, each with its own probability.

Given the probability distribution (C_1, C_2) it is possible to derive explicitly the void distribution $P(\lambda)$ which is the conditional probability that, given an occupied box, it is neighboured at the right by a void of size λ. In Ref.(24) we have presented an approximate calculation but one can actually do it exactly for all the values of the void size (26). The result is given in an iterative form

$$
\left\{
\begin{array}{l}
P(\lambda = 0) = \dfrac{C_2}{\left(1 + \frac{1}{2}C_1\right)\left(C_2 + \frac{1}{2}C_1\right)} \\[4mm]
P(\lambda = 2l + 1) = \dfrac{(1 - C_2)}{2}\, P(\lambda = l) \\[4mm]
P(\lambda = 2l) = \dfrac{(1 + C_2)}{4}\, P(\lambda = l) + \dfrac{(1 - C_2)^2}{4(1 + C_2)}\, P(\lambda = l - 1)
\end{array}
\right. \tag{10}
$$

It is easy to check that the integrated void distribution has the expected behaviour [1]

$$
P(\lambda \geq \Lambda) = F \Lambda^{-D'} \tag{11}
$$

where F is the lacunarity and D' the fractal dimension of the intersection set.

With respect to the use of this distribution as defining the boundary conditions of our elementary growth processes there is an additional fact to consider. Our basic configurations correspond to pairs of sites of which at least one is black, so at the next level of coarse graining they correspond necessarily to a black site. To this site we can directly apply the above void distribution. However if we ask for the probability that a pair of sites is immediately followed by a black (occupied) site the answer is sligthly more

complicated. In fact with pobability $P(\lambda = 0)$ this pair is followed by another pair in which at least one of the two sub-sites is black. So if we consider the probability that the site immediately following the considered pair is black this will require the extra condition that the pair configuration on the right is of type 2 or of type 1 but with the black site on the left. This gives rise to an additional factor $(C_2 + (1/2)C_1)$ and therefore the corrected

probability (for example for $\lambda = 0$) to be used in the calculation of the matrix elements is

$$
P(\lambda = 0) = \dfrac{C_2}{\left(1 + \dfrac{1}{2}C_1\right)} \tag{11}
$$

Note that in Ref.(24) the discussion of this distribution is not very clear but the actual calculation is finally correct and in agreement with the present discussion.

The probability distribution for the size of voids fixes the probability distribution for having a certain boundary condition in the elementary growth process. This implies that the generic matrix element $M_{i,j}$ should be interpreted as the convolution over all possible boundary conditions

$$M_{i,j} \Rightarrow \sum_n P(\lambda_n; C_1, C_2) M_{i,j}(\lambda_n) \qquad (12)$$

Therefore these matrix elements become nonlinear functions of the variables (C1,C2). The iterative equation has then the following structure [24,25]

$$\begin{pmatrix} C_1^{(k+1)} \\ C_2^{(k+1)} \end{pmatrix} = \sum_n P(\lambda_n; C_1, C_2) \begin{pmatrix} M_{1,1}(\lambda_n) & M_{2,1}(\lambda_n) \\ M_{1,2}(\lambda_n) & M_{2,2}(\lambda_n) \end{pmatrix} \begin{pmatrix} C_1^{(k)} \\ C_2^{(k)} \end{pmatrix} \qquad (13)$$

and the corresponding fixed point equation

$$C_1 = \left[1 + \frac{\sum_n P(\lambda_n; C_1, C_2) M_{1,2}(\lambda_n)}{\sum_n P(\lambda_n; C_1, C_2) M_{2,1}(\lambda_n)} \right]^{-1} \qquad (5.12)$$

is now a nonlinear system of equations of infinite order. This can be solved by suitable truncation schemes because the higher orders give exponentially decreasing contributions. The simplest nontrivial method to include the boundary condition fluctuations consists in

assuming that as soon as λ_n is different from zero, the boundary condition is essentially open (in the sense that the next branch can be considered as infinitely far). This approximation is based on the fact that the convergence of the series that define the matrix elements is rather fast so one has to include only a few orders in the calculation. This implies that the considered structure does not change its size appreciably and therefore the relative probabilities within the structure itself are not too sensitive to the distance of the next branch as soon as this is not very close. This we have called the open-closed approximation and it is possible to treat it analytically .

The results of various approximation schemes can be found in Ref.(24) and they show an appreciable degree of systematicity. In particular one can see that the self-consistent treatment of boundary conditions has an important effect. It can also be noted that the agreement with the computer simulations is very good for large values of η and less accurate for smaller values. This can be understood in terms of the screening properties of the Laplace equation that give rise to a faster convergence of the series for large values of η . One should notice however that, strictly speaking the FST results should compare to the box counting dimension that gives an even better agreement with respect to the numerical data reported in Ref.(24).

Finally it should be remarked that the spirit of the present calculation is quite different from Real Space Renormalization not only because one uses a transformation at fixed scale. In fact the size and the number of basic configurations is **strictly fixed** and the calculation is improvable in a systematic way by adding more terms in the series that define the matrix elements of the transformation. In addition growth processes outside the considered cell must be included up the desired level of convergence and the fluctuations of boundary conditions play an important role. These concepts appear natural and necessary in all problems of irreversible growth.

After the introduction of this new method we have developed it along two lines. One is to deepen the conceptual framework of the method and to study its systematicity in

many respects [27]. The other one is to test it on various problems with well known properties. In particular the FST method has been applied succesfully to:
-Fractal and multifractal properties of DLA and DBM in two and three dimensions. Ref.(21,24,25,27,28).
- Percolation in square and triangular lattices. Ref.(29,30).
- Invasion percolation with and without trapping. Ref.(31).
- Ising and Potts Clusters and Droplets. Ref.(32).
- Nature of Cluster aggregation. Ref.(33).

In summary, even if some important questions concerning the self-similarity of growth rules have still to be clarified, the FST method represents, to our knowledge the most succesful and complete theoretical approach to fractal growth.

7. SUMMARY AND OPEN PROBLEMS

The fractal growth models DLA and DBM show how stochastic fractal structures arise from physical processes. They are considered the prototypes of fractal growth models and their role in this field is analogous to the Ising model for critical phenomena.

- Roughly speaking these models give rise to fractal structures that are about the same for different geometries and methods of analysis. There are however small but persistent variations that suggest that the concept of universality is less important for fractal growth with respect to critical phenomena.

- The growth probability shows multifractal features characterized by a universal branch in the spectrum for q>1 and a strongly non universal behavior for the properties of the small probabilities. It is not clear at all that looking at these probability from a multifractal perspective may help in developing theoretical concepts.

- The application of Renormalization Group ideas to these problems turned out to be quite problematic. This is due to various reasons but mainly to the fact that the elimination of the "internal degrees of freedom" is not a natural procedure for these growth processes.

- We have recently proposed an alternative theoretical framework named Fixed Scale Transformation that is based also on a different invariance property with respect to RG. This method allows to understand the origin of fractal structures in these models and to compute its value in a reasonably systematic way. It has also been succesfully applied to various other problems.

- The main open question in the FST approach is that the growth rule of the minimal scale is assumed to hold at all scales. Even if the growth rule itself does not possess any characteristic length this assumption is not obvious. We have performed various studies about this point but a complete understanding has not yet been achieved. The two extreme possibilities are the following. If the growth rule of the minimal scale would strictly reproduce itself at all scales then the present FST method is fully consistent. If on the other hand the growth rule would change with scale up to an asymptotic growth rule one should use this last one in the FST. In this last case the reason one obtains good values for the fractal dimension even using the growth rule of the minimal scale would be the same for which one can obtain good exponent from the exact enumeration of small systems.

REFERENCES

1. B.B.Mandelbrot, "The Fractal Geometry of nature", Freeman, New York (1982).

2.   L.Pietronero and E.Tosatti, "Fractals in Physics", North-Holland, Amsterdam, New York (1986).

3.   L.Pietronero, "Fractals' Physical Origin and Properties", Plenum Publ., New York - London (1989).

4.   L.Pietronero, Physica A $\underline{144}$, 257 (1987); P.H.Coleman, L.Pietronero and R.H.Sanders, Astron. and Astrophys. $\underline{200}$, L32 (1988).

5.   T.A.Witten and L.M.Sander, Phys.Rev.Lett. $\underline{47}$, 1400 (1981).

6.   L.Niemeyer, L.Pietronero and H.J.Wiesmann, Phys.Rev.Lett., $\underline{52}$, 1038 (1984).

7.   See e.g. D.J.Amit, "Field Theory, the Renormalization Group and Critical Phenomena", McGraw-Hill, New York (1978).

8.   H.E.Stanley, Phil. Mag. B $\underline{56}$, 665 (1987).

9.   C.Evertsz, Phys.Rev.B $\underline{41}$, 1830 (1990).

10. C.Evetsz, Ph.D. Thesis, University of Groningen, The Netherlands (1989).

11. P.Meakin and S.Tolman, in Ref.(3), p.137.

12. A.Arneodo, Y.Conder, G.Grasseau, V.Hakin and M.Rabaud, Phys.Rev.Lett. $\underline{63}$, 984 (1989).

13. R.C.Ball and T.A.Witten, Phys.Rev. A $\underline{29}$, 2966 (1984).

14. G.Paladin and A.Vulpiani, Phys.Reps $\underline{156}$, 145 (1987).

15. L.Pietronero and A.P.Siebesma, Phys.Rev.Lett. $\underline{57}$, 1098 (1986).

16. A.Coniglio, C.Amitrano and F.Di Liberto, Phys.Rev.Lett. $\underline{57}$, 1016 (1986).

17. T.Bohr, P.Cvitanovic and M.H.Jensen, Europhys. Lett. $\underline{6}$, 445 (1988).

18. R.Blumenfeld and A.Aharony, Phys.Rev.Lett. $\underline{62}$, 2927 (1989).

19. S.Schwarzer, J.Lee, A.Bunde, S.Havlin, H.E.Roman and H.E.Stanley, Phys.Rev.Lett. $\underline{65}$, 603 (1990).

20. B.B.Mandelbrot and C.J.G.Evertsz, Nature $\underline{348}$, 143 (1990).

21. M.Marsili and L.Pietronero, preprint.

22. M.Plischke and Z.Racz, Phys.Rev.Lett. $\underline{53}$, 415 (1984).

23. T.Nagatani, J.Phys.A $\underline{20}$, L381 (1987); Phys.Rev.A $\underline{36}$, 5812 (1987).

24. L.Pietronero, A.Erzan and C.Evertsz, Physica A $\underline{151}$, 207 (1988).

25. L.Pietronero, A.Erzan and C.Evertsz, Phys.Rev.Lett. $\underline{61}$, 861 (1988).

26. R.R.Tremblay and A.P.Siebesma, Phys.Rev.A $\underline{40}$, 5377 (1989).

27. A.Vespignani and L.Pietronero, Physica A $\underline{168}$, 723 (1990).

28. A.Vespignani and L.Pietronero, Physica A, in print.

29. L.Pietronero, W.R.Schneider and A.Stella, Phys.Rev.B, Rapid Comm., Nov.15, 1990.

30. L.Pietronero and A.Stella, Physica A (1990), in print.

31. L.Pietronero and W.R.Schneider, Physica A (1990), in print.

32. A.Erzan and L.Pietronero, J.Phys.A, in print.

33. L.Pietronero and W.R.Schneider, preprint.

CRACK FORMATION: CROSSOVERS BETWEEN DIFFERENT GROWTH REGIMES AND CRITICAL BEHAVIOR

O. Pla and F. Guinea[†], E. Louis[‡], L. M. Sander['] and P. Meakin[‖]

[†] Instituto de Ciencia de Materiales (CSIC). Facultad de Ciencias. Universidad Autónoma. 28049 Madrid. Spain

[‡] Departamento de Física Aplicada. Universidad de Alicante. Apartado 99 03080 Alicante. Spain

['] Department of Physics. University of Michigan. Ann Arbor MI48109. USA

[‖] Central Research and Development Department. E. I. Du Pont de Nemours Wilmington DE19880-0356. USA

I. Crack formation with growth laws which are not scale invariant

It has been customary to describe various growth and aggregation processes by means of power laws, in which the probability of growth of a point at the surface of the pattern is proportional to a power of the value of a given field at that position. These laws are scale invariant, and, when combined with the randomness inherent to the growth process, they give rise to fractal objects, which are also scale invariant [1-6]. However, we do not expect this to be the most general behavior in nature, and more complex growth laws, with intrinsic length and time scales are likely to occur.

In the present work, we will show how simple concepts allow a qualitative understanding of the shapes formed by means of growth laws which are not scale invariant. We make extensive use of the accepted knowledge of the effects due to scale invariant laws[7].

We concentrate mostly on fracture models[1]. The *field* which describes the growth process is given by the distribution of stresses outside the crack in formation. Despite the vectorial character of the model, the main features are similar to scalar problems, like DLA or dielectric breakdown. The rich phenomenology of fracture processes[8,9] makes specially relevant the study of complex growth laws.

The simplest growth law with an intrinsic scale is tha combination of two scale invariant ones, like $P(T) \propto T \times (1 + (T/T_0)^{\eta - 1})$, where $P(T)$ is the probability that a growth site under stress T will fail at a given instant. The fractal dimension of the cracks formed under homegeneous laws (which correspond to different choices of η) is well understood[4]. The parameter T_0 sets the scale at which one of the laws dominates over the other. It is clear that, by using suitable expansions, more complex laws can written as a combination of many (maybe infinite) power laws, with the required stress scales which determine which is the most relevant power, η.

Growth and Form, Edited by M. Ben Amar *et al.*
Plenum Press, New York, 1991

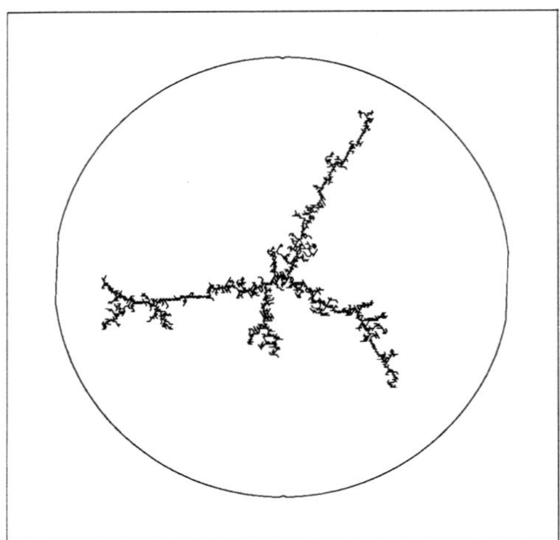

Figure   1

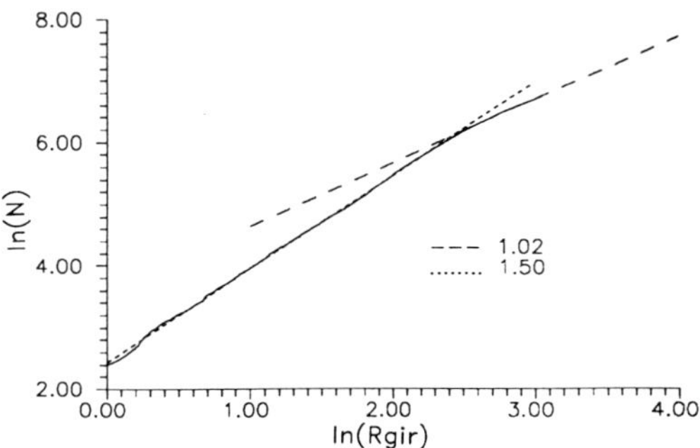

Figure   2

We have performed simulations of crack formation under this law for various choices of η and T_0. In order to understand the results, it is convenient to start with a simple case, like the one depicted in figure (1). There $\eta = 10$, and $T_0 = 0.45$. The system was subjected to an external stress in such way that its value at the boundaries was 0.1. The crack shown has ~ 1400 broken bonds. An analysis of the shape of the crack (averaged over five realizations is shown in figure (2)).

The results are easy to interpret: During the first stages of the growth process, the sresses at the edges of the crack are determined mainly by the applied pressure, which lies well below the internal scale of the growth law, T_0. Then, the shape of the pattern is that associated with the smallest power, $\eta = 1$, and resembles those obtained previously, with a fractal dimension $D \sim 1.5$. As the surface of the crack becomes rougher, and relatively thin branches and tips develop, the stresses at the boundaries increase above T_0 and the higher power in the growth law dominates. The resulting shape has very long and thin branches, with a fractal dimension close to 1. This regime attained in this fashion is self consistent, in the sense that the stresses continue to increase as the branches become longer and thinner. Thus, for long times or large crack sizes, the only relevant part of the growth law is that arising from the $T \rightarrow \infty$ limit.

In order to reach this limit, it is necessary for the stresses at the crack edges to increase indefinitely. The tendency of the boundaries to become rougher and rougher (that even happens in the Eden model, $\eta = 0$) leads to such a situation. Numerical simulations in small systems are hindered by the fact that, as the crack approaches the limits of the sample, stresses relax and tend to be equal to the applied pressure. Thus, the crossover mentioned before tends to be obscured and even reversed when the powers combined in the growth law are similar[6].

Finally, it is worth noting that, when the growth law saturates at high stresses (like $P(T) \propto \tanh(T/T_0)$) an inverse crossover, towards a more dense object, with $D \sim 2$ takes place (see figure (3)).

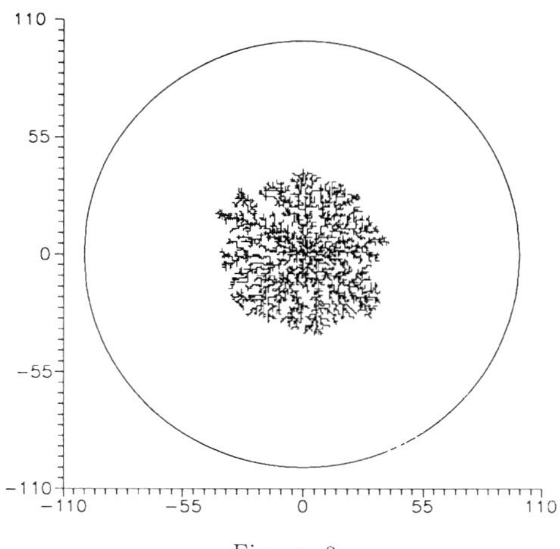

Figure 3

II. Noise and criticality in growth processes

It has been recently proposed that many systems with complex non linear dynamics, when started in complex and random configurations, evolve towards a *self similar critical state*[9-10]. This situation is characterized by being very close to losing equilibrium, in the sense that small perturbations can cause rearrangements of arbitrarily large sizes.

We now present simulations to test whether this hypothesis also applies to growth processes intrinsically out of equilibrium. We study the most common models in the literature, like DLA, dielectric breakdown, and crack formation.

The basis assumption required to explain the existence of a self similar critical state is that the situation of equilibrium first reached by the system is that easiest to destabilize. It cannot be unstable but, there are, presumably, many critical states where the system is stuck, before it can move towards the configurations with lowest energy. In the present context, where the system is always evolving out of equilibrium, the critical state which can be achieved can only refer to the *stationary* situation which describes the evolution at long times. That a stationary state is reached in the growth processes that we consider is well established by numerical calculations: a self similar pattern describes the shape of the system after the initial stages, and the resulting fractal dimension does not change with time.

The way growth takes place is determined by the values of a field at the edges of the pattern, which is, in turn, altered during the evolution of the system. Thus, in order to define the analog of a perturbation or catastrophe in the static case, we can analyze the size of the changes in this field after each growth process. In the spirit of the self critical state hypothesis, we can conjecture that a stationary process is reached when large rearrangements in this field are not too frequent. On the other hand, as soon as alterations which modify the fields over the entire object dissappear, a stationary situation will be obtained. Thus, in order to check this possibility, the simplest procedure is to estimate the distribution of changes in the field which determines the probability of growth as the system evolves.

To increase the accuracy of the calculations, we have analyzed the frecuency of catastrophes when growth takes place at different sites of the boundary of a large, static aggregate. In order to obtain information about dynamical properties, we only need to weight these frecuencies by the probability of growth at each site. The size of the rearrangemets has been defined in two ways: the number of sites around the perturbed position where the field has changed above a given threshold, and the number of iteration steps required to achieve numerical convergence above another threshold. The first definition describes, approximately, the spatial extent of the perturbation induced by the growth step, while the second method is associated with the "time" that the system requires to reach again equilibrium. The results, for aggregates created using the dielectric breakdown algorithm are shown in figure (4).

In both cases, a power law can be defined over many scales, thus confirming the above hypothesis. The aggregates are fractals with $D \sim 1.7$. It is interesting to note that, when the growth law is such tnat more deterministic objects are obtained (and smaller fractal dimensions), the frecuency of large catastrophes decreases.

We have also analyzed quantities closer to real power spectra dissipated as the system evolves. The change in the energy of the field after each growth step (that is, the energy stored in the electrostatic field in the case of DB) is shown in figure (5), along with the corresponding Fourier spectrum. We take the (discrete) time to be proportional to the number of growth steps. While the results are numerically less accurate, a power law can be identified as well.

It is finally interesting to remark that the exponents which describe the different power laws become larger and larger as the patterns which are being formed look more and more deterministic. In other words, the distribution of frecuencies move away from the large catastrophes side of the spectrum. Turning the argument around, it seems likely that stationary regimes cannot be achieved by any choice of exponents, but that there is a minimum threshold required for the existence of such processes. Then, the ubiquous appearance of power laws with exponents close to -1 can be understood.

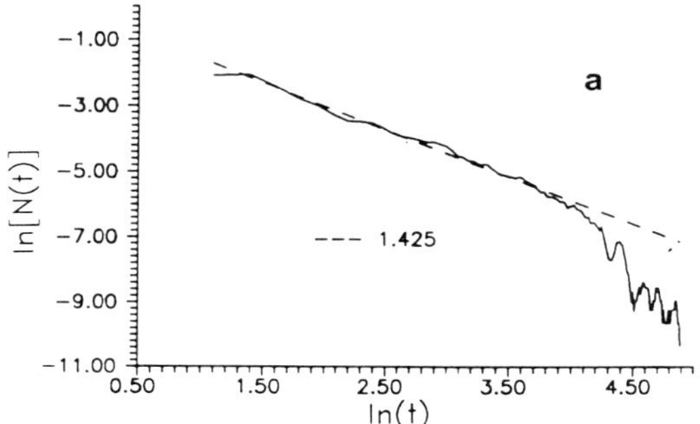

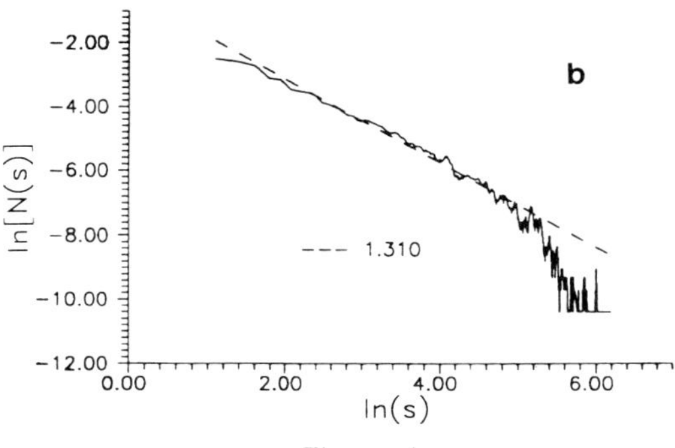

Figure 4

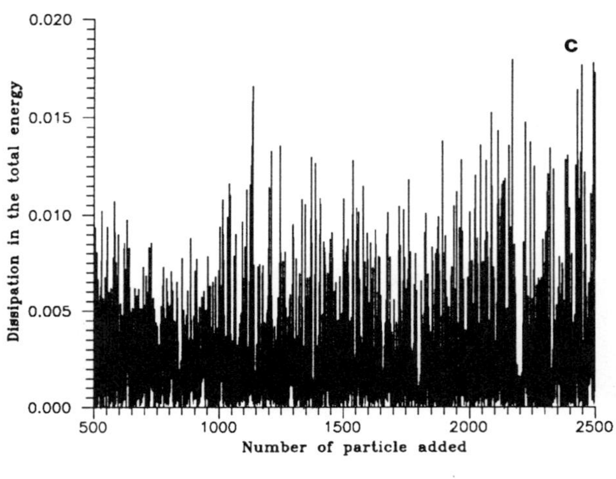

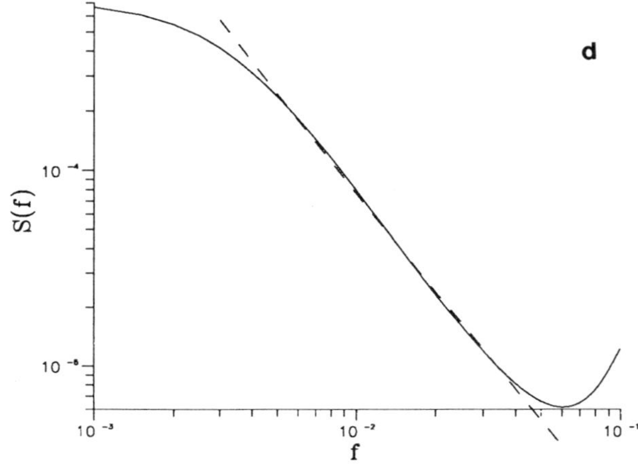

Figure 5

References

[1] E. Louis and F. Guinea, Europhys. Lett. **3** 871 (1987); L. Navas, F. Guinea and E. Louis, J. Phys. A **21** L301 (1988); M. P. López-Sancho, F. Guinea and E. Louis, J. Phys. A **21** L1079 (1988).

[2] H. J. Herrmann, J. Kertész and L. de Arcangelis, Europys. Lett. **10** 147 (1989).

[3] P. Meakin, G. Li, L. M. Sander, E. Louis and F. Guinea, J. Phys A **22** 1393 (1989).

[4] L. Niemeyer, L. Pietronero and H. J. Wiesmann, Phys. Rev. Lett. **52** 1033 (1984).

[5] T. A. Witten and L. M. Sander, Phys. Rev. Lett. **47** 1400 (1981); Phys. Rev. **B27** 5686 (1983).

[6] Part of this work has been reported in P. Meakin, G. Li, L. M. Sander, H. Yan, F. Guinea, O. Pla and E. Louis, *Cargèse Summer School on Pattern Formation Models* (1989), and Phys. Rev. A, to be published.

[7] Y. Termonia and P. Meakin, Nature **320** 6061 (1986).

[8]R. M. Latanision and J. R. Pickens (eds.) **Atomistic of Fracture** Academic Press, New York (1984).

[9]P. Bak, C. Tang and K. Wisenfeld, Phys. Rev. Lett. **59**, 381 (1987), and Phys. Rev. **A 38**, 364 (1988).

[10]P. Bak, K. Chen and M. Creutz, Nature **342**, 780 (1989).

THEORETICAL MODELS FOR CRYSTAL GROWTH FROM ATOM BEAMS

J. Villain

DRF/MDN, Centre d'Etudes Nucléaires de Grenoble
85 X, F-38041 Grenoble Cedex, France

ABSTRACT. *Continuum equations appropriate to describe crystal growth from atom beams are derived in various cases. When desorption is important, the growth is described on very long lengthscales by the Kardar-Parisi-Zhang equation, but should be corrected for shorter lengthscales where surface diffusion is the dominant mechanism. In the absence of desorption, an important effect at sufficiently low temperature comes from the fact that diffusion of incoming atoms on the surface is anisotropic on long lengthscales becaused it is biased by reflexions against terrace edges. As a result, the growth is described by a pseudo-diffusion equation. In the case of a high symmetry surface, (001) or (111), an instability arises. Finally, in the absence of diffusion bias, the growth is described by a nonlinear equation of fourth order with respect to to $\partial/\partial x$ and $\partial/\partial y$, and the appropriate exponents are calculated in a Flory-type approximation.*

1. INTRODUCTION

1.1. Mechanisms of crystal growth

This note is intended to give a simple theoretical description of the macroscopic or mesoscopic aspects (beyond, say, 50 atomic distances) of crystal growth by atom beams. It is of interest to recall first an essential difference with growth from the fluid phase: the destabilising effect of heat or impurity diffusion through the fluid phase is absent, so that the corresponding instabilities (Mullins and Sekerka 1963, Langer 1980, Pelcé 1988, Viscek 1989) do not appear. The interface remains macroscopically planar if it is initially planar. However it becomes rough in the following sense. Let Z be the coordinate normal to the average surface (possibly different from the beam direction z), let X and Y be two rectangular coordinates parallel to the average surface, and let $\vec{R}=(X,Y)$ be a two-dimensional vector, or a (d-1)-dimensional vector in the general case of a d-dimensional space. For an initially planar surface, the height $Z(\vec{R},t)$ will be assumed to be a uniform function of $\vec{R}$ and t, but it has fluctuations due to the fluctuations of the beam. The roughness is characterized by the correlation function

$$G(\vec{R},t) = \left\langle (Z(\vec{R}',t)-Z(\vec{R}'+\vec{R},t))^2 \right\rangle \tag{1}$$

Growth and Form, Edited by M. Ben Amar *et al.*
Plenum Press, New York, 1991

The surface will be said to be rough if $G(\vec{R},t)$ diverges when $\vec{R}$ and t go to infinity. The goal of this note is to study this roughness.

Instead of (1), it would be possible to define a correlation function $\Gamma(\vec{R},\tau;t)$ between the heights at distance $\vec{R}$ in space and τ in time after an irradiation time t. This complication, however, is not extremely useful.

It is of interest to remark that instabilities which are not of the Mullins-Sekerka type can occur in growth by atomic beams, as seen in Subsection 4.4.

1.2. Various modes of growth by atomic beams

1.2.1) The oscillatory mode. The best semiconducting devices or metallic multilayers are grown by molecular beam epitaxy (MBE) at fairly high temperatures, where diffusion is fast. In that case, when the surface is parallel to a high symmetry orientation, the roughness of the surface oscillates in time and exhibits minima (corresponding to the completion of the successive layers) separated by maxima (Fig. 1 a). These oscillations are observed in reflection high energy electron diffraction (RHEED) or by other spectroscopic methods.

This growth mode will not be treated in the present work. The models presented in the next sections are continuous, macroscopic models which are of no use to describe oscillatory growth. Presumably, these models would be in principle applicable on very large lengthscales, which would be unphysical.

We are conscious that most of experimentalists will be disappointed to see that growth in the oscillatory mode is not much studied in the present work. Our main excuse is that it is reasonable to study the simplest problems first. On the other hand, the other types of growth, addressed below, have also been experimentally investigated, and this suggests that their interest is not purely theoretical.

1.2.2) Stepped surface ("Step flow")

From the theoretical point of view, the simplest case is that of a stepped (or vicinal) surface (Fig. 1 b). Such a surface is prepared as a set of large terraces of high symmetry orientation, (001) or (111) in the case of a cubic crystal. These terraces are separated by straight, equidistant, parallel steps. When the beam is switched on, the steps go forward with an average velocity v, and consequently the crystal grows. The beam direction z will be assumed perpendicular to terraces. If ℓ is the distance between steps, the rate of growth in the direction z is

$$\dot{z} = v/\ell \tag{2}$$

1.2.3) High symmetry surface at moderate temperature

The oscillating growth described in § 1.2.1 is observed only at high enough temperatures. At low temperature the atoms have no possibility to move and to look for the lowest energy configuration, so that the resulting object is amorphous rather than crystalline. On the other hand, at moderate temperature, surface diffusion is fast enough to allow the growth of a crystal, but the surface will remain appreciably rough (Fig. 1 c) on lenghscales larger than a temperature-dependent limit, which becomes microscopic at low temperatures. This growth mode might be of interest in the case of certain multilayers in order to avoid volume diffusion. In that case a continuum description is reasonable on a timescale larger than the time τ_0 necessary to complete a layer, even though RHEED oscillations (of period τ_0) may still be present. It is not clear to us whether RHEED oscillations can coexist with roughness.

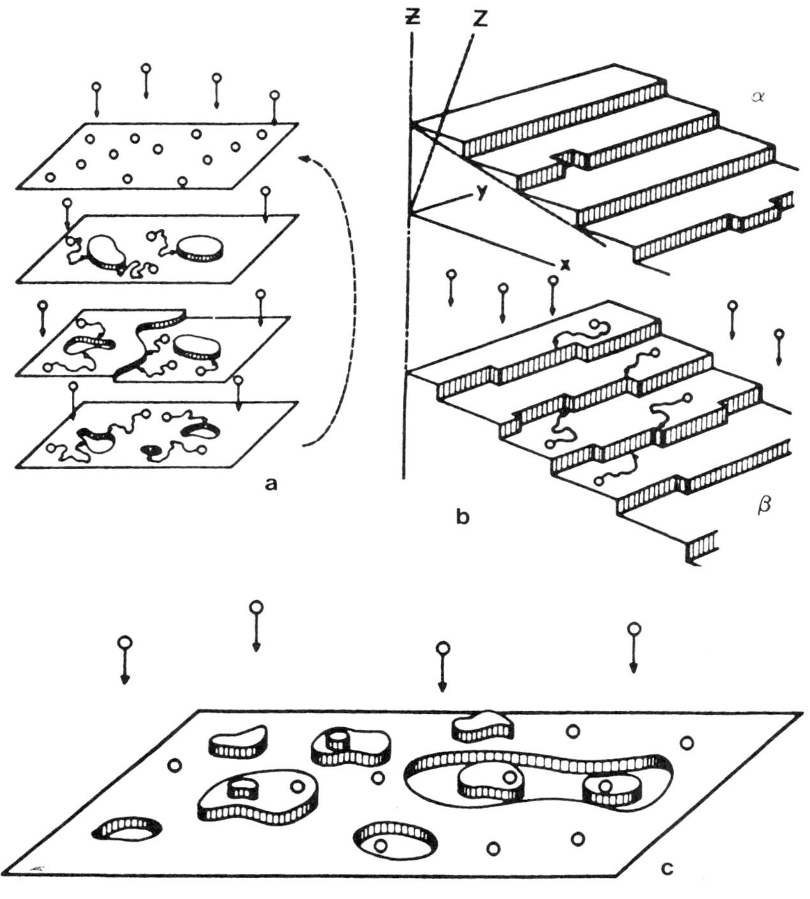

Figure 1

a) high symmetry surface of a crystal growing in an oscillatory way, giving rise to RHEED oscillations. This is the usual MBE procedure. b) A vicinal face of a crystal. α) At equilibrium. β) During growth. c) A high symmetry surface growing in a stationary regime. This regime can be reached when the surface diffusion constant is fast enough to ensure the formation of a crystal rather than an amorphous material, but sufficiently slow to avoid complete disappearance of small terraces when bigger ones begin to form.

<u>1.3. Random effects, inhomogeneity effects and recovery mechanisms</u>

The main purpose of this work is the investigation of the effects of random fluctuations $\delta f(\vec{r},t)$ of the beam intensity $f(\vec{r},t)$. $\vec{r}=(x,y)$ denotes the coordinates perpendicular to the beam direction z. It will be assumed that $\delta f(\vec{r},t)$ is uncorrelated in space and time. Thus

$$\langle \delta f(\vec{r},t)\ \delta f(\vec{r}',t') \rangle = \tau_0\ \delta f^2\ \delta_{\vec{r},\vec{r}'}\,\delta(t-t') \tag{3}$$

where $$\delta f = 1/\tau_0 = \bar{f} \tag{4}$$

is the instantaneous beam fluctuation, which on local scales approximately coincides with the average beam intensity $\bar{f}$. Indeed, in a time τ_0, at a given site, an atom can land or not, with the same probability.

In the absence of atomic motion, the surface would become extremely rough and this would result in amorphisation of the material. The recovery mechanisms which allow for the formation of a good crystal are a) surface diffusion. b) desorption (or "evaporation"). c) volume diffusion. d) formation of vacancies in the superficial layers, which are then incorporated in the bulk. The last two effects will be neglected. The effect (d) would have an effect similar to desorption. Surface diffusion is certainly the dominant effect on short lengthscales (Mullins 1963).

A short calculation will give the order of magnitude of the effect of fluctuations in the beam intensity for a layer of thickness h atoms and size R×R, where R is measured in atomic distances. The average number of atoms deposited (on an initially flat substrate) is R^2h , and its fluctuation is $\sqrt{R^2h}$. The resulting fluctuation in h is therefore

$$\delta h = \sqrt{R^2h/R^2} = \sqrt{h}/R \tag{5}$$

For h=100, the height fluctuation is 0.1 atomic distances on an area of 100×100 atoms. This is not small. The purpose of the next sections is to check whether diffusion is able to heal the surface.

2. THE LINEAR APPROXIMATION

In certain cases the equations which describe growth turn out to be linear and can be solved. Such cases will be studied in Subsections 4.4 and 4.6. It is of interest to recall first the equations which rule the smoothening dynamics of a surface in the absence of a beam, since these classical equations (which are linear in the case of weak fluctuations) are suggestive of the kind of equations we would like to have in the presence of a beam.

<u>2.1. Recovery of a macroscopic profile in the absence of a beam</u>

The problem to be addressed in this section is the following. The surface of a material is, on the average, planar, but has some macroscopic (e.g. sinusoidal) profile, resulting for instance from grooves having been digged on it -a classical experiment (Mullins 1957, 1959, Bonzel et al 1984). How will that surface go back to equilibrium ? The average surface will be assumed to be orthogonal to a high symmetry axis z.

2.1.1) Evaporation dynamics (Mullins 1959)

The chemical potential $\bar{\mu}$ of the vapor will be assumed uniform, so that the evaporation term in the kinetic equation has the form

$$\dot{z}_{ev}\,(\vec{r},t) = B(\mu(\vec{r},t) - \bar{\mu}) \qquad , \qquad\qquad\qquad (6)$$

where the local chemical potential $\mu(\vec{r},t)$ is a function of the shape of the surface at time t in the neighbourhood of $\vec{r}$. The simplest assumption is that it is a function of the partial derivatives of z with respect to x and y, which can be expanded as a power series if the roughness is weak. This assumption can be argued to be correct, except in the case of a crystal surface below its roughening transition, but that case is not relevant in the frame of this article, the other sections of which deal with growing surfaces. A growing surface is, we believe, unavoidably rough. The expansion of μ cannot contain powers of the first derivatives, $(\partial_x z)^p (\partial_y z)^q$, because then (6) would imply that a crystal limited by a plane surface can grow or not according to the orientation of the plane. In reality growth or evaporation depends only on the sign of the difference between the chemical potentials. For the sake of simplicity, only the case of an isotropic surface will be considered, and then the leading term of the expansion is

$$\mu(\vec{r},t) = \text{Const} \times (z''_{xx} + z''_{yy}) = \text{Const} \times \nabla^2 z(\vec{r},t) \qquad\qquad (7)$$

This is a particular form of the Gibbs-Thomson formula, appropriate for variations of weak amplitude. Insertion into (6) yields

$$\dot{z}_{ev}\,(\vec{r},t) = \nu\,\nabla^2 z + \text{Const} \qquad\qquad\qquad (8)$$

with a positive value of ν.

2.1.2) Surface diffusion (Mullins 1959)

The part of $\dot{z}$ which is due to surface diffusion obeys a continuity equation

$$\dot{z}_{dif} = -(j'_x + j'_y) = -\text{div}\,\vec{j}(\vec{r},t) \qquad\qquad\qquad (9)$$

where the current density $\vec{j} = (j_x, j_y)$ is a vector parallel to the average surface direction (not to the local surface). According to irreversible thermodynamics (Landau and Lifshitz 1967)

$$\vec{j}(\vec{r},t) = \text{Const} \times \vec{\nabla}\mu(\vec{r},t) \qquad . \qquad\qquad\qquad (10)$$

Relations (7), (9), (10) yield

$$\dot{z}_{dif}\,(\vec{r},t) = -\,\acute{K}\,\nabla^2(\nabla^2 z) \qquad . \qquad\qquad\qquad (11)$$

If both surface diffusion and desorption are present, both contributions (8) and (11) should be added. The result is a simple equation, which is universal in that sense that the microscopic details of desorption and diffusion are contained in the constants ν and K. Equations (8) and (11) suggest the kind of equations that most of theorists, fond of simple and universal models would like to apply to crystal growth.

2.2. Crystal growth in the linear approximation

If the beam is switched on, the naivest guess is that the local rate of growth of the surface is the sum of (8), (11) and the beam intensity $f(\vec{r},t)$. It will be seen in Section 3 that this is not generally true. However the resulting equation

$$\dot{z}(\vec{r},t) = f(\vec{r},t) + \nu\,\nabla^2 z - K\,\nabla^2(\nabla^2 z) \qquad\qquad (12)$$

turns out to be correct in certain cases. In addition, the solution, to be given below, has certain qualitative features which are also present in the general case.

The solution of (12) is a slight generalisation of the treatment of Edwards and Wilkinson (1982) who treated the case K=0. However, their article is rather difficult to read, and it is of interest to outline the method. The Fourier transform $z_q(t)$ of $z(\vec{r},t)$ satisfies

$$\dot{z}_q(t) = f_q(t) - \alpha(q)z_q(t) \qquad (13)$$

where

$$\alpha(q) = \nu q^2 + Kq^4 \qquad . \qquad (14)$$

(13) is readily integrated, yielding

$$z_q(t) = z_q(0)e^{-\alpha(q)t} + \int_0^t dt' f_q(t')e^{\alpha(q)(t'-t)} \qquad (15)$$

The surface is assumed to be initially flat and orthogonal to the z direction, so that the first term vanishes.

Using (3) and (15), the correlation function of z is found to be

$$\langle z_q(t)z_{-q'}(t)\rangle = \delta_{qq'}\,\tau_o\int_0^t d\tau\,\delta f^2 e^{-2\alpha(q)(t-\tau)} = \delta_{qq'}\,\tau_o^{-1}\,\frac{1-\exp(-2(\nu q^2+Kq^4)t)}{(\nu q^2+Kq^4)} \qquad (16)$$

The Fourier transformation can be inverted, yielding the following results in the limit t=∞.

If $0<\nu<<K$, in 3 dimensions and for long distances,

$$\delta z(r)^2 = \left\langle [z(\vec{r}',t)-z(\vec{r}'+\vec{r},t)]^2\right\rangle \sim 2\pi\tau_o\,\frac{\delta f^2}{\nu}\ln\,(r\sqrt{\nu/K}) \qquad (17)$$

K appears only in the logarithm and does not play an important part. The logarithmic dependence in r is the same as in the original Edwards-Wilkinson model (K=0), and also as in a surface at equilibrium at some effective temperature. Thus the noise δf may be viewed as simulating thermal disorder. This way of thinking can, however, be misleading since the noise does not satisfy any detailed balance formula. Indeed, if $\nu=0$, the K term is unable to evacuate all the energy and formula (16) yields

$$\delta z(r)^2 = \left\langle [z(\vec{r}',t)-z(\vec{r}'+\vec{r},t)]^2\right\rangle \simeq \tau_o\,\frac{\delta f^2}{K}\,r^2 \qquad . \qquad (18)$$

This formula is valid in 3 dimensions for $\nu=0$, or for small ν and short lengths r. It predicts a roughness which is much stronger than thermal roughness. A numerical evaluation is appropriate to get convinced that this result is not in contradiction with the experimental realisation of good layers by MBE. According to (3), $\delta f\approx 1/\tau_o$, where τ_o is the time necessary to complete one atomic layer. The current is given by $\vec{j}=D\rho\vec{\nabla}\mu/T$, where ρ is the density of moving atoms, D their diffusion constant and μ the chemical potential. This yields $K\approx D\rho g/T$, where g is the surface tension. At high enough temperature, D should be of order $1/\tau_1$, where τ_1 is a typical phonon time. The order of magnitude of ρ is the maximum concentration possible without nucleation of terraces. This threshold does not depend very much on the growth rate except if it becomes very slow, and is between 0.01 and 0.1. Thus, taking $g/T\approx 10$, a reasonable evaluation is $K \approx 10^{-10}$ and $\tau_o \approx 1$ second. Then the height variation predicted by (18) is one atomic distance for two points of the surface distant of 10^5 atomic distances. In such a case the continuous approximation is not applicable and formula (18) is not reliable. However, the above numerical evaluation shows that this unreliable formula is not in disagreement with the very smooth surfaces grown by MBE. When the temperature decreases, τ_1 increases drastically and the roughness becomes much stronger. Then the present work becomes of physical interest.

The Fourier transform of (16) in dimension $d \neq 3$ is also of interest. For $\nu \neq 0$ and $d < 3$ the result of the Edwards-Wilkinson model is recovered, namely

$$\delta z(r) \sim r^{-(3-d)/2} \qquad (d < 3) \tag{19}$$

For $\nu = 0$ one finds

$$\delta z(r) \sim r^{-(5-d)/2} \qquad (d < 5) \tag{20}$$

It follows from these formulae that, in all physically relevant cases ($d = 3$ and $d = 2$) the continuum, linear model (12) is consistent with our prejudice that a growing surface should be rough. This property will be found to hold even when nonlinear terms are taken into account.

The above formulae apply in the limit $t = \infty$. Let the limit $r = \infty$ be now considered. In that limit, the Fourier transform of (16) is for $0 < \nu < < K$

$$\left\langle [z(\vec{r}',t) - z(\vec{r}' + \infty, t)]^2 \right\rangle \sim \frac{\pi}{\nu} \tau_0 \delta f^2 \ln(\nu^2 t / K) \tag{21}$$

and for $\nu = 0$

$$\left\langle [z(\vec{r}',t) - z(\vec{r}' + \infty, t)]^2 \right\rangle \sim \tau_0 \delta f^2 (t/K)^{1/2} \tag{22}$$

The crossover from (17) to (21) takes place for $r \approx \xi$, with

$$\xi \approx \sqrt{\nu t} \qquad (\nu \neq 0) \tag{23}$$

and the crossover from (18) to (22) occurs at

$$\xi \approx (K^3 t)^{1/4} \tag{24}$$

Following Kardar's notations, we define exponents z and χ by $\xi \sim t^{1/z}$ and $\delta z(r) \sim r^\chi$. χ is sometimes called α or ζ. Comparison with formulae (21) to (24) yields the following values in $3 = 2 + 1$ dimensions.

	$\nu \neq 0$ (Edwards-Wilkinson model)	$\nu = 0$ (no desorption)
z	2	4
χ	0	1

Below 3 dimensions χ is seen from (20) to be larger than 1 in the absence of desorption. This implies that, below 3 dimensions, the model is not physically acceptable. However, mathematically, it is still consistent.

3. GROWTH WITH DESORPTION

In this Section it will be seen that, in the presence of a beam, the linear equation (12) is generally not acceptable. The nonlinear terms can be explicitly calculated in the case of a stepped surface (Fig. 1 b).

3.1. Case of a stepped surface

The following model will be assumed. An atom which has just landed on the surface diffuses until it finds a step or desorbs. If it reaches a step it does not play any role. The desorption probability obviously depends on the step density $|\vec{\nabla}z|=1/\ell$, where ℓ is the local distance between steps, and z is the coordinate perpendicular to terraces (while Z denotes the direction perpendicular to the average surface). The beam direction is assumed to be the high symmetry orientation z. For short ℓ, the atoms have no time to desorb and the growth rate is

$$\dot{z}(\vec{r},t) = f(\vec{r},t) + \text{diffusion terms.} \tag{25 a}$$

On the other hand, if ℓ is large, all atoms desorb except those which land very close to a step. Therefore, each step has a velocity v=Cf, where C is a constant. Relation (2) then yields

$$\dot{z}(\vec{r},t) = (C/\ell)f(\vec{r},t) + \text{diffusion terms.} \tag{25 b}$$

It is easy to find the appropriate interpolation formula between (25 a) and (25 b). It is convenient to introduce a function of ℓ^{-2} rather than ℓ or $1/\ell$. The interpolation formula is

$$\dot{z}(\vec{r},t) = f(\vec{r},t)\ \varphi(\ell^{-2}) + \text{diffusion terms.} \tag{25 c}$$

The form of the function φ (which will not be explicitly used) has been calculated by Burton, Cabrera and Frank (1951), namely

$$\varphi(1/\ell^2) = 2\ell^{-1}\ \sqrt{\Lambda/\eta}\ \tanh(\frac{\ell}{2}\ \sqrt{\eta/\Lambda}) \tag{26}$$

where η is the evaporation rate and Λ the diffusion coefficient. Expanding φ in (25 c) in a power series around the average orientation yields

$$\dot{z}(\vec{r},t) = f(\vec{r},t)\ \varphi(z'^2_x+z'^2_y) + \text{diffusion terms}$$

$$= f(\vec{r},t)\varphi_0 + f(\vec{r},t)\ \left(z'^2_x-\bar{z}'^2_x+z'^2_y\right)\varphi'_0 + \frac{1}{2}\ f(\vec{r},t)\left(z'^2_x-\bar{z}'^2_x+z'^2_y\right)^2\varphi''_0\ .$$

Here, y is the average step direction, $\bar{z}'_x=1/\bar{\ell}$ is the average slope, and φ_0, φ'_0 and φ''_0 denote the values of φ and its derivatives for $z'_x=\bar{z}'_x$, which can easily be calculated from (26). Introducing the coordinate Z perpendicular to the average surface and neglecting the fluctuations of f except in the first term, the above equation yields

$$\dot{Z}(\vec{r},t) = f(\vec{r},t)\ \varphi_0 + c\ Z'_x + \frac{\lambda}{2}\ Z'^2_x + \frac{\lambda'}{2}\ Z'^2_y + \text{diffusion terms,}$$

where the values of the constants c, λ and λ' can easily be calculated. The constant part of the first term can be eliminated by a translation Z $\to$ Z$-\varphi_0$t, and the second term by a galilean transformation x $\to$ x-ct. Finally

$$\dot{Z}(\vec{r},t) = f(\vec{r},t)\ \varphi_0 + \frac{\lambda}{2}\ Z'^2_x + \frac{\lambda'}{2}Z'^2_y + \text{diffusion terms} \ . \tag{27}$$

The actual beam intensity is renormalized by the factor φ_0. This factor can be omitted if (as generally done) the beam intensity is not directly measured, but deduced from the layer completion time τ_o through (4). On long lengthscales, diffusion terms can be neglected. The above calculation is too crude because it assumes that atoms coming to a step do not play any role. In reality, they can still escape from the step and even from the surface, but these events follow the laws of thermodynamics and are described by equation (8). One finally obtains an anisotropic form of the well-known "KPZ" equation

$$\dot{z}(\vec{r},t) = f(\vec{r},t) + \frac{\lambda}{2} (\vec{\nabla}z)^2 + \nu \nabla^2 z \quad . \tag{28}$$

introduced by Kardar, Parisi and Zhang (1986) to describe the long distance behaviour of the Eden (1958) model. Certain models different from the present one are also represented by (4) on long lengthscales. Certain of these models do exhibit desorption, since particles which land at certain places are not accepted. It is so in the model of Kim and Kosterlitz (1989) and in an exactly soluble, two-dimensional model of Meakin et al. (1986). In the Eden model the justification of (4) is completely different (Kardar et al. 1986). In the Edwards-Wilkinson model, $\lambda=0$ and the ν-term is due to gravity.

The KPZ equation (28) is analogous to a Langevin equation since it contains the random term f. However, it seems preferable to use the word "Langevin equation" only for an equation of the form $\dot{z} = f - \Gamma \, \partial \mathcal{H}/\partial t$. Then (28) is not a Langevin equation because it is not possible to find an effective Hamiltonian $\mathcal{H}$. In the absence of a beam, the λ-term is strictly forbidden, and the equation of motion should be a true Langevin equation.

3.2. High symmetry surface

For a crystal surface of high symmetry at moderate temperature (Fig. 1.3) the growth dynamics is still expected to be represented by an equation of the form (27), with $\lambda=\lambda'$, or by (28) on long lengthscales. However, we have not been able to calculate explicitly λ.

3.3. Properties of the Kardar-Parisi-Zhang equation (28)

They are reviewed in detail elsewhere by more competent specialists (Medina et al. 1989, Wolf 1990) and only a summary will be given here. $\bar{f}$ will be assumed to vanish, since this condition can be fulfilled after a translation $z \rightarrow z-\bar{f}t$.

a) The sign of λ is irrelevant since it can be changed by the transformation $z \rightarrow -z$, $\lambda \rightarrow -\lambda$ which leaves (28) invariant.

b) The nonlinear term in (28) is relevant, at least in 3 dimensions and below 3 dimensions. Indeed the ratio of the nonlinear to the linear (ν-) term in (28) on a lengthscale R scales as

$$\frac{(\vec{\nabla}z)^2}{\nabla^2 z} \approx \frac{(\delta z(R)/R)^2}{\delta z(R)/R^2} = \delta z(R) \quad ,$$

where $\delta z(R) = \sqrt{G(R,\infty)}$ (see formulae (1) and (18)). Since the surface is rough, $\partial z(R)$ goes to infinity with R. This suggests, in fact, that not only the nonlinear term is relevant, but the linear term is irrelevant. The relaxation time τ of a bump or a hole of radius ξ and height h can be calculated from the λ-term alone in the absence of beam, and is given by $h/\tau \approx \dot{h} \approx \lambda(h/\xi)^2$. It follows

$$\tau \approx \xi^2/\lambda h \quad . \tag{29}$$

c) The exponents χ and z defined in Section 2 should be such that the characteristic length ξ and height h associated to a time τ satisfy

$$\xi \sim \tau^{1/z} \qquad (30\ a) \qquad \text{and} \qquad h \sim \tau^{\chi/z} \qquad (30\ b).$$
Insertion into (29) yields

$$\chi + z = 2. \tag{31}$$

d) The exponent χ turns out to be about 0.385 in 3 dimensions (Forrest and Tang 1990).

Let the physical relevance of the KPZ equation (28) in crystal growth be briefly discussed. Firstly, it clearly neglects higher order terms, in particular the K-term of (12). In these terms, z can only appear through it derivatives, if gravity is neglected. Then a discussion analogous to that of point (b) would show that these terms are negligible in the long lengthscale limit. If one wants to describe short lengthscales, it is necessary to reintroduce the K-term (11).

4. GROWTH WITHOUT DESORPTION

At temperatures where MBE is usually performed, desorption is generally negligible at equilibrium. Although this is not necessarily true in the presence of a beam, the limit of vanishing desorption is of physical interest.

4.1. Absence of λ-term in the growth equation

In this Section it will be assumed that all atoms coming from the beam are incorporated in the surface, and that the concentration of vacancies in the bulk is not influenced by the growth process and is just the equilibrium concentration at the temperature of the material. Under these circumstances, the rate of growth $\dot{z}$ of a planar surface (if z is the beam direction, assumed to be also a high symmetry axis) should be independent of the orientation of the surface, therefore independent of $\vec{\nabla}z$. It follows, as noticed by Kariotis (1989) that the coeffecient λ should vanish in (28). Another way to obtain this result is that, if the beam intensity is subtracted, the growth rate should satisfy a continuity equation

$$\dot{z}(\vec{r},t) = f(\vec{r},t) - \text{div } \vec{j}(\vec{r},t) \tag{32}$$

where the current density $\vec{j}$ is perpendicular to z. The λ-term in (28) is not a divergence, and therefore $\lambda=0$.

4.2. Diffusion bias

The next question is whether the linear term of (28) also vanishes. In the absence of a beam, it does, and the dynamics are described by eq. (28) where only higher order derivatives appear. This is a consequence of the detailed balance principle, which obliges the average thermal fluctuation of the current to vanish if $\vec{\nabla}z=0$. A non-vanishing current would indeed imply, through the detailed balance principle, that the particle energy depends on the height z, and this is excluded if gravity is neglected. However, in the presence of a beam, the average current does not vanish. The reason is the following. Assume $\vec{\nabla}z\neq0$, so that the surface has steps (Fig. 2). A number of experimental facts (Fink and Ehrlich 1986, Tsushiya et al 1989) show that freshly landed atoms (which are not thermalized and therefore ignore the detailed balance principle) prefer to be incorporated to the "upper" step than to the lower one. The reason is that incorporation to the lower step would imply that the atom first jumps down to the lower terrace. To do that, the atom should jump over a potential barrier (Fig. 2). This can be understood from the fact

that the number of neighbours is particularly small when the atom passes
through the terrace edge. Experimentally, the effect is strong at room
temperature in usual metals and semiconductors, but does not prevent the
formation of smooth surfaces at around 600 K, so that the order of
magnitude of the potential barrier can be estimated to 300 K. We propose
to call the effect defined in this section ($\vec{j}\neq0$ if $\vec{\nabla}z\neq0$) "diffusion bias".
Its consequences are studied in the next two subsections.

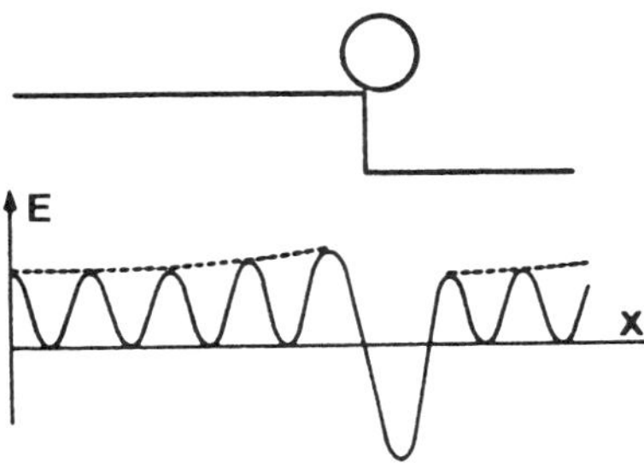

Figure 2. Potential seen (lower part) near a step (upper part) by an atom diffusing on a stepped surface. The potential has a maximum near the terrace edge because an atom at this position (circle) is at maximum distance from its remote neighbours. As a consequence, atoms go more easily to the upper step than to the lower step.

4.3. Case of a stepped surface

In the case of stepped surface, the equation which describes
growth on long lengthscales (longer than the step distance ℓ) can easily
be derived in the case of infinitely high barrier. This approximation can
be expected to be good at low enough temperature and if ℓ is not too
large, otherwise the atoms would try a large number of times to jump the
barrier, and would ultimately succeed with a high probability.

What is needed is an evaluation of the current density, to be
inserted into (32). All landing atoms should go to the upper ledge, and it
will be assumed that they do that in a time τ sufficiently short, so that
the ledge does not move very much in that time. If the atom has landed at
a distance x from the upper edge, its average velocity is x/τ. If the
probability for an atom to have certain values of x and τ is $p(x,\tau)$, the
average current density is

$$|\vec{j}| = \int_0^\ell dx \int_0^\infty d\tau \, p(x,\tau) \, (\bar{f}\tau) \, x/\tau.$$

The factor $(\bar{f}\tau)$ is the average density of particles with a given
τ. The beam intensity fluctuations are neglected in the second term of
(32) which is now being calculated, although they should be taken into
account in the *first* term. The integration over τ is readily done and one
obtains

$$|\vec{j}| = \int_0^\ell dx \, p(x) \, \bar{f} \, x \, ,$$

where $p(x)=1/\ell$ is the probability that an atom has a given value of x. The
final result is

$$|\vec{j}| = \bar{f} \, \ell/2 = \bar{f}/2|\partial z/\partial X| \, ,$$

where X denotes the coordinate perpendicular to the local orientation of
the steps. Coming back to the coordinate x perpendicular to the average
step orientation, the above formula yields

$$j_\alpha = \frac{z'_\alpha\,\bar{f}}{2\left(z'^2_x + z'^2_y\right)} \;. \qquad\qquad (\alpha = x, y)$$

Insertion into (32) yields

$$\dot{z}(\vec{r},t) = f(\vec{r},t) + \nu\,\nabla^2 z(\vec{r},t)\ , \qquad\qquad (33)$$

with
$$\nu = \ell^2\,\bar{f}/2. \qquad\qquad (34)$$

Thus, the linear model of Section 2 is applicable and ν can be explicitly calculated. For short lengthscales, diffusion bias may be less important and the term (11) may dominate. Adding it to (33) yields eq. (12).

4.4. High symmetry surface

In the case of a high symmetry surface, it is not so easy to take diffusion bias into account. Formula (33) would suggest an infinite current. This is in fact not true. The current density in Subsection (4.3) was found to be proportional to the step distance ℓ because the atoms were supposed to diffuse to the upper ledge. This is not true if ℓ is too large because the density of free, diffusing atoms becomes so large that they form their own terraces as in Fig. 1 c. It is reasonable to assume that on a growing surface, all properties are analytic functions of the derivatives of z, as they are for a rough surface near equilibrium (Mullins 1963). In particular

$$\vec{j}(\vec{r},t) = \alpha\,\bar{f}\,\vec{\nabla}z(r,t)$$

where α is a constant and the factor $\bar{f}$ recalls that the coefficient should vanish in the absence of beam. The point is that, according to the discussion of subsection 4.2, α should be positive. Then insertion into (32) yields equation (33) again, but with a negative value of ν!

This implies an instability. Equations (12) and (28), derived in the previous sections, were consistent with a macroscopically planar (since $\chi < 1$), though microscopically rough surface. If $\nu < 0$, the surface cannot be macroscopically planar. To our knowledge, no such instability has been reported, perhaps because experimentalists just try to avoid it.

A more detailed argument and an evaluation of the constant α has been proposed by Villain (1990).

4.5. The case $\nu = 0$

In practice, although diffusion bias is a well-known effect, it seem to be mainly present at rather low temperatures. The fact that it does not prevent the growth of very smooth surfaces at higher temperatures justifies the investigation of the case $\nu = 0$.

The simplest idea (Bruinsma 1990, Bales 1990) would be to use eq. (12) with $\nu = 0$. However, since a nonlinear term is necessary in the case where desorption is possible, nonlinear terms can also be expected to be present in the absence of desorption. We have not been able to produce any real derivation, but the following argument may have some value. One would like to use eq. (33) and an appropriate equation to replace (10) when no chemical potential can be defined. A reasonable Ersatz of (10) is $\vec{j}(\vec{r},t) \sim \vec{\nabla}\rho(\vec{r},t)$, where ρ is the density of adatoms. Then it is reasonable to assume that ρ depends on the surface slope. For a highly symmetric orientation the lowest order expansion should be $\rho=\rho_o+\rho_1(\vec{\nabla}z)^2$. Combining both relations, one arrives at

$$\dot{z}(\vec{r},t) = f(\vec{r},t) - K\,\nabla^2(\nabla^2 z(\vec{r},t)) + \sigma\,\nabla^2(\vec{\nabla}z(\vec{r},t))^2 \tag{35}$$

In this model, the exponents can be calculated in an approximation à la Flory. Firstly a bump or a hole of radius R and height h is healed by the last two terms of (35) after a time τ which satisfies the relation

$$h/\tau \approx \dot{h} \approx (Kh+\sigma h^2)/R^4$$

Hence
$$\tau \approx R^4/(K+\sigma h) \quad . \tag{36}$$

Another relation can be obtained by looking for the time τ necessary for the random noise f to make a bump or a hole of radius R and height h. The number of atoms landing into the area R^{d-1} (if d is the space dimension, usually 3) in the time τ is $\bar{f}\,R^{d-1}\,\tau/\tau_o$, and the fluctuation of this number, responsible for the formation of the bump or hole, is equal to its square root. It should also be equal to the volume R^{d-1} h divided by the atomic volume (assumed to be one). This yields

$$\tau \approx (\tau_o/\bar{f})R^{d-1}\,h^2 \quad . \tag{37}$$

Equating (36) and (37) yields

$$h^2(K+\sigma h) \approx (\bar{f}/\tau_o)R^{5-d} \quad . \tag{38}$$

Therefore the exponent χ is, if $\sigma\neq0$,

$$\chi = (5-d)/3 \quad . \tag{39}$$

In particular, for d=3, $\chi=2/3$. For $\sigma=0$, (38) yields $\chi=(5-d)/2$ in agreement with (20).

Relation (37) cannot be applied to the KPZ equation (28). It can be understood from the physical interpretation given in the present work, that (37) is correct in the case of surface diffusion dynamics and not in the case of desorption dynamics. Indeed, the correct way to handle the problem would be a coarse-graining procedure, in which an initial cell of linear size R would be divided in cells of size bR, then in cells of size b^2R , etc. The fluctuation of the atoms having landed in the initial cell and having remained in that cell is normally expected to depend on all scales. Equation (37) states that they do not. This is (presumably) correct in the case of surface diffusion because the atoms which land in a small cell of size b^pR have no great probability to leave the big, initial cell. In the case of desorption dynamics, on the contrary, desorption is dominated by small cells, because the average value of $|\nabla z|$ in (28) is bigger in small cells. On the other hand, (36) might well be only approximate, in contrast with the KPZ equivalent (29) which is exact.

The above argument is very similar to the one which yields the equilibrium roughness of a domain wall (Villain 1982, Grinstein and Ma 1982) in the random field Ising model (RFIM). The result is the same. The analogy between the RFIM and growth problem was already remarked by Zhang (1988). The analogy between the Eden or KPZ growth and the random *exchange* Ising model is well known (Kardar et al.1986).

A model analogous to (35) has been proposed by Sun et al (1989) with the important difference that the correlation function of f does not satisfy (3), but a similar condition with $\nabla^2 \delta(\vec{r}-\vec{r}')$ instead of $\delta(\vec{r}-\vec{r}')$. The exponents are therefore different.

4.6. A two-dimensional model

Although surface diffusion seems to be the dominant healing mechanism, there are not many numerical studies. The first one seems to be that of Wolf and Villain (1990). This is a one dimensional model in which a freshly landing atom moves by one step (in the direction x of the average surface) if it increases the number of its neighbours, and then does not move any more (Fig. 3).

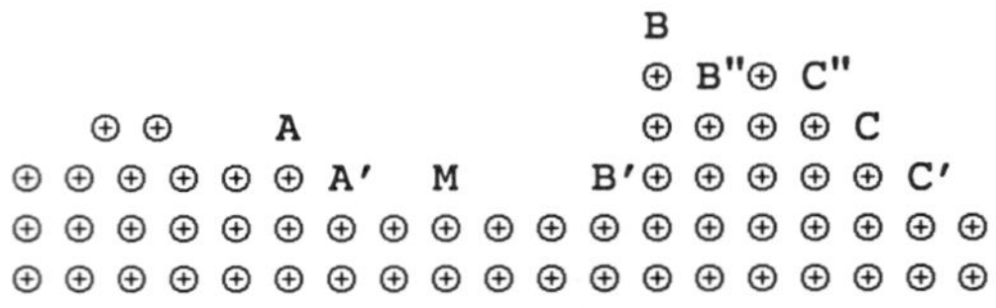

Fig. 3. A one-dimensional model with no evaporation and no vacancy formation. Atoms landing at A',B', B'', C' or M do not move. Atoms landing at A and C'' should go to A' and C respectively. An atom landing at B can choose between between B' and B''.

The numerical results are consistent with equation (35) with $\sigma=0$. It is not quite clear why σ should vanish. Moreover there is no clear derivation of equation (35) in this case. The coefficient K cannot be obtained from the thermodynamical argument of section 2, since there is no hamiltonian, no free energy and no chemical potential in the problem. It is interesting to see that thermodynamic-looking equations can occur in non-thermodynamical problems.

REFERENCES

BALES, G.S., BRUINSMA, R., KARUNASI, X. RUDNICK, X. AND ZANGWILL, A. (1990) to be published in *Science Magazine*.
H.P. BONZEL, E. PREUSS, B. STEFFEN (1984) Appl. Phys. **A 35,** 1 and Surface Sci. **145** , 20 (1984)
BRUINSMA, R. (1990) in "Kinetics of Ordering and Growth at Surfaces" ed. by M. Lagally (Plenum, New York)
BURTON, W.K., CABRERA, N., FRANK, F.C. (1951) Phil. Trans. Roy. Soc. **243,** 299
EDEN, M. (1958) in "Symposium on Information theory in Biology", ed. H.P. Yockey (Pergamon Press, New York) p. 359
EDWARDS, S.F., WILKINSON, D.R. (1982) Proc. Roy. Soc. **A 381,** 17
FINK, H.W., EHRLICH, G., Surface Sci. (1986) **173,** 128
FORREST, B.M., TANG, L.H. (1990) Phys. Rev. Lett. to be published
GRINSTEIN, G., MA, S.K. (1982) Phys. Rev. Lett. **49,** 685
KARDAR, M., PARISI, G., ZHANG, Y.C. (1986) Phys. Rev. Lett. **56,** 889
KARIOTIS, R. (1989) J. Phys. A **22,** 2781
KIM, KOSTERLITZ, J.M. (1989) Phys. Rev. Lett. **62,** 2289
LANDAU, L., LIFSHITZ, E. (1967) *Physique statistique*, Editions Mir, Moscow
LANGER, J.S. (1980) Rev. Mod. Phys. **52,** 1
MEAKIN, P., RAMANLAL, P., SANDER, L., BALL, R.C. (1986) Phys.Rev. **A 34,** 5091
MEAKIN, P., BALL, R.C., RAMANLAL, P., SANDER, L. (1987) Phys.Rev. **A 35,** 5233

MEDINA, E., HWA, T., KARDAR, M., ZHANG (1989) Y.C., Phys. Rev. **A 39**, 3053
MULLINS, W.W. (1957) J. Appl. Phys. **28**, 333
MULLINS, W.W. (1959) J. Appl. Phys. **30**, 77
MULLINS, W.W. (1963) in "Metal Surfaces : Structure, Energetics and Kinetics", Am.Soc. Metals, Metals Park, Ohio, p. 17
MULLINS, W.W., SEKERKA (1963) J. Appl. Phys. **34**, 323 and **35**, 444 (1964)
PELCE, P. (1988) "Dynamics of curved fronts" (Acad. press, New York)
SUN, T., GUO, H., GRANT, M. (1989) Phys. Rev. **A 40**, 1989
TSUSHIYA, M., PETROFF, P.M., COLDREN, L.A. (1989) Appl. Phys. Lett. **54**, 1690
VILLAIN, J. (1982) J. Physique Lett. **43**, L-551
VILLAIN, J. (1990) Submitted to J. Physique
VISCEK, T. (1989) "Fractal growth phenomena" (World Scientific, Singapore)
WOLF, D. (1990) in "Kinetics of Ordering and Growth at Surfaces" ed. by M. Lagally (Plenum, New York)
WOLF, D., VILLAIN, J., to be published in Europhys. Lett.
ZHANG, Y.C. (1988) Preprint

SCALING AND A POSSIBLE PHASE TRANSITION

IN MODELS FOR THIN FILM GROWTH

Leonard Sander and Hong Yan

Department of Physics
The University of Michigan
Ann Arbor, MI, 48109-1120, USA

INTRODUCTION

In recent years there has been a significant interest in simple models for the growth of random rough surfaces. In this note we will report on some progress we have made in a particular sort of model. It is most easily understood as a model for thin film growth. Particles launched from random points bombard a substrate and stick to it and to other particles. To fix our ideas, we give the exact rule for this model (the ballistic aggregation model) written on a lattice: Choose a column at random, and let it grow according to:

$$h(r,t+\tau) = \text{Max}\{ h(r,t) + A, h(r+\delta,t)\} \tag{1}$$

where $h(r,t)$ is the height of the the surface at time t and substrate position r, A is the particle diameter, and δ runs over the nearest neighbors of the column in question. The term involving the neighbors means that the particles can stick on the side of the columns. We will discuss several variants of this kind of growth.

We will review the special sort of symmetry (self-affine scaling) which seems to characterize these surfaces (Family and Vicsek, 1985), and the continuum formulation, the so-called KPZ equation (Kardar, et al.1986) which captures the scaling of many of the models which have been proposed. We give a particularly simple new proof (cf Meakin et al., 1986) for the exact value of one of the scaling exponents in the case of growth on a one-dimensional substrate. We will discuss the weak and strong coupling limits of the theory, and the possibility of a phase transition between the two types of behavior in three and higher dimensions (Yan, et al., 1990). Finally we give numerical results on the solution of the KPZ equation itself in three dimensions to attempt to discuss the possible phase transition in this case.

SELF-AFFINE SCALING

Simulation of the model of Equation (1) and other similar processes (such as the Eden model where each perimeter site is equally likely to grow) gives rise to random rough surfaces whose roughness initially increases with time, and then saturates. Many such simulations have now been done.(see Meakin, et al. 1986, for example).

These results can be most easily summarized by considering $s(r,t) = h(r,t){-}h_0(t)$, where $h_0(t)$ is the spatial average of the height. Then the observations can be systematized by claiming that in steady-state growth in a very large system:

Growth and Form, Edited by M. Ben Amar *et al.*
Plenum Press, New York, 1991

$$G(|\mathbf{r}\text{-}\mathbf{r}'|, |t\text{-}t'|) = <[s(\mathbf{r},t) - s(\mathbf{r}',t')]^2> \sim |\mathbf{r} - \mathbf{r}'|^{2\alpha} F[\,|t\text{-}t'|\,|\mathbf{r} - \mathbf{r}'|^{-z}\,]. \quad (2a)$$

Here $< >$ denotes an ensemble average, and $F(x)$ is a scaling function which is constant for $x<<1$ and which approaches $x^{2\alpha/z}$ for $x>>1$. The exponents α and z are characteristic of the model considered, and have the following interpretations: (i) correlations spread in the surface with rate $t^{1/z}$. (As we will see $z \geq 1$ so that the coupling of the columns makes the correlations spread superdiffusively.) (ii) The typical height difference at any given time is given by $\delta s \sim \delta r^{\alpha}$, with $\alpha < 1$. This is the reason for the name self-affine: the surface is a self-affine fractal (Mandelbrot, 1982). Equation (2), is not completely unexpected since it is what arises from a renormalizable continuum field theory for the surface.

A calculation of the values of α and z is one of the goals of a surface theory. It is very curious that scaling in this form, which is, in principle, eminently measurable, has not yet been measured for any real growing thin film in a convincing way. There are experiments on flow in porous media (Rubio, et al., 1989) which are indirectly related to models of the Eden type where self-affinity seems to have been observed. However, the *values* of the exponents do not agree with our current understanding (see next section), so that this subject is still quite controversial.

We can easily derive the original scaling form of Family and Vicsek (1985) if we assume that if we start at t=0 from a flat surface of width L the dynamics is the same as above. Then:

$$\delta h \sim t^{\alpha/z} \qquad t << L^z \qquad\qquad\qquad\qquad\qquad (2b)$$
$$\sim L^{\alpha} \qquad t >> L^z. \qquad\qquad\qquad\qquad\qquad (2c)$$

Here δh is the rms roughness of the surface averaged over the whole sample and over the ensemble. Growth from a flat surface has two regimes: a transient regime when correlations cover a small part of the surface, and a steady state in which the correlations are limited by the finite size, and where the deposit is many times higher than it is wide. We will find in the sequel that the steady state is the most useful regime to consider in simulations. For the experimentalist, the early time regime is more relevant.

It is also clear that growth in a radial geometry gives rise to a roughly circular deposit of mean radius $R \sim t$ whose mean roughness scales as $R^{\alpha/z}$. The last result follows because the circumference grows as t, and thus the surface consists of many essentially uncorrelated patches; the circular growth is always in the early time regime.

CONTINUUM FORMULATION

Edwards-Wilkinson and KPZ models

It is useful to consider a coarse-grained description of the surface. Suppose we subtract the average growth, h_o, from the outset, and try to formulate a general equation for $s(\mathbf{r},t)$ which could govern the evolution of the surface. We would want to write something like:

$$\partial s(\mathbf{r},t)/\partial t = Q\{s(\mathbf{r}',t)\} + \eta(\mathbf{r},t) \qquad\qquad\qquad (3a)$$

where the functional Q is to be determined, and η is the randomness in flux left over by averaging the shot noise to coarse-grain. It is natural to take η to be an uncorrelated gaussian noise. The functional, Q, may be assumed to depend only on differences of heights, not absolute heights, to be more-or-less local in space and time, and to have no preferred substrate direction. The simplest terms which can be written with these assumptions are:

$$Q = \nu\nabla^2 s + (\lambda/2)|\nabla s|^2 + \cdots \qquad\qquad\qquad (3b)$$

With the first term alone this is the Edwards-Wilkinson model (1982), which is linear, and thus can be solved exactly. The result is $\alpha=1/2$ and $z=2$ for a 1-dimensional substrate, and $\alpha=0$ and $z=2$ for higher dimensions; in fact, in 2+1 dimensions this corresponds to logarithmic growth of the surface thickness ($\delta h \sim (\log L)^{1/2}$ to be precise) and in 3+1 constant δh. It is rather easy to see that a discrete model which yields this upon coarse-graining (Sander, 1985) is

$$h(r,t+\tau) = \text{Ave}\{\; h(r,t) + A,\; h(r+\delta,t)\} \qquad (4)$$

If $\lambda \neq 0$ then we have a non-linear and non-trivial model. The only exact results known in this case are for a 1-dimensional substrate (see next section). The meaning of $\lambda \neq 0$ is that flat surfaces grow at a different rate than sloped ones. For the ballistic model (1), slopes grow faster than flat surfaces because of sticking on the sides. It is widely believed that the essential physics of ballistic growth models and of the Eden model are captured by the KPZ equation without higher order terms.

The exponents α and z are the result of the solution of this non-linear stochastic differential equation, and thus should have the same value for all the models. The simulation evidence for this is relatively strong. It is reasonable to suppose that for small λ we have essentially linear behavior, and that for large λ a different, strong-coupling regime. In fact, in 1+1 dimensions, this is not true; the behavior is always dominated by the non-linear terms except when $\lambda=0$. In higher dimensions the behavior is complex, as we will see below.

Scaling law

If λ is not zero, then the form of the KPZ equation puts a restriction on the values of α and z. This was first pointed out in this context by Meakin et al. (1986). The law, which results from putting a solution of form (2a) into Equation (3) and matching leading powers of δr is:

$$\alpha + z = 2. \qquad (5)$$

The linear model does not obey this rule.

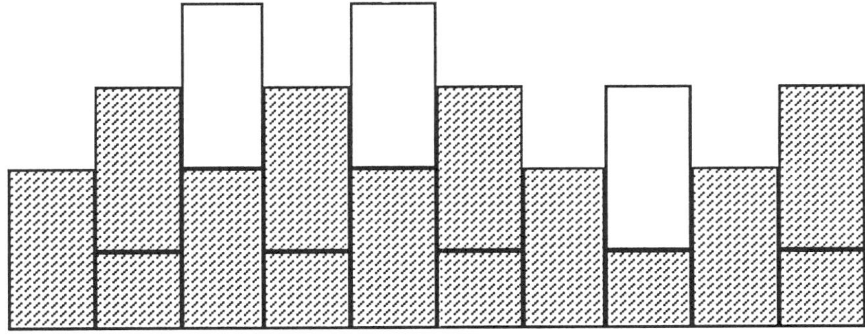

Figure 1. The single-step model. Growth is only possible at local minima of the surface. The white rectangles are growth sites.

EXACT SOLUTION IN THE STEADY STATE IN 1+1 DIMENSIONS

It has been known at least since the KPZ paper that $\alpha=1/2$, $z=3/2$, on a 1-dimensional substrate. It is useful, we think, to present a simple proof of this statement , since the original argument of KPZ is quite involved. The argument is simplest for another model of the ballistic type, where we insist that the steps between adjacent columns are always $\pm A$ and where growth occurs at local minima (Meakin et al.,1986), see Figure 1. This proof completes an argument partially presented in the reference cited.

Consider this model in the steady state. We will calculate $G(x,0)$. From Equation (2) we expect a behavior like $x^{2\alpha}$. Let us suppose (we will prove this in a moment) that each step of $\pm A$ as we proceed along the surface is independent of the proceeding steps. Then the surface is nothing but a random walk, whose mean squared wandering is proportional to the number of steps, i.e., $G(x,0) \sim <(\delta h)^2> \sim |x|$ Thus $\alpha=1/2$. We can use the scaling law to find $z=3/2$. Dhar (1989) has shown this directly using the Bethe ansatz.

It is clear that steps will be uncorrelated if we can show that that all configurations of the steps up and down are equally likely. To this end, consider the master equation for the probability, $P(\{\sigma\},t)$, to be in a given configuration of the steps $\{\sigma\}$ at a given time:

$$\frac{dP}{dt} = \sum_{\{\sigma'\}} W\{\sigma'\rightarrow\sigma\} \, P(\{\sigma'\},t) \; - \; \sum_{\{\sigma'\}} W\{\sigma\rightarrow\sigma'\} \, P(\{\sigma\},t). \quad (6)$$

If we think of this as a matrix equation in which the P's are vectors with an entry for each configuration, then the first term is the off-diagonal process of creating the configuration, and the second is diagonal. There is an entry of a 1 off-diagonal in the matrix for every process which leads to the given surface. But since every maximum in the configuration corresponds to a minimum in the parent the number of off-diagonal elements is the number of maxima. The diagonal element is the number of minima.

In 1+1 dimensions, the number of minima is always equal to the number of maxima. Since the sum of each *row* of the matrix is zero, the steady-state (the eigenvector with eigenvalue zero) has *equal probability* for all configurations. This is identical to the equilibrium state of the infinite temperature Ising model in which there are no spin correlations. This argument should carry over to any model in 1+1 in which the step distribution has a finite second moment, and with spatially uncorrelated noise.

The value $\alpha=1/2$ is probably the most solid result of the theory. It is all the more disturbing that Rubio, et al. found a much larger value from their experimental results, $\alpha\sim0.75$. If this result holds up, then the entire framework of this theory, and its application to real systems may need to be reexamined.

In higher dimensions however, it is *not* necessarily true that the number of minima is equal to the number of maxima. This is what went wrong with the 'proof' in Meakin et al. (1986) that $\alpha = 0$ for 2+1 dimensions which implicitly assumed that the steady-state would have uniform probability for each configuration. It is possible to have a strong-coupling state in higher dimensions with $\alpha\neq0$.

NUMERICAL EVIDENCE FOR A PHASE TRANSITION IN HIGHER DIMENSIONS

In higher dimensions it has long been suspected that both weak and strong-coupling behavior should be possible (Kardar, et al., 1986) and that the transition between the two regimes is, at least in 3+1 dimensions, a non-equilibrium phase transition. There has been, in fact, a good deal of recent activity in models which interpolate between the two limits (e.g., Yan, et al., 1990, Amar and Family, 1990a, Pelligrini and Jullien, 1990, Forrest and Tang, 1990, Guo et al, 1990). We will briefly report on one approach to the problem in this section.

We (Yan et al. 1990) sought a model which would interpolate between the weak and strong coupling limits which are given in Equations (4) and (1) above. An obvious way to do this is to write

$$h(r,t+\tau) = qMax\{h(r,t) + A,h(r+\delta,t)\}+(1-q)Ave\{h(r,t)+A,h(r+\delta,t)\} \quad (7)$$

where q is a random variable which takes on value 1 with probability p and value 0 with probability 1-p. Changing p corresponds to tuning the coupling strength from weak, Edwards-Wilkinson type (p=0) to strong, ballistic aggregation (p=1). That is, as $p \to 0$, effectively $\lambda \to 0$. In order to investigate the model, we did a series of simulations in 2+1 and 3+1 dimensions. Our work is different from that of many other groups in that we are exclusively interested in the steady state, and the value of α: we consider relatively small systems over long times, cf. Equation (2c). Other groups have looked at larger systems in the short time regime, Equation (2b), but this requires great care in avoiding transient effects from filling the first few layers and also passage to the steady state. Our results are summarized in Tables 1 and 2 where we give values of the apparent value of α for 3+1 and 2+1 dimensions.

The abrupt change in behavior in 3+1 dimensions is quite indicative of a phase transition, and this is in agreement with theoretical expectations, and with the work of other authors cited above.

The situation in 2+1 dimensions is extremely confusing and remains controversial. Our results best fit to a continuous change in α, or, alternatively to a new kind of scaling behavior where $\delta h \sim (\log L)^\gamma$ with γ changing continuously from 1 near p=0.2 to $\gamma=1/2$ at p=0. Alternatively, some groups (Tang, et al., 1990) have interpreted this as a crossover of a particularly rapid sort. We cannot, within the confines of this model, distinguish between the two behaviors.

Table 1. Effective roughness exponents α in 3+1 dimensions

L=	10-20	20-40	40-60
p=1	0.29±0.02	0.25±0.02	0.30±0.04
0.6	0.31±0.02	0.27±0.02	0.30±0.05
0.4	0.34±0.01	0.29±0.02	0.22±0.06
0.2	0.20±0.02	0.15±0.03	0.07±0.02
0.0	0.036	0.018	0.010

Table 2. Effective roughness exponents α in 2+1 dimensions

L=	20-40	40-80	80-160	160-320
p=1	0.36±0.01	0.34±0.01	0.36±0.03	0.40±0.08
0.8	0.37±0.01	0.35±0.02	0.36±0.03	
0.6	0.39±0.01	0.37±0.02	0.38±0.05	
0.4	0.40±0.01	0.37±0.02	0.37±0.04	0.36±0.05
0.2	0.29±0.02	0.24±0.03	0.21±0.02	0.23±0.05
0.0	0.14±0.01	0.11±0.01	0.12±0.02	

NUMERICAL SOLUTION OF THE KPZ EQUATION IN 2+1 DIMENSIONS

In order to investigate more closely the situation in 2+1 dimensions we have returned to the KPZ equation in its original form. By doing this we will be able to compare directly with predictions of crossover behavior which are based on continuum arguments.

In particular, Tang et al. (1990) have argued that since 2+1 dimensions appears to be a critical dimension for the continuum theory, we would expect complex behavior near $\lambda=0$. Their prediction is that there is a new length in the theory given by

$$\xi = \xi_0\, e^{8\pi/\varepsilon}, \quad \varepsilon = \lambda^2 D/2\nu^3 \tag{8}$$

where D is the mean squared value of the noise η, and ξ_0 is a microscopic length. The parameter ε is the coupling constant in the perturbation expansion of KPZ. The significance of the new length is that for systems smaller than it the behavior will appear to be weak coupling whereas for larger scales than there will be a crossover to strong coupling.

We have integrated the KPZ equation for various values of ε. Previous work by Amar and Family (1990b) had concluded for a similar calculation that the early time behavior indicated a crossover for all values of ε. We also did early time analysis to get, for very strong coupling $\alpha/z = 0.24\pm0.01$ in agreement with Amar and Family. We do not, however, consider the evidence for a crossover for smaller ε convincing in this regime from our simulations.

We give, in Figure 2, our estimates of effective values of the roughness exponent, α, for this model. This is easily compared with our earlier results on the discrete simulations, and also allows us to test directly for crossover effects as in Equation (8). It is a matter of simple arithmetic to show that with a ξ_0 on the order of the lattice size, our curve for $\varepsilon=4.67$ should already have crossed over if the theory of Tang, et al. is correct. We see no signs of this. However, it is possible that the ε in our numerical work is somehow renormalized from that of the original equation. Our result is, at this point, ambiguous.

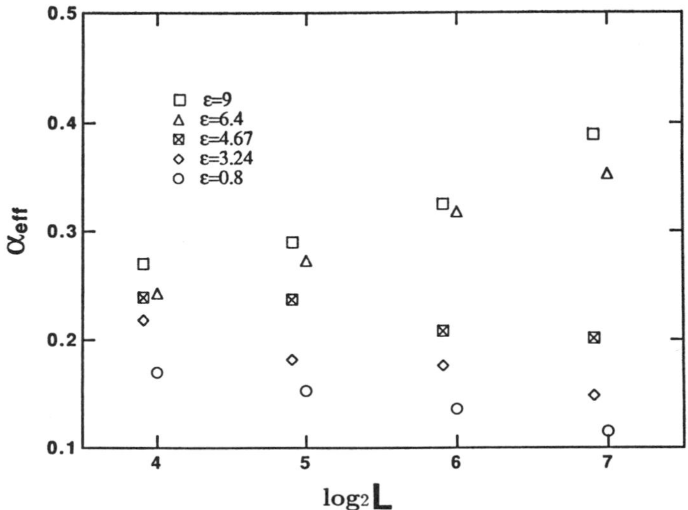

Figure 2. Effective exponent α for systems of various sizes and coupling constants ε.

SUMMARY

In this work we have investigated various aspects of the scaling of random rough surfaces. We find that for an ordinary film there will be a very rapid change in the roughening behavior as we vary the parameters of the theory. For a physical system this might correspond to changing the rate of rearrangements of deposited particles, say, by changing the temperature. This has never been observed in a systematic way. Whether the change corresponds to a true phase transition remains, in our view, an open question. For the experimentalist it is probably not important since a rapid crossover would give the same observable effects.

There are many open questions in this field which would repay future investigation. For example, we have no description of the strong coupling phase of the system, say, in terms of some sort of mean field theory. This would be an interesting thing to try to formulate, and would give us a way to approach properties beyond the simple scaling which we have looked at here.

ACKNOWLEDGEMENTS

We were supported by NSF grant number DMR 88-15908. LMS would like to thank the GPS at Universite Paris 7 for hospitality while some of this work was done.

REFERENCES

Amar, J. and Family, F. 1990a, Phys. Rev. Letters **64**, 543

Amar, J. and Family, F. 1990b, Phys. Rev. A **41**, 3399

Dhar, D. 1989, unpublished

Edwards, S. and Wilkinson, D., 1982, Proc. Roy Soc. London, A**381**, 17

Family, F. and Vicsek, T., 1985, J. Phys. A **18**, L75

Forrest , B. and Tang, L-H., 1990, Phys. Rev. Letters, **64**, 1405

Guo, H. Grossmann, B. and Grant, M., 1990, Phys. Rev. Letters **64**, 1262

Kardar, M., Parisi, G. and Zhang, Y., 1986, Phys. Rev. Lett. **56**, 889

Mandelbrot, B. 1982, *The Fractal Geometry of Nature* (Freeman)

Meakin, P., Ball, R., Ramanlal, P., and Sander, L. 1986, Phys. Rev. A **34**, 5091

Pelligrini, Y. and Jullien, R., 1990, Phys Rev. Letters **64**, 1745

Rubio, M. Edwards, C., Doughtery, A., and Gollub, J., 1989, Phys. Rev. Letters, **63**, 1685

Sander, L., 1986, Proceedings of Symposium on Multiple Scattering of Waves and Random Rough Surfaces,(College Station, PA, July, 1985). Published in ``Multiple Scattering of Waves in Random Media and Random Rough Surfaces'', V.V. Varadan and V.K. Varaden, eds. (Technomic)

Tang, L-H., Natterman, T., and Forrest, B., 1990, Phys. Rev. Letters **65**, 2422

Yan, H., Kessler, D., and Sander, L.M., 1990, Phys. Rev. Letters, **64**, 926

INHOMOGENEOUS GROWTH OF ROUGH SURFACES

Dietrich E. Wolf and Lei-Han Tang[†]

Institut für Festkörperforschung, Forschungszentrum Jülich
POB 1913, D-5170 Jülich, Germany

Abstract. *Growth induced surface roughness is a common phenomenon which apart from its practical importance is interesting because a generally nonlinear dynamics transforms external noise into scale invariant surface fluctuations. It is demonstrated, that investigating macroscopic surface deformations in response to deterministic inhomogeneities of the growth rate provides valuable information on this dynamics.*

INTRODUCTION

Many growth processes at stable planar surfaces show a phenomenon called kinetic roughening: A noisy growth rate is transformed by the surface dynamics into scale invariant surface fluctuations.[1] The growth dynamics in general does not fulfill detailed balance and thus may drive the system into a steady state not characterized by a Boltzmann distribution. This is why one makes a distinction between kinetic and thermal roughening.

To be specific let us consider ballistic deposition of particles onto a flat substrate. As the growth proceeds the surface develops bumps of all sizes up to a characteristic length ξ (parallel to the substrate) that increases with time. The coarsening can usually be characterized by a dynamic exponent[2] z: $\xi \sim t^{1/z}$. By surface roughness one means a power law increase of the variance of the surface height with the distance $R < \xi$ over which the fluctuations are sampled:

$$\Delta h \sim R^{\zeta}. \tag{1}$$

ζ is called the roughness exponent.

In order to get insight into the surface dynamics at work in any particular growth process it is very natural first to study the macroscopic surface deformation in response to an *externally controlled perturbation* of the growth rate. The same surface dynamics

---

[†] Present address: Fakultät für Physik und Astronomie, Ruhr-Universität Bochum, POB 102148, D-4630 Bochum, F. R. Germany

Growth and Form, Edited by M. Ben Amar *et al.*
Plenum Press, New York, 1991

will also determine the roughness in response to *stochastic fluctuations* of the growth rate, and hence the exponents ζ and z.

For instance, we consider an inhomogeneous deposition rate[3]

$$\kappa(\mathbf{r}) = \kappa_0 + \kappa_1 \sum_n \delta(x - L/2 + nL), \tag{2}$$

where $\mathbf{r}$ contains the $(d-1)$ coordinates parallel to the substrate. At equidistant lines perpendicular to the x-axis the deposition may be enhanced ($\kappa_1 > 0$) or reduced ($\kappa_1 < 0$). Two parameters characterize this inhomogeneity: The period L and the strength κ_1. Involuntary variations of the deposition rate are inevitable in experiments, and their effect has actually been observed in molecular beam epitaxy.[4] It should also be possible to produce variations similar to (2) intentionally.

On a mesoscopic level the surface dynamics is described by a stochastic partial differential equation for the height variable $h(\mathbf{r}, t)$. For the classification of growth processes this equation plays a similar role as the Ginzburg-Landau-functional in equilibrium thermodynamics: Irrespective of the microscopic details of the growth process one writes down a phenomenological expansion in the derivatives of h which is compatible with general symmetry considerations. A well-known example is the KPZ-equation proposed by Kardar, Parisi and Zhang:[5]

$$\frac{\partial h}{\partial t} = \nu \nabla^2 h + \frac{\lambda}{2}(\nabla h)^2 + \kappa + \eta. \tag{3}$$

The height increases due to the average deposition rate $\kappa(x)$ whose fluctuations are represented by the noise $\eta(\mathbf{r}, t)$. The Laplacian term with $\nu > 0$ is the simplest way of expressing the tendency of a perturbed surface to become planar again. It also prevents the occurence of singularities ("shock waves") due to the nonlinearity. Frequently the growth velocity depends on the tilt of the surface giving rise to the λ-term. A possible reason[6] could be that the average sticking coefficient for particles arriving at the surface depends on the step density (see Refs.1 for other models leading to a square gradient term). The presence or absence of the nonlinearity is crucial for the surface roughness:[5,7] Growth processes described by (3) with $\lambda \neq 0$ can have different exponents than those belonging to the universality class of the linear theory, $\lambda = 0$.

After switching on the inhomogeneity the surface develops a steady state deformation. In the first approximation this can be calculated by ignoring the noise, just as what one does, for instance, in a continuum description of dendritic growth.[*] Typical profiles predicted for the growth described by (3) are[3]

$$\frac{h(x) - h(0)}{L} = s(\kappa_1) \left(\frac{x}{L}\right)^2 \qquad \text{for} \quad \lambda = 0, \tag{4a}$$

and

$$\frac{h(x) - h(0)}{L} \approx s(\kappa_1) \left|\frac{x}{L}\right| \qquad \text{for} \quad \lambda\kappa_1 > 0, \tag{4b}$$

where $|x| \leq L/2$, and $h(x + L) = h(x)$. $s(\kappa_1)$ is the characteristic slope imposed by the inhomogeneous deposition. Thus one gets direct information about the universality

---

[*] Note however, that in the latter case, the inhomogeneity is controlled by diffusion in the bulk which is itself a dynamical process.

class to which the growth process belongs. This and further examples will be derived below.

Another interesting quantity is the growth velocity $v(\kappa_1, L)$. For $\lambda = 0$ one has

$$v - v_0 = \kappa_1/L, \tag{5a}$$

while for $\lambda \kappa_1 > 0$ in the steady state one finds

$$v - v_0 = \frac{\lambda}{2} s^2(\kappa_1). \tag{5b}$$

v_0 denotes the growth velocity of the unperturbed surface, $\kappa_1 = 0$. Eq.(5b) gives a simple way of measuring λ (see Ref.3 for details).

Finally we ask the question how fluctuations modify the deterministic surface dynamics. It turns out that in the nonlinear case the fluctuations couple to the average profile by renormalizing the system parameters. As a consequence the roughness exponent ζ not only determines the scaling properties of the surface fluctuations but also has a bearing on the average profile.

STEADY STATE PROFILES

As already indicated in the introduction, the significance of investigating steady state profiles and their growth velocity under inhomogeneous growth conditions lies in the fact that one gets information about the universality class to which the growth process belongs. Let us first give three examples that will lead us to formulate more general rules.

First Example : KPZ Equation

Rescaling.– If one neglects the noise η and considers growth rates which are translationally invariant perpendicular to the x-axis, the motion of the surface reduces to a one dimensional problem. A simple way of identifying the expression which determines the macroscopic shape is to formulate the equation of motion in dimensionless quantities $\tilde{h} = (h - v_0 t)/h_0$, $\tilde{x} = x/x_0$, $\tilde{t} = t/t_0$. For instance, (3) then reads

$$\frac{\partial \tilde{h}}{\partial \tilde{t}} = \frac{\nu t_0}{x_0^2} \left[\frac{\partial^2 \tilde{h}}{\partial \tilde{x}^2} + \frac{h_0}{Z} \left(\frac{\partial \tilde{h}}{\partial \tilde{x}} \right)^2 + 2s \frac{x_0}{h_0} \delta \left(\tilde{x} - \frac{L}{2x_0} \right) \right], \tag{6}$$

where (2) has been inserted, and periodic boundary conditions, $h(x + L) = h(x)$, have to be imposed. The parameters Z and s are given by

$$Z = \frac{2\nu}{\lambda}, \quad s = \frac{\kappa_1}{2\nu}. \tag{7}$$

If one asks what the profile will look like on a large scale, $L \to \infty$, it is most convenient to choose the units of length and height as

$$x_0 = L/2, \quad h_0 = sx_0, \tag{8}$$

thereby giving the perturbation a standard form. A practical unit of time is $t_0 = x_0^2/\nu$, so that (6) simplifies to

$$\frac{\partial \tilde{h}}{\partial \tilde{t}} = \frac{\partial^2 \tilde{h}}{\partial \tilde{x}^2} + \frac{Ls}{2Z} \left(\frac{\partial \tilde{h}}{\partial \tilde{x}} \right)^2 + 2\delta(\tilde{x} - 1). \tag{9}$$

From this one can immediately tell that for $L/2 \gg |Z/s|$ the nonlinearity determines the shape of the surface deformation. On the other hand, the case $\lambda = 0$ corresponds to $Z \to \infty$. Then the curvature term in (9) dominates, leading to a macroscopically different profile.

If the nonlinearity is present it also determines the surface deformations caused by the noise, if $\kappa_1 = 0$. The scaling properties of a rough surface imply that the configuration at time t looks statistically similar to the one at a later time $t_0 t$ if it is magnified parallel and perpendicular to the substrate by factors

$$x_0 = t_0^{1/z}, \quad \text{and} \quad h_0 = x_0^{\zeta} = t_0^{\zeta/z}. \tag{10}$$

Hence the nonlinear term in (9) dominates over the curvature term by a factor x_0^{ζ}/Z in the long time limit, provided that $\zeta > 0$.

Solution of (9).– The general steady state solution for $A \equiv Ls/2Z > 0$ is

$$\tilde{h}_+ = (1/A)\ln\cosh(qA\tilde{x}) + q^2 A\tilde{t}, \tag{11a}$$

where $q(A)$ is implicitely given by

$$1 = q\tanh(qA). \tag{11b}$$

Similarly the steady state profile for $A < 0$ is

$$\tilde{h}_- = (1/A)\ln\cos(qA\tilde{x}) - q^2 A\tilde{t}, \tag{12a}$$

with

$$1 = -q\tan(qA). \tag{12b}$$

In the limit $A \to 0$ both $\tilde{h}_+$ and $\tilde{h}_-$ reduce to the parabolic profile (4a), the steady state solution for the linear equation of motion:

$$\tilde{h}_\pm \to \pm(\tilde{x}^2/2 + \tilde{t}). \tag{13}$$

The pile produced by enhanced deposition at $x = L/2$ is symmetric to the groove caused by reduced deposition, see Fig.1.

In the opposite limit, $|A| \to \infty$, this symmetry does not exist. Eq.(11b) yields $q = 1 + O(\exp(-2A))$ so that

$$\tilde{h}_+ \to |\tilde{x}| + (1/2)\lambda s^2 t/h_0, \tag{14}$$

as already noted in (4b) and (5b): If $\lambda\kappa_1 > 0$ the shape of the profile is essentially triangular. As illustrated in Fig.1, noticeable curvature occurs only in a range $|\tilde{x}| \leq A^{-1}$, i.e. $|x| \leq Z/s$.

The case $\lambda\kappa_1 < 0$ is very different: Taking the limit $A \to -\infty$ in (12b) yields $q = -(\pi/2)(A^{-1} - A^{-2}) + O(A^{-3})$. Therefore

$$\tilde{h}_- \to 0, \tag{15}$$

i.e. there is no surface deformation which would grow proportional to L! Upon looking more closely, it turns out that

$$h_- = h_0 \tilde{h}_- \approx Z \ln \cos(\pi x / L) - \pi^2 \nu t Z / L^2, \tag{16}$$

outside the regions $|x \pm L/2| \leq Z/s$ where (16) would diverge logarithmically but the real profile has a cusp (see Fig.1).

Can fluctuations be neglected?– The surface actually fluctuates in the steady state around the macroscopic profiles (13)-(16). On scale L the amplitude of these fluctuations is $\Delta \tilde{h} \sim L^{\zeta-1}$, according to (1) and (8). For both universality classes, $\lambda = 0$ and $\lambda \neq 0$, it is known[1] that the roughness exponent $\zeta < 1$. Therefore the fluctuations can be neglected for the triangular and the parabolic profiles, (13) and (14), and distinguishing them experimentally should be feasible. However the profile (16) with its logarithmic amplitude is completely buried in the noise if $\zeta > 0$. Hence one would have to average over many samples or over many periods L in order to extract the profile. In a computer simulation this is feasible, but in the laboratory it is probably only possible to detect the shape (16) as long as the surface roughness has not yet developed on large scales.

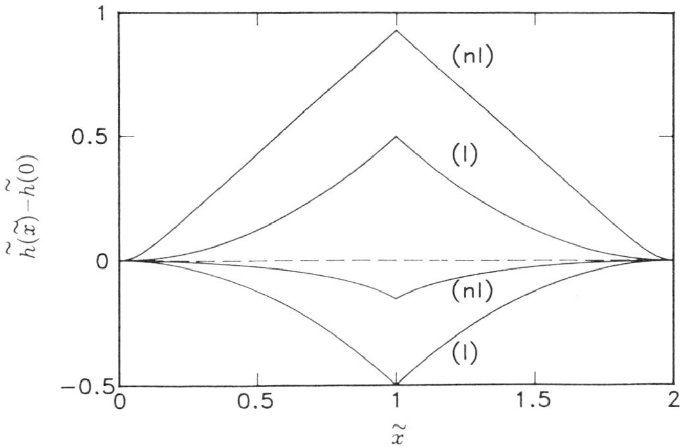

Fig.1. Stationary profiles for the KPZ equation scaled according to (8). (nl): nonlinear case $\lambda > 0$; (l): linear case $\lambda = 0$. The upper two curves are for enhanced, the lower ones for reduced deposition at $\tilde{x} = 1$.

Second Example : Surface Diffusion

As explained e.g. in the contribution by Villain[6] to these proceedings, surface diffusion may lead to a term $-K\nabla^2(\nabla^2 h)$ in the equation of motion (3) for the surface. Therefore, as a second example, we consider the effect of this term in the linear case, $\lambda = 0$. Using the units (8), the equation of motion reads

$$\frac{\partial \tilde{h}}{\partial \tilde{t}} = \frac{\partial^2 \tilde{h}}{\partial \tilde{x}^2} + \left(\frac{2X}{L}\right) \frac{\partial^4 \tilde{h}}{\partial \tilde{x}^4} + 2\delta(\tilde{x} - 1), \tag{17}$$

where $X = \sqrt{K/\nu}$. Obviously the macroscopic profile for $L \to \infty$ is not influenced by the new term. (This conclusion would also be true for $\lambda \neq 0$.) Using the abbreviation

$B \equiv L/2X$, the solution of (17) is

$$\tilde{h} = \tilde{x}^2/2 + \tilde{t} + \frac{1 - \cosh(\tilde{x}B)}{B \sinh B}, \tag{18}$$

which deviates from (13) significantly only for $|\tilde{x} \pm 1| \le B^{-1}$, or $|x \pm L/2| \le X$. The cusp at $x = L/2$ present in the periodic continuation of (4a) is now replaced by a discontinuity in the third derivative.

The fourth order derivative in (17) is also irrelevant for the surface roughness: Rescaling as in (10) shows that this term is small of order $(X/x_0)^2$ compared to the second order derivative. As in the first example, the response to noise is governed by the same term of the equation of motion as the macroscopic response to a deterministic perturbation. Our last example shows that this is not always the case.

A Counter Example

Let us consider the equation of motion

$$\frac{\partial \tilde{h}}{\partial \tilde{t}} = \frac{\nu t_0}{x_0^2} \left[\frac{\partial^2 \tilde{h}}{\partial \tilde{x}^2} + \frac{h_0^3}{Z x_0^2} \left(\frac{\partial \tilde{h}}{\partial \tilde{x}} \right)^4 + 2s \frac{x_0}{h_0} \delta \left(\tilde{x} - \frac{L}{2x_0} \right) \right]. \tag{19}$$

Using (8) shows that the nonlinear term determines the macroscopic surface deformation. For $C \equiv s^3 L/Z \gg 1$ one will again get a triangular steady state profile as in (4b), but now the growth velocity increases by an amount $v - v_0 = (\nu/Z)s^4$ rather than quadratically in s as in (5b). For $C \ll -1$ one will get no macroscopic response proportional to L.

For the surface roughness, however, the nonlinear term in (19) is irrelevant. With (10), this term is rescaled be a factor $x_0^{3\zeta - 2}$ relative to the second order derivative. Without the nonlinearity the roughness exponent is[8] $\zeta = (3 - d)/2$. Hence $3\zeta - 2 < 0$, so that the nonlinear term will be an irrelevant perturbation which does not change the exponents of the linear theory.

General Rules

Suppose, experimentally one finds that the macroscopic response to an inhomogeneous growth rate can be ascribed to a term in the equation of motion which contains a product of N_2 derivatives of h where the sum of the orders of the derivatives is equal to N_1.[*] Under (8) this term scales like $x_0^{N_2 - N_1}$. All other terms in the equation of motion must be rescaled by a smaller factor. If they contain n_1 derivatives and are nonlinear of order n_2, it follows that

$$n_2 - n_1 < N_2 - N_1. \tag{20}$$

Our previous example shows that in principle any of these terms can be crucial for the roughness, if

$$n_2\zeta - n_1 > N_2\zeta - N_1. \tag{21}$$

For any realistic deposition process at a stable surface the roughness exponent has a value between 0 and 1. Hence it follows from (20), (21) that

$$n_2 < N_2, \quad n_1 < N_1, \quad \zeta < (n_1 - N_1)/(n_2 - N_2) < 1. \tag{22}$$

---

[*] For instance, $N_1 = 2$, $N_2 = 1$ for the first term on the right-hand-side de of (19), $N_1 = 4$, $N_2 = 4$ for the second term.

The above naive power counting argument shows that even if the term describing the macroscopic response does not coincide with the one governing the fluctuations, it gives upper bounds for the possible values of n_1, n_2 and the roughness exponent.

RENORMALIZATION OF SYSTEM PARAMETERS

So far we have neglected the fluctuations. Their influence will be discussed in this section for the inhomogeneous KPZ-equation (3).

Coupling Between the Average Profile and the Fluctuations

It follows from (3) that the noise averaged profile $H(x,t) \equiv \langle h(\mathbf{r},t) \rangle$ satisfies the equation

$$\frac{\partial H}{\partial t} = \nu H'' + \frac{\lambda}{2} \langle (\nabla \delta h)^2 \rangle + \frac{\lambda}{2} H'^2 + \kappa, \tag{23}$$

while the fluctuations $\delta h = h - H$ obey

$$\frac{\partial \delta h}{\partial t} = \nu \nabla^2 \delta h + \lambda H' \delta h' + \frac{\lambda}{2}((\nabla \delta h)^2 - \langle (\nabla \delta h)^2 \rangle) + \eta. \tag{24}$$

Here, a prime denotes partial differentiation with respect to x. Because of the nonlinearity the fluctuations and the average profile are coupled. Let us expand

$$\langle (\nabla \delta h)^2 \rangle = c_0 + c_1 H + c_2 H' + c_3 H'' + c_4 (H')^2 + ... \tag{25}$$

Obviously $c_1 = 0$, as a uniform translation of H has no effect on δh. Also a uniform tilt, $H' = a$, does not influence the value of $\langle (\nabla \delta h)^2 \rangle = b$ which will be constant under these conditions. The transformation $\hat{h} = \delta h + bt$ and $\hat{x} = x - at$ reduces (24) to the ordinary KPZ-equation (3) without changing $\langle (\nabla \delta h)^2 \rangle = \langle (\nabla \hat{h})^2 \rangle$. Therefore $c_2 = 0$ and $c_4 = 0$, and the effect of the fluctuations should simply be a renormalization of ν and κ in (23). Thus one expects that the imposed slope s and the parameter Z that determines the size of the curved part of the macroscopically triangular profile (14) will deviate from the constants (7) due to the influence of the fluctuations.

Numerical Results

We have examined[3] s and Z in the single-step model in two space dimensions.[9,10] The growth starts from a flat ("zig-zag") substrate parallel to the (11)-direction of a square lattice. Particles (squares) can be added provided that the surface length remains constant. Periodic boundary conditions are imposed. To gain a computational advantage we employed a parallel updating scheme on the two checkerboard sublattices:[11] Eligible growth sites on a given sublattice are filled simultaneously with a probability $1/2$. Before the growth inhomogeneity is switched on we let the surface evolve until its roughness is fully developed. Then the growth probability above a fixed substrate site is changed to a value $p \neq 1/2$ rendering the growth process inhomogeneous. For reduced deposition, $p < 1/2$, we found a macroscopically triangular profile, implying that $\lambda < 0$. This can be understood easily, since the number of growth sites decreases upon tilting the surface.

We checked if s is linear in $(p - 1/2)$ which should be proportional to κ_1 in (2). This would be expected according to (7), if fluctuations are irrelevant. Although the simulation for small $(p - 1/2)$ is hampered by strong finite size effects (the curved part

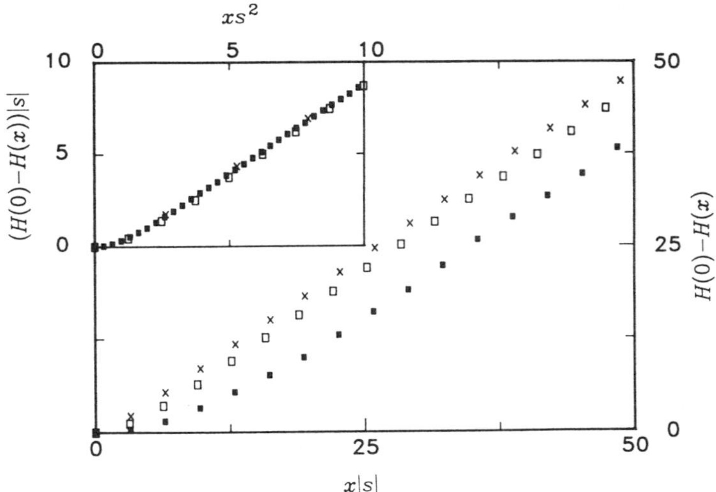

Fig.2. Rounded parts of the stationary profiles scaled in two different ways for different slopes and system sizes (see Ref.3). A data collapse (inset) is obtained only if one assumes that $\nu/\lambda \propto s^{-1}$.

of the profile becomes large making an accurate measurement of s difficult), our data can roughly be fitted by a power law $s \approx \kappa_1^{2.8}$. This indicates that the fluctuations indeed renormalize the system parameters.

The evidence for a renormalization of Z is much clearer. Figure 2 shows the rounded part of the stationary profile at three different values of s. Finite size corrections are negligible for the L-values considered. The x-coordinate is scaled by s^{-1} in accordance with (11a). The rounding extends to larger values of xs as s decreases, thus preventing the data collapse which would be expected if Z were constant. A data collapse is achieved by a different scaling of both x and $H(0) - H(x)$, in the way shown in the inset. All three curves fit well to (11a) by assuming

$$Z = 2\nu/\lambda = (1.7 \pm 0.3)/s. \tag{26}$$

A Scaling Argument Explaining $Z \sim |s|^{\zeta/(\zeta-1)}$

Equating $|s|$ to the typical slope fluctuations $\delta h/R \sim (R/a)^{\zeta-1}$ one gets a characteristic distance

$$R(s) = a \mid s \mid^{1/(\zeta-1)} \tag{27}$$

up to which the fluctuations are capable of washing out the imposed asymptotic slope. a is the lattice constant. A data collapse for systems of different s is expected if one scales the x-coordinate by R and the surface height by $sR = a(R/a)^\zeta$. This leads to a scaling Ansatz for the average profile

$$H(x, L, t, s) - H(0, L, t, s) = sRF(x/R, L/R, t/R^z). \tag{28}$$

For any *nonlinear* theory of surface growth, (27) and (28) offer a new way of determining the roughness exponent through the introduction of a tunable characteristic length R. In the present case, R is simply a measure of the size of the rounded part of the average profile. The data collapse in Fig.2 implies that $\zeta = 1/2$, which is the known value for this model in two space dimensions.[9,10]

Our finding of the scaling (28) and the s-dependence (26) of ν/λ is in perfect agreement with the scaling properties of the KPZ-equation. Like any continuum theory describing the nonlinear behavior of a discrete model the KPZ-equation depends on the lower cutoff or coarse graining length l. Hence there is a whole family of possible coefficients for a given lattice model, with[5]

$$Z(x_0) = x_0^\zeta Z(1), \tag{29}$$

depending on the choice of $l(x_0) = x_0 l(1)$. If one wants to map two configurations of the discrete model with imposed characteristic lengths R_1 and $R_2 = x_0 R_1$ onto each other by rescaling, one must choose different coarse grainings, $l_2 = x_0 l_1$. Hence, with (27), the appropriate continuum description involves different coefficients, $Z(s) \sim |s|^{\zeta/(\zeta-1)}$. Of course, this argument is valid only if the coarse graining length $l \ll R$.

CONCLUSION

We have shown that macroscopic surface profiles obtained from inhomogeneous deposition provide useful information about the phenomenological equations governing the growth. In some cases like the KPZ-equation the surface deformation can be ascribed to the term which is most relevant also for the surface response to noise. In general, a naive power-counting argument shows that the term governing the deterministic response puts only bounds on the possible terms responsible for the surface roughness on large scales. The notion of scale-dependent coefficients in nonlinear growth processes has been illustrated by a computer simulation of the single-step model. It shows that the roughness exponent also describes characteristic features of the average surface deformation.

We would like to thank J. Kertész, T. Nattermann, L. Sander, T. Vicsek and J. Villain for stimulating discussions. This work was supported by the Deutsche Forschungsgemeinschaft within SFB 341. Part of the computer simulations were done at the supercomputer center HLRZ Jülich.

REFERENCES

[1] For recent reviews see: T. Vicsek, *Fractal Growth Phenomena*, (World Scientific, Singapore, 1989); J. Krug and H. Spohn, in "Solids Far From Equilibrium: Growth, Morphology and Defects, ed. by C. Godrèche (Cambridge University Press, 1990); D.E. Wolf, in "Kinetics of Ordering and Growth at Surfaces", ed. by M. Lagally (Plenum, New York, 1990)

[2] F. Family and T. Vicsek, J. Phys. A**18**, L75 (1985)

[3] D. E. Wolf and L.-H. Tang, Phys. Rev. Lett. **65**, 1591 (1990); L.-H. Tang and D. E. Wolf, to be published

[4] J. E. Parmeter, R. Kunkel, B. Poelsema, L. K. Verheij, and G. Comsa, in Proceedings of the 11th International Vacuum Congress and 7th International Conference on Solid Surfaces, Cologne, West Germany, 1989, to be published

[5] M. Kardar, G. Parisi and Y.-C. Zhang, Phys. Rev. Lett. **56**, 889 (1986); E. Medina, T. Hwa, M. Kardar, and Y.-C. Zhang, Phys. Rev. A **39**, 3059 (1989)

[6] J. Villain, this volume

[7] S. F. Edwards and D. R. Wilkinson, Proc. R. Soc. A **381**, 17 (1982)

[8] P. G. de Gennes, Faraday Symposium **5**, 61 (1971)

[9] P. Meakin, P. Ramanlal, L. M. Sander, and R. C. Ball, Phys. Rev. A **34**, 5091 (1986)

[10] M. Plischke, Z. Rácz and D. Liu, Phys. Rev. B **35**, 3485 (1987)

[11] B. Forrest and L.-H. Tang, Phys. Rev. Lett. **64**, 1405 (1990)

CROSSOVER PHENOMENA IN KINETIC ROUGHENING

Lei-Han Tang and Thomas Nattermann

Fakultät für Physik und Astronomie
Ruhr-Universität Bochum
D-4630 Bochum
Federal Republic of Germany

Bruce M. Forrest

Institut für Theoretische Physik
Universität Heidelberg
D-6900 Heidelberg
Federal Republic of Germany

INTRODUCTION

Many growth processes in nature involve, on the molecular level, random attachment and detachment of particles at the interface. One such example is the growth of an amorphous film by the sputtering technique, where energetic atoms land on a cold substrate at the end of their ballistic flights. In the absence of rapid surface diffusion, fluctuations in the incoming flux of particles (known as "shot-noise") lead to accumulation of excess material at parts of the surface, thereby create surface height fluctuations whose amplitude may diverge with the size of the substrate. Such growth-induced roughening can be different from thermal roughening due to the difference in either the noise spectrum or, perhaps more importantly, the way a growing surface responds to a given type of noise in comparison with its equilibrium counterpart.

To understand universal features of the kinetic roughening phenomenon, one is tempted to construct phenomenological models which capture the hydrodynamic behavior of a growing surface. Following such an approach, Kardar, Parisi, and Zhang (KPZ) (1986) proposed a nonlinear stochastic equation for the local growth of a suitably coarse-grained surface height $h(\mathbf{x}, t)$ above a d-dimensional substrate plane in the form,

$$\partial h / \partial t = \nu \nabla^2 h + \frac{\lambda}{2} (\nabla h)^2 + \eta(\mathbf{x}, t). \tag{1}$$

Here an average constant growth velocity has been taken out. The noise term η is usually assumed to have neither spatial nor temporal correlations and to satisfy a Gaussian distribution with a variance D, i.e.,

$$\langle \eta(\mathbf{x}, t) \eta(\mathbf{x}', t') \rangle = 2 D \delta^d (\mathbf{x} - \mathbf{x}') \delta(t - t'). \tag{2}$$

where d is the surface dimension. Without the nonlinear term, Eq. (1) is simply the Langevin equation for the relaxation dynamics of an equilibrium surface above the roughening temperature T_R (Chui and Weeks, 1978). In this case the coefficient ν is related to the stiffness constant of the surface. KPZ (1986) argued that inclusion of the nonlinear term is essential in order to describe certain growth processes such as the Eden model (Eden, 1958). In the case of a restrcited solid-on-solid growth model (Kim and Kosterlitz, 1989), such a term can arise from a reduction in the number of growth sites for a tilted surface.*

Growth and Form, Edited by M. Ben Amar *et al.*
Plenum Press, New York, 1991

The linear Langevin equation at $\lambda = 0$ can be easily solved for any given noise spectrum (de Gennes, 1971). In particular, under the assumption (2), one obtains the well-known logarithmic roughness for a two-dimensional interface at $T > T_R$, and the diffusive relaxation of height fluctuations characterized by a *dynamic exponent* $z = 2$. Full solution to the nonlinear problem, on the other hand, has so far eluded analytical means and minds. Nevertheless, one can gain some understanding concerning the importance of the nonlinear term by setting up a suitable perturbative scheme at small values of λ. This was first done by Forster, Nelson, and Stephen (FNS) (1977) for the related Burgers equation (an equation for $-\nabla h$) using dynamic renormalization group (RG) techniques. The method was later adapted to the surface growth problem by KPZ (1987) (see also Medina *et al*, 1989). Some of the results from this approach will be discussed below.

Computer simulation has become an indispensible tool in physics nowadays and the study of the kinetic roughening phenomenon has not been an exception. In most growth simulations the task of developing necessary computer codes is straightforward, yet with a bit of thinking valuable information complementary to analytical results can be gained from the exercise. One of the observations coming out of the numerical studies is that a growing surface can evolve spontaneously into a critical steady state which possesses scaling properties in space and in time. More precisely, it was found that the amplitude Δh and the lifetime τ of height fluctuations across a sufficiently large distance ξ grow as power-laws of ξ, i.e., $\Delta h \sim \xi^\zeta$ and $\tau \sim \xi^z$. Here ζ is known as the *roughness exponent*. Also of interest is a third exponent $\beta \equiv \zeta/z$ which describes how the width of the surface increases with time starting from a flat substrate.

Early numerical studies were focused on the determination of ζ and z for the class of growth models which are believed to be described by (1) and (2) (e.g., Meakin *et al*, 1986; Wolf and Kertész, 1987; Plischke *et al*, 1987; Kim and Kosterlitz, 1989). The exponent identity

$$\zeta + z = 2$$

predicted based on a symmetry of (1) under temporally uncorrelated noise (Medina *et al*, 1989) agreed well with simulation results. In addition, there is now a general consensus over the value $\zeta(1) = 1/2$ and $z(1) = 3/2$ for a one-dimensional interface. However, the exact values of $\zeta(d)$ and $z(d)$ in higher dimensions are not known. Recent accurate simulation data of Forrest and Tang (FT) (1990a) yielded

$$\beta(2) = 0.240 \pm 0.001, \qquad \beta(3) = 0.180 \pm 0.005.$$

These numbers deviate significantly from either the expression

$$\zeta(d) = 1/(d+1), \qquad z(d) = (2d+1)/(d+1)$$

of Wolf and Kertész (1987), or the conjecture

$$\beta(d) = 1/(d+2), \qquad \zeta(d) = 2/(d+3), \qquad z(d) = 2(d+2)/(d+3)$$

of Kim and Kosterlitz (1989), both were based on simulations of much smaller system sizes than those of FT. The FT result for $\beta(2)$ has recently been confirmed by direct simulations of the KPZ equation (Moser *et al*, 1990).

Here we address the question of dynamic crossover between growth regimes characterized

---

* It should be noted, however, that the applicability of (1) to any particular experimental system has not been demonstrated. In fact, Villain (1990) has argued that the gradient-squared nonlinearity is not allowed if there is no desorption of particles in an epitaxial growth of thin films. Interestingly, most surface growth models studied in computer simulations seem to comply with (1) except at special parameter values where λ vanishes accidentally. Exceptions were found only recently after the principle of forbidding the square-gradient nonlinearity was elucidated (Wolf and Villain, 1990). This confirms the belief that, in a rough phase and on sufficiently large length scales, the gradient-squared nonlinearity dominates all other terms which are compatible with the local surface growth condition (Kardar *et al*, 1986).

by different roughness and/or dynamic exponents. Our study was motivated by an on-going controversy regarding the existence of a kinetic phase transition in the physically interesting case $d = 2$ (Amar and Family, 1990; Yan et al, 1990; Guo et al, 1990a; Pellegrini and Jullien, 1990). The discussion also contributes toward the clarification of the origin of a smaller effective roughness exponent observed in some growth simulations (Chakrabarti and Toral, 1989; Guo et al, 1990a). We begin with a general discussion of the extended scaling hypothesis for a kinetic phase transition. This general consideration is applied to Eq. (1) at small values of the nonlinear coupling parameter, where a crossover between the $z = 2$ dynamics in the linear regime to a different dynamical behavior in the nonlinear regime is analyzed. With the help of the dynamic RG method of FNS, we obtain explicit predictions concerning the form of the crossover scaling function and the dependence of the crossover length on growth parameters. These analytical results are then compared with simulation data on a deposition and evaporation model. We conclude with a discussion of some unsettled issues and a summary of results.

EXTENDED SCALING HYPOTHESIS FOR A KINETIC PHASE TRANSITION

In surface growth models we identify different phases by their scaling properties in the hydrodynamic limit. These scaling properties can be conveniently examined and compared through the measurement of appropriate correlation functions. For example, one may consider the mean-square width of a surface of linear size L growing from a flat substrate at time $t = 0$,

$$w^2(L,t) \equiv \left\langle \overline{[h(\mathbf{x},t) - \overline{h}\,]^2} \right\rangle = \frac{1}{2}\left\langle \overline{[h(\mathbf{x},t) - h(\mathbf{x}',t)]^2} \right\rangle.$$

Here the angular brackets denote average over different realizations of the noise and the overline bars denote average over all surface sites in a given configuration. Computer simulations show that $w^2(L,t)$ usually obeys a scaling form (Family and Vicsek, 1985),

$$w^2(L,t) \sim L^{2\zeta} F(t/L^z), \tag{3}$$

where the scaling function $F(x) \sim x^{2\zeta/z}$ for $x \ll 1$ whereas it approaches a constant for $x \gg 1$. Equation (3) simply reflects the fact that a growing surface has no intrinsic length or time scale apart from the lattice spacing. The width of the surface in the rough phase is only restricted by the smaller of the two: the $relaxation$ $length$ $\xi(t) \sim t^{1/z}$ and the linear system size L. A $kinetic$ $phase$ $transition$ takes place if $\{\zeta, z\}$ change their values from, say, $\{\zeta_A, z_A\}$ to $\{\zeta_B, z_B\}$ as certain growth parameter p is varied. In general one should allow for the possibility of a third set of exponents $\{\zeta_C, z_C\}$ at the transition (Kertész and Wolf, 1989).

How does one reconcile the three possibly different set of exponents with a continuous transition? To answer this question let us recall that usually an intrinsic length ξ_c emerges as the transition point is approached. In the case of ordinary critical phenomena, ξ_c is simply the correlation length which diverges at the critical point. On length scales much smaller than ξ_c the system behaves as if it is at the critical point, while normal state behavior is expected on length scales much larger than ξ_c. One should, however, take into account that under growth conditions, what we termed "normal state" is often by itself critical, in that it too possesses infinite range spatial and temporal correlations. Thus the scaling forms which describe the ordinary critical phenomena should be modified accordingly. In addition, since we are talking about connecting different power laws rather than moving away from criticality, it appears more reasonable to call ξ_c the $crossover$ $length$.

The conclusion is that, on either side of the transition, (3) is still correct asymptotically, i.e., for length scales much larger than ξ_c. Below this length scale one sees a different scaling which characterizes the transition point. The two behaviors can be conveniently incorporated in an extended scaling form of the type (Kertész and Wolf, 1989; Wolf, 1990),

$$w^2(L,t) \sim \xi_c^{2\zeta_C} f_i(L/\xi_c, t/\tau_c), \tag{4}$$

where $\tau_c \simeq \xi_c^{z_C}$ is the $crossover$ $time$. The subscript i stands for either A or B. To describe how fast the crossover length increases as the growth parameter p approaches its value p_c at the transition, one may introduce a crossover exponent ϕ such that

$$\xi_c(p) \sim |p - p_c|^{-1/\phi}.$$

Equation (4) expresses the fact that $w^2(L, t)$ is now determined by the ratios of three relevant lengths of the problem: the crossover length ξ_c, the linear system size L, and the characteristic relaxation length $\xi(t)$. The prefactor in front of f_i is needed in order to recover the simple scaling form (1) at the transition. The latter in addition requires that

$$f_i(a, b) \simeq a^{2\zeta_C} g(b/a^{z_C})$$

for $a, b \ll 1$. In the opposite limit $a, b \gg 1$ one should recover the asymptotic scaling in the two phases, i.e.,

$$f_i(a, b) \simeq a^{2\zeta_i} g_i(b/a_i^z).$$

Compare this result with (3) we obtain, asymptotically,

$$w^2(L, t) \sim \begin{cases} \tau_c^{2\beta_C} (t/\tau_c)^{2\beta_i} & \text{if } t/\tau_c \ll (L/\xi_c)^{z_i}; \\ \xi_c^{2\zeta_C} (L/\xi_c)^{2\zeta_i} & \text{if } t/\tau_c \gg (L/\xi_c)^{z_i}. \end{cases} \tag{5}$$

If one of the arguments of f_i is much smaller than unity and the other much bigger than unity, f_i depends only on the smaller of the two.

A special case of interest is when the surface at the transition is logarithmically rough ($\zeta_C = 0$). For simplicity let us consider only the two-length problem, ξ_c and ξ, with the latter being either $\xi(t)$ in the case $\xi(t), \xi_c \ll L$, or L in the case $L, \xi_c \ll \xi(t)$. Guo $et\ al$ (1990b) wrote down a crossover scaling form

$$w^2(\xi) \sim \tilde{g}(\xi/\xi_c) + \ln \xi_c. \tag{6}$$

To obtained the logarithmic behavior at small values of ξ they required $g(x) \simeq \ln x$ for $x \ll 1$. The constant term in (6) is needed in order to cancel out the ξ_c dependence from the first term in this limit. A power-law form at $\xi \gg \xi_c$ requires $g(x) \sim x^{2\zeta_i}$ for $x \gg 1$. The behavior of g at intermediate values of x remains arbitrary.

There actually exists a second crossover scaling form which also connects a logarithmic to a power-law roughness (Tang $et\ al$, 1990). It has the following appearance,

$$w^2(\xi) \sim \ln \xi_c \left[G(\xi/\xi_c) + \ln \ln \xi_c \right], \tag{7}$$

where $G(x) \simeq -\ln(-\ln x)$ at $x \ll 1$ and $\sim x^{2\zeta_i}$ at $x \gg 1$. Unlike (6), Eq. (7) predicts that the amplitude of the asymptotic power-law, when ploted against the scaled variable $x = \xi/\xi_c$, grows with ξ_c. We shall see below that only (7) is supported by the RG analysis of the KPZ equation at $d = 2$.

RENORMALIZATION GROUP ANALYSIS IN (2+1) DIMENSIONS

If the KPZ equation is valid not only in the hydrodynamic limit, but also as a continuum equation for a class of growth processes at smaller length and time scales, then the crossover behavior which occurs at intermediate length scales should also be universal. RG analysis of the KPZ equation can yield partial or full expression for the crossover scaling functions discussed above.

FNS (1977) showed in a one-loop RG calculation that $d = 2$ is the marginal dimension for the related Burgers equation, below which even the slightest nonlinearity will dominate on sufficiently large length scales. The picture changes at $d > 2$, where a sufficiently weak nonlinearity is irrelevant and a transition takes place as the strength of the nonlinear term, measured by the dimensionless parameter $g_B = (a_B/\pi)^{2-d} D_B \lambda_B^2 / \nu_B^3$, is increased above some critical value. Here a_B is the lower cut-off length of the continuum equation (1). In the following discussion we shall focus on the $d = 2$ case, where the $\lambda = 0$ fixed point is marginally unstable.

In this situation there is a crossover between the logarithmic roughness at small length scales and power-law roughness at large length scales when g_B is sufficiently small. A corresponding crossover occurs in the dynamic behavior of the system.

Our discussion starts with the following expression for the mean-square surface width,

$$w^2(t, L) \simeq \int_{2\pi/L \leq |\mathbf{k}| \leq \pi/a} \frac{d^d \mathbf{k}}{(2\pi)^d} \frac{D}{\nu k^2} [1 - \exp(-2\nu k^2 t)]. \tag{8}$$

where we have assumed that the surface is flat at $t = 0$. Equation (8) is obtained from (1) and (2) at $\lambda = 0$. In a one-loop calculation we assume that (8) is valid also in the nonlinear case, but the coefficients D and ν now depend on the length scale $b a_B$ in the way determined by the RG flow equations (FNS, 1977),

$$d\nu/dl = K_d g(2 - d)\nu/4d, \qquad d\lambda/dl = 0, \qquad dD/dl = K_d g D/4. \tag{9}$$

Here $K_d^{-1} = 2^{d-1} \pi^{d/2} \Gamma(d/2)$, and $l = \ln b$. Equations (9) are valid when the dimensionless coupling parameter $g \equiv (b a_B/\pi)^{2-d} D \lambda^2/\nu^3$ is sufficiently small (weak-coupling).

In the case $d = 2$, ν is not renormalized in the one-loop approximation, and D has the same scale dependence as g. Thus the right-hand side of the last equation in (9) is given by $D^2 g_B/(8\pi D_B)$. A simple integration yields

$$D(b) \simeq D_B(\ln \xi_c)/\ln(\xi_c/b), \tag{10}$$

where

$$\xi_c = \exp(8\pi/g_B) \tag{11}$$

gives the crossover length in units of a_B.

We can now evaluate (8) by letting $\nu = \nu_B$ and $D = D(b)$. For simplicity we let $L \to \infty$. The correct b to use now depends on both k and t: $b = \pi/(k a_B)$ if $\nu_B k^2 t \gg 1$ and $b = \xi(t) \simeq \sqrt{\nu_B t}/a_B$ if $\nu_B k^2 t \ll 1$. The latter case is irrelevant since there the integration is effectively cut off by the term in the square brackets. We now approximate this term by a step function which vanishes for $k < 1/[\xi(t) a_B]$. Performing the integration then gives

$$w^2(t) \simeq \frac{D_B \ln \xi_c}{\pi \nu_B} \ln \frac{\ln \xi_c}{\ln \xi_c/\xi(t)}. \tag{12}$$

The singularity of (12) at $\xi(t) = \xi_c$ is due to lack of a strong-coupling fixed point at a finite g. In fact, (12) is expected to be valid only if $\xi(t)/\xi_c$ is sufficiently small. By connecting (12) to an asymptotic power law one obtains (7). Note that the amplitude of w^2 contains an additional factor D_B/ν_B.

The bare nonlinear coupling parameter g_B is expected to vary smoothly with actual growth parameters, such as p^+ and p^- in the model described below. The exponential form (11) suggests that the crossover length ξ_c should increase very rapidly when these growth parameters are varied so as to decrease g_B. For example, reducing g_B from 10 to 1 brings ξ_c from about 10 to about 10^{11}, a change of ten decades in the crossover length!

COMPUTER SIMULATION

A solid-on-solid growth model was introduced by Forrest and Tang (1990a) to study kinetic roughening in the KPZ family. The model describes growth along the body-diagonal direction of a hypercubic lattice, thus known as the hypercube-stacking model (HSM). It is equivalent to the single-step model of Meakin $et\ al$ (1986) and Plischke $et\ al$ (1987) when $d = 1$. In the (2+1)-dimensional case, with a different definition of the surface height, the model can also be interpreted as a growth algorithm for the dense-packing face-centered-cubic structure along the (111) direction.

A typical surface configuration in the $d = 2$ case is shown in Fig. 1(a), along with its projection onto the (111) substrate plane [1(b)]. It is evident that the three types of faces of the surface, i.e., (100), (010), and (001), are projected onto 60° rhombi in three different orientations. A growth event corresponds to depositing a cube at an inward corner of the surface. The move is performed with a probability p^+ each time such a corner is sampled. Similarly, evaporation events take place at outward corners of the surface with a probability p^-.

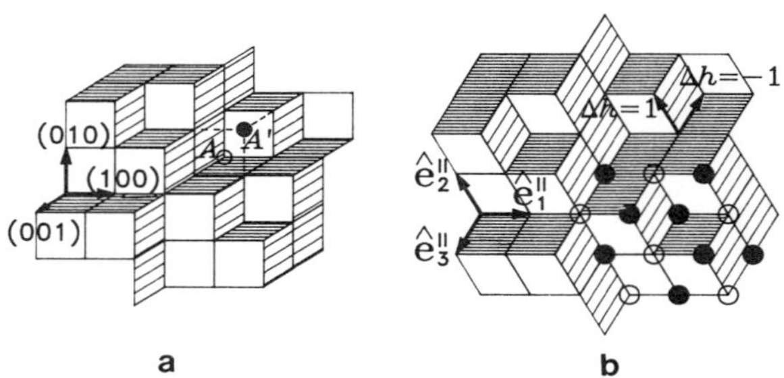

Fig. 1. The hypercube stacking model at $d = 2$. (a) The surface of a cube stack with only (100), (010), and (001) faces exposed. The corner labelled A is an adsorption site, and A' a desorption site. (b) A rhombus tiling obtained by projection onto a (111) substrate plane. Open ($s = 1/2$) and solid ($s = -1/2$) circles in (b) indicate Ising spins on vertices (of part) of the tiling. A rhombus edge corresponds to a satisfied antiferromagnetic bond (see Forrest and Tang, 1990a, b).

A major computational advantage of the HSM is that the surface height configuration can be mapped to a d-state Potts model on the d-dimesional substrate lattice via the tiling representation. This allows for very efficient memory usage in simulations, especially in the two-dimensional case, where each substrate site requires only one-bit of memory space. In addition, the growth rules described above can be easily expressed in terms of spin-flips which can then be implemented through simple bit operations. Using a sublattice updating scheme the whole algorithm can be very efficiently vectorized on machines such as the Cray. On a Cray-XMP/416 simulations of systems of 11520^2 and 2×192^3 substrate lattice sites in two and three dimensions have been performed (FT, 1990a). For details of the mapping and the simulation method the reader is referred to Forrest and Tang (1990b).

In the following we compare the predicted scaling (7) with numerical simulation data on the HSM at $d = 2$. Time is measured in units of sweeps of the whole surface. We studied a system of $N = L^2 = 5760^2$ surface sites at $p^+ = 1/2$ and various values of p^-, up to $t = 4096$. Saturation of the surface width at this size is estimated to take place at $t \geq 5 \times 10^5$. The need for having such a large system size is obvious, as it will prevent a third length scale to complicate the analysis. Figure 2 shows the simulation data $w^2(2t) - w^2(t)$ versus t on a log-log scale. In this way a logarithmic $w^2(t)$ is represented by a constant on the plot, while a power-law behavior has the usual appeerence of a straight line at a finite slope. From the figure we see that data at $p^- = p^+ = 1/2$ shows logarithmic scaling as expected for a thermally rough surface. The dashed line gives the coefficient of $\ln t$ using the exact result of Blöte and Hilhorst (1982) for the stiffness constant, assuming $z = 2$. At $p^- = 0$ a simple power-law scaling is observed down to very early times. In between the data show a crossover behavior with a changing effective exponent.

To check the crossover picture quantitatively, we shift each data set at $t \geq 4$ horizontally and vertically on the plot so as to achieve a data collapse. The result is shown in Fig. 3 for $1/16 \leq p^- \leq 3/16$. This procedure also defines the crossover length $\xi_c(p^-)$ and the scaling amplitude $A(p^-)$ in terms of the amount of horizontal translation $2\log_{10}\xi_c(p^-)$ and vertical translation $\log_{10} A(p^-)$, respectively. Of course we only know the relative magnitude of these

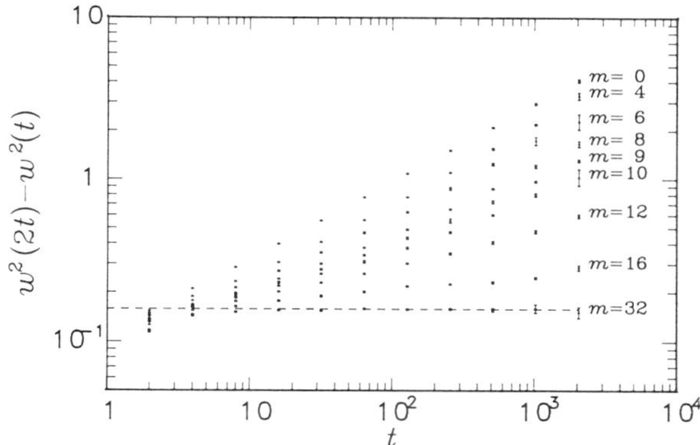

Fig. 2. Simulation data for the mean-square surface width w^2 versus time t, starting from a flat substrate at $t = 0$. Here $p^+ = 1/2$ and $p^- = m/64$.

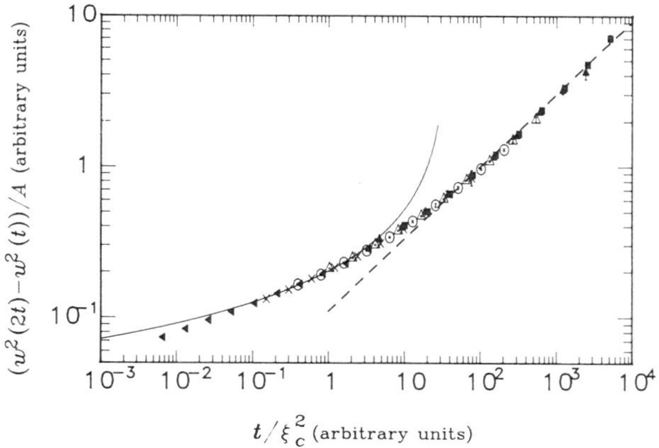

Fig. 3. Data collapse achieved by a pure translation of each data set shown in Fig. 2 at $t \geq 4$. The amount of translation defines ξ_c and A. The solid line gives the RG result at small values of the scaled variable.

quantities. The operation becomes ambiguous when the curvature of a data set becomes too small, which is the case at larger values of p^-. The solid line represents the one-loop result for

$$G(\sqrt{2u}) - G(\sqrt{u}) \simeq -\ln(1 + \ln 2/\ln u)$$

at small $u = \xi^2(t)/\xi_c^2 \sim t/\xi_c^2$. The dashed line indicates an asymptotic power-law dependence at $\beta = 0.24$.

DISCUSSION AND CONCLUSION

The good data collapse seen in Fig. 3 shows that the seemingly decreasing effective exponent β_{eff} with increasing p^- (at least up to $p^- = 3/16$) is due to a crossover effect. Moreover, the data collapse curve gives a numerically determined form for the scaling function G in terms of t, which matches well with the solid line at small values of the scaling variable. To provide further check on the scaling form (7), we plotted A against ξ_c on a semi-log scale, as shown in Fig. 4. The rapid increase of ξ_c with decreasing $\kappa = (p^+ - p^-)/p^+$ is evident: a 30% decrease in κ increases ξ_c by more than 30 times! The straight line in the figure gives the dependence of A on ξ_c assuming D_B/ν_B to be a constant. Our data thus show that D_B/ν_B depends only weakly on p^-.

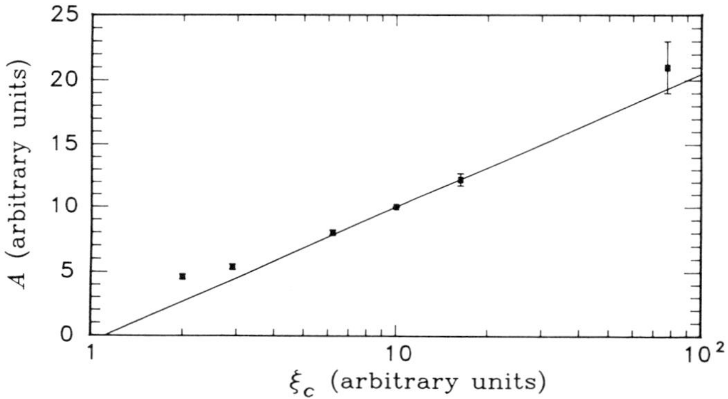

Fig. 4. Crossover length ξ_c versus the scaling amplitude A. Relative uncertainties on ξ_c (not shown) are comparable to those on A.

One might still argue that the above analysis does not rule out the possibility of a phase transition for p^- closer to p^+. Our opinion is that although the data on the HSM provide the best support so far for the crossover picture, it is ultimately impossible to settle down the issue from a numerical study. The matter is likely to be complicated by the additional fact that most lattice models contain an intrinsic length scale which is often nonuniversal (Wolf, 1990). In the best cases this length scale is of the order of a few lattice constants. In other cases it may be much bigger. Only above this length scale can one hope to see the type of universal behavior predicted based on the KPZ equation. It may be useful in some cases to devise a better continuum equation which includes additional terms. These terms may be irrelevant in the truly hydrodynamic limit, but they may change the scaling behavior at intermediate length scales which computer experimentalists have to live with. An example along this line is provided by the crossover effect due the discrete height of lattice planes. A systematic RG treatment of the effect below the thermal roughening temperature has been presented by Noziéres and Gallet (1987) and by Hwa *et al* (1990).

In summary, we have demonstrated the importance of extended scaling in analyzing kinetic phase transitions. We discussed in detail the crossover from logarithmic to power-law scaling which occurs in the weak-coupling regime of the KPZ equation at $d = 2$. RG calculations yielded quantitative information concerning the functional form of the crossover. We established that the observed varying effective exponent in a large-scale simulation of the HSM is due to a very slow crossover of the KPZ equation at $d = 2$. Our results may shed some light onto the recent controversy over the nature of kinetic phase transition(s) in various $(2+1)$-dimensional growth models. Given the ineffectiveness of the nonlinear term in driving the system into the strong-coupling regime at $d = 2$, it would be interesting to explore the effect of adding other terms to the continuum KPZ equation, which may modify the crossover behavior discussed here.

From the experimental point of view, the type of crossovers in kinetic roughening discussed here may be relevant for the growth of a crystal under a weak driving force, in which case quasi-thermal equilibrium properties of the surface are maintained on small length scales while kinetic roughening effects prevail on sufficiently large length scales. Given the exponentially large crossover length for the KPZ equation at small values of the bare-coupling parameter, it is conceivable that even a real crystal may not be able to see its hydrodynamic exponents.

ACKNOWLEDGEMENTS

We wish to thank D. E. Wolf for numerous valuable discussions. T. N. and L. T. are supported in part by the DFG through SFB 237. One of us (T. N.) acknowledges a grant from the German Israeli Foundation for Scientific Research and Development.

REFERENCES

Amar J. G. and Family F., 1990, Phys. Rev. Lett. **64**, 543; Phys. Rev. A **41**, 3399.
Blöte H. W. J. and Hilhorst H. J., 1982, J. Phys. A **15**, L631.
Chakrabarti A. and Toral R., 1989, Phys. Rev. B **40**, 11419.
Chui S. T. and Weeks J. D., 1978, Phys. Rev. Lett. **40**, 733.
Eden M., 1958, *in* "Symposium on Information Theory in Biology," H. P. Yockey ed., Pergamon Press, New York, p.359.
Family F. and Vicsek T., 1985, J. Phys. A **18**, L75.
Forrest B. M. and Tang L.-H., 1990a, Phys. Rev. Lett. **64**, 1405.
Forrest B. M. and Tang L.-H., 1990b, J. Stat. Phys. **60**, 181.
Forster D., Nelson D. R., and Stephen M. J., 1977, Phys. Rev. A **16**, 732.
de Gennes P. G., 1971, Faraday Symposia, No. 5, 61.
Guo H., Grossmann B., and Grant M., 1990a, Phys. Rev. Lett. **64**, 1262.
Guo H., Grossmann B., and Grant M., 1990b, Phys. Rev. A **41**, 7082.
Hwa T., Kardar M., and Paczuski, 1990, MIT preprint.
Kardar M., Parisi G., and Zhang Y.-C., 1986, Phys. Rev. Lett. **56**, 889.
Kertész J. and Wolf D. E., 1989, Phys. Rev. Lett. **62**, 2571.
Kim J. M. and Kosterlitz J. M., 1989, Phys. Rev. Lett. **62**, 2289.
Meakin P., Ramanlal P., Sander L. M., and Ball R. C., 1986, Phys. Rev. A **34**, 5091.
Medina E., Hwa T., Kardar M., and Zhang Y.-C., 1989, Phys. Rev. B **39**, 3053.
Moser K., Wolf D. E., and Kertész J., 1990, to be published.
Noziéres P. and Gallet F., 1987, J. Physique (France) **48**, 369.
Pellegrini Y. P. and Jullien R., 1990, Phys. Rev. Lett. **64**, 1745.
Plischke M., Rácz Z., and Liu D., 1987, Phys. Rev. B **35**, 3485.
Tang L.-H., Nattermann T., and Forrest B.M., 1990, Phys. Rev. Lett. **65**, 2422.
Villain J., 1990, J. Physique (France) (in press); This volumn.
Wolf D. E., 1990, *in* "Kinetics of Ordering and Growth at Surfaces," M. Legally ed., Plenum, New York.
Wolf D. E. and Kertész J.,1987, Europhys. Lett. **4**, 651.
Wolf D. E. and Villain J., 1990, Europhys. Lett. (in press).
Yan H., Kessler D., and Sander L. M., 1990, Phys. Rev. Lett. **64**, 926.

MISCELLANEOUS SUBJECTS

THREE SHORT STORIES ON CHIRAL STRUCTURES IN CONDENSED MATTER

Yves Pomeau

Laboratoire de Physique Statistique
24, rue Lhomond, 75231 Paris Cedex 05
France

ABSTRACT

Chirality plays a crucial role in many instances in condensed matter physics. The three examples of this role that I present pose well defined questions and answers to these questions will be proposed. An unifying thema of these stories is perhaps that the chiral information may be transported from the molecular to the (quasi) macroscopic level. The first example concerns the observation of crystals growing in the form of spirals on Langmuir monolayers. I argue that this might be due to an uneven distribution of impurities in the growing crystal, this one being itself related to the molecular structures through the lack of symmetry of the Wulff's plot for crystals of chiral material. The second example (a joint work with J. Lega) is related to the helical structures observed by N. Mendelson on strings of a mutant of Bacterium Subtillis. It is possible to explain this helicity as resulting from a buckling of the cell wall under forces depending on the helical structure of the molecules. This provides a way for transferring the genetic information on chirality from molecular to large scales. The third story (a joint work with P. Coullet and J. Lega) is about nonvariational effect in the dynamics of Bloch walls in ferromagnets. Those Bloch walls have an helicity depending on their preparation, but with a definite sign, if the magnetic interaction is not too anisotropic. Then under some conveniently chosen external stress (a rotating magnetic field) it is possible to move those Bloch walls in a direction depending on the sign of their helicity. This is a typical nonequilibrium phenomenon, because for gradient flow systems, the dynamics of the wall between two phases (here the two orientations of the magnetization) is independent on the wall structure and is a function of the (free) energy difference between the two sides of the wall only.

---

[*] This talk was presented at the Nato Advanced Study Institute on "Growth and Form: Nonlinear aspects" in Cargese(France), July 17-27, 1990.

Growth and Form, Edited by M. Ben Amar *et al.*
Plenum Press, New York, 1991

1. SPIRAL CRYSTALS IN LANGMUIR MONOLAYERS[+]

McConnell et al. have [2] observed the growth of spiral crystals in 2D Langmuir monolayers. These crystals are made of an optically active substance and for a given handedness of the molecules all spirals turn in the same sense, left or right, depending on the sign of the optical activity of the molecules. Various explanations have been proposed for this phenomenon. It has been suggested [3] that this is because chiral molecules (or even nonchiral molecules in a chiral background) diffuse at an angle with the concentration gradient, this being in agreement with geometrical symmetries, and the angle between the concentration flux with the concentration gradient should change sign under a global mirror symmetry operation.

Another explanation [4] makes use of the Wulff's construction. This gives [5] the equilibrium shape of a crystal, under the constraint that its surface energy is the lowest for a prescribed angular dependence of the capillary constant. It does not seem to be possible to explain spiral crystals by using Wulff's construction, as this yields convex crystals only. Perhaps nonconvex equilibrium shapes could follow from negative line tension for some orientations. But this would give anyway an unique equilibrium shape up to an overall dilation, although the spirals observed in the experiments have a variable number of turns, depending presumably on the duration of the growth. Moreover it does not seem possible to explain in this way the observed bending of the crystal axis along the spirals (see below). Indeed the molecules making those crystals have a permanent dipole moment perpendicular to the layer. This leads to a new term in the Gibbs expansion of the free energy as a function of the size of the system:

Usually this expansion reads (for such two dimensional objects) as:

$$E = e\,A + \sigma\,L + \ldots,$$

where A is the area of the crystal, e its energy per unit area, σ the line tension of the crystal/melt interface, and L the perimeter of the crystal. As shown by de Gennes, when one takes into account for the electrostatic interaction a new term is to be inserted between the area (eA) and the line (σL) term, it is of order $L\ln L$ (and thus in between the area, of order L^2 and the perimeter in the large L limit). It is however unlikely that this new term has anything to do with the spiral crystals: it represents a repulsive force between the dipoles, although the spiral shape clearly increases this electrostatic energy with respect to a straight crystal. More generally, it is reasonnable to believe that the logarithm, as often in condensed matter physics can be seen as a kind of constant, except for anomalously large crystals and/or coefficients.

[+] The interested reader may find a more detailed exposition of this question in ref.[1].

Other constraints on any explanation of these spirals are the two following experimental facts. First the orientation of the crystal rotates with the spiral, so that the crystal may be seen as permanently bent [6]. Also, to have converging spirals as observed, the local radius of curvature should decrease as the crystal grows.

This permanent bending of a growing monocrystal can be explained as follows: suppose that during its growth more impurities are incorporated on one side of this crystal than on the other, as seen from the tip. Substitution of an impurity in place of a regular atom in a lattice leads to local compression or dilation, depending whether the impurity is larger or smaller than a regular atom. The mathematical description of this phenomenon is the same as for the bending of a nonuniformly heated crystal presented in §7 of [7]. Without trying to solve the full mathematical problem, one may reason as follows to get order of magnitude estimates: The relevant quantity one is looking at is the curvature of the bent monocrystal. This curvature is proportional, in the weak bending limit, to the gradient of concentration of foreign molecules in the lattice. Let **grad**c be this gradient, boldface being for vectors. Thus a phenomenological law for the bending is:

$\kappa = \alpha$ **grad**c, where α is a coefficient and **grad**c the average value of the gradient across the crystal, κ being directed along the normal to the shape of the crystal.

Measuring everything in atomic units, one would get a radius of curvature of order of a molecular length (say 10 Angstroms at most) for a concentration gradient of order of one molecule per unit cell of the crystal (again a microscopic length). Thus measuring c in dimensionless relative concentrations (c=average number of foreign atoms per unit cell of the crystal), one gets that α is dimensionless too. This means that, up to numbers of order 1, the curvature of the crystal is of the same order as the gradient of dimensionless concentration of impurities in the crystal. Thus a small amount of impurities only is sufficient to produce the observed bending with a radius of curvature much bigger than any molecular length.

To explain spiral crystals by this phenomenon of permanent bending however, it remains to understand other experimental facts:

(i) why the nonequilibrium situation of a concentration gradient in the solid is stable over long periods of time? This is likely because impurities diffuse very slowly in a solid. However one should remember that these crystals "float" on a liquid/vapor interface. Thus it is possible perhaps to change the impurity concentration in the 2D solid by letting impurities diffuse into - or eventually react with- the underlying liquid. This could be done by changing the physicochemical conditions (pH,...) of this liquid. The radius of curvature of the crystals would increase as the impurities are incorporated in the underlying fluid.

(ii) How does the impurity gradient build up as the crystal grows, in relation with the optical activity of the molecules? This is what is explained below.

The theory of the growth of solids at the expense of a supercooled melt introduces [8] two conditions at the interface. First Stefàn's condition for the conservation of mass of the impurity: in the liquid phase the impurity diffuses according to Fick's law, then the moving solid/liquid boundary is a source (or a sink) for the concentration field, because the equilibrium concentration is usually not the same in the solid and liquid phases. The other boundary condition at the interface says that the local curvature of this boundary depends on the local impurity concentration through a condition given by the Gibbs-Thomson theory. This is where the optical activity enters.

Suppose that the tip of the growing crystal is axissymetric. Then by this Gibbs-Thomson relation there is a difference between the impurity concentration on the two sides of the tip if the angular dependence of the surface (actually line-) tension has no mirror symmetry with respect to the direction of growth, and this is precisely what happens if the crystal is made of molecules lacking this symmetry.

It is indeed possible to formulate mathematically this remark although probably not terribly useful: the final formulae will depend on parameters as the solid/liquid line tension that are likely extremely difficult if not impossible to measure. We assumed a symmetric shape of the crystal although there is probably a complicated interplay between the assymmetry of the crystal shape and the concentration field in the liquid to give finally a non symmetric concentration field in the solid. This lack of symmetry comes from the formulation of the boundary conditions through the Gibbs-Thomson relation and the angular dependence of the line tension. Until now we have assumed implicitely that the crystal will grow along a direction fixed with respect to the local crystal axis. This may have various explanations but fits well with the property [8.b] that the velocity of growth of needle crystals is very sensitive to the anisotropy in the surface tension. Finally the image of the growth that we gave implied the tip only in an attempt to make the overall picture hopefully clearer, whereas dendrites as the parabolae predicted by the theory by Ivantsov [8] grow by deposition of material all along the sides of the needle crystal. A theory more elaborate than the one we have presented could probably take this into account, without altering the basic mechanism that we propose, because in a sense the middle part of the crystal comes everywhere from the growing tip.

Our analysis explains why the spirals grow by converging. The growing crystal gets rid of the impurities in the fluid phase that becomes more concentrated nearby. Thus the growing crystal will meet a more and more concentrated impurity field and thus will bend more and more.

This permanent bending of growing monocrystals coming from impurity gradients in the solid could be perhaps observed in other physical conditions. One could think of imposing this gradient by a steady laminar flow of melt with impurities more or less perpendicular to the direction of growth. In this case only a crystal growing parallel or antiparallel to the external flow would not bend, one of those two possible directions being probably stable and the other unstable. It is also possible that the bending related to the growth process of a needle crystal in a gradient leads to an instability during the growth even in an homogeneous and nonchiral medium: as the needle grows, it generates a gradient of the concentration field in its vicinity that could generates an instability against bending, but this is clearly not what is observed in the experiments of reference [2] since spirals generated by this instability should have at random one helicity or the other.

Finally, as kindly suggested by P. Oswald at the conference, it is perhaps possible to grow three dimensional dendrites with helicity, by using for instance cholesteric liquid crystals.The mechanism for spiralling would be then similar to the one presented here, but for the fact that it happens in 3D, where probably there is a better control of everything than in 2D Langmuir layers.

2. AN EXAMPLE OF HELICAL STRUCTURE IN BIOLOGY[++]

Bacteria use to split in two (mitosis) at each new generation. N. Mendelson [10] has observed a mutation of Bacillus subtilis where there is no splitting at the mitosis and the cells keep linked together. Those mutants generate strings of cells that may even reach macroscopic length (centimeters). Mendelson observed too that those strings take the shape of helicoidal coils and that the sign of the helicity as well as it pitch is fixed under given physicochemical conditions of growth. This is a clear indication that the "macroscopic" helicity of those strings is a direct consequence of the chirality of biomolecules, a property discovered by Pasteur. Now I will be interested in the way in which this molecular genetic "message" of chirality is transferred from the molecular up to the macroscale.

It is known from experiments that the chirality of the strings is entirely due to the wall of the cells. For this mutant, this wall may be seen as a cylinder made of solid material. We shall even assume that this tube is physically made of a regular 2D square lattice, even though it is known to have a much more complicated structure, but our approach is macroscopic in a sense and does not pay attention to every detail. Furthermore this idealization by a square lattice has been already used for this specific problem [10].

The general idea of our approach is as follows: we assume that the square lattice is oriented with a family of generatrices parallel to the axis of the cylinder. Thus the lattice may be seen as made of circles of molecules encircling the cylinder, each circle interacting with two neighbours: one above and one below if the cylinder is supposed to be vertical. Then I assume that those circles are rigid (this approximatioon is shown to be correct in

---

[++] A more complete version of this joint work with J. Lega is to be appear in the Comptes-Rendus de l'Académie des Sciences [9].

[9]) and look for the setting that minimizes their interaction energy. This interaction energy has indeed a molecular origin and I show that certain components of it are typical of chiral molecules. Then, if this chiral part of the energy is large enough, the equilibrium configuration of pairs of neighbouring circles is twisted in a chiral way. Finally, by extending this to the whole stack of circles making the cylinder, its equilibrium form may be chiral [9], again if a certain chiral component of the interaction between molecules is large enough.

Supposing that the molecules are points, the chirality of those molecules does appear in set of those point-molecules lacking mirror symmetry. This requires nonplanar sets made of four molecules at least. Thus I shall consider the interaction energy between set of four neighbours on the square lattice. If one restricts oneself to sets where each point is the nearest neighbour of at least another point in the same set, they are a priori eight geometrically different sets of four neighbours by supposing that the mirror images are different and that the interaction is not the same in the vertical and in the horizontal direction. I shall deal with two types of chiral interaction: $U_{c,1}$ between three molecules on a circle and another in the circle below (for instance) and $U_{c,2}$ between the four molecules at the vertex of a square unit on the lattice. It happens that $U_{c,1}$ leads to a distortion of the lattice with respect to the orientation of the cylinder, but by leaving it straight, although $U_{c,2}$ may lead to a helical shape of the cylinder if it is strong enough.

To make things more precise, let $r_{i,j}$ be the position of a molecule on the square lattice(possibly deformed), the discrete index i labels the azimuthal position on the horizontal circles and the index j labels the circle. The argument of the interaction energy $U_{c,1}$ is thus $(r_{i-1,j}, r_{i,j}, r_{i+1,j}, r_{i,j+1})$. As the circles are assumed rigid, one can to minimize energy $U_{c,1}$ only by changing the position of the molecule $r_{i,j+1}$ with respect to the rigid triad $(r_{i-1,j}, r_{i,j}, r_{i+1,j})$. But once the position of this molecule has been found to minimize this energy, one do the same for all the other particles on the circle (j+1) by putting them at the same minimum by a convenient horizontal rotation around the axis of the cylinder. This rotation however does generate a circle labelled (j+1) that has generally not the same radius as the circle of label j, although it should, because those circles are assumed to be rigid. This can be cured by changing by a small amount the diameter of the circle labelled (j+1) to bring it back to the prescribed value. The final result is that, under $U_{c,1}$, the particles with the same azimuthal index i are generally no more along the same generatrix of the cylinder at equilibrium and winds up around it, but again without changing its general shape, so that it remains a straight cylinder.

I claim now that turning on the interaction $U_{c,2}$ may lead to a transition (a buckling transition in technical terms) of a straight to a helical cylinder. With the above choice of coordinates, this total interaction energy is the sum over all unit squares of the lattice of the energy $U_{c,2}$ $(r_{i,j}, r_{i+1,j}, r_{i,j+1}, r_{i+1,j+1})$, and we suppose that this energy is chiral in the sense that it does change value in a mirror symmetry of the set of points in its argument. Indeed we shall retain the part of $U_{c,2}$ that is odd in this symmetry. We shall even chose a specific form for this interaction, by saying that it is proportional to the algebraic volume of the tetrahedron built on the four points $(r_{i,j},\ r_{i+1,j},\ r_{i,j+1}, r_{i+1,j+1})$. Indeed this is a rather crude approximation, in obvious contradiction for instance with the physical requirement that this interaction should tend to zero when the particles split apart. This is not very important because we will consider small deviations away from the planar structure and this approximation by the algebraic volume may be seen then as a first term in the expansion of the chiral part of $U_{c,2}$ near this planar configuration where it is obviously zero.

Now we shall determine the change in the chiral part of the interaction energy related to a small and rather arbitrary deformation of the cylinder. Let Δ_j be the vector $(r_{i,j}-r_{i+1,j})$ and Δ_{j+1} be the vector $(r_{i,j+1}-r_{i+1,j+1})$. Those two vectors have the same length because they are drawn on two rigid circles. We consider a situation where the cylinder is only slightly deformed, so that Δ_j and Δ_{j+1} are almost identical. The geometrical transformation mapping the pair $(r_{i,j},\ r_{i+1,j})$ into the pair $(r_{i,j+1},\ r_{i+1,j+1})$ may be written as the product of a translation T of small amplitude in the plane of the circle labelled j and of a small rotation around an axis in the same plane and of vector ω. With those assumptions Δ_{j+1} is equal to $\Delta_j + \omega x \Delta_j$. Then the volume of the tetrahedron $(r_{i,j},\ r_{i+1,j}, r_{i,j+1},\ r_{i+1,j+1})$ is equal to the triple product :

$$V = (\Delta_j,\ \Delta_j + \omega x \Delta_j,\ r_{i,j} - r_{i,j+1}) = (\Delta_j,\ \Delta_j + \omega x \Delta_j,\ e_z + T)$$

where e_z is the lattice unit along the generatrices of the straight cylinder. We assumed the chiral interaction to be proportional to V. From the linearity of the triple product with respect to each of its arguments, V is equal to $(\Delta_j,\ \Delta_j + \omega x \Delta_j, T)$. The interaction we are looking at is actually the interaction between rigid circles, and it will be proportional to the mean value of the volume V along a circle. Supposing that the lattice unit is much less than the radius of this circle, one may consider the position label i as continuous and so obtain

$$<V> = \frac{a^2}{2}\ \omega.T,\ \text{where } a \text{ is the lattice unit length.}$$

The change in the interaction energy due to a deformation of the cylinder is the sum of this last contribution plus two others, expressing that the straight cylinder is at equilibrium in the absence of chiral interactions. Said otherwise, this energy must have other positive definite contributions that are quadratic in the deformation. Thus the second variation of this energy reads:

$$\delta^2 E = \alpha' T^2 + \beta' \omega^2 + 2\gamma' \omega.T\ ,$$

with α' and β' positive. The last rectangular term $(2\gamma' \omega.T\ )$ only is typical of the chirality and is thus proportional to the average volume $<V>$ computed before. If $\gamma^2 > \alpha'\beta'$, then the quadrtatic form $\delta^2 E$ is not positive definite and the straight cylinder is unstable against a buckling instability leading ultimately to an helical "macroscopic"shape.

This provides a possible explanation to the observed spiral structures of the strings of bacteria described at the beginning of this section. The quantity $\dfrac{\gamma'^2}{\alpha'\beta'}$ is dimensionless, and the coefficients α', β' and γ' are mean values of an energy of interaction along a circle , thus they all are proportional to the perimeter of this circle, and the ratio $\dfrac{\gamma'^2}{\alpha'\beta'}$ is the square of the ratio of the chiral part of the interaction energy (between four neighbours for instance) to its nonchiral part. In principle one could derive the long range part of this interaction, as a function of polarizability parameters of the molecules. The final result would be quite akward, and probably of no much real use, because it would depend on parameters poorly known for real molecules and because one expects the chiral interaction to be mostly short ranged, depending on steric forces between molecules with a complicated shape. All this leads to the reasonnable assumption that the ratio $\dfrac{\gamma'^2}{\alpha'\beta'}$ is generally of order 1, and eventually bigger than 1 for some molecules.

However, even if the ratio $\dfrac{\gamma'^2}{\alpha'\beta'}$ is bigger than 1, this is not enough to explain the observed helical shape of the string of bacteria. Indeed the change in the pair of circles minimizing the energy $\delta^2 E$ has still a rotation invariance: the direction of either of the vectors $\mathbf{T}$ or ω is arbitrary in a plane. Then this model with an interaction between two circles only would predict a rather random global structure, with each pair of circles having its own random orientation for the vector T for instance. As shown in reference [9], this degeneracy disappears under the effect of interactions between next nearest neighbouring circles, and the corresponding shape of the cylinder in its lowest energy state is a spiral, as observed. An order of magnitude estimate shows that in some sense there is no loss of helicity (in order of magnitude) when one goes from the molecular scales (the ratio $\dfrac{\gamma'^2}{\alpha'\beta'}$) to the macroscopic scale (pitch of the cylinder): it remains of the same order of magnitude in dimensionless units.

To end this section, it is interesting to notice that if one considers a plate drawn on the square lattice, but now with free boundaries (that would be a piece of the cell wall), then the chiral interaction leads to a change in the elastic energy of first order in a deformation parameter with respect to the perfectly planar shape. This means that the equilibrium shape of such a plate is chiral (say like a piece of propeller) and nonplanar.

This remark gives also an opportunity to emphasize a point that was alluded to before: the chiral part of the interaction should be not invariant under the interchange of the horizontal and vertical direction of the square lattice. If it were so, the buckling of a square piece of the lattice could not be chiral, because by an in-plane rotation of $\dfrac{\pi}{2}$, one would change the sign of the chirality!

3. NONEQUILIBRIUM DYNAMICS OF BLOCH WALLS[+++]

Consider a ferromagnet with a direction of easy magnetization lying in a plane of easy magnetization too. This means that in a perfectly organized magnetized state, the spontaneous magnetization will point in one of the two opposite directions along the easy magnetization axis, and that pertubations of this will send the magnetization to another orientation, but still in the easy plane. In most real uniaxal magnets, the magnetization is not uniform , and points in either of the two opposite preferred directions and so called "domain walls" draw the border between regions of opposite magnetization. In the core of those walls, the magnetization interpolates continuously between those two magnetizations(but for lattice effects that I shall neglect). By imposing the local magnetization to stay in the easy magnetization plane, one still finds two possible wall structures: in the Néel wall, the magnetization stays in the preferred direction, but decays to zero before to grow in the opposite direction. On the contrary, in the Bloch wall, the magnetization keep more or less a fixed modulus, but rotates continuously from up to down in the easy magnetization plane. Néel wall is preferred at large in-plane anisotropies, whereas the Bloch wall is preferred at low anisotropies.

The helicity enters into the game because a Bloch wall has helicity: when the magnetization turns in the easy magnetization plane from up to down, it may do so in either of the possible sense on a circle, and each sense corresponds to a given helicity of the wall structure.

The way in which nonequilibrium comes into the story is not completely obvious. When a wall (Bloch or Néel) drifts in a ferromagnet, this is usually interpreted by saying that the magnet lower its free energy by replacing (at constant rate if the wall move at constant speed) domains with the bad orientation by the good one. This is what happens for instance in an external magnetic field: good/optimal domains parallel to the external field expand at the expense of bad domains antiparallel to the magnetic field. This optimization of the free energy is of course rather independent of the helicity of the Bloch wall, if such a wall separates the two domains. Below I show a possible situation where the direction of motion of a Bloch wall depend on its helicity, something that could not be found in a variational system(=close to equilibrium), where this direction is dictated by an optimization principle.

To write the equation for this problem, I follow an idea used first (as far as I know) by Ginzburg[12]and that goes as follows. Let us represennt the local magnetization in the easy magnetization plane as a complex number χ, that may be thus a function of time and/or space. The modulus of this number is the strength of the magnetization and its angle measures its orientation. Below the Curie point, this magnetization takes at

---

[+++] This is a shortened version of a joint work[11]with Pierre Coullet and J. Lega, submitted for publication.

equilibrium anyone of the two possible opposite values that I will take as + or -1 by a convenient choice of unit. Forgetting for the moment the anisotropy in the easy plane, one finds that a natural equation for the dynamics of magnetization is of the relaxation (or variational)type and reads:

$$\frac{d\chi}{dt} = \chi - \chi^*\chi^2,$$

where χ^* is the complex conjugate of χ. This equation predicts that any initial condition will keep a constant angle (because it is invariant under a change of phase) and relax to the modulus 1, after a time of order 1, again by a proper choice of time scale. The anisotropy in the easy plane is introduced by adding to this equation a non phase invariant term. The most simple such term is $\gamma\chi^*$, γ coefficient measuring the in plane anisotropy, so that the proposed dynamical equation is:

$$\frac{d\chi}{dt} = \chi - \chi^*\chi^2 + \gamma\chi^*,$$

Depending on the initial value of the angle, solutions of this equations tends to large times to either of the symmetric stable equilibria $\chi = \pm i\,(1+\gamma)^{1/2}$. There are also two unstable equilibria $\chi = \pm(1-\gamma)^{1/2}$ and the unstable "paramagnetic" state $\chi = 0$. Notice that this requires $\gamma^2 < 1$, as we shall assume.

In this theory the anisotropy is measured by the coefficient γ. A small (/large)γ means a small (/large) anisotropy. To describe walls in this theory, one has to consider a nonuniform χ. The physics says that the magnetization tends to be as uniform as possible. This is represented by adding a diffusion term to the dynamical equation: diffusion tends to make even the distribution of magnetization in space. This leads to the new equation (a partial differential equation now):

$$\frac{\partial\chi}{\partial t} = \chi - \chi^*\chi^2 + \gamma\chi^* + \frac{\partial^2\chi}{\partial x^2} + h, \tag{1}$$

This is written in 1D (coordinate x) and the length unit has been chosen to set to 1 the diffusion coefficient. Furthermore the effect of a possible external magnetic field was introduced through the last term on the right hand side, h, that is a number proporttional to the intensity of this field. This equation has a variational structure(as well as its predecessors) in the sense that it can be written in the form:

$$\frac{\partial\chi}{\partial t} = -\frac{\delta H}{\delta\chi^*},$$

where $\dfrac{\delta H}{\delta \chi^*}$ is the Fréchet derivative of the functional

$$H[\chi] = \int dx \; [\tfrac{1}{2}|\chi|^4 - |\chi|^2 - \tfrac{\gamma}{2}(\chi^2+\chi^{*2}) + |\tfrac{\partial \chi}{\partial x}|^2 - h \, (\chi+\chi^*)] \; .$$

This equation shows that the (free) energy H tends to values as low as possible in the relaxation dynamics of the system toward its equilibrium state. The stable uniform states are indeed minima of the integrand of H, for uniform solutions and h=0.

Domain walls are stationary solutions of (1) with h=0, joining two domains where the magnetization takes different stable values, that are, for instance $\chi=(1+\gamma)^{1/2} e^{i\phi}$, with, say $\phi=0$ at $x=-\infty$ and $\phi=\pi$ at $x=+\infty$. Depending on the value of γ, this problem may have [12] either one solution, the stable Néel wall at large γ or three solutioons at small γ: the now unstable Néel wall and the stable left- and rigth handed Bloch walls. The transition is continuous and occurs at $\gamma=\tfrac{1}{3}$. The Bloch wall solution is [12]:

$\chi_0=X_0+iY_0$ (2),

   where $X_0=(1+\gamma)^{1/2}$ th(ax), $Y_0=\pm(1-3\gamma)^{1/2}$sech (ax) and $a^2=2\gamma$.

The left-right hand symmetry is thus the complex conjugation in this formalism, and it is also equivalent to a change of sign of the quantity a. Thus the Bloch walls have the same kind of symmetry as a corkscrew and are images of each other in a mirror. This leads quite naturally to the following idea. Consider an uniform magnetic field turning at constant speed while staying in the easy magnetization plane. The effect of this field on a Bloch wall will be a bit like a constant rotation on a corkscrew, that is a displacement at constant speed in a direction depending on the handedness of the screw. From this we expect a constant drift for Bloch wall solutions of (1) with a time dependent h representing a rotating field. With our representation this is realised by putting for h in (1) $h_0 e^{i v t}$.

The equation so obtained, that reads:

$$\frac{\partial \chi}{\partial t} = \chi - \chi^*\chi^2 + \gamma\chi^* + \frac{\partial^2 \chi}{\partial x^2} + h_0 e^{ivt}, \tag{3}$$

is time dependent and does not seem to have any simple solution, even uniform in space. So I shall limit myself to a perturbative approach, in the small h_0 limit. To show my point, I shall take as unperturbed solution of (3), anyone of the two Bloch wall solutions(that are written as χ_0) in equation (2), and then expand a solution of (3) in powers of h_0.

At the first oder in h_0, one finds small oscillations with the frequency v around the

Bloch wall solution (2). Those small amplitude oscillations are described by the solution of (3) that reads at zeroth and first order as:

$$\chi = \chi_0 + h_0[\alpha_s \sin(vt) + \alpha_c \cos(vt) + i\,\beta_s \sin(vt) + i\beta_c \cos(vt)],$$

where the α's and β's are real functions of x defined by the solution of a set of coupled inhomogeneous linear equations, deduced by a straightforward algebra from (3):

$$\Lambda_+\alpha_s - 2X_0Y_0\,\beta_s = -v\alpha_c \qquad\qquad (4.a),$$
$$\Lambda_+\alpha_c - 2X_0Y_0\,\beta_c = v\alpha_s - 1 \qquad\quad (4.b),$$
$$\Lambda_-\beta_s - 2X_0Y_0\,\alpha_s = -v\beta_c - 1 \qquad\quad (4.c),$$
$$\Lambda_-\beta_c - 2X_0Y_0\,\alpha_c = v\beta_s \qquad\qquad (4.d),$$

where the linear operators Λ_+ and Λ_- are defined by their action on a test function $\Gamma(x)$ as:

$$\Lambda_+ \Gamma(x) = (1+\gamma-3X_0^2-Y_0^2)\Gamma + \frac{d^2\Gamma}{dx^2}$$

and

$$\Lambda_- \Gamma(x) = (1-\gamma-X_0^2-3Y_0^2)\Gamma + \frac{d^2\Gamma}{dx^2} \ .$$

Counting the number of free parameters in (4), one sees that they have the right number of constraints when one imposes to their solution to decay far away from the Bloch wall, that is at x tending either to + or to minus infinity. This does not prove however that (4) has an unique solution. If two (or more) solutions of (4) existed, the difference between those two solutions would be the amplitude of small oscillations of the Bloch wall, in the absence of external forcing. This is clearly impossibe, because of the relaxational character of the original equation. Furthermore, it is possible to prove that, at least for small frequencies this equation has certainly a solution, although this not an easy result.

By continuing the expansion of the solution of (3) to higher orders in powers of h_0, one meets, as often in this sort of expansion, a solvability condition. This is because, in the formal expansion of the solution of (3), one solves at the n-th order a linear problem in the form:

$$\Omega\chi_n = (\text{right hand side}),$$

where the r.h.s. is a known finite algebraic combination of the lowest order contributions $\chi_{n-1}, \chi_{n-2}, \chi_{n-3}, \ldots$ where the index n is for the order in small parameter h_0, and Ω is a linear operator. For n even, χ_n as a function of time is decomposed into Fourier component $e^{invt}, e^{i(n-2)vt}, e^{i(n-4)vt}, \ldots$ and a time independent part, say $\chi_{n,0}$. This time independent part is the solution of a linear equation of the form:

$$\Omega\chi_{n,0} = (\text{right hand side})_0 \qquad\qquad (5),$$

which is actually an ordinary differential equation with respect to the variable x, and for the unknown function $\chi_{n,0}$. This equation cannot be solved as written, because Ω has a non trivial kernel for zero frequency functions (the explicit writing of this operator is given below). In technical terms, this means that there exists a non zero function of x, $\chi_{n,k}$ such that $\Omega\chi_{n,k}=0$. Assuming now(and this can be made more precise, but at the price of a lot more work) that everything is like in ordinary algebra (i.e. that Ω is a numerical square matrix, and χ_n a vector, instead of a differential operator and a function repectively), the equation (5) has no solution in general, because the "matrix" Ω has zero as an eigenvalue. Before to consider in more details what to do next, I will (hopefully) make all the above considerations more precise by giving explicitely the form of the operator Ω when restricted to zero frequency perturbations. This is indeed nothing but the right hand side of (1), linearized near the Bloch wall solution χ_0:

$$\Omega\chi_{n,0}=\chi_{n,0}-2\,|\chi_0|^2\chi_{n,0}-\chi_0^2\chi^*_{n,0}+\gamma\chi^*_{n,0}+\frac{d^2\chi_{n,0}}{dx^2},$$

and for the complex conjugate:

$$\Omega\chi^*_{n,0}=\chi^*_{n,0}-2\,|\chi_0|^2\chi^*_{n,0}-\chi^{*2}_0\chi_{n,0}+\gamma\chi_{n,0}+\frac{d^2\chi^*_{n,0}}{dx^2}.$$

As announced this operator has a zero mode, because by putting in those two equations $\chi_{n,0}=\frac{d\chi_0}{dx}$, and the complex conjugate relation, one gets zero, because the corresponding expressions are nothing but the derivative with respect to x of the r.h.s. of (1) and of its complex conjugate. This is a rather familiar phenomenon in applied mathematics and it occurs whenever one linzearizes an autonomous nonlinear equation as (1) [autonomous means that its coefficients are independent of x].

The attentive reader will have noticed that the same solvability problem would appear already at the *first* (instead of *second* at finite frequency) order in h, at zero frequency v. This is because then this first order solution requires already the inversion of the operator Ω at zero frequency. Then the physics tells us what happens: a constant external magnetic field is imposed to the system and the wall (whatever its internal structure) will drift in order to replace the domains antiparallel to this external magnetization by domains parallel to it. This means that the Bloch wall solution $\chi_0(x)$ becomes actually a moving wall solution(for h_0 small at least) of the form $\chi_0(x-ut)$. The velocity of the wall, u, is a free parameter till now, and it is small, as one expects the wall velocity to vanish at zero h. Indeed one expects the same in the other problem: time dependent h(t), and second order perturbation. Without going into all the details, that are quite tedious, one assumes in the time dependent problem that u is second order in h_0 and constant. This constant takes care of the solvability problel, that is of the non trivial kernel of Ω, by cancelling the kernel component of the r.h.s. of (5).

This yields finally the following result, for the dominant contribution to u (again in the time dependent problem):

$$u=h_0^2\frac{N}{D},\ \text{with:}$$

$$N = \int_{-\infty}^{+\infty} dx \left\{ (\alpha_c \beta_c + \alpha_s \beta_s) \frac{d(X_0 Y_0)}{dx} + \frac{1}{4}[3(\alpha_c^2 + \alpha_s^2) + \beta_c^2 + \beta_s^2] \frac{dX_0^2}{dx} \right\}$$

$$+ \int_{-\infty}^{+\infty} \frac{dx}{4}[3(\beta_c^2 + \beta_s^2) + \alpha_c^2 + \alpha_s^2] ,$$

and

$$D = \int_{-\infty}^{+\infty} dx \left[\left(\frac{dX_0}{dx}\right)^2 + \left(\frac{dY_0}{dx}\right)^2 \right],$$

which shows that, in the small h_0 limit, the velocity u is proportional to the square of h_0 times a numerical function of v and γ. This function is such that u is odd with respect to v, in agreement with the qualitative considerrations presented before: the motion of a corkscrew changes direction when the sense of rotation is changed. Moreover, by looking at the same expression for the Néel wall solution, without helicity (which can be done in the above expression for u by setting $Y_0 = 0$ and putting for X_0 its expression for a Néel wall) one finds by simple symmetry considerations that the velocity u is then equal to zero, as expected.

It is more delicate to show that u is of order v as it tends to zero. This is because physically, in this low frequency limit the wall (Bloch or Néel) makes large excursion at first order in h_0: the external magnetic field is then quasistatic, and a constant magnetic field would give a constant velocity of order h_0; thus the amplitude of the displacement in a slow time-periodic external field is about of the order of this velocity times a period, that is of order $\frac{h_0}{v}$. Nevertheless one can show that, on top of those "large" first order(in h_0) fluctuations of a Bloch wall without mean drift, there is a mean drift at second order, proportional to v , that changes sign with the helicity of the Bloch wall.

All those predictions have been tested [11] directly and successfully by solving numerically the model equation (3), with an equilibrium Bloch wall as initial condition. Concerning the possibility of an experimental observation of this displacement of Bloch walls in an externally rotating magnetic field, and without discussing this in any detail, I can only stress that such a displacement speed, of second order in an oscillating external field has been observed already by Schlömann and coll.[13], but with a geometry completely different of the one I have considered (in particular, the Bloch walls move in a direction independent on their handedness). However the connection with the present problem is that the velocity measured by Schlömann et al. [13] is of order $v\, h_0^2$, as the one we predict. Thus it seeems to be reasonable to expect that the velocity found here will be of the same order of magnitude for a real magnet as the one measured in [13] with the same h_0 and v, which leaves some hope to have it observable.

REFERENCES

[1] Y.Pomeau, Europhys. Letters, $\underline{3}$, 1201(1987)

[2] R.M. Weis and H.M.McConnell, Nature,$\underline{310}$, 47(1984)

[3] J.S. Langer, Physica Scripta (Nobel Symposium)$\underline{T9}$,119(1985)

[4]H.M.McConnell, D.Keller and H.Gaub, J.Phys. Chem., $\underline{90}$,1717(1986)

[5] L.Landau et E.Lifshitz "Physique Statistique", editions Mir (Moscou), 1961

[6] V.T.Moy, D.J. Keller, H.E. Gaub and H.M. McConnell, J.Phys. Chem. $\underline{90}$, 3198 (1986)

[7] L.Landau, E.Lifshitz, "Théorie de l'élasticité", editions Mir, Moscou (1967)

[8.a]J.S.Langer , Rev. of Mod. Phys. ,$\underline{52}$, 1(1980)

[8.b]D.Kessler, J. Koplik and H.Levine, Adv. in Phys., $\underline{37}$,255(1988)

[9] Y.Pomeau, J. Lega, Comptes rendus de l'Académie des Sciences, to appear (1990)

[10] N.H. Mendelson, J.J. Thwaites, Comments on Theoretical Biology, 1989, vol 1, p 217-236, and references therein

[11] P.Coullet, J. Lega and Y. Pomeau "Dynamics of Bloch Walls", preprint, submitted for publication

[12] L.N. Bulaevskii and V.L. Ginzburg, Sov. Phys. JETP **18**, 530 (1964)

[13] E. Schlömann and J.D. Milne, IEEE Trans. Magn. **Mag-10**, 791 (1974); E. Schlömann, IEEE Trans. **Mag-11**, 1051 (1975)

ON THE GROWTH AND FORM OF DISLOCATION PATTERNS

D. WALGRAEF†
Service de Chimie-Physique, Université Libre de Bruxelles
B-1050, Brussels, Belgium

N.M.GHONIEM
Mechanical, Aerospace and Nuclear Engineering Department
University of California, Los Angeles, CA 90024, USA

ABSTRACT

Defects are known to play an important role in the macroscopic behavior of equilibrium and non equilibrium structures. They are able to disorganize spatio-temporal patterns, to trigger transitions between patterns of different symmetries, but also to form various kinds of microstructures. In particular, the occurence of spatial instabilities leading to the formation of defect patterns is overwhelming in solids driven away from thermal equilibrium by physico-chemical constraints. Some of these instabilities have been recently studied within the framework of dynamical models for the defect densities. The basic properties of these models which take into account the motion and interaction of defects are reviewed. It is shown by a specific example, namely the ordering of vacancy loops in irradiated materials, how the diffusion and nonlinear interactions trigger the formation of dislocation patterns.

INTRODUCTION

Several aspects of defect properties have recently been studied in non-equilibrium systems. Their role in the disorganization of spatio-temporal structures and in the transitions between structures of different symmetries has been analyzed in various contexts [1-2]. However, in the case of high defect densities such as for example in deformed or irradiated solids, topological defects may acquire a collective behavior and particular types of ordering phenomena may occur. Effectively, when crystalline materials are driven away from thermal equilibrium by external constraints (monotonic or cyclic loadings, corrosion, laser or particle irradiation,...), they display several types of instabilities associated with the ordering of defect populations which modify their macroscopic properties [3-4]. Typically, the defect densities increase significantly and homogeneous defect distributions become unstable. The resulting microstructures are associated with the inhomogeneity of deformation, with strain localization, inhomogeneous irradiation-induced swelling, etc. They also act as initiating centers for micro-crack nucleation, influence crack propagation and void lattice formation. Therefore defect ordering can have very practical implications.

---

† Senior Research Associate, National Fund for Scientific Research (Belgium).

Growth and Form, Edited by M. Ben Amar *et al.*
Plenum Press, New York, 1991

"

Furthermore these phenomena are related to strong nonequilibrium conditions imposed on the material, and they cannot usually be interpreted by classical thermodynamical or mechanical concepts. Hence, new theoretical tools are needed to describe and understand these and many other aspects of today's materials science and solid state physics. For example, methods related to the explicit use of genuine nonequilibrium techniques, nonlinear dynamics and instability theory are now applied to several studies of materials subjected to agressive mechanical, thermal or chemical environments [3]. For example, since it appears that defect microstructures result from a dynamical equilibrium between different processes, kinetic models were proposed to describe this collective behavior by taking into account the motion (diffusion, transport, ...) and nonlinear interactions (annihilation, pinning, clustering,...) between defects [3-4]. We will discuss here the basics of a dynamical description of the collective behavior of defect populations in driven solids.

Such a description is usually based on reaction-diffusion schemes. It is now understood that, when several defect populations with different mobilities interact via sufficiently highly nonlinear processes, pattern forming instabilities are expected [5]. These instabilities have been, and are still, investigated in various fields including hydrodynamics, chemistry, biology and nonlinear optics. Besides the determination of the critical wavelength , it is also essential to investigate the geometry, the symmetries, and the stability ranges of the selected structures. As extensively discussed elsewhere , the difficulties of the post-bifurcation analysis lie in the fact that the complexity of the dynamics does not allow, in general, the attainement of analytic solutions for the different variables. However, near the instability or bifurcation points, the dynamics can be reduced to much simpler forms by taking advantage of the time and space scale separation between stable and unstable modes and by projecting the dynamics on its unstable manifold. The resulting slow mode dynamics which governs the system evolution on its longest time scale becomes similar to the time dependent Landau-Ginzburg dynamics describing phase transitions in equilibrium systems. This description leads then to amplitude equations for the patterns and allows the derivation of their phase dynamics. Pattern selection and stability may then be discussed in this framework as shown in several contributions of this volume.

In driven materials, it is only recently that the problem of pattern selection, stability, symmetry, or of the influence of the underlying lattice , has been adressed along these lines. Preliminary results are promising, however fundamental questions raised by the anisotropies, the three-dimensional character of the samples and the time evolution of the constraints remain open.

REACTION-TRANSPORT DYNAMICS FOR DEFECT POPULATIONS

Metals and alloys under physico-chemical constraints such as deformation, corrosion, particle or laser irradiation, present several types of microstructures including dislocation patterns (slip, kink or shear bands, Lüders bands, dislocation cells,...), vacancy loops, bubbles, cavity and void lattices. These structures which originate in the spatial organization of point or line defect populations have a strong influence on the macroscopic properties of the materials and on their resistance to external constraints. Hence, the understanding of the formation, selection and stability properties of these defect patterns is of primary importance both from fundamental and from technological points of view.

Several attempts have recently been made to describe these phenomena via numerical simulations of the molecular dynamics or cellular automata types, but also in the framework of kinetic models for the defect populations [3-6]. We will describe

here some basic aspects of these models which are based on the fundamental elements of the collective behavior of each defect population, namely :
(1) their motion (diffusion for point defects such as interstitials and vacancies, glide, climb and cross-slip for dislocations),
(2) their nonlinear interactions which correpond for example to recombination of point defects, capture or emission of point defects by microstructures and defect creation mechanisms induced by the external constraint.

Defect Motion

Different types of motion are observed, according to the nature of defects considered. Point defects such as interstitials and vacancies diffuse in the crystal and the corresponding diffusion coefficient are computed by using energetic considerations. It turns out that their relative value depends on temperature; in the case of 316 steel, for example, we have [10]:

$$\frac{D_v}{D_i} = 6.10^2 exp(-1eV/k_B T)m^2 s^{-1}$$

where D_v is the vacancy diffusion coefficient, D_i is the interstitial diffusion coefficient, k_B is the Boltzmann's constant and T the absolute temperature. Hence, according to the temperature of the material, they may differ by orders of magnitude.

The motion of dislocations corresponds to glide, climb or cross-slip. Glide is the fastest and is typical of the plastic regime when dislocations move almost freely on well-defined crystalline planes. Dislocations of opposite Burgers vectors move in opposite directions. In real materials, there are also obstacles to dislocation motion : impurities, stacking faults, grain boundaries, dislocation clusters forming the so-called forest of immobile dislocations. Line defects may also have very different mobilities, which is a perequisite for pattern forming instabilities in reaction-diffusion systems.

Defect Creation and Multiplication

Point defects are always present in crystalline solids at finite temperature but much larger amounts of such defects are created under external constraints. Under irradiation, ballistic and cascade effects provide creation mechanisms for interstitials and vacancies. The clustering or collapse of point defects leads to the formation of interstitial or vacancy loops. Dislocation creation, on the other hand, is a direct consequence of the deformation process, but the increase of the dislocation density is mainly due to multiplication effects such as the Frank-Read or Bardeen-Herring mechanisms [7].

Defect Interactions

The process mentioned above are essentially linear. The nonlinearities of the dynamics are mainly due to defect interactions. In fact, a few of them dominate the defect dynamics. A particularly important phenomenon is the pair annihilation of defects of opposite nature (annihilation of dislocations of opposite Burgers vectors, recombination of interstitials and vacancies) which leads to quadratic nonlinearities. Other quadratic nonlinearities are related to the formation of dislocation dipoles or to the capture of point defects by dislocations and loops. Higher order nonlinearities may be associated with the pinning of free dislocations by clusters of the nearly immobile dislocations of the forest [5].

Several dynamical descriptions of plastic instabilities and dislocation pattern formation have been based on the effects described above. Since they are discussed in detail in the literature [3-5], we will rather illustrate here the collective behavior of point defects and vacancy loops in irradiated metals and alloys and describe the dynamics of their spontaneous ordering.

VACANCY LOOP ORDERING IN IRRADIATED MATERIALS

The ordering of vacancy loops occurs frequently in metals and alloys irradiated at moderate doses and high temperatures [8]. The uniform distribution of loops which are created by cascade collapse becomes unstable beyond some threshold related to various parameters such as the irradiation dose, the kinetic damage rate, the bias in the migration of point defects to loops and network dislocations, etc... A minimal model has been proposed by Murphy to describe the dynamics of defect populations in metals and alloys under particle irradiation[9-10]. It is based on the rate theory of radiation damage originally developed by Bullough, Eyre and Krishan [10], and expanded further by Ghoniem amd Kulcinski [12] to include the dynamics of point defects in the fully dynamic rate theory. The network dislocations which are also present in the material are assumed to have a constant uniform distribution. The effect of interstitial loop formation is included in the network density. By taking into account the basic mechanisms cited above, the kinetic equation for the defect concentrations are written as follows, where c_v corresponds to vacancies, c_i to interstitials, ρ_L and ρ_N to vacancy loops and network dislocation densities :

$$\partial_t c_i = K - \alpha c_i c_v + D_i \nabla^2 c_i - D_i c_i (Z_{iN}\rho_N + Z_{iL}\rho_L)$$

$$\partial_t c_v = K(1 - \nu) - \alpha c_i c_v + D_v \nabla^2 c_v - D_v(Z_{vN}(c_v - \bar{c}_{vN})\rho_N + Z_{vL}(c_v - \bar{c}_{vL})\rho_L)$$

$$\partial_t \rho_L = \frac{1}{|\vec{b}|r_L^0}[(\nu K - \rho_L(D_i Z_{iL}c_i - D_v Z_{vL}(c_v - \bar{c}_{vL})))] \tag{1}$$

where K is the displacement damage rate and ν the cascade collapse efficiency, α is the recombination coefficient, $\vec{b}$ the Burgers vector, r_L^0 the mean vacancy loop radius and $Z_{.,.}$ the bias factors which will be approximated by $Z_{iL} = Z_{iN} = 1+B$ and $Z_{vL} = Z_{vN} = 1$. $\bar{c}_{vN}$ and $\bar{c}_{vL}$ are the thermally emitted vacancies from network dislocations and vacancy loops. The various coefficients appearing in these rate equations may be computed theoretically or related to measurable quantities [12].

<u>Linear Stability Analysis</u>

On writing the rate equations (1) in a dimensionless form [12], the linear evolution matrix of inhomogeneous perturbations of the uniform steady state (x_i^0, x_v^0, x^0) is easily determined and simply given, in Fourier space, by :

$$\begin{bmatrix} \omega + \mu(1 + x^0) + x_v^0 + q^2\bar{D}_i & x_i^0 & \mu x_i^0 \\ x_v^0 & \omega + 1 + x^0 + x_i^0 + q^2\bar{D}_v & x_v^0 - \bar{x}_{vL} \\ \mu x^0 & -x^0 & \omega + \Delta \end{bmatrix} \tag{2}$$

where

$$\lambda_v = D_v Z_{vN}\rho_N, \quad \bar{D}_. = D_./\lambda_v, \quad \alpha/\lambda_v = \gamma, P = \gamma K/\lambda_v, \quad \tau = \lambda_v t,$$

$$x = \frac{\rho_L}{\rho_N}, \quad x_i = \gamma c_i, \quad x_v = \gamma c_v, \quad \tau_0 = br_L^0\rho_N\gamma$$

$$\mu = \frac{Z_{iN}D_i}{Z_{vN}D_v} = (1 + B)\frac{D_i}{D_v}. \tag{3}$$

and

$$\mu x_i^0 = x_v^0 - \bar{x}_{vN}, \qquad \nu P = x^0(\bar{x}_{vL} - \bar{x}_{vN}) = x^0 \Delta \qquad (4)$$

It turns out that the characteristic equation has two negative roots while the third one may change its sign and becomes positive when

$$\mu A + q^2 \bar{D}_i [A + 1 + B - \frac{\nu P}{\Delta^2} B(x_v^0 - \bar{x}_{vL})] + q^4 \bar{D}_i \bar{D}_v \leq 0 \qquad (5)$$

where $A = 1 + x_i^0 + \frac{x_v^0}{\mu} + \frac{\nu P}{\Delta}$

Hence, a pattern forming instability occurs when

$$\frac{B}{\nu}\Big|_c = [1 + \sqrt{\frac{\Delta}{\nu P}}]^2 = [1 + \sqrt{\frac{\rho_N}{\rho_L^0}}]^2 \qquad (6)$$

for a critical wavelength given, in unscaled units, by:

$$\lambda_c = 2\pi [\frac{D_v(\bar{c}_{vL} - \bar{c}_{vN})}{(1 + B)\nu K \rho_N}]^{1/4} \qquad (7)$$

We see that it decreases with increasing network dislocation density, cascade collapse efficiency and damage rate. On the other hand, its temperature dependence is more difficult to asses since D_v is an increasing function of the temperature while $(\bar{c}_{vL} - \bar{c}_{vN})$ is a decreasing function of the temperature and its global behavior may vary from material to material. As an example, consider 316 SS steel irradiated at 500^0 C with a displacement damage rate of 10^{-6} dpa s^{-1} [12]. The critical wavelength is nearly 1.24 μm for solution-annealed material with a typical dislocation density of 10^{13} m^{-2}. The wavelength is smaller for cold-worked material, on the order of 0.39 μm for a dislocation density of 10^{15} m^{-2}.

Since the orientation and the number of wavevectors underlying the possible structures are not fixed by this linear analysis, a nonlinear analysis in the post bifurcation regime is needed to discuss the emergence of stable patterns.

The various coefficients which represent defect creation, annihilation, and migration to dislocations and vacancy loops may be obtained from experimental data or theoretical analysis. K is the displacement damage rate and ϵ the cascade collapse efficiency. D_i and D_v are the diffusion coefficients, α the recombination coefficient, b the length of the Burgers vector. $Z_{,.}$ are the bias factors and $\bar{c}_{v,.}$ the thermally emitted vacancies from the microstructures.

AMPLITUDE EQUATIONS FOR DEFECT MICROSTRUCTURES

In the systems considered here, the fluctuations of vacancy and interstitial concentrations evolve much more rapidly than vacancy loop density fluctuations. Hence the dynamics may be reduced through a multiple scale analysis [13] or the adiabatic elimination of the fast variables [15] and the slow mode or order parameter-like variable will be associated with the vacancy loop density. On expanding the point defect concentrations as power series in the vacancy loop density one obtains [13] :

$$\partial_t \sigma(\vec{x}, t) = [\epsilon - \xi_0^2(q_c^2 + \nabla^2)^2]\sigma(\vec{x}, t) + v\sigma^2(\vec{x}, t) - u\sigma^3(\vec{x}, t) \qquad (8)$$

where $\sigma(\vec{x}, t) = x(\vec{x}, t) - x^0$, $b = B/\nu$, $\frac{b - b_c}{b_c} = \epsilon$, $\xi_0^2 \propto (x_0)^{-1/2}$, $v = 2/(x_0)^{3/2}$ and $u = 2/(x_0)^{5/2}$.

This equation is characteristic of Turing-like instabilities and has been derived in several contexts, such as reaction-diffusion and hydrodynamic systems. For $b > b_c$, its stable solutions of this Landau-Ginzburg type of dynamics correspond to :

(1) roll or wall structures associated with spatial modulations of the order parameter (here the vacancy loop density) in one direction. They appear via a second orderlike transition, or supercritical bifurcation.

(2) rodlike hexagonal or triangular structures appearing via a first orderlike transition (subcritical bifurcations) defined by the following amplitude equations:

$$\tau \dot{A}_i = [\epsilon + \frac{4\xi_0^2}{q_c^2}(\vec{q}_i.\vec{\nabla})^2]A_i + v A_{i-1}^* A_{i+1}^* - 3u[|A_i|^2 + 2\sum_{j \neq i}|A_j|^2]A_i \qquad (9)$$

where $\sigma(\vec{x},t) = \sum_{i=1}^{3} A_i(\vec{x},t)e^{i\vec{q}_i.\vec{x}} + c.c.$, with $|\vec{q}_i| = q_c$, and $\vec{q}_1 + \vec{q}_2 + \vec{q}_3 = 0$
The stable steady state is given by $|A_i| = A = \frac{1}{30u}[v + \sqrt{v^2 + 60u\epsilon}]$.

(3) bcc lattices or filamental structures of cubic symmetry, also associated with a subcritical bifurcation and defined similarly to hexagonal structures but with six pairs of wavevectors. In this case, the corresponding steady state is then given by:

$$\sigma(\vec{x}) = A[\cos \frac{q_c}{\sqrt{2}}x \cos \frac{q_c}{\sqrt{2}}y + \cos \frac{q_c}{\sqrt{2}}y \cos \frac{q_c}{\sqrt{2}}z + \cos \frac{q_c}{\sqrt{2}}z \cos \frac{q_c}{\sqrt{2}}x] \qquad (10)$$

with $A = \frac{1}{33u}[v + \sqrt{v^2 + 33u\epsilon}]$

When the bifurcation parameter is increased, the 2d and 3d structures may in turn become unstable (the hexagonal structure for $\epsilon > 4v^2/3u$ and the bcc structure for $\epsilon > 3v^2/u$). Hence, between threshold and $3v^2/u$, bcc dislocation structures should be expected while above this limit the structure should consist of regularly spaced planes of maximum density.

The bifurcation diagram for bcc and planar wall structures is sketched in fig.1.

In the case of an anisotropic diffusion of interstitials as in hcp materials where the mobility of interstitials is much larger in the basal planes than between these planes, it is easy to show with the same method [16], that the stable patterns for vacancy loops correspond to planar arrays with planes of maximum density parallel to the planes of high interstitial mobility in agreement with experimental observations [17]. From a more general point of view, pattern formation in systems with uniaxial anisotropy may be described by the following dynamics for the order-parameterlike variable :

$$\partial_t \sigma(\vec{x},t) = [\epsilon - \xi_{||}^2(q_c^2 + \nabla^2)^2 + \xi_\perp^2 \nabla_\perp^2]\sigma(\vec{x},t) + v\sigma^2(\vec{x},t) - u\sigma^3(\vec{x},t) \qquad (11)$$

where $\nabla_\perp^2 = \nabla_y^2 + \nabla_z^2$, $y0z$ being the plane perpendicular to the easy axis $0x$.

As discussed in [18], pattern selection results, in this case, from the competition between the anisotropy which favors wall structures with walls perpendicular to the easy axis, and the nonlinearities which favor cellular structures. It may also be shown that the stability of the selected patterns is affected by the anisotropy [19]. In particular, as shown in fig.2, the anisotropy increases the stability range of the favored wall structures.

436

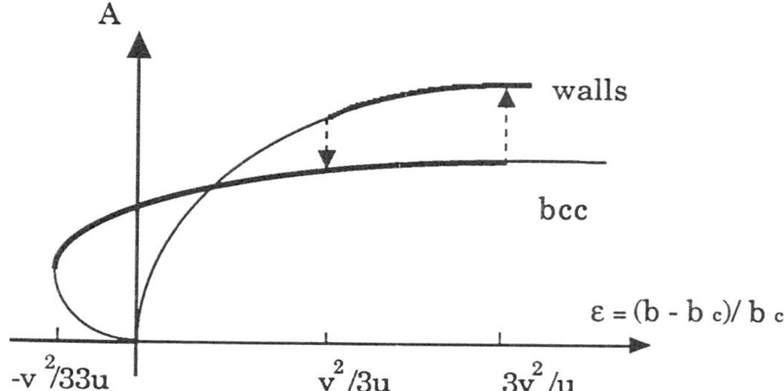

Figure 1 . Schematic bifurcation diagram associated with the amplitude equation for the microstructures showing the transition and the hysteresis loop between bcc and wall structures (heavy lines represent stable states, and thin lines represent unstable states).

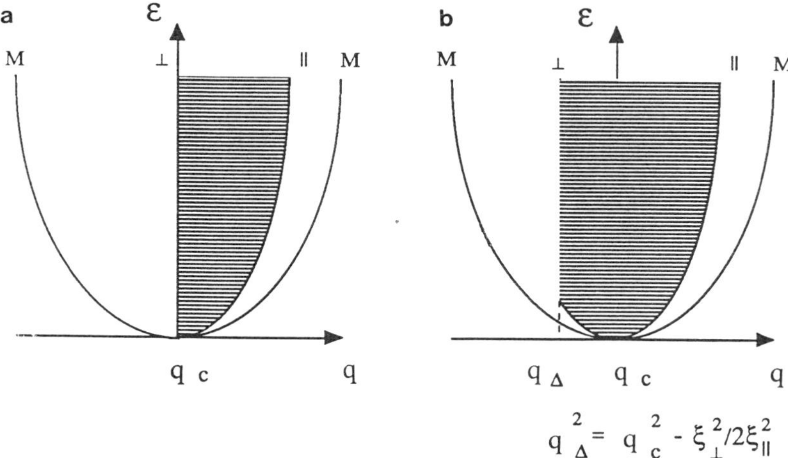

$$q^2_\Delta = q^2_c - \xi^2_\perp/2\xi^2_\parallel$$

Figure 2 . Stability range of wall structures described by eq.11 in the (ϵ, q) plane in (a) isotropic and (b) anisotropic systems (in the hatched regions, wall structures are stable versus long wavelength phase fluctuations).

An interesting situation also occurs when there are two preferred directions for the unstable wavevectors, i.e. for the following slow mode dynamics:

$$\partial_t \sigma(\vec{x},t) = [\epsilon - \xi_0^2(q_c^2 + \nabla^2)^2 - \kappa \nabla_x^2 \nabla_y^2]\sigma(\vec{x},t) + v\sigma^2(\vec{x},t) - u\sigma^3(\vec{x},t) \qquad (12)$$

In this case, the linear part of the dynamics selects walls perpendicular to the x or y directions, or square structures built on the two preferred wavevectors. However, for u constant, this structure is ruled out by the nonlinearities [19]. Hence, the resulting structure is expected to correspond to coexisting domains of walls perpendicular to the x or y directions, as in labyrinth structures [5].

The result of the present analysis obtained in the weakly nonlinear regime beyond pattern forming instabilities need of course to be confronted with detailed experimental investigations and to at least 2-D numerical simulations. It is interesting to note that, in the case of irradiated materials, recent experimental observations by Jäger et al.[8] indicate that spatial microstructure modulation is a general phenomenon under ion-irradiation conditions where cascade are produced. Over a limited range of conditions, Jäger observed that the wavelength was insensitive to temperature, dose rate, and type of primary knock-on atom. Dislocation loops and tangles were found to be arranged in planar arrangements (walls), with a wavalength of 0.03 to 0.06 μm. While the model discussed here is in general agreement with these experimental observations, it should be viewed as a step towards a generalized theoretical explanation of the nature of microstructure ordering under irradiation.

CONCLUSION

It has been shown that the coupling between reaction and transport may induce pattern forming instabilities in defect distributions in driven crystalline materials. For example, uniform distributions of point defects such as interstitials and vacancies in irradiated materials may become unstable and lead to the precipitation of solid solutions, to the nucleation of voids and of void lattices. In this framework, it is shown that vacancy loop ordering occur under very general conditions in irradiated metals and alloys. It mainly results fom the different mobilities and bias in the migration of point defects to line defects such as vacancy loops or network dislocations. Structures with different symetries may be simultaneously stable beyond the primary bifurcation. For example, when the diffusion and interactions of point defects are isotropic the maxima of the vacancy loop density may either correspond to bcc lattices or planar arrays. Hence, these structures could be in non parallel orientations, i.e. with a structure different from the structure of the host lattice.On increasing further the displacement damage rate bcc lattices become unstable and a first-orderlike transition should occur to planar structures. In the case of anisotropic interstitial diffusion, planar structures should be the rule. Hence, since the symmetry of the defect structures is a crucial issue in irradiated materials [20], the present discussion shows that a careful study of the post-bifurcation regime is needed to test the relevance of particular kinetic models to the interpretation of experimental observations.

Furthermore a coherent description of materials instabilities associated with the spatio-temporal organisation of defects will hopefully lead to a deeper understanding of these phenomena. Due to the strong nonequilibrium conditions under which they occur, classical mechanical or thermodynamical considerations are not sufficient and we need an important input from nonlinear dynamics and instability theory. Effectively, despite the huge complexity of defect dynamics, even in the case of phenomenological models, valuable information can be obtained via the reduced dynamics near instability points leading to a possible description of the pattern selection and stability properties in the post-bifurcation regime. Hence, by the combination of the results of bifurcation analysis, amplitude equation formalism and numerical simulations we expect significant breakthroughs in the understanding and prediction of the effects of materials instabilities on the macroscopic behavior of driven or degrading solids.

Acknowledgements. Financial assistance through a NATO grant for international collaboration in research (890489) is gratefully acknowledged.

References

1. J.Lega, in "Patterns, Defects and Materials Instabilities," D. Walgraef and N.M. Ghoniem, eds., Kluwer, Dordrecht, 1990.
2. F.Busse and L.Kramer, "Nonlinear Evolution of Spatio-Temporal Structures in Dissipative Continuous Systems," Plenum, New York, 1990.
3. G.Martin and L.P.Kubin, "Nonlinear Phenomena in Materials Science," Transtech, Aedermannsdorf (Switzerland), 1988.
4. D.Walgraef and N.M.Ghoniem, "Patterns, Defects and Materials Instabilities," Kluwer, Dordrecht, 1990.
5. D.Walgraef and E.C.Aifantis, Res Mechanica **23** (1988), p. 161.
6. a)G.Martin, Phys.Rev. **B30** (1984), p. 1424.
6. b)K.Krishan, Radiat.Eff. **66** (1982), p. 121.
7. D.Hull and D.J.Bacon, "Introduction to Dislocations," 3rd edition, Pergamon, Oxford, 1984.
8. W.Jäger, P.Ehrhart and W.Schilling, in "Nonlinear Phenomena in Materials Science," G.Martin and L.P.Kubin eds., Transtech, Aedermannsdorf (Switzerland), 1988, p. 279.
9. S.M.Murphy, Europhys.Lett **3** (1987), p. 1267.
10. S.Murphy, in "Nonlinear Phenomena in Materials Science," G. Martin and L.P. Kubin eds., Transtech, Aedermannsdorf (Switzerland), 1988, p. 295.
11. R.Bullough, B.L.Eyre and K.Krishan, J.Nucl.Mat. **44** (1975), p. 121.
12. N.M.Ghoniem and G.L.Kulcinski, Radiation Effects **39** (1978), p. 47.
13. D.Walgraef and N.M.Ghoniem, Phys.Rev. **B39** (1989), p. 8867.
14. A.C.Newell, in "Lectures in Applied Mathematics," vol.15, M.Kac, ed., American Mathematical Society, 1974, p. 157.
15. H.Haken, "Advanced Synergetics," Springer, Berlin, 1983.
16. N.M.Ghoniem and D.Walgraef, in preparation.
17. J.H.Evans, Mater.Sci.Forum **15-18** (1987), p. 869.
18. D.Walgraef and C.Schiller, Physica **D25** (1987), p. 423.
19. D.Walgraef, in "Nonlinear Phenomena in Materials Science," G. Martin and L.P. Kubin eds., Transtech, Aedermannsdorf (Switzerland), 1988, p. 77.
20. R.W.Cahn, Nature **329** (1987), p. 284.

RAYLEIGH-TAYLOR INSTABILITY OF A THIN LAYER

M. Fermigier, L. Limat, E. Wesfreid, P. Boudinet, C. Ghidaglia and C. Quilliet

Laboratoire de Physique et de Mécanique des Milieux Hétérogènes
Ecole Supérieure de Physique et de Chimie Industrielles
10 rue Vauquelin, 75231 Paris CEDEX 05, France

INTRODUCTION

In this communication, we report some recent results[1,2] obtained in the study of the Rayleigh-Taylor instability of a thin layer. This interfacial instability exhibits interesting "growth and form" effects, in which the non-linearities play an essential role: formation of two-dimensional patterns of different symmetries, transition between patterns (sometimes involving front propagation phenomena), secondary instabilities...

The Rayleigh-Taylor instability[3-5] is a gravitational instability occuring when there is an adverse density stratification in a fluid, for instance at the interface separating two fluids of different densities, the upper fluid being heavier than the lower one. Most of the available studies of its non-linear development deal with the case of thick layers, that is very important for practical applications: rising of magmas in geophysics[4], or laser implosion of fusion targets[5] for instance. Very few studies, based essentially on numerical simulations[6,7], have investigated the case in which one of the layers is very thin, its thickness e being in particular small compared to the capillar length $l_c = \sqrt{\rho g/\gamma}$ (as usual, $\rho = \rho_{up} - \rho_{down}$ designates the density difference, g the gravity and γ the surface tension). This case is also of practical importance, for instance in the genesis of two-phase flow in situations of film boiling[8,9].

EXPERIMENT - GENERAL OBSERVATIONS

In our experiment, a drop of silicone oil (viscosity η=1000 cP, density ρ=0.97 g/cm^3, surface tension γ=21 dynes/cm) is firstly spread by gravity on an horizontal glass plate yielding a viscous pancake, approximately 30 cm in diameter and a fraction of millimeter thick. Secondly, the instability is started by quickly inverting the plate. The development of the instability is then monitored by video recording or by photographs taken at fixed intervals. Figure (1) is a typical example obtained after 600 s. The deformation of the interface is revealed by means of a dye previously mixed with the oil. The thicker parts appear as dark spots on the picture. As can be seen on this figure, different patterns are obtained: axisymmetric patterns (concentric rings), axisymmetric patterns whose rings break into peaks (often with a sixfold symmetry), hexagonal patterns and finally, lines which we sometimes call "rolls", in analogy with the rolls observed in convective instabilities.

These patterns are in fact transient structures growing from initial perturbations of the interface. For instance the axisymmetric structures generally grow from the initial dimple created by specks of dusts on the interface. In a similar way, the "rolls" are initiated by the thickness gradient at the boundaries of the pancake. The evolution of the pattern seems then to involve three processes: (i) local growth of the amplitude (ii) spreading by an increase of their

Growth and Form, Edited by M. Ben Amar *et al.*
Plenum Press, New York, 1991

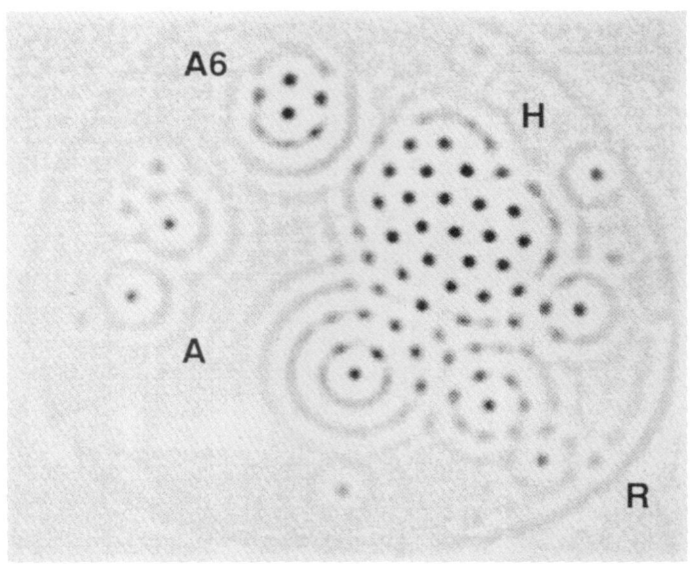

Fig. 1. Typical structures occuring in the instability. Thicker parts of the fluid appear as dark spots. (A) axisymmetric structures, (A6) axisymmetric structures developing a sixfold symmetry, (H) hexagonal pattern, (R) "rolls" or "lines" structures.

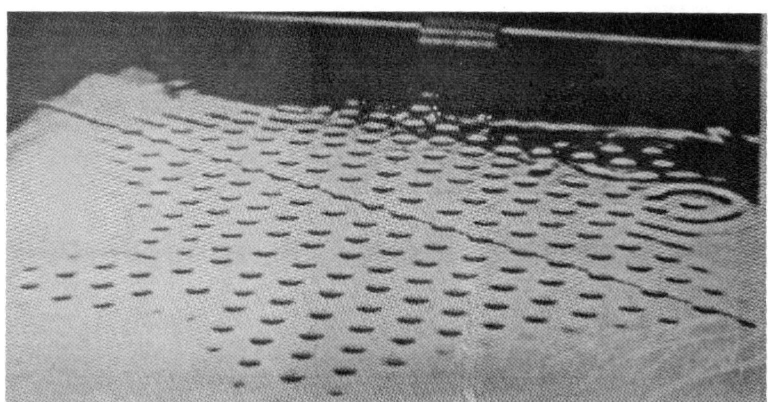

Fig. 2. Perspective view of the experiment seen from above, through the glass plate: a more or less regular hexagonal lattice of pendant drops connected by a very thin film is left below the plate in the final stage of the experiment.

spatial extent (iii) transition towards the hexagonal symmetry. After a complicated history, one finally gets a more or less ordered hexagonal system of pendant drops (see figure 2). In the latest stage, depending on the initial thickness, these drops can exhibit secondary instabilities: drop falling, drop coalescence, falling after coalescence...

EVOLUTION EQUATION - LINEAR GROWTH

The evolution equation governing the thickness variation $\zeta(\mathbf{r},t)=e(\mathbf{r},t) - e_0$ can be derived rather easily in the lubrication approximation, the slope of the interface of order e_0/l_c being assumed to be small[1,6,7] (This approximation neglects inertial effects and assumes that the horizontal dependance of the velocity field is negligible compared to the vertical one):

$$\frac{\partial \zeta}{\partial t} + \frac{1}{3\eta}\nabla.[(e_o+\zeta)^3\nabla(\rho g\zeta+\gamma\Delta\zeta)] = 0 \tag{1}$$

This is the mass conservation equation, in which the mass flux is proportional to the pressure gradient $\nabla P=\nabla[\rho g\zeta+\gamma\nabla^2\zeta]$. This pressure gradient involves an hydrostatic contribution $\rho g\zeta$ associated to the destabilizing gravity, and a capillary one dependant upon the curvature of the interface. The fact that the "mobility" of the fluid is proportional to the cube of the thickness $e^3=(e_0+\zeta)^3$, results from the structure of the flow that reduces to a half-Poiseuille one. After linearizing this equation, one gets the dispersion relationship of Fourier modes $\exp[iqx+\sigma t]$:

$$\sigma = \frac{e_0{}^3}{3\eta}[\,\rho g q^2 - \gamma\, q^4]\tag{2}$$

The variation of σ versus q is suggested below:

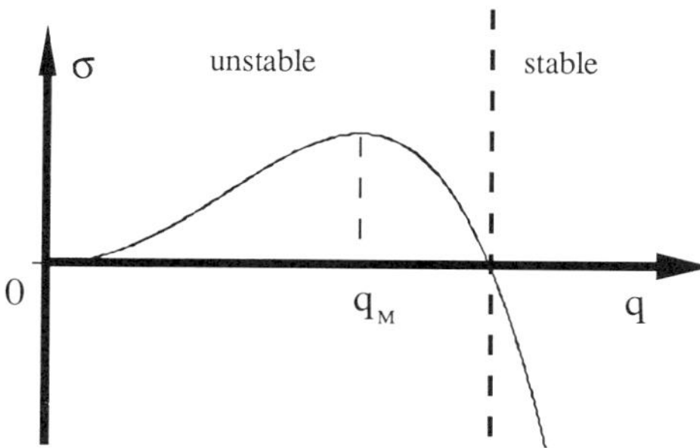

A large band of Fourier modes are unstable, the maximum growth rate being obtained for a wave number $q_M = \sqrt{\rho g/2\gamma}$. The associated wave-length $\lambda_M = 2\pi \sqrt{2}\, l_c \approx 13.2$ mm is consistent with that observed in our experiments (of order 12.5 mm for the most regular patterns).

443

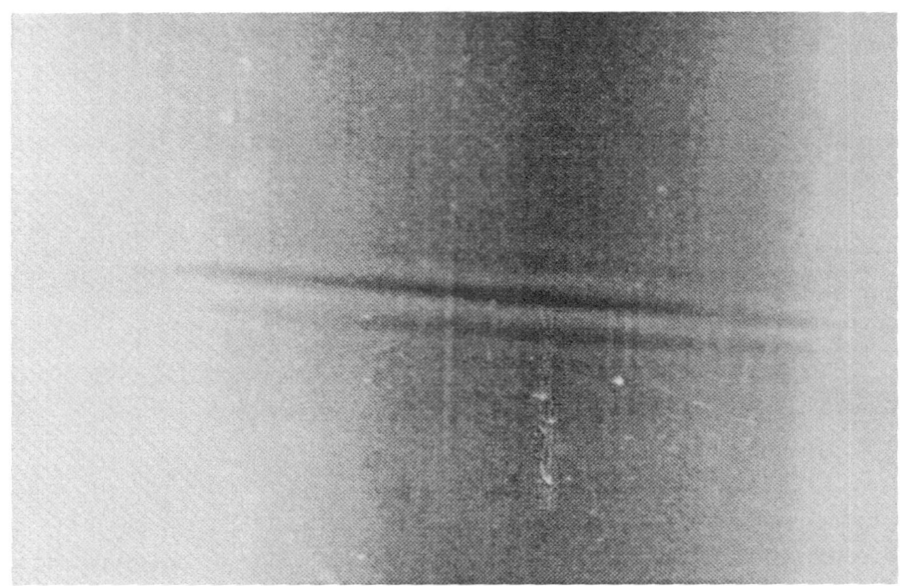

Fig. 3.   Instability initiated by a single wire stretched through the fluid layer.

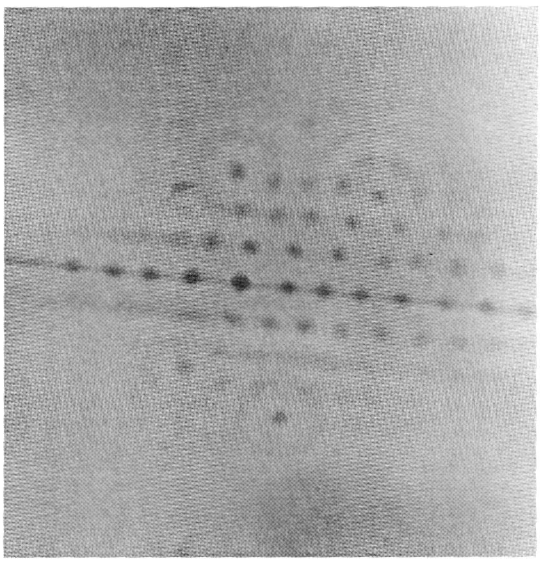

Fig. 4 Transition of the roll system of fig. 3 towards an hexagonal symmetry.

TRANSITION ROLLS-HEXAGONS

We have investigated the time evolution of the patterns, starting from well defined, controled initial conditions. In a first experiment, a thin metal wire is stretched across the oil layer after the spreading is completed. The wire has a diameter slightly larger than the oil thickness, and induces an initial perturbation which decays exponentially away from the wire, with a characteristic length equal to l_c. Once the glass plate is inverted, "rolls" developed parallel to the wire (see fig.3). These structures do not remain one-dimensional: after a while, the lines break in peaks of an hexagonal pattern (fig. 4).

This tendency of our system to form an hexagonal pattern can be roughly explained[1] by expanding eq. (1) over the Fourier modes. After some calculations, the evolution equation of the amplitude $A_q(t)$ of a mode q is given by:

$$\frac{dA_q}{dt} = (2q^2-q^4)\, A_q + 3 \sum_{q_a, q_b} (q_a \cdot q)(2-q_a^2) A_{q_a} A_{q_b} \delta(q - q_a - q_b)$$

$$+ 3 \sum_{q_a, q_b, q_c} (q_a \cdot q)(2-q_a^2) A_{q_a} A_{q_b} A_{q_c} \delta(q - q_a - q_b - q_c) \quad + \;.... \tag{3}$$

where the wave numbers q and the time t have been nondimensionalized, using $1/q_M$ and $1/\sigma(q_M)$ as units of length (in the horizontal direction) and of time, and where the A_q are normalized by the initial thickness e_o. The presence of a second-order term in this equation is to be noted, and means that our system is not invariant by amplitude reflection. Usually in the physics of interfacial instabilities[10], this property tends to favor the occurence of an hexagonal symmetry. This tendency can be made more obvious by considering the growth of a pair of modes $\pm q_1$, with a perturbation of small amplitude on the two other pairs of modes $\pm q_2$ and $\pm q_3$ as suggested on figure (5).

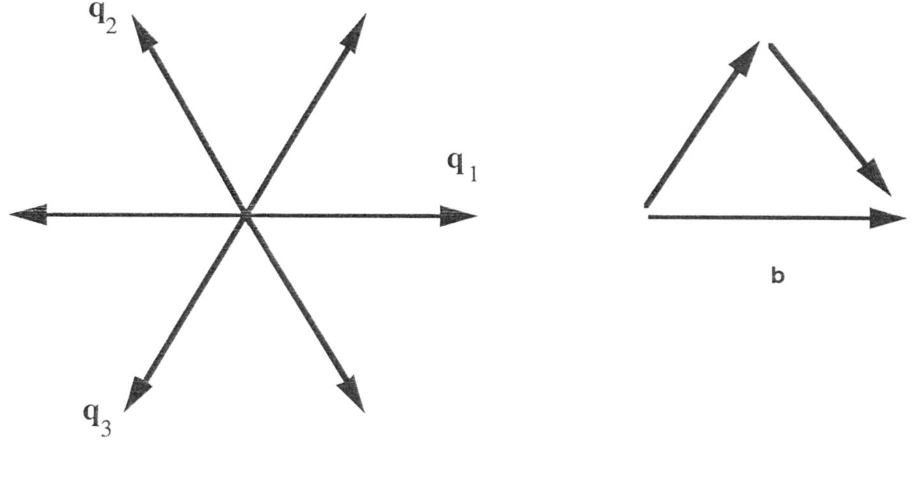

Fig. 5.   (a) The three pairs of modes involved in our simplified discussion of the transition rolls-hexagons. (b) Selection rule associated to the second order non-linearity.

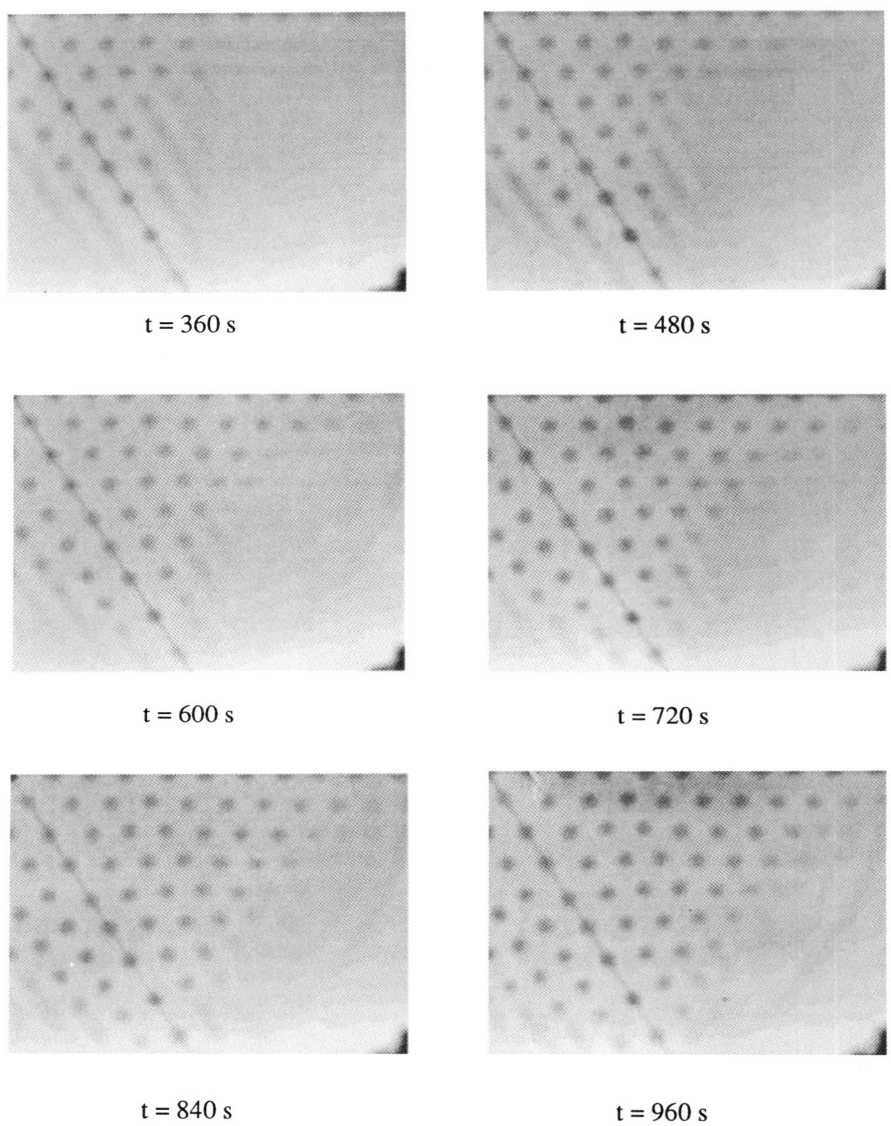

Fig. 6. Time evolution of the instability initiated by two wires crossed at 60°.

This situation can also be understood as the competition between a roll pattern of amplitude $A_R(t) = A_{q1} - A_{q2} = A_{q1} - A_{q3}$ and an hexagonal pattern of small amplitude $A_H(t) = A_{q2} = A_{q3}$. For simplicity, the amplitudes are taken to be real, and only the most unstable wave-number $q_M = 1$ is taken into account. At second order, eq. (3) reduces to:

$$\frac{dA_R}{dt} = A_R \qquad - 3\, A_R A_H$$

$$\frac{dA_H}{dt} = A_H + 3A_H^2 + 3A_H A_R \tag{4}$$

In these equations, it is clear that the second order interaction suggested on figure (5 -b) tends to amplify the growth of the hexagonal perturbation, and to damp that of the roll pattern. We believe that this effect is the basic mechanism involved in the transition roll-hexagon .

ROLLS, SQUARE AND HEXAGONAL PATTERNS

The tendency of our system to develop a pattern with an hexagonal symmetry is enhanced if the initial perturbation is created by two wires crossed at 60° (fig. 6). As in the previous experiment two sets of "rolls" appear beside each wire. When the two sets of lines cross they create a perfect hexagonal pattern.

We have also try to force a pattern of square geometry by means of two wires crossed at 90°. The final result is presented on fig. (7) . The hexagonal symmetry is again clearly favored, but the square one seems to survive locally in some regions of the pancake. The "stability" (metastability is perhaps a better term) of this structure seems to be intermediate between that of the "rolls" and that of the hexagonal pattern.

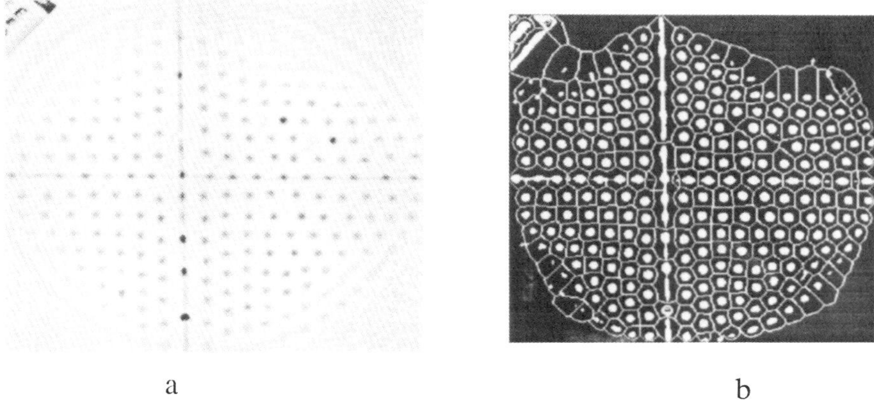

a   b

Fig. 7. (a) final structure obtained with two wires crossed at 90°. (b) skeletonization obtained from image treatment. In some regions the square structure seems to survive.

A conclusion consistent with this observation can be obtained thoretically, by calculating the growth of the three patterns using a perturbation method, the initial amplitude ε being treated as a small parameter. At third order, this gives[1]:

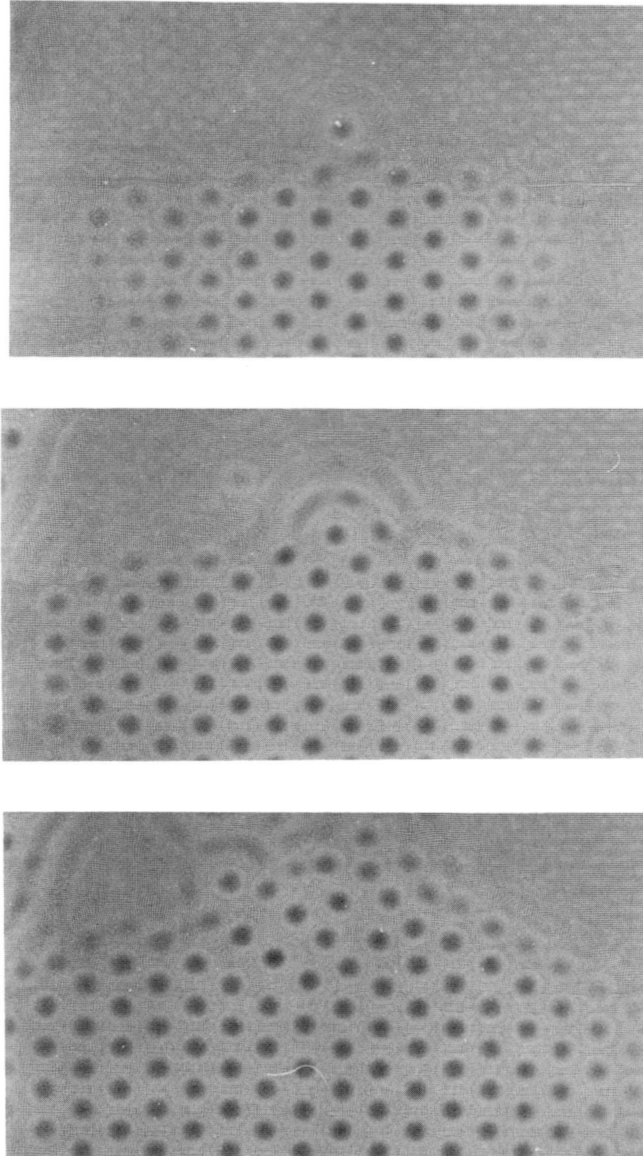

Fig. 8. Time evolution of a front between an hexagonal pattern and a region of unperturbed thickness. The hexagonal pattern has been forced by means of an array of needles. The deformation of the interface is revealed by observing a ruled screen across the oil layer. The time interval between pictures is 240 s.

$$A_R \approx \varepsilon\, e^t \qquad\qquad - 3\,\varepsilon^3\, [\, e^{3t} - \frac{11}{8}\, e^t\,] \qquad\qquad + \ldots\ldots$$

$$A_S \approx \varepsilon e^t \qquad\qquad + \frac{9}{8}\,\varepsilon^3\, e^t \qquad\qquad\qquad + \ldots \qquad (5)$$

$$A_H \approx \varepsilon\, e^t + 3\,\varepsilon^2\, (e^{2t} - e^t) + \frac{1}{5}\,\varepsilon^3\, [\, 6e^{3t} - 96e^{2t} + \frac{1629}{8}\, e^t\,] + \ldots$$

where A_R, A_S and A_H are respectively the amplitudes of the "roll", square and hexagonal patterns. We see on these equations that the "rolls" are damped at third order, while the hexagonal pattern is amplified at second and third order. As announced above, the case of the square pattern is intermediate: at third order its growth is practically that obtained in the linear theory. It is also interesting to note that an hexagonal pattern of negative amplitude would be damped at second order. This means that an hexagonal lattice of "spikes" is amplified, while a lattice of "holes" would be damped. This situation results from the non-invariance by amplitude reflection of our system.

FRONT PROPAGATION PHENOMENA

Another typical feature of this instability appears on figs. (3,4,6). The patterns do not grow uniformly in space, but are in fact localized, their spatial extent increasing with time. This effect is particularly obvious on figure (6), where the "rolls" as well as the hexagonal systems exhibit a spreading dynamics. In analogy with what is observed in other instabilities[11], we suggest that this spreading may involve the propagation of a front[12-14] between a stable state (rolls or hexagons) and an unstable one (domain of unperturbed thickness). This idea can be precised by comparing the Swift Hohenberg model of convection

$$\frac{\partial u}{\partial t} = [\varepsilon - (1+\Delta)^2]u - u^3 \qquad (6)$$

to eq. (1) rewritten nondimensionally as:

$$\frac{\partial \zeta}{\partial t} = [1 - (1+\Delta)^2]\zeta - \nabla.[(3\zeta + 3\zeta^2 + \zeta^3)\nabla(2\zeta + \Delta\zeta)] \qquad (7)$$

As a first guess, we suggest that the front propagations in our experiment could be governed by the same linear mechanisms involved in the SH model (marginal stability case I), but with the control parameter ε being always equal to 1. This has allowed us to calculate the wave-length λ_L left behind the front, and the velocity of the front: $\lambda_L \approx 0.95\lambda_M$ and $v^* = 0.38e_o^3\rho^2 g^2/\eta\gamma$ ($v^* = 4.6$ in adimensional units of eq. 7). The orders of magnitude obtained experimentally are consistent with this hypothesis. For instance, for the most regular patterns λ_L is of order 12.5 mm, that seems to be slightly smaller than our estimate for λ_M, $\lambda_M \approx 13.2$ mm , based on the value of γ given above. The theoretical value of v^* is difficult to determine because of its strong cubic dependance upon the layer thickness, that we did not measure accurately. The orders of magnitude $v^* \approx 0.05$ to 0.1mm/s are however consistent with the experiment. More accurate measurements, and a numerical study of front propagation in the case of equation (7) are under progress[15].

An interesting feature of our experiment is the possibility to observe two-dimensional aspects of front propagation. An example is given on figure (8), where a localized hexagonal pattern has been forced by means of an array of needles. At long time, it seems that the front parallel to one of the pair of wave-vectors is unstable and breaks in two fronts perpendicular to the two other pairs. The problem of two-dimensional front propagation has been scarcely studied, the only available study being that of Schiller[14]. We believe that our experiment could constitute an interesting tool of investigation of these problems.

ANNULAR PATTERNS

The theoretical discussion of the development of instabilities in annular geometry is in general more difficult. We have developed in our case a simplified approach[1] based on the Fourier decomposition of the solutions of the linear problem. These solutions are of the kind:

$$\zeta_{q,n}(r,\theta,t) = J_n(qr) \cos(n\theta) \, e^{\sigma t} \qquad (8)$$

where J_n are the Bessel functions of the first kind, σ being again given by eq. (2). The Fourier content of these solutions corresponds to a continuous distribution of wave-vectors $\mathbf{q}$ of constant modulus q, which amplitude $a_n(\phi)=\cos(n\phi)$ depends on the azimuthal angle $\phi=(\mathbf{x},\mathbf{q})$. The same geometrical construction as that used for the hexagonal pattern (fig. 5-b) gives the second order coupling between the amplitudes $a_n(\phi,t)$ in a second order development[1]. After some calculations, one is led in this case to the conclusion that the patterns amplified at second order are of the kind:

$$\zeta = J_o(qr) + \alpha_1 J_6(qr)\cos(6\theta) + \alpha_2 J_{12}(qr)\cos(12\theta) + \ldots \qquad (9)$$

As mentioned more above, structures of this kind are very often nucleated by specks of dust falling on the interface. In their initial stage of evolution, only the J_0 component appears, while both the amplitude and the spatial extent of the pattern are growing. Later, a J_6 component may appear, very often by interference between two neighboring J_0 patterns. Qualitatively, this process is slower than the equivalent rolls-hexagons transition. We believe that the longer life-time of the axisymmetric pattern (J_0) is related to the fact that contrary to the "roll" pattern, it is amplified at second order. After the J_6 component, a J_{12} component appears, and so on. An example is given on fig. 9.

A careful analysis of the evolution of this kind of structure is under progress. We mention that sometimes, structures of five-fold or even seven-fold symmetry may also appear. An example is available on fig. 1.

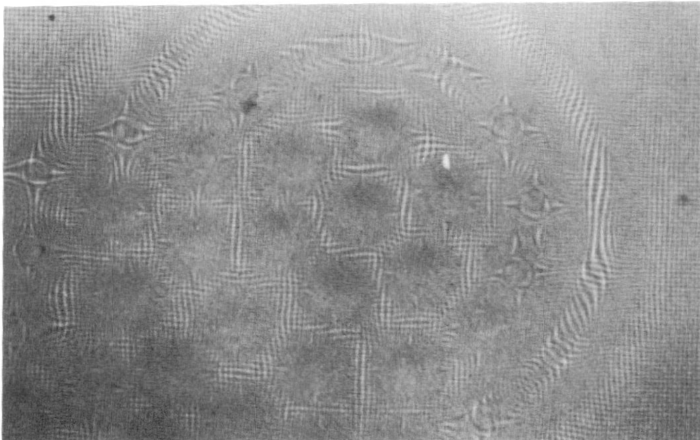

Fig. 9. Pattern obtained after transition of an axisymmetric structure towards the hexagonal symmetry. As in fig. 7, the thickness evolution is revealed by observing a ruled screen across the oil layer. Note the existence of 12 maximas on the second annulus, associated to the J_{12} component.

CONCLUSION

The Rayleigh-Taylor instability of a thin layer constitutes a nice example of a growth process driven by non-linearities. The observed behavior is very rich, involving a great variety of self organization phenomena: formation of two-dimensional patterns, transition between different symmetries, front propagation in two dimensions... The main mechanisms have been recognized, and more precise studies involving quantitative measurements, numerical simulations and theoretical calculations are now under progress.

Acknowledgment: we have benefitted from very helpful discussions with M. Brachet, H. Brand, G. Dewell, C. Mitescu, I. Mutabazi, J. C. Nataf, J. Prost, W. Van Saarloos and D. Walgraef. We also acknowledge helpful criticisms of the manuscript from R. Goodwin.

REFERENCES

1) M. Fermigier, L. Limat, J. E. Wesfreid, P. Boudinet and C. Quilliet, to appear in J. Fl. Mech.
2) M. Fermigier, L. Limat, J. E. Wesfreid, P. Boudinet, M. Petitjean, C. Quilliet and T. Valet, Phys. Fl. A $\underline{2}$, 1517 (1990).
3) S. Chandrasekhar, "Hydrodynamic and hydromagnetic stability", Dover (1981)
4) J. A. Whitehead and D. S. Luther, "J. Geophys. Res. $\underline{80}$, 705 (1975); J. A. Whitehead, Ann. Rev. Fl. Mech. $\underline{20}$, 61 (1988)
5) D. H. Sharp, Physica D $\underline{12}$, 3 (1984)
6) T. P. Hynes, "Stability of thin films", PHD Thesis, Cambridge Univ. (1978)
7) S. Yiantsios and B. G. Higgins, Phys. Fl. A $\underline{1}$, 1484 (1989).
8) J. Berenson, Int. J. Heat Mass Transfer, $\underline{5}$, 985 (1962)
9) J. Sainson, "Contribution à l'étude des transitions rapides de phase", Thesis, Université Paris 6 (1989); V. Hakim and J. Vannimenus, to be published.
10) E. Buzano and M. Golubitzky, Phil. Trans. Roy. Soc. (London) A $\underline{308}$, 617 (1983)
11) G. Ahlers and D. S. Cannell, Phys. Rev. Lett. $\underline{50}$, 1583 (1983)
12) G. Dee and J. S. Langer, Phys. Rev. Lett. $\underline{50}$, 383 (1983)
13) W. Van Saarloos Phys. Rev. A $\underline{37}$, 211 (1988); $\underline{39}$, 6367 (1989)
14) C. Schiller, "Modelisation of microstructures in metals", Thesis, Université Libre de Bruxelles (1989).
15) C. Mitescu, L. Limat and E. Wesfreid, Com. to 43rd APS Meeting, New-York (1990), Bull. Am. Soc. $\underline{35}$, 10, 2277 (1990).

STABLE LAWS IN BREATH FIGURES ON A

ONE-DIMENSIONAL SUBSTRATE

I. Yekutieli

Service de Physique du Solide
  et de Résonance Magnétique
CEN-SACLAY
91191 Gif-sur-Yvette Cedex, France

Breath figures are ensembles of droplets formed by condensation of water vapor on cold non-wetting surfaces. Mist on a window is an everyday example of this phenomenon. The formation of breath figures involves two mechanisms: growth of individual droplets by continual condensation of water and coalescences of pairs of droplets, when they touch, into single droplets. Extensive studies of breath figures[1,3] show that the radius of a droplet grows as a power law in time between coalescences

$$r_i(t) \approx t^\alpha \tag{1}$$

where the exponent α is governed by the mode of condensation of water vapor into droplets. When two droplets touch, they coalesce very rapidly into one droplet. The new droplet is centered on the center of mass of it's two parent droplets and assuming that the form of a droplet does not depend highly on its radius, the radius of the new droplet is given by conservation of mass

$$r_{1+2} = \left(r_1^{\,3} + r_2^{\,3}\right)^{1/3} \tag{2}$$

At the beginning of such an experiment initial droplets nucleate on the substrate. They are so small that there are hardly any coalescences observed. When the droplets grow to radii of the order of the separation between them, many coalescences are then observed. In this regime, which persists for a long time, the ensemble of droplets seems self similar in time. The distribution of droplet radii is scaling in time. The coverage of the substrate, i.e., the fraction of it covered by droplets, which is null in the begining, attains a constant value in time, hinting that the whole breath figure is indeed self-similar in time.

Computer simulations of breath figures using eqs.(1,2)[2,3,5] show that these equations define a reasonable model for breath figures. For droplets on a one dimensional substrate the coverage stabilizes at 0.79 . In Fig. 1 we show results of a simulation of droplet growth in one dimension: distributions of gaps between droplets (g), droplet diameters (d) and distances between neighboring droplet centers (h) at different times well into the scaling regime superposed together.

We have simplified this last model to a bare minimum which still resembles breath figures. We take all droplets to be equal in size at all times. When two droplets

Growth and Form, Edited by M. Ben Amar *et al.*
Plenum Press, New York, 1991

coalescence they form a droplet of the same size as theirs, centered in the middle between them. This can be viewed as a purely geometric model by looking at the evolution of the centers of the droplets in time. Starting from points distributed on the substrate, at each step one looks for the two closest points and replaces them by a single point in the middle between them. For this model one can look at the distribution of distances between neighboring points and the coverage (where one takes the diameter of the droplets equal to the shortest distances between any two points). In the one dimensional version of this model the points partition the line into intervals. Simulations show that in the long time limit this model is driven toward a stable distribution of intervals scaling in time, implying also a constant coverage. In Fig. 2 we plot the scaled distributions of intervals at different times in the scaling regime. The scaled distribution of droplet diameters is a δ-function at 1, and the distribution of gaps between droplets is the same as the distribution of intervals only shifted by 1 to the left. The constant coverage achieved is 0.72. The

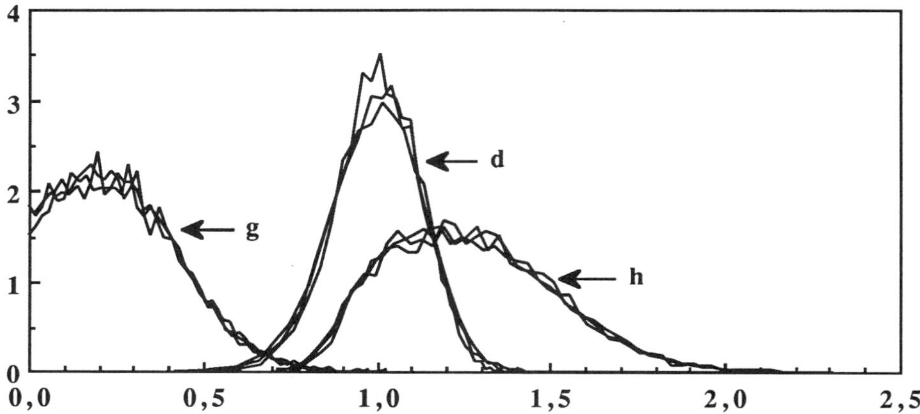

Figure 1. Distributions of gaps between droplets (g), droplet diameters (d) and intervals (h) scaled by the average diameter at 3 different times in the scaling regime.

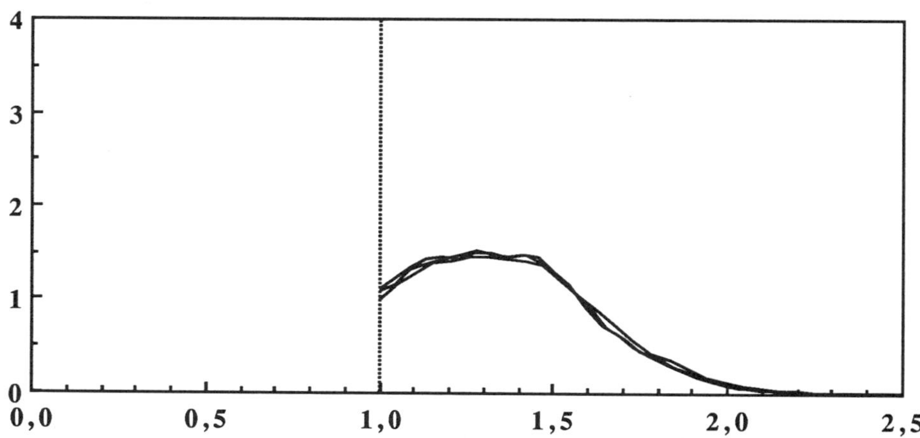

Figure 2. Distribution of intervals (h) scaled by the minimal interval at 3 different times in the scaling regime.

curve in Fig. 2 resembles the curve for the distribution of intervals in Fig. 1, showing that this simple model is not very different from the more complicated droplet model. Note that in both cases almost all gaps are smaller than the smallest droplet, signifying almost jammed-states (with allusion to the parking problem[6] where cars of equal length are parked randomely along a street with no overlap until the street is jammed).

For an analytical description of this model, we assume, as a first approximation, that lengths of neighboring intervals are uncorrelated. We define $n_t(h)dh$ as the number of intervals of length in $(h,h+dh)$ and h_0 as the shortest interval. If during a time increment dt we coalesce all intervals in (h_0,h_0+dh_0), the evolution of $n_t(h)$ is given by

$$n_{t+dt}(h)=n_t(h)+\frac{2n_t(h_0)dh}{N_t}[n_t(h-\tfrac{1}{2}h_0)\theta(h-\tfrac{3}{2}h_0)-n_t(h)] \tag{3}$$

the prefactor of the square brackets giving the probability of an interval to be next to an interval h_0, the terms inside the brackets corresponding to gains and losses in $n_t(h)$ as a result of coalescences and N_t the total number of intervals. Iteration of this recurrence relation leads to a scaling distribution in the long time limit which differs considerably from the simulation results. This is a consequence of long range correlations between interval lengths developing in the system. Higher order approximations can be taken[4,5] to improve the description of this model.

We have seen that the evolution of a system of "droplets" by growth and coalesence toward a self-similar regime is a phenomena general to a large class of systems resembling breath figures. In all the cases studied, the system is driven by means of local rules towards a global stable statistical distribution. Currently we are looking into yet even simpler models of one-dimensional breath figures where the first-order approximation is exact and two-dimensional versions of these models.

REFERENCES

1. D. Beysens and C. M. Knobler, Phys. Rev. Lett. 57:1433 (1986).
2. D. Fritter, C. M. Knobler, D. Roux and D. Beysens, J. Stat. Phys. 52:1447 (1988).
3. A. Steyer, P. Guénoun, D. Beysens, D. Fritter, C. M. Knobler, submitted to Europhys. Lett.
4. B. Derrida, C. Godrèche and I. Yekutieli, Europhys. Lett. 12:385 (1990).
5. B. Derrida, C. Godrèche and I. Yekutieli, preprint.
6. A. Rényi, Publ. Math. Inst. Hung. Acad. Sci. 3:109 (1958).

ORDERING PROCESS IN THE DIFFUSIVELY COUPLED LOGISTIC LATTICE

Claudine V. Conrado and Tomas Bohr

The Niels Bohr Institute
Blegdamsvej 17
DK-2100 Copenhagen

INTRODUCTION

Coupled map lattices have been studied in the last years as an attempt to bridge the gap between low-dimensional chaotic systems and spatially extended systems showing turbulent behaviour [1, 2, 3, 4]. For recent reviews see [5, 6]. In a chaotic coupled map lattice a large number of maps, each of which can generate chaotic behaviour, are coupled together as a coarse model of turbulence. From this approach we can hope to learn something about the characteristic length and time scales in "spatio-temporally chaotic" systems and about the emergence of ordered states in strongly fluctuating systems.

In the present work we shall study *transient* behaviour in a coupled map lattice and, in particular, how the transient times and their distribution depend on the system size. Low-dimensional chaotic repellors are well known: many dynamical systems have chaotic dynamics that decays into simpler behaviour after a characteristic transient or escape time. The relation of the escape time to the chaotic invariants of the repellor is rather well understood [7] but in spatially extended systems not much is known. General arguments have been given, why these times should be exponentially (or even hyperexponentially) large in the system size [8, 9]; in the example studied here it grows only algebraically and can be understood well in terms of diffusing defects as discussed below. This is in agreement with the types of behaviour observed in [10].

In the particular example discussed here, the turbulent state decays into a state with long range order superimposed on small amplitude chaotic fluctuations. The ordering process turns out to be very analogous to the ordering in thermodynamic systems at first order phase transitions [11], but with certain interesting differences. Here even finite systems and systems in one spatial dimension order completely and this is due to the limited noise - of "nonequilibrium" nature - present in the ordered states of the map lattice in contrast to the thermodynamic case.

The specific coupled map lattice which we have used is defined in terms of fields $x_n(\vec{r})$ defined on a D-dimensional lattice with periodic boundary conditions indexed by $\vec{r}$ and discrete time n [12]. The dynamic equation for x consists of 1) a local transformation given by the map $f(x)$ and 2) diffusion - or averaging - among the nearest neighbors:

$$x_{n+1}(\vec{r}) = (1 - \epsilon)f(x_n(\vec{r})) + \frac{\epsilon}{2D}\sum_{\vec{r}'} f(x_n(\vec{r}')). \tag{1}$$

The local map $f(x)$ is chosen to be the logistic map

$$f(x) = 1 - ax^2. \tag{2}$$

Growth and Form, Edited by M. Ben Amar *et al.*
Plenum Press, New York, 1991

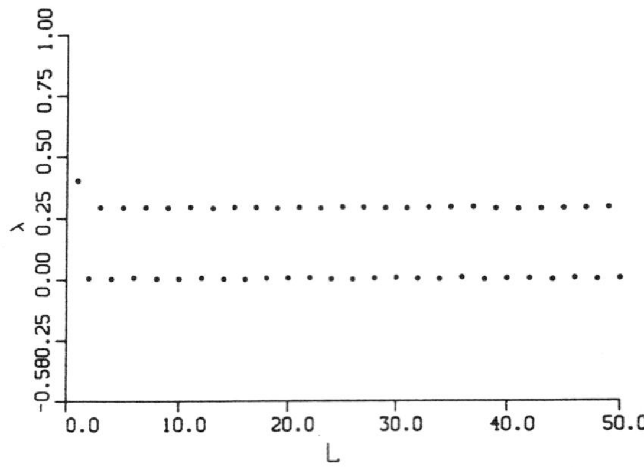

Figure 1. Size dependence of the maximum Lyapunov exponent λ for $a = 1.80$ in the one-dimensional model. These results were obtained for a fixed number of time iterations for all L's from a random initial condition.

Several aspects of this model has been studied in one and two dimensions. For a small coupling ϵ, it was found that the system has distinct phases characterized by different spatiotemporal patterns, depending on the value of the parameter a in (2). When a is increased from the accumulation point $a_c = 1.4055...$ of the single map [13], we observe a transition sequence from a frozen random state to pattern selection and to fully developed spatiotemporal chaos [8, 12]. In the pattern selection phase, the system is spatially ordered with selected domains of small size and the time evolution is less chaotic in the sense of a smaller Lyapunov exponent. The behaviour of the two-dimensional model for small coupling is found to be essentially the same. In particular for $\epsilon = 0.1$, the simplest domains (size 1) are selected resulting in a checkerboard-like pattern in two dimensions and a zigzag-like pattern in one dimension [8, 12]. For large coupling, the two-dimensional system has a turbulent state with characteristic length and time scales varying as powers of $(a - a_c)$ [4]. Systems with sizes smaller than this characteristic length become ordered while systems with larger sizes display "incoherent" chaos or turbulence.

In this work, we investigate the behaviour of one and two-dimensional systems for varying lattice size L in the region of the parameter space where the simplest pattern is selected, namely $\epsilon = 0.1$ and $1.74 \leq a \leq 1.90$. The aim is to study the characteristic transient time for the turbulent repellor at varying system sizes. We are interested in the existence and stability of the selected pattern for large lattices, for which loss of correlation between points far away from each other could lead to a disordered or incoherent state.

RESULTS AND DISCUSSION

To describe and distinguish the observed chaotic states, we calculate Lyapunov exponents of the coupled map lattice (see e.g. [5, 6]). The maximum Lyapunov exponent λ is given in terms of the growth rates χ_i at time i of a tangent vector by [14]

$$\lambda = \lim_{n \to \infty} \frac{1}{n} F_n, \tag{3}$$

where

$$F_n = \sum_{i=1}^{n} \log \chi_i. \tag{4}$$

458

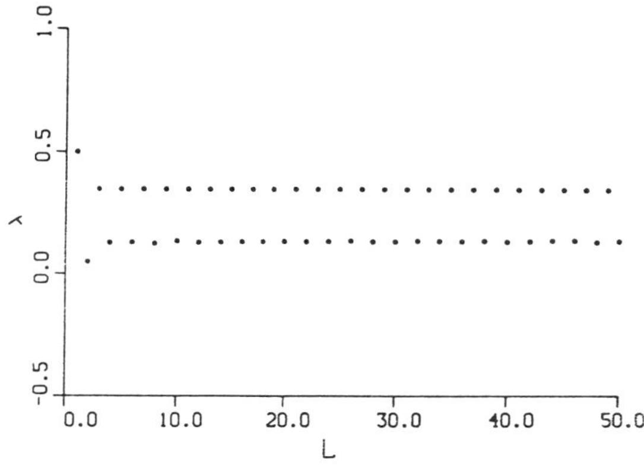

Figure 2. Same as Fig. 1 for the two-dimensional model with $a = 1.85$.

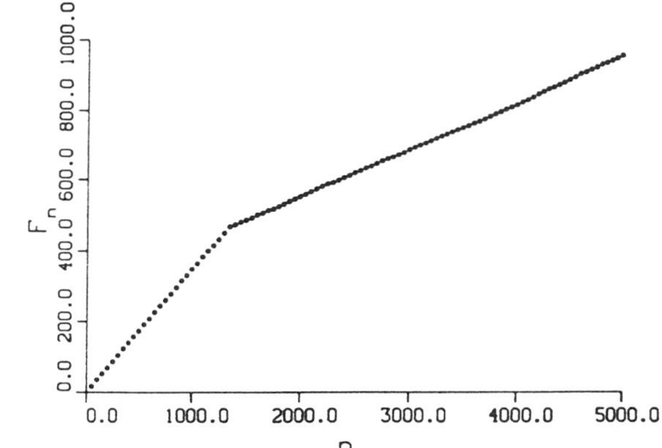

Figure 3. F_n (see eq. (4)) as a function of the discrete time n for the $L = 30$ two-dimensional lattice and $a = 1.85$ starting from a random initial condition.

In Fig. 1, we show $\lambda(L)$ for the one-dimensional system and $a = 1.80$ and in Fig. 2 for the two-dimensional system and $a = 1.85$. For each L, we start from some random initial condition and iterate a number of time steps in order to obtain the selected pattern. We can see the alternate behaviour for odd and even lattices due to the presence in odd lattices of a chaotic defect separating domains with opposite phase, which gives positive contributions to λ [12]. As can be seen from (3), λ is given by the slope of F_n in (4). This quantity is shown in Fig. 3 for the $L = 30$ two-dimensional lattice for $a = 1.85$. The break in the curve happens at the point at which the lattice becomes ordered. The ordered state has a positive Lyapunov exponent, but smaller than that of the transient chaotic state, which, as seen from Fig. 3, is rather well-defined[1].

It is interesting to notice that λ is constant for increasing (even) L as Figs. 1 and 2 show (this was checked for one-dimensional lattices for values of L up to 200 and in the two-dimensional case for values up to 100). From this result, it seems that, independently of the

---

[1] The change in the Lyapunov exponent coincides with the disappearance of the last defects. This of course does not mean that the state is perfectly ordered. Indeed, at least in two dimensions, the "ordered state" has a positive Lyapunov exponent. We are currently in the process of investigating the properties of this state in more detail.

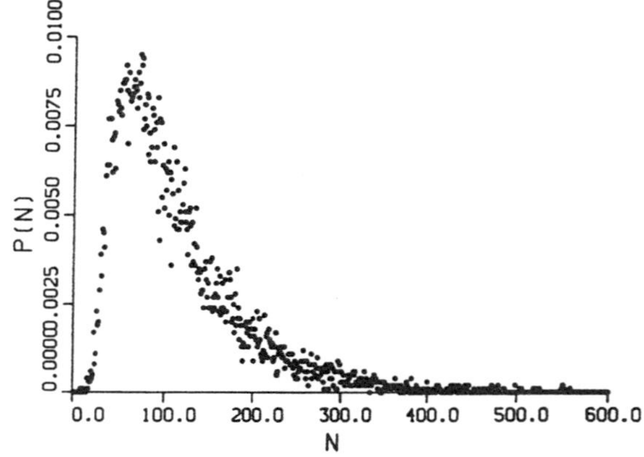

Figure 4. Distribution $P(N)$ of the number of time steps N required to obtain the selected pattern for the $L = 6$ one-dimensional lattice and $a = 1.80$. The number of different initial conditions is $N_i = 10000$.

size of the lattice, we will always obtain the selected pattern in that region of the parameter space if we iterate a sufficient number of times.

In order to understand the ordering process we computed the number N_m of time steps necessary to obtain the selected pattern as a function of the lattice size. Starting from many random initial conditions we found that this number, which is obtained from curves like the one in Fig. 3, fluctuates enormously. For the one-dimensional system it obeys a distribution $P(N)$ as given in Fig. 4 for $L = 6$ and $a = 1.80$. We can see a rapid increase from zero in the beginning of the curve but a slow exponential decay.

The mean value N_m obtained from $P(N)$ varies with L as shown in Fig. 5. According to that curve we find the following behaviour

$$N_m(L) \sim AL^B, \tag{5}$$

where $A \sim 2.0$ and $B \sim 2.0$. This time scale ($time \sim (distance)^2$) is expected in problems involving random walkers and it agrees with the picture of pattern formation through collisions and annihilations of defects whose motion is diffusive [8]. In Fig. 6, we show the behaviour of α with the lattice size, α being defined through the relation

$$P(N) \sim e^{-\alpha N}, \tag{6}$$

which fits the tail of the distribution $P(N)$. From Fig. 6, we have

$$\alpha(L) \sim A'L^{-B'}, \tag{7}$$

where $A' \sim 1.0$ and $B' \sim 2.0$.

The fact that, within the numerical accuracy, $B = B' = 2$ suggests a simple model to describe the late stages of the ordering process, namely a particle inside a one-dimensional box of size L with absorbing walls in $x = 0$ and $x = L$ obeying the diffusion equation[2]:

$$\dot{u}(x,t) = \nu\nabla^2 u(x,t) \tag{8}$$

with

$$u(x = 0, t) = u(x = L, t) = 0. \tag{9}$$

---

[2]This picture seems to describe the process satisfactorily at large times since even though the initial number of defects is large, their number rapidly diminishes. For very few defects we then look at only one and approximate the presence of the others by absorbing fixed walls.

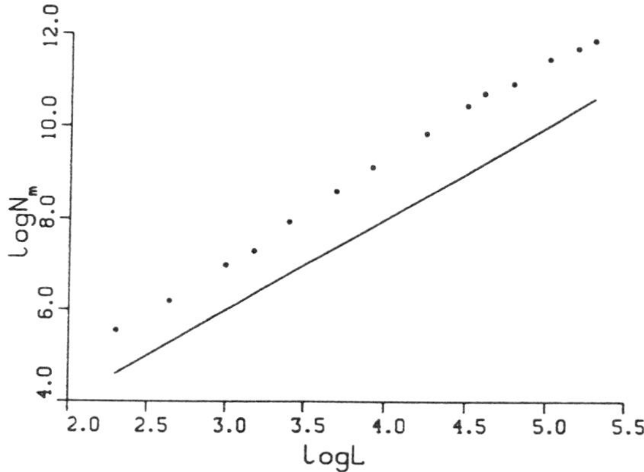

Figure 5. Log(N_m), the mean value obtained from the distribution $P(N)$, as a function of log(L) for one-dimensional lattices and $a = 1.80$. The straight line with slope 2 is also shown.

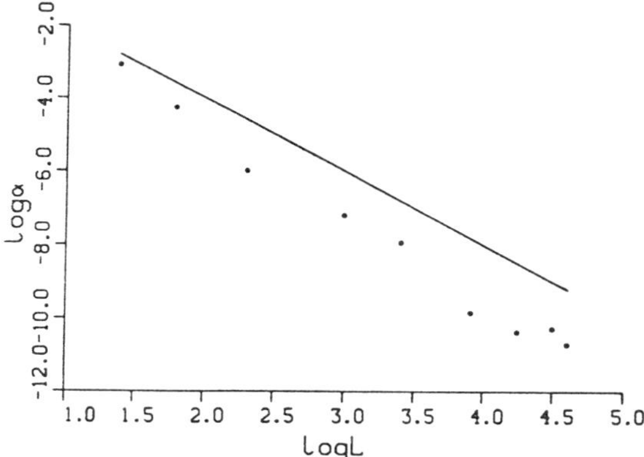

Figure 6. Log(α) (see eq. (6)) as a function of log(L) for the one-dimensional model with $a = 1.80$. The straight line with slope -2 is also shown.

If we assume a uniform distribution for the particle inside the box at $t = 0$, the solution of (8) with boundary conditions (9) is given by

$$u(x,t) = \frac{4}{\pi L} \sum_{m=0}^{\infty} \frac{1}{2m+1} e^{-\nu \frac{(2m+1)^2 \pi^2}{L^2} t} \sin \frac{(2m+1)\pi}{L} x. \tag{10}$$

Let us now look at $p(t)$ defined as follows:

$$p(t) = -\frac{d}{dt} \int_0^L u(x,t) dx. \tag{11}$$

It gives the variation in time of the probability of finding the particle inside the box or the distribution function of absorbing times of the particle, which, for large times, should approximate the distribution $P(N = t)$. Using (10) we find

$$p(t) = \frac{8\nu}{L^2} \sum_{m=0}^{\infty} e^{-\nu \frac{(2m+1)^2 \pi^2}{L^2} t} \tag{12}$$

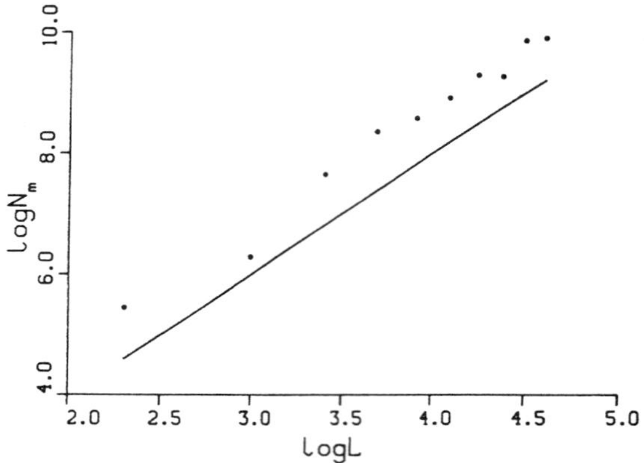

Figure 7. Same as Fig. 5 for two-dimensional lattices and $a = 1.85$.

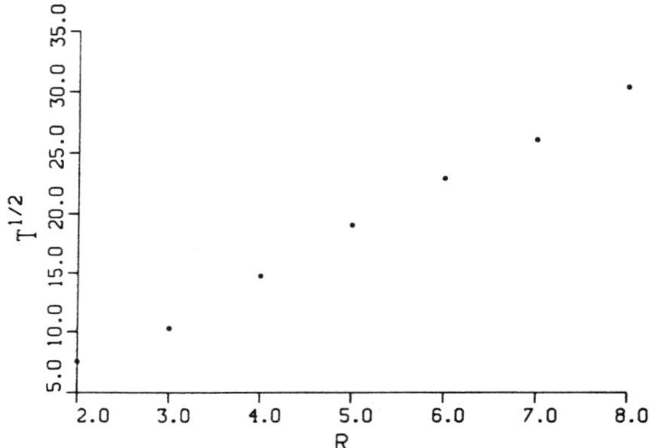

Figure 8. Square root of the average disappearance time T of created defects in the $L = 20$ two-dimensional ordered lattice with $a = 1.85$ as a function of their typical size R. The surface-tension coefficient (see eq. 15) is $\sigma = 0.032$.

and for large times, namely $t \gg L^2/\nu$, this expression becomes

$$p(t) \sim \frac{8\nu}{L^2} e^{-\nu \frac{\pi^2}{L^2} t}, \tag{13}$$

which indeed has the exponential form (6) with $\alpha \sim L^{-2}$ as found in (7). One can further estimate the diffusion constant from the prefactor in (7).

The average time t_m for absorption of the particle is given by

$$t_m = \int_0^\infty t p(t) dt, \tag{14}$$

since $p(t)$ is normalized. We find

$$t_m = \frac{1}{12\nu} L^2, \tag{15}$$

which has the form (5). The value of the prefactor should not be taken seriously since we can describe only the late stages of ordering by (8). It differs from the value estimated from (7) by a factor of approximately 2.

For the two-dimensional model, the computation of $P(N)$ is very difficult due to the computational time required in order to have a reasonable statistics. However, we show in Fig. 7 N_m as a function of L for that model (average taken only over a few values) and from the curve it seems that the L^2 behaviour is again obtained. In the two-dimensional case defects in the pattern formation are replaced by one-dimensional strings [12] with the same diffusive behaviour, which is expected to lead to the L^2 scaling.

It is interesting to discuss, once the pattern is selected, how created defects propagate through the system. In the one-dimensional case they just diffuse until they meet and annihilate and the original pattern is recovered. In the two-dimensional case we have the same behaviour if walls (opened strings) separating domains with opposite phase are created. However, we can also create islands (closed strings) and due to effects of surface tension (not present in the one-dimensional model) the time T of disappearance of those islands of typical size R follows the expected law $T \sim R^2$ as can be seen in Fig. 8 for the $L = 20$ two-dimensional lattice. From the slope of the line we can obtain the effective surface-tension coefficient σ for our coupled map model, this coefficient being defined through the relation

$$R = (2\sigma T)^{\frac{1}{2}}. \tag{16}$$

One could imagine that the average time required to obtain the selected pattern should be smaller for two-dimensional systems due to the effects of surface tension discussed above. This is not true since configurations with walls passing through the entire system give rise to very long transient times and essentially one-dimensional behaviour.

ACKNOWLEDGEMENTS

We would like to thank G. Grinstein and J. Villain for helpful discussions. CVC was supported by CNPq-Conselho Nacional de Desenvolvimento Científico e Tecnológico (Brazil) through a scholarship. TB was supported by Novo Nordisk through a Hallas-Møller scholarship.

REFERENCES

[1] Kaneko, K., Prog. Theor. Phys. 72, 480 (1984); 74, 1033 (1985).

[2] Chaté, H. and Manneville, P., Physica 32D, 409 (1988).

[3] Bohr, T., Grinstein, G., He, Y. and Jayaprakash, C., Phys. Rev. Lett. 58, 2155 (1987).

[4] Bohr, T. and Christensen, O. B., Phys. Rev. Lett. 63, 2161 (1989).

[5] Crutchfield, J. P. and Kaneko, K., Directions in Chaos Vol.1, ed. Hao Bai-Lin, World Scientific (1987).

[6] Bohr, T., Applications of Statistical Mechanics and Field Theory to Condensed Matter, ed. A. R. Bishop, D. Baeriswyl and J. Carmelo, Plenum.

[7] Tél, T., Directions in Chaos Vol.3, ed. Hao Bai-Lin, World Scientific (1990).

[8] Kaneko, K., Physica 34D, 1 (1989).

[9] Kaneko, K. and Crutchfield, J. P., Phys. Rev. Lett. 60, 2715 (1988).

[10] Kaneko, K., Phys. Lett. 149A, 105 (1990).

[11] Gunton, J. D., Phase Transitions and Critical Phenomena Vol. 8, ed. C. Domb and J. L. Lebowitz.

[12] Kaneko, K., Physica 37D, 60 (1989).

[13] Bergé, P., Pomeau, Y. and Vidal, C., Order within Chaos, Wiley, (1984).

[14] Eckmann, J. -P. and Ruelle, D., Rev. Mod. Phys. 67, 617 (1985).

FRACTIONAL STATISTIC and PLANAR BROWNIAN WINDING

Stéphane Ouvry

Division de Physique Théorique[1]
IPN, Orsay Fr-91406, and
LPTPE, Tour 16
Université Paris 6, France

The computation of the second virial coefficient of a gas of anyons and the derivation of the probability $\mathcal{P}(n)$ for a planar brownian curve to wind n times around a given point are considered. The joint probability $\mathcal{P}(n, A)$ for the curve to wind n times and enclose an algebraic aera A is also studied and a lower bound on the arithmetic aera of the curve is given.

1. INTRODUCTION

Two-dimensional systems of charged flux tubes, called anyons, define non standard statistics that has been interpreted by some authors[1,2] as exotic or fractional, since they interpolate between bosons and fermions. The main step[1] was to recognize through a reformulation of the problem of quantum statistics in terms of path integrals the importance of the topological properties of multiconnected configuration spaces in spatial dimensions lower or equal to 2 (see Wu in ref. (2) for a braid group formulation). Charged flux tubes yield a simple quantum mechanical realization[2] of these statistics. From a point of view of field theory, such objects can be described semiclassically since they arise as classical solutions of a Higgs model in 2+1 dimensions with topological Chern-Simons terms[3]. It has been shown recently[4] that the anyon Hamiltonian follows from the quantum action of charged particles coupled to a statistical gauge field

---

[1] *Laboratoire des Universités Paris 11 et Paris 6 associé au CNRS*

Growth and Form, Edited by M. Ben Amar *et al.*
Plenum Press, New York, 1991

with a purely topological Chern-Simons dynamics. However the complete derivation of the statistical mechanics of a gas of anyons is still an open problem. A mean field approximation (valid for a high anyonic density) exhibits an anyonic superconductivity behaviour near the fermionic statistics. It is not a surprise that both braid group and Chern-Simons theories play an important role in this context since the realization by Witten[5] of Jones polynomials (that caracterize braid groups) in terms of correlation functions in the topological Chern-Simons field theory.

In the following [6,7] we first propose a simple method to compute the second virial coefficient for a low density gas of anyons. We then compute the probability distribution for a closed planar brownian curve to wind n times around a prescribed point and enclose a given algebraic area. Interestingly enough, this quantity is shown to be related to the second virial coefficient of a gas of anyons.

The material presented here is basically contained in a review presented at the 4th Annecy meeting on Theoretical Physics (March 1990)[8].

We consider N identical particles of mass m_o and charge e carrying a flux tube ϕ and living in a two-dimensional space. The N-body Hamiltonian reads

$$2m_o H = \sum_{i=1}^{N} (\vec{p}_i - \alpha \vec{A}_i)^2 \tag{1}$$

where $\vec{A}_i = \sum_{j \neq i} \vec{k} \times \vec{r}_{ij}/r_{ij}^2$ and $\vec{k}$ is the unit vector perpendicular to the plane. One has the property that $\vec{A}_i$ can be gauged away by a singular gauge transformation $\vec{A}_i = \vec{\partial}_i \sum_{k<l} \theta_{kl}$.

In the two-body case the relative Hamiltonian takes the form

$$m_o h_r = -\partial_r^2 - \frac{1}{r}\partial_r - \frac{1}{r^2}\partial_\theta^2 + 2i\frac{\alpha}{r^2}\partial_\theta + \frac{\alpha^2}{r^2} \tag{2}$$

Parametrizing the relative wave function $\psi(r,\theta) = \exp(im\theta)f(r)$ where m is even for bosons and odd for fermions leads to the Schrödinger equation

$$m_o E f(r) = (-\partial_r^2 - \frac{1}{r}\partial_r + (m-\alpha)^2 \frac{1}{r^2})f(r) \tag{3}$$

The spectrum of h_r is continuous and $\psi(r,\theta) = (\frac{k}{2\pi})^{1/2} e^{im\theta} J_{|m'|}(kr)$ where $m' = m - \alpha$. One can gauge away the gauge potential in the Hamiltonian. The gauge transformed wave function now reads

$$\psi'(r,\theta) = e^{-i\alpha\theta}\psi(r,\theta) = (\frac{k}{2\pi})^{1/2} e^{im'\theta} J_{|m'|}(kr) \tag{4}$$

It is a solution of the free Hamiltonian but with a shifted angular momentum m', which is no more an integer. It follows that under the interchange of the two particles the wavefunction picks up a phase $\exp(i\pi\alpha)$, a characteristic of fractionnal statistics. Thus in the singular gauge the theory is free (the magnetic field vanishes everywhere whereas in the regular gauge it is infinite on each particle) but the interaction shows up in the multivaluedness of the wave function. This feature is directly generalizable to the N-body case.

If an harmonic potential term $m_o(\omega r/2)^2$ is added the spectrum becomes discrete and a standard calculation[1] yields in the bosonic case (m even) the spectrum (energy, degeneracy) $((2j + 1 + \alpha)\omega, j + 1)$ and $((2j + 1 - \alpha)\omega, j)$ where j is a positive or null integer and α has been assumed to lie in the interval $[0, 2]$ without loss of generality (in the fermionic case, m odd, α has to be shifted by 1). This spectrum indeed interpolates between the Bose $\alpha = 0$ and Fermi $\alpha = 1$ cases. If one adds[6] an external orthogonal magnetic field B to the system, one obtains the energy levels $(2n + 1 + (|m| \pm \alpha)(1 \pm B/B'))eB'/2m_o$ with n integer ≥ 0 and where the $\pm$ signs refer to $m \geq 0$ or $m < 0$ accordingly $(B'^2 - B^2 \equiv (2m_o\omega/e)^2)$. In that case the infinite degeneracy of the Landau levels is completely lifted.

2. THE SECOND VIRIAL COEFFICIENT OF A GAS OF ANYONS

The equation of state for a real gas expanded in powers of the density ρ reads $PV = \frac{N}{\beta}(1 + a_2\rho + a_3\rho^2 + \cdots)$ where the a_i's are the virial coefficients. The computation of $a_2(T)$ alone requires the knowledge of relative two body interactions. It is defined as $a_2 = -b_2/b_1^2$ where $Vb_2 = 2Tr\exp(-\beta H_2) - [Tr\exp(-\beta H_1)]^2$ and $Vb_1 = Tr\exp(-\beta H_1)$. This expression follows from a low density approximation[9] for the grand partition function.

For a gas of free bosons or fermions one has $a_2 = \mp\frac{\pi\beta}{2m_o}$. Therefore

$$a_2 = \mp\frac{\pi\beta}{2m_0} - 2V\frac{\Delta Tre^{-\beta H_2}}{[Tre^{-\beta H_1}]^2} \tag{5}$$

where $\Delta Tre^{-\beta H_2}$ stands for the interacting part of the two-body partition function. Giving a non ambiguous meaning to $\Delta Tre^{-\beta H_2}$ is not not a trivial task since the spectrum is continuous and thus a regularization is needed (for example putting the system in a box as in ref.(10)). However we have seen that in the presence of an harmonic well regulator the spectrum becomes entirely discrete. The point is that in the limit ω goes to 0 (which ensures the transition to a continuous spectrum) one can obtain directly a finite result for the second virial coefficient, as follows[6].

In the case of an harmonic regulator, since $Tre^{-\beta H_1} = Tre^{-\beta H_{CM}}$ and since in the limit $\omega \to 0$ one has that $Tre^{-\beta H_1} = \frac{m}{2\pi\beta}V$ one can rewrites (5) as

$$a_2^\omega = \mp \frac{\pi\beta}{2m_o}(1 \pm 8\Delta Z^\omega) \tag{6}$$

Here ΔZ^ω stands for the interacting part of the regularized relative two-body partition function $Z^\omega(\alpha)$. The latter is defined as $Z^\omega(\alpha) = \sum_{j=0}^{\infty}\left((j+1)e^{-\beta\omega(2j+1+\alpha)} + je^{-\beta\omega(2j+1-\alpha)}\right)$ and reads

$$Z^\omega(\alpha) = \frac{1}{2}\frac{\cosh\beta\omega(\alpha-1)}{\sinh^2\beta\omega} \tag{7}$$

ΔZ^ω must vanish when the electromagnetic interaction between the anyons is switched off thus it must be defined as (boson, fermion)

$$\Delta Z^\omega = Z_\omega(\alpha) - Z_\omega(0) \tag{8a}$$

$$\Delta Z^\omega = Z_\omega(\alpha+1) - Z_\omega(1) \tag{8b}$$

Thus the second virial coefficient can be computed as $a_2 = \lim_{\omega\to 0} \mp\frac{\pi\beta}{2m_o}(1 \pm 8\Delta Z^\omega)$ that leads straightforwardly to the known result[10]

$$a_2 = -\frac{\pi\beta}{2m_o}(1 + 2\alpha(\alpha-2)) \tag{9a}$$

$$a_2 = \frac{\pi\beta}{2m_o}(1 - 2\alpha^2) \tag{9b}$$

Another point is that it is possible to recover this result in the continuum, by computing the two-point Green's function $G(\vec{r},\vec{r}',\beta) = < \vec{r}'|exp(-\beta h_{rel})|\vec{r}>$ or $G(\vec{r},\vec{r}',\mathcal{M}) = \int_0^\infty d\beta G(\vec{r},\vec{r}',\beta)exp - (\beta\mathcal{M}^2)$ (its Laplace transform). In terms of Green's functions, the second virial coefficient takes the form (details of this computation are given in ref.(6)): $a_2 = \mp\frac{\pi\beta}{2m_o}\left(1 \pm 4\int d^2\vec{r}(\Delta G(\vec{r},\vec{r}) \pm \Delta G(\vec{r},-\vec{r}))\right)$.

3. THE WINDING OF PLANAR BROWNIAN CURVES

From now on we are going to use an approach developped by Edwards [11] in the context of polymer modelisation, namely exploit the properties of the brownian winding in a functional scheme. Probabilistic problems will turn out to be linked to quantum mechanical spectrum determinations.

The Wiener integral representation of the transition probability for a brownian particle to start from $\vec{r}'$ and end at $\vec{r}''$ after a "time" τ is

$$P(\vec{r}'',\vec{r}') = \frac{1}{2\pi\tau}\exp-\frac{(\vec{r}''-\vec{r}')^2}{2\tau} = N\int_{\vec{r}(0)=\vec{r}'}^{\vec{r}(\tau)=\vec{r}''}\exp\left(-\frac{1}{2}\int_0^\tau \dot{\vec{r}}^2(s)ds\right)[D\vec{r}] \tag{10}$$

where N is a normalization factor. In these units the average end to end square distance between $\vec{r}'$ and $\vec{r}''$ is $\langle [\vec{r}(\tau) - \vec{r}(0)]^2 \rangle = 2\tau$.

Let us first ask the question what is the probability for a closed brownian curve to wind n times around a given point? We start from the path integral transition probability $P(\vec{r}'', \vec{r}')$ and impose a constraint expressing that the total angle wound at time τ around O is $2\pi n$, namely $n = \frac{1}{2\pi} \int_0^\tau \dot{\theta} ds$ where θ is the polar angle between $\vec{r}'$ and $\vec{r}''$ around the point O. After inserting the Kronecker constraint $\delta_{n, \frac{1}{2\pi} \int_0^\tau \dot{\theta} ds} = \int_0^1 d\xi \exp i\xi 2\pi(n - \frac{1}{2\pi} \int_0^\tau \dot{\theta} ds)$ we get for $P(\vec{r}', \vec{r}', n)$

$$N \int_0^1 d\xi e^{i2\pi\xi n} \int_{\vec{r}(0)=\vec{r}'}^{\vec{r}(\tau)=\vec{r}'} [D\vec{r}] e^{-\int_0^\tau \left(\frac{\dot{r}^2(s)}{2} + i\xi\dot{\theta} \right) ds} \tag{11}$$

The "action" appearing now in the path integral describes a particle of unit charge and mass moving in a vortex field localized at the origin and carrying a flux $\phi = -2\pi\xi$. If we assume that the brownian curve can wander everywhere in the plane, the resulting probability for it to wind n times around the origin reads $P(n) = \int_{plane} P(\vec{r}', \vec{r}', n) d^2\vec{r}'$ which indeed involves the partition function $Z(\xi)$ of a particle in a vortex with for "temperature" $1/kT = \tau$. Thus one has

$$P(n) = N \int_0^1 d\xi e^{i2\pi\xi n} Z(\xi) \tag{12}$$

It is striking that the corresponding Hamiltonian can be viewed as the two body relative Hamiltonian of a system of anyons carrying both an electric charge and a magnetic flux and interacting with an harmonic force. As we have seen above, the partition function $Z(\xi)$ diverges and a suitable regularisation is to assume that the particle is attracted to the point O by an harmonic force. The total partition function (boson + fermion) then follows from (8) and reads

$$Z_\omega(\xi) = \frac{\cosh \tau\omega(\xi - \frac{1}{2})}{2 \sinh \tau\omega \sinh(\frac{\tau\omega}{2})} \tag{13}$$

By a proper normalization $\mathcal{P}_\omega(n) = P_\omega(n) / \sum_{n=-\infty}^{n=+\infty} P_\omega(n)$ one finally obtains

$$\mathcal{P}_\omega(n) = \frac{2\tau\omega \sinh(\frac{\tau\omega}{2})}{\cosh(\frac{\tau\omega}{2})(\tau^2\omega^2 + 4\pi^2 n^2)} \tag{14}$$

a result already known in the literature[12].

When ω is infinitesimal $\mathcal{P}_\omega(n)$ behaves as $\frac{\tau^2\omega^2}{4\pi^2 n^2}$ and $\mathcal{P}_\omega(0)$ as $1 - \frac{\tau^2\omega^2}{12}$. In the limit $\omega \to 0$ one thus has $\mathcal{P}(0) = 1$. This result reflects the fact that for $\omega = 0$, the particle is no more attached to the origin, thus a typical closed curve will have a vanishing winding number. The set of curves with a non vanishing winding number being of zero measure, it thus follows that $\mathcal{P}(0) = 1$. It would be interesting to distinguish in the $n = 0$ sector, curves which do not enclose the origin from curves which do enclose the origin but an equal number of times clockwise and anticlockwise. Such a distinction can not however be reached within the scope of this analysis [13]. However when ω is very small one gets an $1/n^2$ behaviour for the probability of n windings around a given fixed point.

Our results can be given a simple interpretation if we adopt the following point of view: the integration $\int d^2\vec{r'}$ used in calculating $P(n)$ means counting all the closed curves that begin at every point $\vec{r'}$ and wind n times around the origin. Equivalently, it means considering all the closed curves beginning at a given fixed point, and counting all the points of the plane that are wound around n times by those curves. Thus, for each closed curve, we can define different winding sectors, each one being labelled by the winding number n of any point inside it. Calling S_n the arithmetic area of the n-sector, it follows that $\lim_{\omega \to 0} P_\omega(n) \propto < S_n >$ where one averages over all possible closed curves of a given average length. It follows that $< S_n >= \frac{c}{n^2}$ $(n \neq 0)$, where c is a proportionality constant yet to be determined.

One way is to calculate the probability $\mathcal{P}_\omega(n, A)$ for the curve to enclose a given algebraic area A and wind n times around a given point. Imposing simultaneously the constraints $\delta(n - \frac{1}{2\pi} \int_0^\tau \dot\theta ds)$ and $\delta(A - \frac{1}{2} \int_0^\tau \vec{r} \times \dot{\vec{r}}.\vec{k} ds)$ (for the latter one uses the identity $2\pi\delta(x) = \int_{-\infty}^{+\infty} \exp(i\lambda x)d\lambda$) we have to compute the partition function of a charged particle moving in a vortex field (flux : $-2\pi\xi$) placed at the origin and a uniform magnetic field, $+\lambda$, perpendicular to the plane. Moreover, we use the same harmonic well regulator as before. Thus all what is needed is the spectrum determined in section 2. Following the same approach as for $\mathcal{P}_\omega(n)$, we get

$$P_\omega(n, A) = N \int_0^1 d\xi \int_{-\infty}^{+\infty} d\lambda \exp(i2\pi\xi n - i\lambda A) Z_\omega(\lambda, \xi) \tag{15}$$

Defining $N \, X_\omega(n, A) = P_\omega(n, A)$ the normalized probability thus reads $\mathcal{P}_\omega(n, A) = X_\omega(n, A)/\mathcal{N}$ where $\mathcal{N} = \sum_n \int dA X_\omega(n, A) = \frac{\pi}{\sinh \omega\tau} \coth \frac{\omega\tau}{2}$.

We can calculate exactly the quantity $X_\omega(n, A)$ as well as its limit $X(n, A)$ when $\omega \to 0$. Details of this rather tedious computation are given in Ref.(7). We can show that, for $n \neq 0$, the quantity $X(n, A)$ is finite and positive and it becomes infinite for $n = 0$, that is in agreement with the discussion given above.

Moreover, in the limit $\omega \to 0$, an interesting relationship holds:

$$\sum_{n=-\infty}^{+\infty} nX(n,A) = \frac{1}{\tau}A\mathcal{P}(A) \qquad (16)$$

where $\mathcal{P}(A)^{14}$ is the normalized probability for the closed brownian curve to enclose A. Note that this equation has at the level of the partition function the following counterpart

$$\lim_{\omega \to 0}\left[\frac{\partial}{\partial \xi}Z_\omega(\lambda,\xi)|_{\xi=0} = -\frac{2\pi}{\tau\mathcal{N}}\frac{\partial}{\partial \lambda}Z_\omega(\lambda,\xi=0)\right] \qquad (17)$$

Following (16), we can define the new quantity $< S(n,A) > = \tau\frac{1}{\mathcal{P}(A)}X(n,A)$ which satisfies $\sum_{n=-\infty}^{+\infty} n < S(n,A) > = A$. It follows $< S(n,A) >$ is the mean value of the arithmetic area of the n-sector ($n \neq 0$), the total algebraic area enclosed by the curve being fixed and equal to A. Note that we can calculate[7], A being fixed, the average of the total arithmetic area $\sum_{n\neq 0} < S(n,A) >$ enclosed by the curve, except for the zero-winding sector. Thus the use of simultaneous conditions on n (winding number) and A (algebraic area) gives access to some information on the arithmetic area enclosed by the brownian curve. Finally, averaging $< S(n,A) >$, over A allows to compute $< S_n > = \int dA\mathcal{P}(A) < S(n,A) >$ leading to $c = \tau/2\pi$: the mean area of the n-sector ($n \neq 0$) is completely determined.

References

1. J.M. Leinaas and J.Myrheim, Nuovo Cimento B37 (1977) 1; J.M. Leinaas, Nuovo Cimento A 47 (1978) 1; M. G. G. Laidlaw and C. M. de Witt, Phys. Rev. D3 (1971) 1375

2. F. Wilczek Phys. Rev. Letters Vol 49 (1982) 957; for a recent review on the subject see R. Mackenzie and F.Wilczek, Int. J. Mod. Phys. 3 (1988) 2827; Y.S.Wu, Proc. 2nd Symp. Foundations of Quantum Mechanics, Tokyo 1986, p 171-180 and Phys. Rev. Letters Vol 53 (1984) 111

3. H.J. de Vega and F.A. Schaposnik Phys. Rev. Letters Vol 56 (1986) 2564; S.K. Paul and A. Khare Phys. Lett. 174B (1986) 420

4. G.Semenoff Phys.Rev. Letters Vol 61 (1988) 517; Y-G. Chen, F. Wilczek, E. Witten and B.I. Halperin, Intern. J. Mod. Phys. B3 (1989) 1001

5. E. Witten Comm. Math. Phys. 121 (1989) 351; V.F.R. Jones Bull. AMS 12 (1986) 103

6. A. Comtet, Y. Georgelin and S. Ouvry J. Phys. A : Math. Gen. **22** (1989) 3917-3925; when spin effects are included see T. Blum, R. C. Hagen and S. Ramaswamy

Phys. Rev. Lett 64 (1990) 709

7. A. Comtet, J. Desbois and S. Ouvry J. Phys. A : Math. Gen. **23**(1990) 3563

8. S. Ouvry, Nucl. Phys. B (Proc. Suppl.) 18B (1990) 250

9. B. Khan and G. E. Uhlenbeck, Physica 5 (1938) 399

10. D.P. Arovas, R. Schrieffer, F. Wilczek and A. Zee Nucl. Phys. B251 (1985) 117; J.S. Dowker, J. Phys. A18 (1985) 3521

11. S.F. Edwards Proc. Phys. Soc. **91** (1967) 9

12. F. W. Wiegel 1977 J. Chem. Phys. **67** n^o2 469-472

13. This problem has been recently solved by M. Yor (private communication)

14. P. Levy, Processus stochastiques et mouvement brownien, Paris 1948; M. G. Brereton and C. Butler C 1987 J. Phys. A : Math. Gen. **20** 3955-68; D. C. Khandekar and F. W. Wiegel F W 1988 J. Phys. A : Math. Gen. **21** 563; B. Duplantier 1989 J. Phys. A : Math. Gen. **22** 3033-3048

CRYSTALLIZATION AND CONVECTION IN COOLING MAGMA CHAMBERS

Geneviève Brandeis

Laboratoire de Dynamique des Systèmes Géologiques
Institut de Physique du Globe de Paris
4, Place Jussieu. 75252 Paris Cedex 05 - France

INTRODUCTION

Rocks that we see at the Earth's surface are crystalline assemblages and some of them result from solidification of a liquid called magma. Undoubtedly, the most spectacular flows of magma in Nature are lava flows that occur over kilometers on the slopes of volcanoes during eruptions (Williams and McBirney, 1979). Although several km^3 are sometimes erupted during an eruption, much larger volumes of lava are stored under volcanoes in reservoirs that are called magma chambers. Such reservoirs are situated at shallow depth (few km) under the Earth surface.

There are several difficulties to overcome for geophysicists that study solidification problems in these liquids. First, very few direct observations can be made on these processes because most chambers are buried deep into the Earth. Second, cooling occurs on very long time-scales compared to the life of a geophysicist. For example, it takes ~1 year to grow a crystal in quasi-equilibrium conditions to its natural size of 1mm. It also takes ~30 years to solidify a 100m thick lava lake (Helz, 1987) that cools at the Earth surface (this is a small chamber that has been formed during a volcanic eruption by filling up a depression). Another specific problem in geophysics is the large dimensions of the object that is studied. For example, the chamber under the Kilauea Volcano (Hawaii) has been inferred to have an horizontal dimension of ten kilometers and a vertical dimension of a few kilometers (Ryan et al, 1981). For all these reasons, there cannot be any direct methods of observations of the solidification processes in geology. Therefore, theoretical models have to be coupled with numerical and fluid dynamical experiments. It should also be mentioned that, in order to understand the cooling of these reservoirs of large dimensions, the problem of solidification is not studied microscopically.

The understanding of the dynamics of magma chambers during solidification is essential to the prediction of volcanic eruptions (Blake, 1984; Tait et al, 1989) and to the interpretation of complex structures observed in fossil magma chambers. These were active magma chambers some millions years ago. They are now completely solidified and erosion has brought them to the Earth's surface, for the great luck of geologists. The study of these reservoirs is also very important as these rocks are often associated with ore deposits such as platinum. For example, the largest platinum deposit of the world is observed in the Bushveld (South

Africa), which is the largest fossil chamber on Earth. The mechanisms of deposition that form these layers are still not clearly understood.

In this paper, the solidification and processes that occur on large scales during cooling of magma chambers are discussed. Beforehand, we need to know what is the liquid we are dealing with and what are its properties. The paper is divided into three parts. In the first part, a brief introduction to magmas is made and their specific characters are outlined. In the second one, cooling characteristics and possible mechanisms occurring during solidification are described. In the last part, I illustrate the specific characters of magmas in discussing the interaction of crystallization with convection in cooling magma chambers.

A BRIEF INTRODUCTION TO MAGMAS

A magma is a molten rock and can be either entirely liquid or a mixture of liquid and crystals, and possibly gases. It is a multi-component system, essentially made of oxides. The composition is not fixed and varies from a place to another. In decreasing order of abundance, a magma is made of SiO_2, Al_2O_3, MgO, FeO, CaO, Na_2O, K_2O and other minor constituents. The major component is SiO_2, whose concentration varies enormously from one magma to another, from 44% to 73% (McBirney, 1984).

Physical properties

Magma physical properties depend strongly on composition. The temperature of fusion T_L (or liquidus) varies between 1400°C and 900°C for different compositions, the SiO_2-rich magmas having the lowest temperatures T_L. Furthermore, addition of only slight amounts (<2%) of H_2O drastically reduces the melting point by ~100°C (Shaw, 1965).

For the fluid dynamical processes that occur in magma chambers, two properties are essential. Magmas are very viscous fluids, since their dynamic viscosities μ can vary between 10 to 10^6 Pa.s, depending on composition (Ryan and Blevins, 1987). The more viscous ones are the more enriched in SiO_2. The density also varies with composition, but less strongly than viscosity, between 2.2-2.9g/cm^3 for the liquid and 2.5-3.3g/cm^3 for the crystals (Murase and McBirney, 1973). Both density and viscosity vary with temperature.

Crystallization behavior

Since magmas are multi-component systems, it can be expected that the liquidus (T_L, temperature at which the first crystalline phase appears) is not equal to the solidus (T_S, temperature at which the last liquid crystallizes). The melting range (T_L-T_S) is large and ~200°C. Magmas crystallize different mineral phases, and phase diagrams are rather complex owing to the large number of chemical components present (Elhers, 1972). Crystals are of different mineralogical types; most silicate minerals exhibit the property of solid solution such that in the same mineralogical configuration, the composition can vary. Thus, it is expected that crystallizing magmas produce different residual liquids, depending on their initial composition.

Physical properties, such as density and viscosity, change drastically in the melting range. In particular, viscosity depends strongly on crystal content. A magma can be considered as a solid, from a dynamical point of view, when it reaches a critical temperature T_C for which crystal content is ~65% (Marsh, 1981). Note that T_C is about ~100°C below T_S or half temperature between T_L and T_S. This is important for the dynamics of the chamber. It must also be noted that a magma is a Newtonian fluid when it is above the liquidus, and becomes non-newtonian for a crystal content > ~25% (Shaw, 1969).

Magmatic crystals are small, compared to the dimensions of the reservoir. A typical size is 1 mm for crystals grown in quasi-equilibrium conditions They are mostly faceted crystals. Dendritic and spherulitic crystals can be found in nature, however they are rather rare. Nucleation occurs rather heterogeneously on preexisting grains (Kirkpatrick, 1977; Dowty, 1980).

<u>What is the shape and the size of the reservoir ?</u>

A magma chamber is a 3-D reservoir where the magma, either liquid or partially crystallized, is stored. When the aspect ratio height/length is small, the chamber is called a sill. When the aspect ratio is large, it is called a dyke. In the following, only sills will be considered. The height of a sill can vary from a few meters up to 10 km, while the lateral extension can vary from a few hundred meters to hundreds of kilometers. The volume of the chamber varies from few km^3 up to ~10^6 km^3 for the largest intrusion, i.e. the Bushveld (Wager and Brown, 1968). Although the walls can be highly irregular, the chamber can be considered as a first approximation as a 1-D system (vertical dimension or depth).

COOLING CHARACTERISTICS OF MAGMA CHAMBERS

Although the volume of the chamber can be enormous, we suppose as a first approximation that the time needed to fill these chambers is negligible compared to the cooling time. Thus, the system is considered as closed. Most magma chambers are observed in the shallow part of the crust where rocks are entirely solid, and cooler than the magma. Therefore, the magma is cooled from all sides. Note that there is no heating from bottom. Since the magma is hotter than the surrounding rocks, cooling is a transient process. The only equilibrium state is when the magma has reached the temperature of the matrix, and has, thus, completely solidified.

Suppose that the magma is initially isothermal at a temperature T_0 $\geq T_L$. Its crystal content is initially 0. Immediately after intrusion, steep thermal gradients develop rapidly by conduction in the magma and the surrounding rocks. The temperature at the contact between the matrix and the magma is instantaneously equal to $(T_0+T_r)/2$ where T_r is the matrix temperature (Jaeger, 1968). Hence, the contact is at a subsolidus temperature and magma crystallizes along the margins. Since the magma is not a pure material, there is a given thickness over which the magma is partially crystallized. It is defined as the mush zone, in which the crystal content varies, say, between 1% to 99%. Associated with crystal content variations, there are strong viscosity variations in this zone, between that of the magma and that of a solid (>10^{18}). Suppose that for a while heat transfer is by conduction. Due to the importance of latent heat release, the mush zone is of small thickness (0.1-10m) compared to that of the whole reservoir (Brandeis and Jaupart, 1987). This mush zone moves inward into the interior of the chamber as cooling proceeds.

Since the magma is much hotter than the surrounding rocks and the reservoir is of large thickness, it can be expected that convection occurs in the chamber (Bartlett, 1969). The understanding of convection in magma chambers has been considerably modified by the recent introduction of fluid dynamics in geology. However, due to the complexities of the highly non-linear physics, no definite picture can yet been drawn for reasons that are explained in the following.

The intensity and the patterns of the convective fluid motions depend on two dimensionless numbers, i.e. the Prandtl number and the Rayleigh number. The Prandtl number is equal to

$$Pr = \nu / \kappa \qquad (1)$$

where ν is kinematic viscosity and κ thermal diffusivity. Since magmas are viscous fluids, Pr is large ($>10^3$). The temperature profile is not the classical one of Rayleigh–Bénard (cooled from above, heated from below), however, we can still define a Rayleigh Number as:

$$Ra = \rho.g.\alpha.\Delta T_0.d^3 / (\mu.\kappa) \qquad (2)$$

where ρ is density, α thermal expansion coefficient, ΔT_0 the potentially driving temperature contrast (i.e. difference between the initial temperature T_0 and the critical temperature T_c), d the thickness of the reservoir, and μ the dynamic viscosity of the magma at T_0. With typical magmatic values ($\alpha=5\times10^{-5}{}^{\circ}C^{-1}$, $\kappa=10^{-6}m^2s^{-1}$), the Rayleigh number is always very large. It is $> 10^{16}$ for a low viscosity magma ($\mu=10Pa.s$) in a reservoir of thickness d=1km, assuming $\Delta T_0=100^{\circ}C$. Convection is surely 3-D time-dependent (chaotic), and perhaps turbulent. However, convection has not been studied in these regimes, neither in the laboratory nor numerically, and the patterns of convection cannot be predicted unmistakably (Krishnamurti, 1970). Furthermore, note that Ra is $> 10^{14}$ for $\Delta T=1^{\circ}C$. Does it mean that only $1^{\circ}C$ temperature contrast can lead to turbulent convection ? It seems paradoxical that such a small temperature contrast can lead to turbulence. We will discuss this paradox in the last section.

Consider the development of convection in the chamber. The upper boundary layer goes unstable and descending plumes are generated. As an exemple, Fig.1 shows photographs from a laboratory experiment illustrating the development of a field of plumes at the upper boundary of a layer of fluid cooled suddenly (from Jaupart and Brandeis, 1986). These experiments were performed in a high-Pr and constant viscosity fluid. There are no rising plumes and the upwards motions are diffuse. In

Fig.1. Photograph of developing thermal convection in a layer of silicone oil cooled from above and below (from Jaupart and Brandeis, 1986). At the top boundary, cooling generates cold plumes wich move downward.

the chamber, heat is transferred in the boundary layers by conduction.
The interior is well-mixed and the interior temperature is homogeneous
and evolves with time, due to the cooling of the chamber. Furthermore,
these experiments have shown the existence of a stagnant layer at the
bottom of the reservoir. The mush in the bottom layer moves inward by
conductive cooling.

However, strong changes of viscosity at the margins, due to crystallization, strongly modify the patterns obtained at the same Rayleigh
number in a constant viscosity fluid. Recently, convection at high Ra
number with variable viscosity fluid has been studied theoretically and
experimentally in geophysics in order to understand convection in the
mantle (Richter et al, 1983; Jaupart and Parsons, 1985). Note that this
physics is highly non-linear. The upper boundary layer is divided in two
parts (Fig.2): the lower unstable one, which is able to participate to
convection and where viscosity varies no more than a factor of about 10
and the upper stable one which behaves rigidly and where the viscosity
contrasts are larger. Thus, the driving temperature contrast is much
smaller than the thermal contrast between the critical temperature and
the initial temperature (Fig.2). These studies were performed for the
classical Rayleigh-Bénard profile and much remains to be learned on
convection in variable viscosity fluid in magma chambers boundary
conditions. In particular, theory has shown that, for turbulent
convection, there is a scaling between Nu, the Nusselt number (ratio of
total flux over conductive flux at same ΔT) and Ra (Townsend, 1964;
Deardoff et al, 1969). That is, Nu is proportional to $Ra^{1/3}$. This law is
widely used in studies of convection in magma chambers, however, precise
heat flux measurements have still to be made in order to establish its
validity in high-Ra convection in high-Pr fluids.

Furthermore, these previous studies were made on non-crystallizing
fluids, hence there are two other major complexities in the problem of
cooling magma chambers. First, the kinetics of crystallization has to be
considered, since it is important to compare the time-scale for convective instability and that for crystallization (Brandeis and Jaupart,
1986). They can be of the same order of magnitude, and in that case there

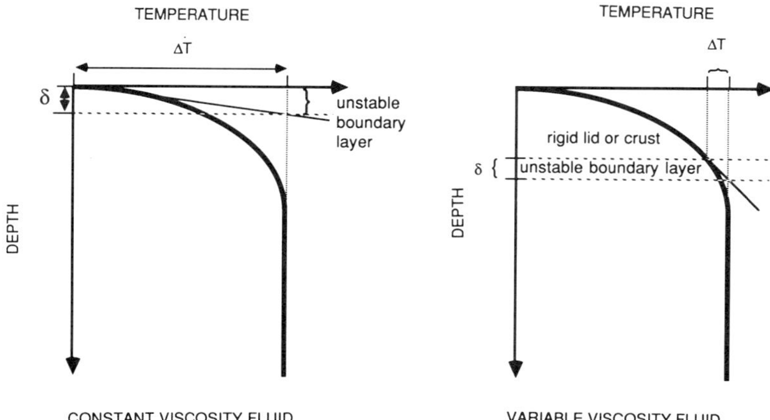

Fig. 2. Structure of the upper unstable boundary layer in case of a
constant viscosity fluid and a variable viscosity fluid. In the
second case, only part of the boundary layer is unstable. Note that
the driving thermal contrast ΔT is greatly reduced.

is some interaction between crystallization and convection. This process
is discussed in the next section. Second, since magmas are not pure materials, residual liquids are of different composition than the initial
liquid, and thus, have a different density. This may lead to development
of compositional convection (McBirney et al, 1985; Nilson, 1985) which
may explain the different chemical compositions of the rocks observed in
fossil chambers. At the bottom of the chamber, crystallization contrac-
tion is such that light residual liquid may escape from the mush. It
mixes with the parental liquid in the interior and the liquid composition
slowly evolves. Thus, the rocks that crystallize from it, can be expected
to have different compositions. This mechanism involves convection in a
porous medium as well as compositional changes associated with crystallization. The physics is also highly non-linear. Several studies have
been made on crystallization at the bottom of the chamber (Huppert and
Worster, 1985; Worster, 1986; Tait and Jaupart, 1989). Furthermore, many
metallurgical studies have been made on the unidirectional solidification
of an alloy (Hurle et al, 1983, Glicksman et al., 1986, for example).

The problem of compositional convection will not be discussed any
further in this paper, nor other secondary processes, such as crystal
settling or compaction in the mush that can occur during solidification.

INTERACTION BETWEEN CRYSTALLIZATION AND CONVECTION IN MAGMA CHAMBERS

How is the picture of high-Ra chaotic convection modified by crystallization ? In particular, can the interior temperature, noted T_i, decrease to the critical temperature T_c (temperature under which viscosity
is considered as infinite), by convection or does convection stop before

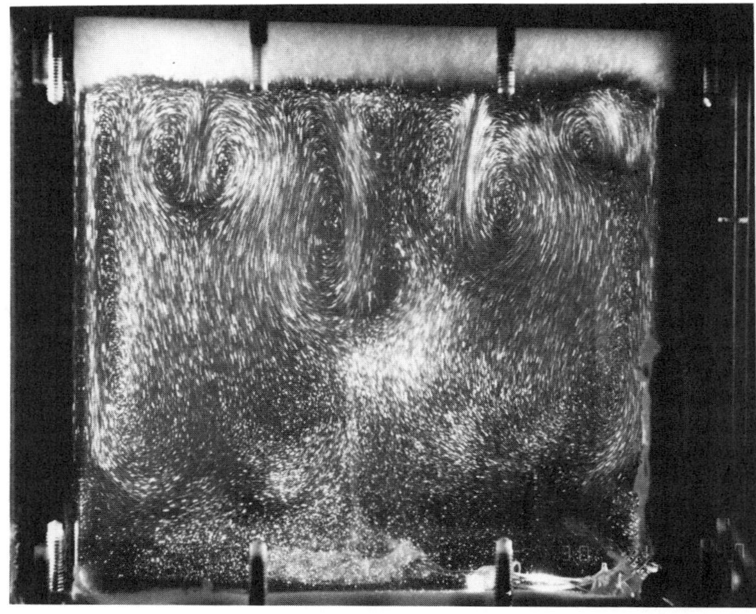

Fig. 3. Photograph of convection in a layer of paraffin cooled from above
(from Brandeis and Marsh, 1989). Note the crystallized part, i.e,
the crust (in white) and the downgoing plumes below the crust

reaching T_c? What is the effective driving temperature contrast ? Is convection turbulent or not ? All these questions are currently the subjects of strong debates in the geological literature (Kerr et al, 1989; Marsh, 1989; Brandeis and Marsh, 1989).

In theory, fluid motions can occur as long as the fluid has not reached the critical temperature T_c. However, take a closer look at the structure of the upper unstable boundary layer. In the upper boundary layer, there are strong temperature, and thus viscosity gradients. Crystallization develops in this layer, and only part of the upper boundary layer can participate to convective motions. In theory, for low viscosity magmas, the unstable part of the upper boundary layer where this factor of 10 increase in viscosity corresponds to ~25% crystals. However, it takes a finite amount of time to achieve complete crystallization.

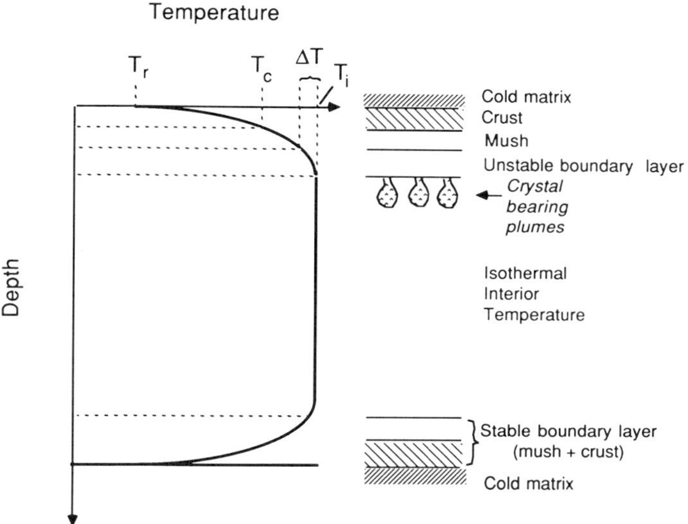

Fig.4. Temperature profile versus depth in a cooling magma chamber. The interior is isothermal at a temperature $T_i \sim T_L$, thus, ΔT is small. The unstable boundary layer develops below the mush and the crust.

Furthermore, it also requires a finite amount of time to destabilise part of this layer. Thus, the two respective time-scales have to be compared. This requires the knowledge of the kinetics of crystallization (Brandeis and Jaupart, 1986). Although it is difficult to perform measurements on real systems, due to complex phase diagrams and long times to crystallize, Brandeis and Jaupart (1987) have shown that non-dimensional analysis of the problem associated to data of crystal sizes observed at the margins of dikes, can give access to kinetic rates.

Using realistic kinetic rates and the concept of marginal convective instability, numerical experiments have shown that the effective temperature contrast driving convection is always smaller than ΔT_0, and depends on the initial viscosity and temperature of the magma (Brandeis and Jaupart, 1986). In low viscosity magmas, convective breakdown occurs before the completion of crystallization and involves partially crystallized magma. The convective regime is thus characterized by descending crystal-bearing plumes. For high viscosity magmas, crystallization completely stabilizes the upper boundary layer and the driving contrast is very small, hence thermal convection is weak.

Thus, the whole issue is whether or not some undercooled melt and/or crystals can participate to the convection. Several fluid dynamical experiments on crystallizing fluids have recently been performed in order to predict the long time-behavior of the cooling of the reservoir. In the first one, the analog fluid is paraffin (Brandeis and Marsh, 1989). Paraffin crystals are fine (hair-like) and small (~1mm) dendrites. The liquid layer is initially slightly above the liquidus and suddenly cooled from above. When the fluid is superheated, vigorous convection sets in but quickly decreases in intensity. As in the previous experiments, there is a field of convective plumes (Fig.3). Convection becomes very sluggish when the liquid reaches the liquidus. Due to the crystallization behavior, convection is restricted to the liquid part only and no crystal-bearing plumes have been seen. The time-scale for convection to decrease the interior temperature to the liquidus is equal to:

$$\tau = 14.3 \ d^2 \ \kappa^{-1} \ Ra^{-1/3} \tag{3}$$

where Ra is the calculated Rayleigh number with $\Delta T = T_0 - T_L$ (Jaupart and Brandeis, 1986; Brandeis and Marsh, 1989). Marsh (1989) and Brandeis and Marsh (1989) have suggested that the vigorous convection is restricted to magmas above the liquidus in the chamber. Using equation (3), the characteristic convective time-scale is equal to $\sim 10^8$ s for a low viscosity magma in a 1km deep reservoir. This time-scale is small compared with the total solidification time, thus, vigorous convection plays a negligible role in the cooling history of magma chambers.

Another experiment has been performed with the same boundary conditions but with aqueous isopropanol solutions (Kerr et al., 1989, 1990). The crystallization behavior of this fluid is quite different from that of paraffin. The crystals are coarse and dendritic. Furthermore, the Prandtl number is small. In these experiments, they observe turbulent convection until complete solidification. When the interior temperature reaches the liquidus, there are some disequilibrium processes associated to convective motions that allow this temperature to be below the liquidus. However, the driving temperature difference is very small ($<1°C$), once the fluid is below the liquidus. They also observe coarse crystals that form at the bottom and release light fluid, however, there are no crystal-bearing plumes in these experiments as predicted by Brandeis and Jaupart (1986).

How can we use the results of these experiments to magma chambers, since the crystallization morphologies of these analog fluids and that of real magmas are quite different? What we can surely say is that vigorous convection with large temperature contrasts is restricted to temperature above the liquidus and represents a short time in the cooling history of the chamber. There might be some crystal bearing-plumes inside the chamber, however, the driving temperature contrast remains small. Convection is part of an overall intimate balance between phase equilibria, crystal growth and heat transfer. The final picture of the temperature profile in a chamber can be represented on Fig.4 with T_i very close to T_L. We do not consider here compositional convection which may destabilise part of the lower boundary layer. Whether or not convection may be turbulent with such small temperature contrasts is still unclear. Crystal bearing plumes remain to be seen in laboratory experiments before the debate on turbulent convection can be closed.

CONCLUSION

In this paper, I have outlined the specific characters of magmas. The study of the solidification problems in geophysics has to be combined with numerical and fluid dynamic experiments. Since these problems are highly non-linear, development of computers in the future will undoubtedly improve our understanding of the solidification processes. The introduction of fluid dynamics in geology has considerably modified the understanding of solidification in magma chambers. Laboratory experiments are very helpful in gaining new insights on the problem. However, they may be limited by the difficulty in finding analog fluids which have the principal characteristics of magmas. In particular, similar crystal morphologies are probably essential in this problem.

<u>Acknowledgements</u>

This paper is the result of the fruitful collaboration with Claude Jaupart in Paris (IPGP) and with Bruce Marsh in Baltimore (JHU).

REFERENCES

Bartlett, R.W., 1969, Magma convection, temperature distribution and differentiation, <u>Am. J. Sci</u>, 267:1067-1082

Blake, S., 1984, Volatile oversaturation during the evolution of silicic magma chambers as an eruption trigger, <u>J. Geophys. Res</u>., 89:8237-8244

Brandeis, G., and Jaupart C., 1986, On the interaction between convection and crystallization in cooling magma chambers, <u>Earth Plan. Sci. Lett</u>., 77:345-361

Brandeis, G., and Jaupart C., 1987, The kinetics of nucleation and crystal growth and scaling laws for magmatic crystallization, <u>Contrib. Miner. Petrol</u>., 96:24-34

Brandeis, G., and Marsh, B.D., 1989, The convective liquidus in a solidifying magma chamber : a fluid dynamic investigation, <u>Nature</u>, 339:613-616

Deardorff, J.W., Willis, G.E., and Lilly, D.K., 1969, Laboratory experiments of non-steady penetrative convection, <u>J. Fluid Mech</u>., 35: 7-31

Dowty, E., 1980, Crystal growth and nucleation theory anf the numerical simulation of igneous crystallisation, <u>in</u> "Physics of magmatic Processes", Princeton University Press, 419-485

Ehlers, E.G., 1972, "The interpretation of geological phase diagrams", Freeman & Company, San Francisco

Glicksman, M.E., Coriell, S.R., and McFadden, G.B., 1986, Interaction of flows with the crystal melt interface, <u>Ann. Rev. Fluid Mech</u>, 18:307-335

Helz, R.T., 1987, Differentiation behavior of Kilauea Iki lava lake, Kilauea Volcano, Hawaii : an overview of past and current work, <u>in</u> Magmatic processes. Physochemical principles, Mysen, B.O., ed, <u>Geoch. Soc. Spec. Pub</u>. 1:241-258

Hurle, D.T.J., Jakeman, E. and Wheeler A.A., 1983, Hydrodynamic stability of the melt during solidification of a binary alloy, <u>Phys. Fluids</u>, 26:624-626

Huppert, H.E., and Worster, M.G., 1985, Dynamic solidification of a binary alloy, <u>Nature</u>, 314:703-707

Kerr, R.C., Woods, A.W., Worster, M.G., and Huppert, H.E., 1989, Disequilibrium and macrosegregation during solidification of a binary melt, <u>Nature</u>, 340:357-362

Kerr, R.C., Woods, A.W., Worster, M.G., and Huppert, H.E., 1990, Solidification of an alloy cooled from above. Part II : Non-equilibrium interfacial kinetics, <u>J. Fluid Mech</u>., 217:331-348

Kirkpatrick R.J., 1977, Nucleation and growth of plagioclase in the Kilauea Volcano, Makaopuhi and Alae, lava lakes, Hawaii, <u>Geol. Soc. Amer. Bull</u>, 88:78-84

Krishnamurti, R., 1970, On the transition to turbulent convection, Part 2: The transition to time dependent flow, <u>J. Fluid. Mech.</u> 42:309-320

Jaeger, J.C., 1968, Cooling and solidification of igneous rocks <u>in</u> "Basalts : The Poldervaart treatise on rocks of basaltic composition, vol. 2", Hess. H.H. and Poldervaart Arie, eds, , New York, John Wiley & Sons, 503-536

Jaupart, C. and Parsons, B., 1985, Convective instabilities in a variable viscosity fluid cooled from above. <u>Phys. Earth Planet. Int.</u>, 39:14-32

Jaupart C., and Brandeis G., The stagnant bottom layer of convecting magma chambers, <u>Earth Plan. Sci. Let.</u>, 80:183-199

McBirney, A.R., 1984, "Igneous petrology", Freeman Cooper & Co, San Francisco, 504pp,

McBirney, A.R., Baker B.H., and Nilson, R.H., 1985, Liquid fractionation. Part 1. Basic principles and experimental simulations, <u>J. Vol. Geoth. Res.</u>, 24:1-24

Marsh, B.D. 1981, On the cristallinity, probability of occurence, and rheology of lava and magma, <u>Cont. Mineral. Petrol</u>, 78, 85-98

Marsh, B.D., 1989, On convective style and vigour in sheet-like magma chambers, <u>J. Pet.</u>, 30:479-539

Murase, T., and McBirney, A.R., 1973, The properties of some common igneous rocks and their melts at high temperature, <u>Geol. Soc. Am. Bull.</u>, 84:3563-3592

Nilson, R.H., 1985, Countercurrent convection in a double-diffusive boundary layer, <u>J. Fluid Mech.</u>, 160:181-210

Richter, F.M., Nataf, H.C., and Daly, S.F., 1983, Heat transfert and horizontally averaged temperature of convection with layer viscosity variations. <u>J. Fluid Mech.</u>, 129:173-192

Ryan, M.R., and Blevins, J.Y.K., 1987, The viscosity of synthetic and natural silicate melts and glasses at high temperatures and 1 bar (100000 Pascals) Pressure and higher pressure, <u>U.S.G.S. Bull.</u>, 1764

Ryan, M.R., Koyanagi, R.Y., and Fiske, R.S., 1981, Modelling the three dimensional structure of a macroscopic magma transport systems : Application to Kilauea Volcano, Hawaii, <u>J. Geophys. Res.</u>, 86:7111-7129

Shaw, H.R., 1965, Comments on viscosity, crystal settling, and convection in granitic magmas : <u>Am. J. Sci.</u>, 263:120-152

Shaw, H.R., 1969, Rheology of basalt in the melting range, <u>J. Pet.</u>, 10: 510-535

Tait, S.R., Jaupart, C. and Vergniolle S., 1989, Pressure, gas content and eruption periodicity of a shallow, crystallising magma chamber. <u>Earth Planet. Sci. Lett.</u>, 92:107-123

Tait, S.R., Jaupart, C., 1989, Compositional convection in viscous melts, <u>Nature</u>, 338:571-574

Townsend, A.A., 1964, Natural convection in water over an ice surface, <u>Q. J. R. Meteor. Soc</u>. 90:248-259

Wager, L.R., and Brown, G.M., 1968, "Layered igneous rocks", Edinburgh, Oliver and Boyd

Williams, H., and McBirney, A.R., 1979, "Volcanology", Freeman Cooper, San Francisco

Worster, M.G., 1986, Solidification of an alloy from a cooled boundary, <u>J. Fluid Mech.</u>, 167:481-501

Nonlinear Systems far from Equilibrium

and Missing Route to "Living State"

Yasuji Sawada, Tomoaki Itayama and Mika Sato

Research Institute of Electrical Communication
Tohoku University
Sendai 980   Japan

ABSTRACT

Process of regeneration of a hydra from dissociated cell
aggregate was studied. The process induces cell-sorting
period, cavity-formation period and structure forming
period. Simultaneously, nerve network formation takes
place, apparently induced by head structure formation.
The aggregate changes from initially disordered cell
assembly to an ordered animal within a few days.   Measured quantities related to the transition are presented.
Difference between the living state and known   dynamical
states of nonlinear system far from equilibrium is discussed.

INTRODUCTION

What is the "living state"? This question has repeatedly
been asked in the history of science. Most recent waves in
physics community were triggered by irreversible
thermodynamics[1] in 70's followed by a storm of reseach in
nonlinear dynamics in 80's. It turned out, however, the state
of reseach for twenty years in this direction is still far
from understanding of living states. A question arises how
much the nonlinear physics can contribute to understanding
living states.

Hydra is known to have a strong ability for
regeneration[2]. A small hydra regenerates from a tissue taken
from an hydra body. Hydra regenerates even from an aggregate
of cells dissociated from hydra animals. Hydra has a simple
body plan. The body is essentially a tube-shaped stomach with
an opening at the top around which five or six tentacles are
regularly arranged. Hydra has relatively few kind of cells;
endodermal and ectodermal epitherial cells, which constitute

single internal and external cell walls of the body, nerve cells, undifferentiated interstitial cells, nematocytes which are found only in tentacles, and so on.

It is to be noted that the initial aggregate is a random, disorganized cell ensemble where the original positional information is completely destroyed. Each cell is living in a sense, when it is dissociate and aggregated. But the cluster of cells just aggregated is not living as an animal. Through the course of regeneration the system recovers its life. By studying the process of regeneration one might be able to observe the dynamics of achieving a living state of an animal.

In this paper we review recent progress in the study of regeneration of hydra from a dissociated cell aggregate carried out in our laboratory[3,4,5,6], and comment on our present understanding of the living state.

MATERIALS AND METHODS

The animals used were Hydra magnipapillata, standard wild- type strain 105, obtained from Dr. T. Sugiyama(National Institute of Genetics, Mishima, Japan). They were cultured in modified "M" solution(Muscatine and Lenhoff, 1965; Sugiyama and Fujisawa, 1977) at a constant temperature of $18^{\circ}C$ and fed on newly hatched brine shrimp nauplii, 6 days a week. Experimental animals were starved 1 day before use.

Dissociation and aggregation were accomplished by the method of Flick and Bode(1983). Groups of animals or tissue pieces were soaked for about 15 min in hyperosmotic medium and Dissociated by repeated pipetting. After 2 min, the cell suspension was filtered through a nylon mesh with 53 micron opening. Dissociated cells collected in fresh dissociation medium were concentrated by low-speed centrifugation. All precedures upto this step were carried out at $4^{\circ}C$, and then the aggregated cells were incubated at room temperature. This is the time origin of regeneration. The medium was gradually diluted to the normal medium as the regeneration progresses.

Nerve cells were visualized by an immunocyto-chemical method using antibodies which specifically stain nerve cells containing particular neuropeptides. Antibodies to RF-amide and Oxytosin-Vassopressin[7] were used in the present experiments. The average number density of nerve cells and the average connection number between the nerve cells were counted by eyes.

REGENERATION PROCESS

A) <u>Regeneration of Body Structure</u>

The regeneration process from a dissociated cell aggregate to a complete hydra may be divided into three stages[8]. [Fig.1]

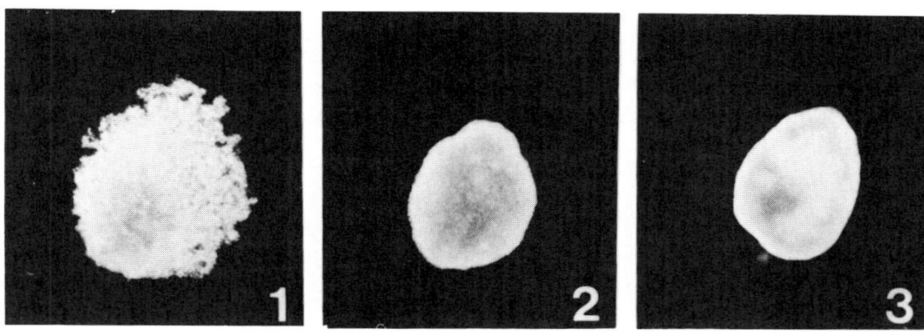
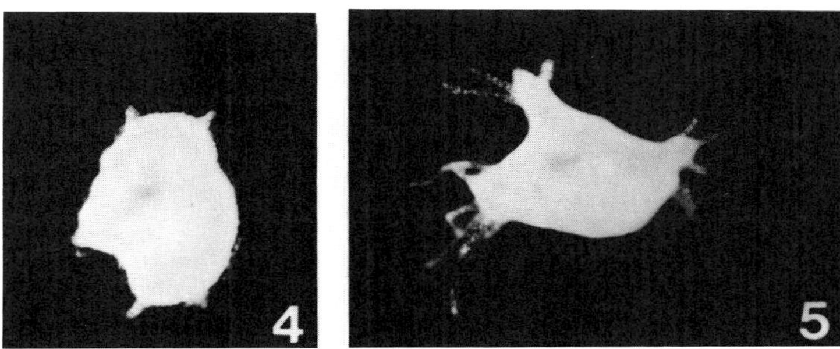

Fig.1 Series of photographs of an regenerating aggregate. Each taken at (1) 0hrs, (2) 4.5hrs, (3) 20.5hrs, (4) 2days, (5) 3.5 days from starting of regeneration.(Figure taken from reference 8)

In the first stage, cell sorting takes place among the ectodermal and endodermal epitherial cells. Through this process the ectodermal cells moves to the surface of the aggregate and form a monolayer, which is backed by the mono-cell layer of the endodermal cells, and the surface of the aggregate becomes smooth. [Fig.1(2)] The time needed for an aggregte to accomplish this stage is about 6 hrs.

The second stage is the cavity formation period.[Fig.1(3)] When the double cell layer is formed, the internal space is isolated from the external world, and some chemical potential difference, such as osmotic pressure, may be established between them. It is often observed that a clump of the cells which have been left unused for forming the double layer are vomited from inside the aggregate. By the end of this stage, 24hrs after regeneration star⁺ed, the aggregate is an approximately isotropic and homogeneous baloon of double cell layer.

In the third stage, many things happen. The scenario depends on what are the composition of cells composing the aggregate. Here we describe only the case of an aggregate in which cells from whole animals are unselectively assembled. Tentacles start appearing after two or three days almost randomly over the surface of the aggregate.[Fig.1(4)] Then the aggregate is elongated along an axis which will be the future body axis of a normal hydra.[Fig.1(5)] A mouth opening appears later at the apex of the axis. The tentacles which are located far from the opening are absorbed in the body and disappear. The tentacles which happened to be in the vicinity of the opening are gradually arranged around it.

B) Regeneration of Control Sytem

It appeared from the measurement that most of nerve cells visualized by the antibody used in the present experiments are thrown away from the aggregate in the cell-sorting and cavity-formation stages. Then the observed density starts increasing from about 50hrs. due to differentiation of the interstitial cells into nervre cells. The increment is most steep around 90hrs., then the tendency gradually saturates.[Fig.2]

Simultaneously, it was observed that the average number of connection of a nerve cells to the others increases.[Fig.3] The increment of the connection is somewhat earlier than that of the density. It might mean that the nerve cells has a tendency to have connection as soon as they are differentiated. The maximum number of connection is about five, suggesting the nearest neighbor interaction as average, due to the two dimensional arrangement of the network.

The electrical activity of the nerve systems and the epitherio-muscular cells were measured by micro glass electrodes.[Fig.4] The aggregate does not show any activity clearly distinguishable from noise for about 10 hrs from the starting time of the regeneration. Then the activity increases rapidly. At about 20hrs one finds decrease of the activity, which may be attribute to the the osmotic pressure

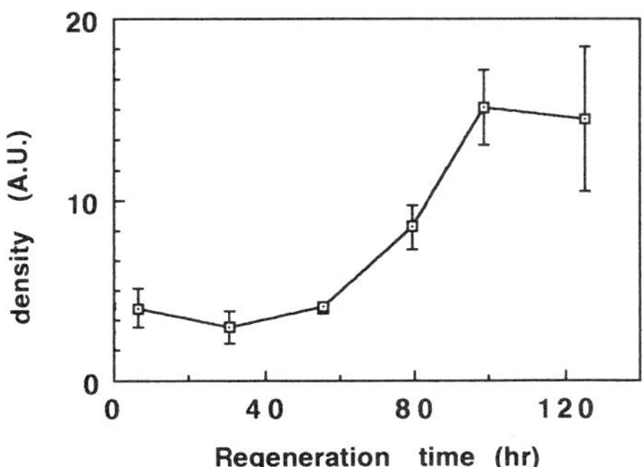

Fig.2　Number density of nerve cells as a function of regenerating time. (Figure taken from reference 7)

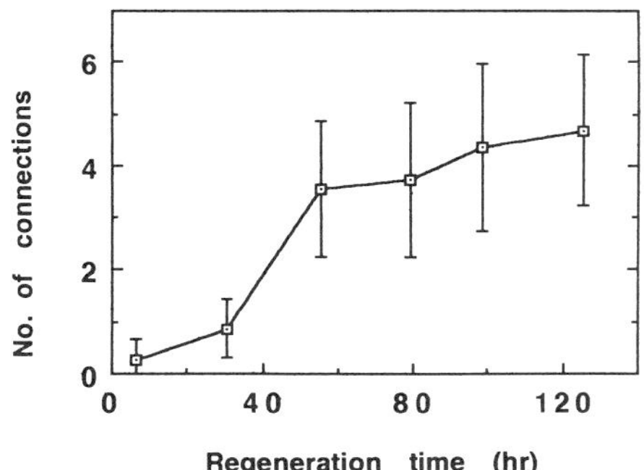

Fig.3 The average connection number of a nerve cell with the others as a function of regenerating time (Figure taken from reference 7)

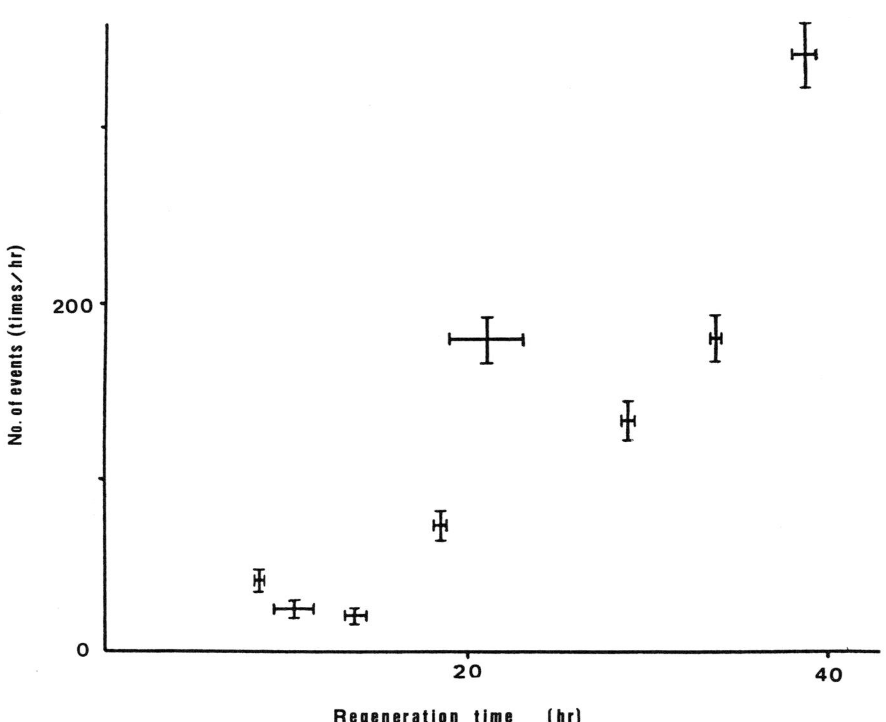

Fig.4 Number of events per hour of the electric pulses measured by a micro glass electrode as a function of regenerating time. (Figure taken from reference 9)

decrease of the aggregate due to the expelling of the internal cell clumps. Recent two points measurements have revealed that the correlation of the electrical activity at two differents points also has a sharp rise at about 15 hrs[9].

DISCUSSION

The experimental results of regeneration of a primitive animal from a dissociated cell aggregate suggested following observations.

Living state is hierarchical, as it has been well recognized. A disorganized biological system, whose constituents are individually in a living state, can be organized into a new living state one step higher in the hierarchy. Living state in a hierarchical step may be characterized by a self-organized structure, such as a digesting sac and a mouth in the present case, and by a self-organized information generator used for the control of the structure. The structure in a higher hierarchical step means ordered distribution of function of the constituents. Each function of the constituents in a lower level of the hierarchy is again achieved by a self-organized structure and information generator, necessary for supporting the living state of the constituents. The function of the constituents differs from one to the other, once they interacts each other to form a living state of a higher hierarchy.

The ordinary bifurcation route of nonlinear systems with small degree of freedom driven far from equilibrium, such as Rayleigh-Benard convection system, is;

spatial structure-->temporal oscillation-->chaos-->turbulence

There are some similarities and differences between this scenario and the regeneration scenario of biological systems. In the physical systems two kinds of structures have mainly been discussed. One is spatially periodic structure such as convectional structure and chemical structure. The other is growth structure such as dendritic crystals. Although the latter structure is localized, the boundary is constantly expanded. The pattern formation of biological systems has often been understood in terms of nonlinear reaction diffusion mechanism[6]. Yet, a stable localized structure which is seen in the biological system is not understood. It is not common to the nonlinear systems in the laboratory except for the localized convectional structure recently observed in binary mixtures and liquid crystals[10].

The activity of a nerve cell is the result of dynamics of excitable medium which is closely related to the limit cycle and chaos. The most of the nonlinear systems driven further in the laboratory would bifurcate to spatio-temporal chaos and then turbulence. Although information is constantly produced in a turbulent system, it is dissipated to heat. On the other hand the information produced by the nerve system is used to control the structure.

Therefore, the route of bifurcation in nonlinear systems to living state is missing, even if we forget for a moment the hierarchical structure which is essential to the living state. It would be interesting if one could find some nonlinear sytems which goes through a localized structure formation which functions by the temporal signals produced by an oscillatory and following bifurcations.

REFERENCE

1. P.Glansdorff and I.Prigogine, _Thermodynamic Theory of Structure, Stability and Fluctuation_(Wiley,1981)

2. See for example, H.M.Lenhoff(ed.), _Hydra:Research Method_ (Plenum,1983)

3. H.Shimizu and Y.Sawada, Transplantation Phenomena in Hydra:Cooperation of Position Dependent and Structure Dependent Factors Determines the Transplantation Result, Dev. Biol. _122_, 113(1987)

4 M.Sato and Y.Sawada, Regulation in Numbers of Tentacles of Aggregated Hydra Cells, Dev. Biol. _133_, 119(1989)

5 H.Ando, Y.Sawada, H.Shimizu and T,Sugiyama, Pattern Formation in Hydra Tissue without Developmental Gradients, Dev. Biol. _133_, 405(1989)

6. M.Sato, H.Bode and Y.Sawada, Patterning Processes in Aggregates of hydra Cells Visualized with the Monoclonal Antibody, TS19, Dev. Biol. _141_, 412(1990)

7. T.Itayama and Y.sawada, Neural Network Formation and Activity of an Aggregate of Diisociated Hydra Cell. In _Cooperative Dynamics in Complex Physical Systems_ (ed. H.Takayama, Springer-Verlag, 1988)p.262

8. Y.Sawada, T.Itayama and S.Mika, _Physics of Living State_ (Ed. Y.Sawada and T.Musha, Ohmusha to appear)

9. T.Itayama and Y.Sawada, Development of Contractile Motion and the Electrical Activity in Regenerating Hydra Aggregates, submitted

10.Recent literatures may be found in _Advances in Fluid Turbulence_, Proc. 8th Int. Conf. Center for Nonlinear Studies, Physica D, _37(1989)_

ARGOUL, Francoise , Centre de Recherche Paul Pascal, Domaine Universitaire
33405 Talence Cedex, France

ARNEODO, Alain , Centre de Recherche Paul Pascal, Domaine Universitaire
33405 Talence Cedex , France

ASSENHEIMER , Michel , The Weizmann Institute of Science 76100 Rehovot , Israel

BEN AMAR , Martine , Laboratoire de Physique Statistique , Ecole Normale Supérieure ,
24 rue Lhomond, 75231 Paris Cedex 05 , France

BILGRAM , Jorg , Lab.fur Festkorperphysik , Eigenossische Technische Hochschule ,
Honggerberg, CH-8093 Zurich , Suisse

BILLIA , Bernard , Lab.Physique Cristalline , Case 5 , Faculté St Jérome , 13397 Marseille
Cedex 13 , France

BOHR , Thomas , Nordita , Blegdamsvej , 17, 2100 Copenhagen , Danemark

BRANDEIS , Geneviève , Lab. Dynamique des Systèmes Géologiques , Univ. Paris 7,
4 Place Jussieu , 75252 Paris Cedex 05, France

BENISTY, Henri, Thomson LCR, Domaine de Corbeville, 91404 Orsay, Cedex

BRENER, Efim , Institute of Solid State Physics, Academy of Sciences of the USSR,
142432 Chernogolovka, Moscow District

CHEVRIER , Joel , Centre de Recherche sur les Mécanismes de Croissance Cristalline ,
case 913 , Campus de Luminy , 13288 Marseille Cedex , France

CLADIS , Patricia , AT&T Bell Laboratories , Rm 10448 , 600 Mountain Avenue , Murray
Hill , NJ 07974-2070 , USA

CLAVIN , Paul , Laboratoire de Recherche en Combustion , Université de Provence , St
Jérome , Case 252 , 13397 Marseille Cedex 13 , France

CONRADO , Claudine , Nordita, Blegdamsvej, 17 , 2100 Copenhagen , Danemark

COUDER , Yves , Laboratoire de Physique Statistique, Ecole Normale Supérieure , 24 rue
Lhomond, 75231 Paris cedex 05, France

DAVAILLE , Anne, Laboratoire de Dynamique des Systèmes Géologiques, I.P.G.,
Tour 14-15 ; 4, Place Jussieu , 75252 Paris cedex 05, France

ECKLER, Institut fur Raumsimulation, Deutsche Forschungsunstalt fur Luft und Faunfarht
D-5000 Koln 90, Germany

EMSELLEM , Virginie , Ecole Normale Supérieure, L.P.S. , , 24, rue Lhomond , 75231
Paris cedex 05, France

FAVIER , Jean Jacques , Laboratoire d'Etudes de la Solidification, Dept. de la Métallurgie ,
Centre d'Etudes Nucléaires de Grenoble , 85 X , 38041 Grenoble cedex, France.

GALLET , François , Laboratoire de Physique Statistique , Ecole Normale Supérieure ,
24 rue Lhomond , 75231 Paris cedex 05, France

GARCIA YBARRA , Pedro , U.N.E.D., Dept. Fisica Fundamental , Apdo 60141
28080 Madrid , Espagne

GLICKSMAN , Martin, Material Engineering Dept. Rensselaer Polytechnic. Inst. Troy ,
New York 12180 , USA

GOLLUB , J.P., Physics Dept. , Hoverford College , Hoverford, Pa 19041   USA

GUINEA , Franciscio , Departemento de Fisica Solido, Universidad Autonoma,
Cantoblanco , 28049 Madrid , Espagne

HAKIM , Vincent , Laboratoire de Physique Statistique, Ecole Normale Supérieure ,
24, rue Lhomond , 75231 Paris cedex 05, France

HOWATH , Viktor , Institute for Technical Physics, HPS , P.O. Box 76 , Budapest, 1325
Hongrie

HUPPERT , Herbert , Institute of Theoretical Geophysics, Department of Applied
Mathematics and Theoretical Physics, 20 Silver St. , Cambridge, CB3 9EW, UK

HURLIMANN , Erich , Lab. fur Festkorperphysik , E.T.H. , CH-8093 Zurich , Suisse

JENSEN , Nordita, Blegdamsvej, 17 ; 2100 Copenhagen , Danemark

KADANOFF, Leo, The James Franck Institute, The University of Chicago, 5640 South
Ellis Avenue, Chicago, Illinois 60637, USA

KARMA , Alain, Physics Department, Northeastern University , Boston , MA 02115 ,
USA

KESSLER , David, Dept. of Physics, University of Michigan , Ann Arbor, MI 48109,
USA

KUROWSKY , Pascal , Groupe de Physique des Solides, Université Paris VII , 4 Place
Jussieu , 75251 Paris cedex 05, France

KURZ , Wielfried , Laboratoire de Métallurgie Physique ; 34, Ch. de Bellevue, CH-1007
Lausanne , Suisse

LANGER   , James , Institute for Theoretical Physics, University of California, Santa
Barbara, CA 93106, USA

LEVINE , Herbert , University of California, Institute for Non Linear Science , La Jolla,
California 92093 , USA

LIMAT , Laurent , Laboratoire d'Hydrodynamique et Mécanique Physique, ESPCI ; 10, rue
Vauquelin, 75231 Paris cedex 05, France

LIPSON , Stephen , Technion Israel Institute of Technology, Dept. of Physics , Technion
City, Haifa 32000 , Israel

MERON , Ehud , The Weizmann Institute of Sciences , 76100 Rehovot , Israel

MIKHAILOV , Alexander , Laboratory for Condensed - Systems Theory and Synergetic , Dept. of Physics , Moscow State University , 11 7234 Moscow, USSR

MISBAH , Chaouqi , G.P.S. Tour 23 ; 2, Place Jussieu , 75251 Paris cedex 05, France

MOREAU , René , Madylam , ENSHMG , BP 95 ; 38402 St Martin d'Heres cedex, France

MUSHOL, Martin, City College of New York, Phys. Dept, 138 Street & Convent Avenue, New York, NY 10031

MULLER-KRUMBHAAR , Heiner , Inst. fur Festkorperforschung , Der Kernforschungsunlage, D5170 Julich , RFA

MUZY, Jean Francois, Centre de Recherche Paul Pascal, Domaine Universitaire, 33405 Talence Cedex, France

NOZIERES, Philippe, Collège de France, 11 Place M. Berthelot, 75231 Paris cedex 05

OSWALD, Patrick, Ecole Normale Superieure de Lyon, 46 allée d' Italie;
69364 Lyon cedex 07, France

OUVRY, Stéphane, Division de Physique Théorique, IPN, Orsay Fr-91406

PELCE, Pierre, Laboratoire de Recherche en Combustion, Université de Provence, St Jérome, Case 252, 13397 Marseille cedex 13, France

PIETRONERO, L., Dept. di Fisica, Univ. di Roma, " La Sapienza", Piazzale A.Moro, 0185 Roma, Italie

POCHEAU, Alain, Laboratoire de Recherche en Combustion, Université de Provence, St Jérome, Case 252, 13397 Marseille cedex 13, France

POMEAU, Yves, Laboratoire de Physique Statistique, Ecole Normale Supérieure, 24 rue Lhomond, 75231 Paris cedex 05, France

RAPPEL, Wouter-Jan, Institute for Non Linear Science, U.C. San Diego, La Jolla, CA 92093, USA

ROLLEY, Etienne, Laboratoire de Physique Statistique, Ecole Normale Supérieure, 24 rue Lhomond, 75231 Paris cedex 05, France

SANDER, Leonard, Groupe de Physique des Solides, Université de Paris VII, Tour 23, 2 Place Jussieu, 75251 Paris cedex 05, France

SAWADA, Yasuji, Research Institute of Electrical Communication , Tohoku University, Sendai 980 Japan

SEARBY, Geoffrey, Laboratoire de Recherche en Combustion, Université de Provence, St Jérome, 13397 Marseille cedex 13, France

SUN, Jiong, Laboratoire de Recherche en Combustion, Université de Provence, St Jérome, 13397 Marseille cedex 13, France

TABELING, Patrick, Laboratoire de Physique Statistique, Ecole Normale Supérieure, 24 rue Lhomond, 75231 Paris cedex 05, France

TANG, Leihan, Fakultat fur Physik und Astronomie, Ruhr-Universitat Bochum, 4630 Bochum, FRG

TANVEER, Saleh, Ohio State University, Mathematics Department, Columbus, Ohio 43210 USA

TEMKIN, Dimitri, I.P. Bardin Institute of Ferrous Metals, Moscow - USSR

TRIVEDI, Rohit, Ames Lab. US DOE and the Dept.of Material Sc. and Engineering, Iowa State University, AME, IA 50011, USA

VAN SAARLOOS Wim, Institute-Lorentz, University of Leiden, P.O. Box 9506, 2300 RA Leiden, The Netherlands

WALGRAEF Daniel, Service de Chimie Physique, Université libre de Bruxelles, B-1050 Bruxelles, Belgique

VILLAIN, Jacques, DRF/MDN, Centre d' Etudes Nucléaires de Grenoble, 85 X, F-38041 Grenoble cedex, France.

WOLF, Dietrich, IFF, Forschungszentrum Julich, POB 1913, D-5170 Julich, Germany

YAN, Hong, University of Michigan, Ann Arbor, MI 48109, USA

YEKUTIELI, Iddo, SPSRM, Orme des Merisiers, CEN Saclay, 91191 Gif sur Yvette Cedex, France